T0329792

Optical Fibre Sensors

Optical Fibre Sensors

Fundamentals for Development
of Optimized Devices

Edited by

Ignacio Del Villar
Ignacio R. Matias

IEEE Press Series on Sensors
Vladimir Lumelsky, Series Editor

IEEE PRESS

WILEY

Published by John Wiley & Sons, Inc., Hoboken, New Jersey.

Published simultaneously in Canada.

For general information on our other products and services or for technical support, please contact our Customer Care Department within the United States at (800) 762-2974, outside the United States at (317) 572-3993 or fax (317) 572-4002.

Wiley also publishes its books in a variety of electronic formats. Some content that appears in print may not be available in electronic formats. For more information about Wiley products, visit our web site at www.wiley.com.

Library of Congress Cataloging-in-Publication Data:

Names: Del Villar, Ignacio, 1978- editor. | Matias, Ignacio R., 1966–
editor. | Institute of Electrical and Electronics Engineers.
Title: Optical fibre sensors : fundamentals for development of optimized
    devices / edited by Ignacio Del Villar, Ignacio R.
    Matias.
Other titles: Optical fibre sensors (John Wiley & Sons)
Description: Hoboken, New Jersey : Wiley-IEEE Press, [2021] | Series: IEEE
    Press series on sensors | Includes bibliographical references and index.
Identifiers: LCCN 2020020787 (print) | LCCN 2020020788 (ebook) | ISBN
    9781119534761 (cloth) | ISBN 9781119534778 (adobe pdf) | ISBN
    9781119534792 (epub)
Subjects: LCSH: Optical fibre detectors.
Classification: LCC TA1815 .O696 2021 (print) | LCC TA1815 (ebook) | DDC
    621.36/92–dc23
LC record available at https://lccn.loc.gov/2020020787
LC ebook record available at https://lccn.loc.gov/2020020788

Cover Design: Wiley
Cover Image: © MirageC/Getty Images

Set in 9.5/12.5pt STIXTwoText by SPi Global, Pondicherry, India

Contents

List of Contributors

Francesco Baldini
Institute of Applied Physics "Nello
Carrara" (IFAC)
National Research Council (CNR)
Florence
Italy

Francesco Chiavaioli
Institute of Applied Physics "Nello
Carrara" (IFAC)
National Research Council (CNR)
Florence
Italy

Marco Consales
Optoelectronics Group
Department of Engineering
University of Sannio
Benevento
Italy

Brian Culshaw
Department of Electronic and
Electrical Engineering
University of Strathclyd, Glasgow
Scotland, UK

Andrea Cusano
Optoelectronics Group
Department of Engineering
University of Sannio
Benevento
Italy

Jean Carlos Cardozo da Silva
Graduate Program in Electrical and
Computer Engineering
Federal University of Technology
-Paraná Brazil

Ignacio Del Villar
Department of Electrical, Electronic
and Communications Engineering
Public University of Navarre
Pamplona
Spain

César Elosua
Department of Electrical, Electronic
and Communications Engineering
Public University of Navarre
Pamplona
Spain

José Rodolfo Galvão
Graduate Program in Electrical and
Computer Engineering
Federal University of Technology
-Paraná Brazill

Ambra Giannetti
Institute of Applied Physics "Nello
Carrara" (IFAC)
National Research Council (CNR)
Florence, Italy

Sillas Hadjiloucas
Department of Biomedical Engineering
University of Reading
Reading, UK

Arthur H. Hartog
Worthy Photonics Ltd
Winchester, UK

T. Hien Nguyen
Photonics and Instrumentation
Research Centre
City University of London
London, UK

Alessandra Kalinowski
Graduate Program in Electrical and
Computer Engineering
Federal University of Technology
-Paraná Brazil

Kamil Kosiel
Łukasiewicz Research Network –
Institute of Electron Technology
Al. Lotników 32/46, 02-668 Warsaw
Poland

Alayn Loayssa
Department of Electrical, Electronic
and Communications Engineering
Public University of Navarre
Pamplona, Spain

Diego Lopez-Torres
Department of Electrical, Electronic
and Communications Engineering,
Public University of Navarre,
Pamplona, Spain

Dajuan Lyu
National Engineering Laboratory for
Fibre Optic Sensing Technology
(NEL-FOST)
Wuhan University of Technology
Wuhan, China

Cicero Martelli
Graduate Program in Electrical and
Computer Engineering
Federal University of Technology
-Paraná Brazil

Ignacio R. Matias
Institute of Smart Cities
Public University of Navarre
Pamplona, Spain

Carlo Molardi
School of Engineering
Nazarbayev University
Astana
Kazakhstan

Talita Paes
Graduate Program in Electrical and
Computer Engineering
Federal University of Technology
-Paraná Brazil

Marco N. Petrovich
Optoelectronics Research Centre
University of Southampton
Southampton, UK

Marco Pisco
Optoelectronics Group
Department of Engineering
University of Sannio
Benevento, Italy

Armando Ricciardi
Optoelectronics Group
Department of Engineering
University of Sannio
Benevento, Italy

Mateusz Śmietana
Institute of Microelectronics and
Optoelectronics
Warsaw University of Technology
Koszykowa
Warsaw, Poland

Tong Sun
Photonics and Instrumentation
Research Centre
City University of London
London, UK

Daniele Tosi
School of Engineering, Nazarbayev
University, Astana, Kazakhstan
and
Laboratory of Biosensors and
Bioinstruments
National Laboratory Astana
Astana, Kazakhstan

Minghong Yang
National Engineering Laboratory
for Fibre Optic Sensing Technology
(NEL-FOST), Wuhan University of
Technology
Wuhan, China

Acknowledgment

As editors, we would like to express our gratitude to all the members of Wiley-IEEE Press for their assistance and help.

Also special thanks to the wonderful team of authors that have written the chapters of the book. To our collaborators in the Public University of Navarra: Alayn Loayssa, Diego López, and César Elosua, we must add a list of prestigious authors that cover multiple countries all over the world: Minghong Yang and Dajuan Lyu, from the National Engineering Laboratory for Fibre Optic Sensing Technology (NEL-FOST) Wuhan University of Technology, Wuhan (China); Daniele Tosi and Carlo Molardi, from the Nazarbayev University, School of Engineering, Astana (Kazakhstan); Arthur H. Hartog, from the Worthy Photonics Ltd, Winchester (UK); Cicero Martelli, Jean Carlos Cardozo da Silva, Alessandra Kalinowski, José Rodolfo Galvão, and Talita Paes, from Universidade Tecnológica Federal do Paraná (Brasil); T. Hien Nguyen and Tong Sun, from the Photonics and Instrumentation Research Centre, City, University of London (UK); Armando Ricciardi, Marco Consales, Marco Pisco, and Andrea Cusano, from the Optoelectronics Group, Department of Engineering, University of Sannio, (Italy); Francesco Chiavaioli, Ambra Giannetti, and Francesco Baldini, from the Institute of Applied Physics 'Nello Carrara' (IFAC), Sesto Fiorentino (Italy); Sillas Hadjiloucas, from the Department of Biomedical Engineering, University of Reading (UK); Kamil Kosiel and Mateusz Śmietana, respectively from the Łukasiewicz Research Network-Instytut Technologii Elektronowej in Warsaw (Poland) and from the Institute of Microelectronics and Optoelectronics in the Warsaw University of Technology (Poland); Marco N. Petrovich, from the University of Strathclyde, Royal College Building, Glasgow, Scotland, (UK); and Brian Culshaw, from the Optoelectronics Research Centre, University of Southampton, (UK).

We would like also to thank our families and friends, because without their support this project would not have been possible.

Finally, just as members of this optical fibre sensor community, we want to thank the dedication to all those who pioneered this more than half a century ago and to those who will continue to do so, because this road is made by walking and, fortunately, the goal is every closer.

About the Editors

Ignacio Del Villar, PhD, is an Associate Professor in the Electrical, Electronic and Communications Engineering Department at the Public University of Navarra, Spain, where he teaches on electronics and industrial communications. He is a member of the IEEE and an Associate Editor of different journals. In addition, he has participated in multiple research projects and co-authored more than 150 papers, conferences, and book chapters related to fibre-optic sensors.

Ignacio R. Matias, PhD, is the Scientific Director of the Institute of Smart Cities and Professor of the Electrical, Electronic and Communications Department at the Public University of Navarra, Spain. He was one of the Associate Editors who founded the *IEEE Sensors Journal*, promoting fibre-optic sensors since then through conferences, special issues, awards, books, etc. He has co-authored more than 500 book chapters, journal and conference papers related to optical fibre sensors. He is currently member-at-large at the IEEE Sensors Council AdCom.

1

Introduction

Ignacio R. Matias[1] and Ignacio Del Villar[2]

[1] Institute of Smart Cities, Public University of Navarre, Pamplona, Spain
[2] Department of Electrical, Electronic and Communications Engineering, Public University of Navarre, Pamplona, Spain

The optical telegraph, invented in 1791 by Claude Chappe, consisted of a network of stations that allowed the transmission of information at a speed of one symbol in two minutes between Paris and Lille (i.e. 230 km) [1]. Each station monitored, with the aid of a telescope, the character that was represented with a wooden semaphore in the previous station. This system was widely used for about 50 years because it was much faster than sending messages by letter, but it required direct vision between each couple of consecutive stations. Consequently, bad weather, or simply the night, prevented its utilization. These are the main reasons why with the invention of the electrical telegraph, a system based on a guided electrical signal, the utilization of the optical telegraph came soon to an end.

However, in parallel to the invention of the electrical telegraph, in 1841, the path towards optical guiding was started with an important discovery by two French researchers, Jean Daniel Colladon and Jacques Babinet, who independently demonstrated that it was possible to guide light in a curved waveguide [2]. Colladon proved this with light rays trapped in a water jet by total internal reflection, whereas Babinet did the same in a bent glass rod.

Another breakthrough occurred in 1966, when Charles Kao (he received the Nobel Prize in Physics in 2009) and George Hockham published a work demonstrating that the attenuation in optical fibres available at the time was caused by impurities, rather than fundamental physical effects such as scattering. They

Optical Fibre Sensors: Fundamentals for Development of Optimized Devices, First Edition.
Edited by Ignacio Del Villar and Ignacio R. Matias.
© 2021 The Institute of Electrical and Electronics Engineers, Inc.
Published 2021 by John Wiley & Sons, Inc.

pointed out that fibres with low loss could be manufactured by using high-purity glass [3, 4]. This idea was proved in the North American company Corning in 1970, with the development of an optical fibre with losses lower than 20 dB/km. Soon afterwards, in 1977, losses were reduced to such a point that General Telephone and Electronics could carry live telephone traffic, 6 Mbit/s, in Long Beach, California, whereas the Bell System could transmit a 45 Mbit/s fibre link in the downtown Chicago phone system. Since that year optical fibre has become the most widely used guided medium in the twentieth century, mainly thanks to the huge bandwidth it presents compared with other guided communication media such as twisted pair and coaxial cable.

Optical communication is the main application of optical fibre. However, there is a second domain where this structure can be used: sensors. Despite the impact of optical fibre in the domain of sensors not being as big as in communications, their presence in the global market cannot be neglected. Indeed, it is the natural and ideal platform in terms of integrating the sensor in the communication system.

Optical fibre sensors (OFSs) can be classified in many different ways. The main classification concerns to the location where the light is modulated, existing in two groups: extrinsic and intrinsic OFSs. In both cases there is a parameter (physical, chemical, biological, etc.) that modulates light. However, the difference is that in an extrinsic OFSs light is guided to the interaction region, extrinsic to the optical fibre, where light is modulated, and after this modulation light is collected again in the optical waveguide, whereas in an intrinsic OFS light is always guided by the optical fibre. In Figure 1.1 the difference between an intrinsic and an extrinsic OFS

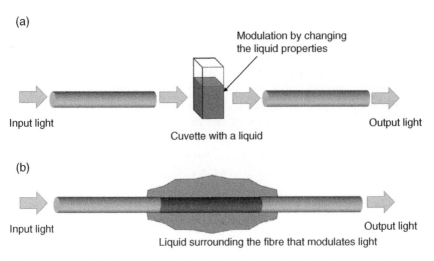

Figure 1.1 (a) Extrinsic sensor: light is modulated outside of the fibre. (b) Intrinsic sensor: light is modulated while it is transmitted through the fibre.

is shown. In the case of an extrinsic sensor, light is modulated outside of the fibre by a liquid (its properties may change as a function of temperature, for instance), whereas in the case of the intrinsic sensor, a fibre has been spliced to two other fibres (one input and one output fibre), which allows an enhanced interaction with the outer medium. In this case, a liquid modulates the light at the same time it is being transmitted through the fibre.

Probably the first OFS was the fibrescope. In 1930 Heinrich Lamm, a German medical student, assembled a bundle of optical fibres to carry an image. His purpose was to use the device for obtaining images of inaccessible parts of the body. He tried to patent the device, but John Logie Baird and Clarence W. Hansell had patented a similar idea some years before. The quality of the images that Lamm obtained was not good, but he is the first researcher that experimentally achieved this breakthrough in the history of optical sensors. Afterwards, in 1954, the Englishman Harold H. Hopkins and the Indian Narinder S. Kapany presented results of better quality on the same principle [5].

Some years later, in 1967, the first effective demonstration of a fibre-optic sensor, the Fotonic sensor, was published [6]. The device was also based on a fibre bundle. However, this time the arrangement was different. Some of the fibres emitted light, and some others did not. The fibre bundle illuminated a surface in front of the fibre, and some part of light was coupled to the fibres that did not transmit light. The amount of light reflected back depended on the distance between the fibre bundle end and the surface. Consequently, the device could be used as a displacement sensor (Figure 1.2).

This type of sensor was the basis for the commercialization of the MTI Fotonic sensor. In the 1980s, the MTI 2000 version allowed monitoring vibration and displacement. Nowadays it is still sold under the version MTI 2100, which is the same concept but with improved characteristics such as the ability to operate in cryogenic, vacuum, high pressure, or in high magnetic field and harsh environments. The resolution has also been improved from 1 nm in the MTI 2000 to 0.25 nm with the MTI 2100 and frequency response from direct-coupled (dc) to 150 kHz in the MTI 2000 up to dc-500 kHz in the MTI 2100.

The concept used in the Fotonic sensor was also the basis for detection of intracranial pressure by using a surface that is a diaphragm that can be deformed by the action of pressure. Depending on the pressure, the surface is deformed, and in this way, the light coupled back to the receiving fibre is modulated. The commercialized device was called Camino ICP Monitor.

Interferometric fibre sensors emerged in the 1970s, the most successful one among them being the optical fibre gyroscope (OFG) (see Figure 1.3). The basic principle was very simple. Light from a laser is split by a beam splitter and enters the fibre on both ends. Both beams go out of the fibre and a photodetector receives them. Thanks to the Sagnac effect, both beams interfere constructively and

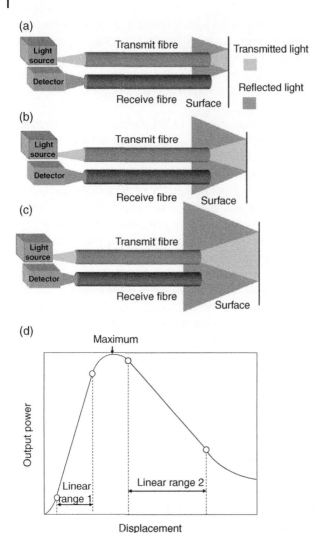

Figure 1.2 (a–c) Fotonic sensor setup with a fibre bundle composed of one transmitting and one receiving fibre: (a) with the surface too close and hence only a small part is coupled back to the receiving fibre; (b) with the surface at the optimal position for a highest coupling; and (c) with the surface too far and hence a great part of light is lost and not coupled to the receiving fibre. (d) MTI 2100 diagram showing the power detected as a function of the distance (the maximum is obtained when the distance is neither too big nor too small).

(a)

(b)

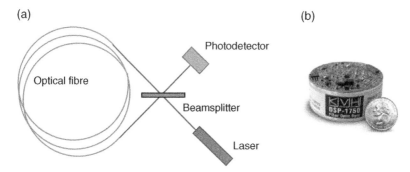

Figure 1.3 (a) Simplified setup: light from a laser is split by a beam splitter and enters the fibre on both ends. The two beams go out of the fibre and the photodetector receives them. Due to the Sagnac effect, both beams interfere constructively and destructively depending on the rotation speed of the device. (b) Commercial optical fibre gyro with a size comparable to a coin (from KVH website).

destructively depending on the rotation speed of the device. The first publication dates from the year 1976 [7]. Since that moment the device has been improved with additional elements such as polarization control, but the initial concept is still maintained. The true benefit of the OFG over traditional spinning-mass gyros is that it has no moving parts. As a result, OFGs are faster, tougher, more reliable and demand far less maintenance. That is why they have become an essential component in platform stabilizing systems, for example, for large satellite antennas, in missile guidance, in subsea navigation, and in aircraft stabilization and navigation, and a host of other applications [8]. It moves about 1000 million US$ per year according to MarketsandMarkets: Fibre Optics Gyroscope Market by Sensing Axis (1, 2, and 3), Device (Gyrocompass, Inertial Measurement Unit, Inertial Navigation System, and Attitude Heading Reference System), Application, and Geography – Global Forecast to 2022.

Based on the acousto-optic effect, it was possible also to develop hydrophones, OFSs that could detect acoustic waves when immersed in water. One of the first approaches was based on interferometry [9], by combining the signals transmitted by an optical fibre that was not immersed in water with the signal reflected at the end facet of another optical fibre immersed in water. By exciting an acoustic wave in front of the fibre immersed in water, it was possible to observe variations in the detected signal. Though it has not been a commercial success like OFG, this application still attracts interest, and the utilization of a Fabry–Pérot cavity (i.e. a coating on the end facet of the optical fibre immersed in water) allows avoiding the use of the reference fibre because in this way an interferometric pattern in the optical spectrum is generated. The setup is depicted in Figure 1.4, and a commercial device is available at the company Precision Acoustics. Its immunity from

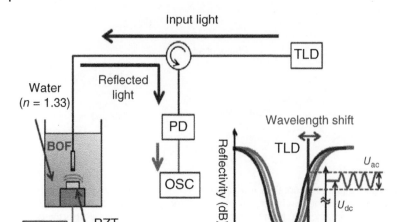

Figure 1.4 Optical setup for a Fabry–Pérot hydrophone [10]. OSC is oscilloscope, PD photodiode, PG pulse generator, PZT piezoelectric transducer, and TLD tunable laser diode. *Source*: Reproduced with permission of Elsevier.

electromagnetic radiation makes it particularly suited for high-frequency measurements in hostile fields.

As we can see, this property was also included in the Fotonic sensor and is one of the key advantages of optical fibres in general. However, in order to make a fibre optical sensor the first option of an end user, more advantages are required compared with the rest of sensors in the market. In the case of the OFG, the key property was that it was not necessary to use moving parts, which means long duration and fast response.

A second OFS success was the measurement of current and voltage with the aid of the Faraday effect [11, 12]. As an example, ABB has developed a commercial device called fibre-optic current sensor (FOCS), which can be used instead of magnetic systems due to its exceptional accuracy and reliability. It can measure uni- or bidirectional DC currents of up to 600 kA with an accuracy of ±0.1% of the measured value (Figure 1.5).

Strain gauges are another well-known application where optical fibres can be used. The first work was published in 1978 [13]. SOFO, from the company Smartec, is a commercial example that can be used for surface mounting or embedding in concrete and mortars. It is ideal for long-term structural deformation monitoring and presents a 20-year track record in field applications.

(a) **Faraday effect principle**

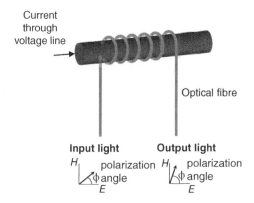

Current
through
voltage line

Optical fibre

Input light   **Output light**
H   polarization *H*   polarization
$\angle\phi$ angle $\angle\phi$ angle
E   *E*

(b) **ABB fibre optic current sensor**

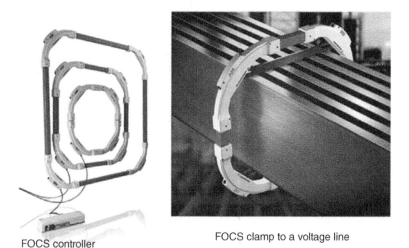

FOCS controller

FOCS clamp to a voltage line

Figure 1.5 (a) Basic principle of optical fibre sensors: the polarization of the input light in and optical fibre is rotated by the action of the magnetic field generated around a line transmitting current. (b) Commercial ABB FOCS sensor.

In addition, the invention of optical fibre Bragg gratings (FBGs) in 1978 [14] widened even more the possibilities of OFSs in terms of detection of strain, because the path was open to include multiple Bragg gratings in the same optical fibre, each one operating at a different wavelength, and to use a multiplexing technology (developed in parallel back in 1980 [15]), to analyse each signal separately. This can be used to monitor strain at multiple points in aircrafts, tunnels, etc., in what is typically called structural health monitoring [16]. The first commercial Bragg grating sensors were available in 1995, and since that moment many companies have commercialized their own FBGs.

However, despite it being possible, unlike electronic gauges, to include multiple strain OFSs in the same wire and despite strain OFSs being less sensitive to vibration or heat and far more reliable than electronic gauges, they have not achieved a commercial success comparable with the OFG. Here we can see a good example of the problem that faces OFSs: there is an electronic competitor, the metallic strain gauge, that nowadays is more widespread than optical fibre gauges because engineers are more familiarized with electronic technology. Like OFSs, electronic sensors have also become popular thanks to another technology, electronics, and to the vast utilization of copper wire for communications. Moreover, the computer, the basic unit in the information technology era, is also based on electronics. All this has made it possible for electronic sensors to nearly monopolize the domain of sensors. Therefore, it is necessary to find applications where optical fibre makes a difference compared with the electronic counterpart.

In this sense, it is important to consider the advantages and disadvantages of optical fibre. The main good points of optical fibre are [17, 18]:

- Small size (its diameter is typically around 100 μm, which allows embedding in many structures) and lightweight.
- Low losses, which allow remote sensing.
- Anti-electromagnetic interference and anti-radio-frequency interference.
- No electrical biasing is required to guide light, so the resulting sensors are passive, which is very relevant in environments with an explosion risk.
- High bandwidth, which allows multiplexing and multi-parameter sensing,
- Distributed sensing in optical fibre communication lines: it is possible to develop modulation techniques that allow physical quantities to be measured along the fibre itself.

However, there are also important concerns [18], which are being progressively solved as the technology matures:

- Cost
- Complexity in interrogation systems
- Unfamiliarity of the end user with the technology

By taking a look at these properties, it is easy to understand why the most successful type of OFS, in terms of covering the sensors market, is distributed sensing. First, it is possible with optical fibre to make distributed measurements over distances up to several tens of kilometres, an ability that is unique to fibre optics. A second advantage of those mentioned above is the small diameter of optical fibre, which allows embedding it in tunnels, bridges, or concrete constructions [19, 20], and, once installed, the initial cost is compensated with a continuous monitoring of variables such as strain, temperature, or vibration, an operation that may last years and that does not affect the optical fibre it is embedded in. This explains its success in the following domains:

- Civil engineering: leakage of dams and river embankments, monitoring of cracks in bridges and other concrete structures; structural health monitoring of large civil projects; and fire monitoring and safety alarms for roads, subways, tunnels, etc. [21].
- Petrochemical: detection of oil and natural gas transmission pipelines or storage tank leaks; temperature monitoring of oil depots, oil pipes, and oil tanks; and detection of fault points.
- Power cable: detection and monitoring of surface temperature of power cable and location of accident points; temperature monitoring of power plants and substations; and detection of fault points and fire alarms.
- Aerospace: monitoring of aircraft pressure, temperature, fuel level, and landing gear status; temperature and strain monitoring of composite skins; and measurement of stress and temperature of aircraft jet turbine engine systems [22].

Distributed sensing technology can be classified in two groups: quasi-distributed (multiplexing FBGs like in Figure 1.6 are a good example) and distributed sensing [24]. A comparison between both technologies is presented in Figure 1.7. With quasi-distributed sensing, discrete points can be monitored, whereas with distributed sensing changes in any point in the optical fibre path length can be detected. Effective gauge lengths of the order of 1 m are common, and there are some that go to even shorter discrimination lengths [8]. Regarding purely distributed sensing, the first works date from the 1980s [25, 26], and since that moment up to now, the utilization of Rayleigh, Brillouin, and Raman scattering for remotely detecting changes in a parameter at a specific point has been widely explored [22, 27, 28].

The optical time-domain reflectometer (OTDR) is the typical commercial device, though there are many types of detectors such as the example presented in Figure 1.7c for sensing an acoustic field. The basic principle of this type of device is the injection of a series of optical pulses into the fibre and the further detection of light that is scattered or reflected back from points along the fibre. These points may be splices, failures, or even changes introduced by variables such as temperature (in this last case the system can be used to detect a fire), strain, or vibration. Since that moment many companies have focused on distributed sensing, such as Omnisens, Sensornet, Silixa, Fotec, Luna, OptaSense, or Future Fibre Technologies, just to mention a few.

In 2017 their market was more than 1 billion US dollars, and it is expected to grow at a 10.4% annual growth rate through 2026. The main application is

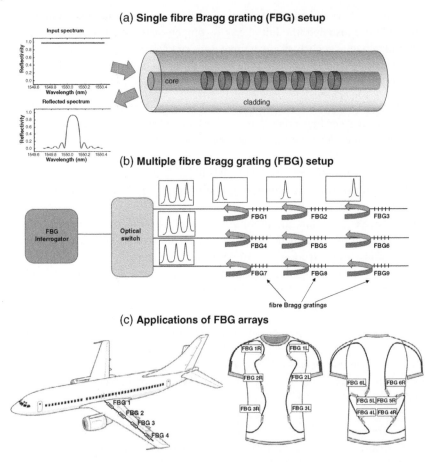

Figure 1.6 (a) Single fibre Bragg grating. (b) Multiple fibre Bragg gratings in a multiplexed system monitored with an interrogator. (c) Applications of FBG arrays for monitoring strain in different points of an aircraft and for developing a smart textile [23].

the oil and gas vertical segment, which occupied a 60.9% share of the global distributed fibre-optic sensor market in 2015. But also pipelines, intrusion detection and security, transport, and infrastructures are other important domains where this technology is used.

Consequently, it can be concluded that optical fibre distributed sensors, along with the gyro, are the two most successful OFSs and both cases can serve as an example to follow towards new commercial opportunities.

In addition to this, OFS research during the last years has focused on two important fields: the fabrication of specialty fibres, where the main breakthrough took place in 1996 with the first microstructured optical fibres [29], and the

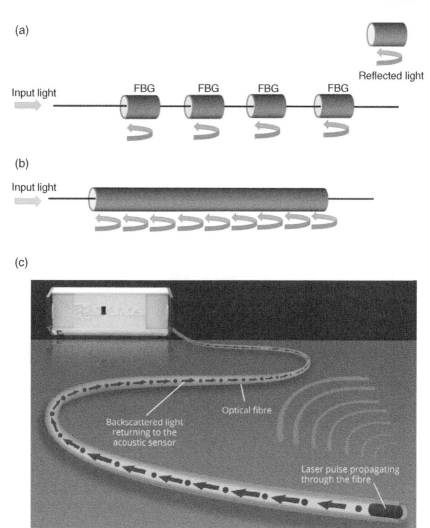

Figure 1.7 (a) Quasi-distributed sensing (with FBGs). (b) Distributed sensing. (c) Commercial distributed sensor for detection of an acoustic field from Silixa.

improvement of nanodeposition techniques [30, 31], which has permitted OFSs to be used in the domain of gas, chemical, and biological sensors [32–37]. The explanation is simple. The optical fibre transmits light, and light can be modulated by parameters that affect the guidance of light through the optical fibre such as strain, temperature, or surrounding medium refractive index. Consequently, if the deposition of a material on the optical fibre modulates the transmission of light

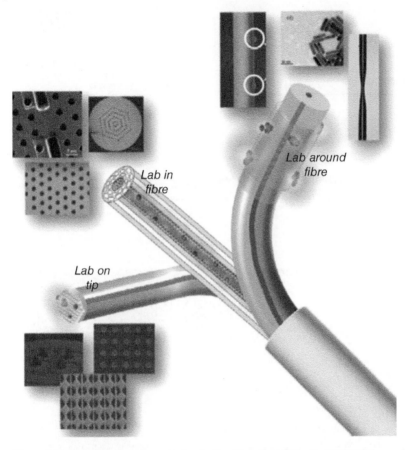

Figure 1.8 Combination of nanotechnology with optical fibre. Deposition of nanostructures: around the fibre, inside the fibre (in the holes of holey fibre), and on the tip of optical fibre (in a probe in reflection configuration). *Source*: With permission from [40] © Wiley.

through the fibre, sensitive materials will modulate the transmission of light through the optical fibre as a function of almost any parameter (i.e. any environmental variables, chemical or biological species, etc.). Moreover, nanotechnology is evolving so much that a lab on fibre can be developed with optimized sensitivity to one or several parameters [38] (Figure 1.8). This positions OFSs in the strategic field of nanophotonics. Even an array of nanoantennas has been deposited on the tip of an optical fibre to enhance the Raman scattering detection [39].

Other important fields are human structural health monitoring, also called biomechanics [41–43], or the development of optical fibre composed of new materials. In this sense, it is well known that the use of plastic optical fibre (also called polymer optical fibre) is an economic alternative to silica fibre in optical communications. Though it has disadvantages, such as higher losses, this technology can be transferred also to optical fibre sensing for the same purpose: low-cost solutions. In addition, the polymer itself presents some different mechanical and thermal properties, which allows new possibilities for the development of multi-parameter sensors, new modulation schemes, and embedded systems for several target applications (e.g. textiles, composite and concrete integration) [44]. The success of plastic optic fibre suggests the exploration of other materials that can be used for harsh environment applications, as it will be shown in one of the chapters of the book. Moreover, OFS technology has evolved so much that even a spider silk optical sensor has been developed for detection of chemical vapours [45]. So the question arises as to why not even metamaterials could be used in OFSs [46].

Considering the current research lines of OFS technology and the commercial devices that are already available, this book will aim for providing the reader with the key concepts towards transforming research into final products. The success of distributed OFSs and the gyro must be followed by others along the twenty-first century, and to this purpose we will combine basic concepts, such as the elements that compose an OFS setup and how light propagates through optical fibre, along with the latest progress of OFSs in multiple important domains of the modern society.

To this purpose, the book will be divided into these sections:

Chapter 2 offers the basics for understanding light propagation in optical fibre: single-mode and multimode fibres under both a geometric optics and wave theory perspective. As a special and challenging case, the propagation through microstructured optical fibre will also be discussed. Finally, some ideas on propagation of light through specialty optical fibres optimized for sensing will be presented.

Chapter 3 describes the key elements that are necessary for an OFS setup (i.e. the optical source, the detector, light coupling, splices, etc.).

Chapters 4 and 5 present different detection techniques: intensity modulation, polarimetric sensors, phase modulation (interferometers), wavelength modulation, and detection based on Rayleigh, Raman, and Brillouin scattering.

Chapters 5–7 focus on applications of distributed sensing: structural health monitoring, biomechanical sensing, and the gas and oil industry (this group is the most successful domain of application of OFSs).

Chapters 8–11 present other important domains such the application of nanotechnology towards improving the performance of OFSs, gas and volatile organic compound sensors, chemical sensors, and biosensors.

Finally, Chapter 12 addresses the important topic of interaction of light with matter with a biomedical perspective. Chapter 13 shows detection in harsh environments, one of the domains where optical fibre can more successfully compete with other technologies. Chapter 14 concludes the book with a thorough analysis of the future trends of OFSs.

References

1 Dilhac, J.-M. (2001). The telegraph of claude chappe: an optical telecommunication network for the 18th century. *IEEE Conference on the History of Telecommunications,* St. John's, Newfoundland, Canada (25–27 July 2001).

2 Johnston, W.K. (2004). The birth of fibreoptics from light guiding. *J. Endourol.* 18 (5): 425–426.

3 Kao, K.C. and Hockham, G.A. (1966). Dielectric-fibre surface waveguides for optical frequencies. *Proc. Inst. Electr. Eng.*: 1151–1158.

4 Natarajan, V. (2010). The 2009 nobel prize in physics: honoring achievements in optics that have changed modern life. *Resonance* 15 (8): 723–732.

5 Hopkins, H.H. and Kapany, N.S. (1954). A flexible fibrescope, using static scanning. *Nature* 173 (4392): 39–41.

6 Menadier, C. and Kissinger, C. (1967). The fotonic sensor. *Instrum. Control Syst.* 40: 114–120.

7 Vali, V. and Shorthill, R.W. (1976). Fibre ring interferometer. *Appl. Opt.* 15 (5): 1099–1100.

8 Culshaw, B. and Kersey, A. (2008). Fibre-optic sensing: a historical perspective. *J. Lightwave Technol.* 26 (9): 1064–1078.

9 Bucaro, J.A., Dardy, H.D., and Carome, E.F. (1977). Fibre-optic hydrophone. *J. Acoust. Soc. Am.* 62: 1302–1304.

10 Kim, K.S., Mizuno, Y., and Nakamura, K. (2014). Fibre-optic ultrasonic hydrophone using short Fabry-Perot cavity with multilayer reflectors deposited on small stub. *Ultrasonics* 54 (4): 1047–1051.

11 Rogers, A.J. (1973). Optical technique for measurement of current and high voltage. *Proc. Inst. Electr. Eng.* 120 (2): 261–267.

12 Rogers, A.J. (1979). Optical measurement of current and voltage on power systems. *Electr. Power Appl.* 2 (4): 120–124.

13 Butter, C.D. and Hocker, G.B. (1978). Fibre optics strain gauge. *Appl. Opt.* 17 (18): 2867–2869.

14 Hill, K.O., Fujii, Y., Johnson, D.C., and Kawasaki, B.S. (1978). Photosensitivity in optical fibre waveguides: application to reflection filter fabrication. *Appl. Phys. Lett.* 32 (10): 647–649.

15 Nelson, A.R., McMahon, D.H., and Gravel, R.L. (1980). Passive multiplexing system for fibre-optic sensors. *Appl. Opt.* 19 (17): 2917–2920.

16 Kinet, D., Mégret, P., Goossen, K.W. et al. (2014). Fibre Bragg grating sensors toward structural health monitoring in composite materials: challenges and solutions. *Sensors (Switzerland)* 14 (4).

17 Elosua, C., Elosua, C., Arregui, F.J. et al. (2017). Micro and nanostructured materials for the development of optical fibre sensors. *Sensors* 17 (10): 2312.

18 Rajan, G. (ed.) (2015). *Optical Fibre Sensors: Advanced Techniques and Applications.* Taylor & Francis.

19 Barrias, A., Casas, J.R., and Villalba, S. (2016). A review of distributed optical fibre sensors for civil engineering applications. *Sensors (Switzerland)* 16 (5).

20 Ye, X.W., Su, Y.H., and Han, J.P. (2014). Structural health monitoring of civil infrastructure using optical fibre sensing technology: a comprehensive review. *Sci. World J.* 2014.

21 Li, H.N., Li, D.S., and Song, G.B. (2004). Recent applications of fibre optic sensors to health monitoring in civil engineering. *Eng. Struct.* 26 (11): 1647–1657.

22 Bao, X. and Chen, L. (2012). Recent progress in distributed fibre optic sensors. *Sensors (Switzerland)* 12: 8601–8639.

23 Lo Presti, D., Massaroni, C., Schena, P.S.E., Formica, D., Caponero, M.A., and Di Tomaso, G. (2018). Smart textile based on FBG sensors for breath-by-breath respiratory monitoring: tests on women. *MeMeA 2018 – 2018 IEEE International Symposium on Medical Measurements and Applications, Proceedings*, Rome, Italy (11–13 June 2018), pp. 1–6.

24 Miller, M. (2017). Distributed sensing environments, in industrial environments. *Cabling.* https://www.cablinginstall.com/connectivity/fibre-optic/article/16467703/distributed-sensing-cable-in-industrial-environments (accessed 21 March 2020).

25 Rogers, A.J. (1981). Polarization-optical time domain reflectometry: a technique for the measurement of field distributions. *Appl. Opt.* 20 (6): 1060–1074.

26 Hartog, A.H. (1983). A distributed temperature sensor based on liquid core optical fibres. *J. Lightwave Technol.* LT-1 (3): 498–509.

27 Thévenaz, L. (2010). Brillouin distributed time-domain sensing in optical fibres: state of the art and perspectives. *Front. Optoelectron. China:* 3–1.

28 Ukil, A., Braendle, H., and Krippner, P. (2012). Distributed temperature sensing: review of technology and applications. *IEEE Sensors J.* 12 (5).

29 Knight, J.C., Birks, T.A., Russell, P.S.J., and Atkin, D.M. (1996). All-silica single-mode optical fibre with photonic crystal cladding. *Opt. Lett.* 21 (19): 1547–1549.

30 Nalwa, H.S. (2002). *Handbook of Thin Films, Volume 1: Deposition and Processing of Thin Films.* Elsevier.

31 Sharma, A.K., Jha, R., and Gupta, B.D. (2007). Fibre-optic sensors based on surface plasmon resonance: a comprehensive review. *IEEE Sensors J.* 7 (8): 1118–1129.

32 Chiavaioli, F., Baldini, F., Tombelli, S. et al. (2017). Biosensing with optical fibre gratings. *Nanophotonics* 6 (4): 663–679.

33 Elosua, C., Matias, I.R., Bariain, C., and Arregui, F.J. (2006). Volatile organic compound optical fibre sensors: a review. *Sensors* 6 (11): 1440–1465.

34 Gupta, B.D., Srivastava, S.K., and Verma, R. (2015). *Fibre Optic Sensors Based on Plasmonics.* World Scientific Publishing Co.

35 Caucheteur, C., Guo, T., and Albert, J. (2015). Review of plasmonic fibre optic biochemical sensors: improving the limit of detection. *Anal. Bioanal. Chem.* 407 (14): 3883–3897.

36 Wang, X.-D. and Wolfbeis, O.S. (2013). Fibre-optic chemical sensors and biosensors (2008–2012). *Anal. Chem.* 85 (2): 487–508.

37 Hodgkinson, J. and Tatam, R.P. (2013). Optical gas sensing: a review. *Meas. Sci. Technol.* 24: 012004.

38 Ricciardi, A., Ricciardi, A., Crescitelli, A. et al. (2015). Lab-on-fibre technology: a new vision for chemical and biological sensing. *Analyst* 140 (24): 8068–8079.

39 Amythe, E.J., Dickey, M.D., Bao, K.J. et al. (2009). Optical antenna arrays on a fibre facet for in situ surface-enhanced raman scattering detection. *Nano Lett.* 9 (3): 1132–1138.

40 Vaiano, P., Vaiano, P., Carotenuto, B. et al. (2016). Lab on fibre technology for biological sensing applications. *Laser Photonics Rev.* 10 (6): 922–961.

41 Roriz, P., Carvalho, L., Frazão, O. et al. (2014). From conventional sensors to fibre optic sensors for strain and force measurements in biomechanics applications: a review. *J. Biomech.* 47 (6).

42 Roriz, P., Frazão, O., Lobo-Ribeiro, A.B. et al. (2013). Review of fibre-optic pressure sensors for biomedical and biomechanical applications. *J. Biomed. Opt.* 18 (5).

43 Al-Fakih, E., Osman, N.A.A., and Adikan, F.R.M. (2012). The use of fibre bragg grating sensors in biomechanics and rehabilitation applications: the state-of-the-art and ongoing research topics. *Sensors (Switzerland)* 12 (10).

44 Bilro, L., Alberto, N., Pinto, J.L., and Nogueira, R. (2012). Optical sensors based on plastic fibres. *Sensors (Switzerland)* 12 (9).

45 Thevenaz, L., Tow, K.H., Chow, D., and Vollrath, F. (2016). Spider silk thread as a fibre optic chemical sensor. *SPIE Newsroom* 10 (2): 10–12.

46 Ubeid, M.F. and Shabat, M.M. (2014). Reflected power and sensitivity of a D-shape optical fibre sensor containing left-handed material. *Sens. Lett.* 12 (11): 1628–1632.

2

Propagation of Light Through Optical Fibre

Ignacio Del Villar

Department of Electrical, Electronic and Communications Engineering, Public University of Navarre, Pamplona, Spain

In 1841 Colladon focused sunlight with a lens onto a water tank. The light spectacularly illuminated the water jets squirting out through the holes of the tank. The light rays were trapped in the fluid along the curving arc of the water until the jets broke up in sparkles of light [1]. What was the basis for this phenomenon?

Water was acting in that case as a waveguide (i.e. a medium that is used to propagate light), and this propagation of light through a specific medium can be explained in two ways: geometric optics and wave theory. The former is an approximation of the latter, and it requires simple mathematics to be solved. However, it can be used only for those cases where the medium that guides light is thick (typically starting from 200 μm [2, 3]).

Oppositely, wave theory can be applied to all cases and allows analysing rigorously the propagation of light through the optical fibre. However, it is more computationally demanding.

Now both models will be explained in detail.

2.1 Geometric Optics

According to Snell's law of refraction, light coming from an input medium with refractive index n_1 is transmitted in an output medium of refractive index n_2 with

Optical Fibre Sensors: Fundamentals for Development of Optimized Devices, First Edition.
Edited by Ignacio Del Villar and Ignacio R. Matias.
© 2021 The Institute of Electrical and Electronics Engineers, Inc.
Published 2021 by John Wiley & Sons, Inc.

an angle θ_2 that is related to the original angle of incidence θ_1 through the following equation:

$$n_2 \sin(\theta_2) = n_1 \sin(\theta_1) \tag{2.1}$$

Figure 2.1a describes the transmission of light in the second medium with a different angle, which is known as refraction.

It is important to consider that only part of the light is transmitted in the second medium because another part of the light is reflected. The equation that describes this is

$$I = R + T \tag{2.2}$$

where I is the input optical power, R is the reflected power, and T is the transmitted power.

In the special case analysed in Figure 2.1a, the refractive index of the input medium n_1 is higher than the refractive index of the output medium n_2. Consequently, according to Eq. (2.1), the angle θ_2 is higher than the angle θ_1.

If the angle θ_1 is progressively increased, there will be a moment when the angle θ_2 will be 90° (see Figure 2.1b). This is called the critical angle θ_1, because above this angle all light is reflected if no losses are considered in the two media involved. This is also called total internal reflection (TIR), and it is the basic principle of light propagation in an optical fibre and also of the phenomenon of surface plasmon resonance (SPR), which can be used for sensing and will be discussed in the next chapters.

In order to understand TIR, first it is important to know the different parts of a standard single-mode fibre (S-SMF), which is one of the most widely used types of optical fibre.

The S-SMF structure in Figure 2.2 is composed of several cylindrical layers. The three outer layers (i.e. the buffer, the strengthening fibres, and the jacket) are used

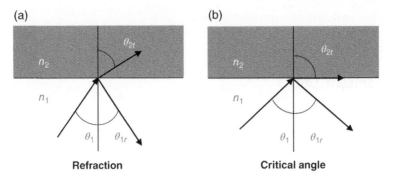

Figure 2.1 (a) Refraction. (b) Total internal refraction at the critical angle.

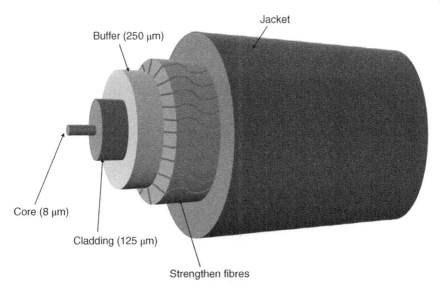

Figure 2.2 Different cylindrical layers and its typical diameters in an optical fibre: the core and the cladding are responsible for light propagation, the buffer is a rigid plastic layer used to protect the core and the cladding, the strengthening fibres are a flexible layer that protects the core and the cladding against crushing, and the jacket is used to cover all the structure.

for protecting the inner layers (the core and the cladding), which are responsible for light propagation.

The core and the cladding are typically made of silica, though other materials can also be used, such as plastic, semiconductor, etc. In any case, what is common for all optical fibres is that the core presents a higher refractive index than the cladding (in the case of silica, this can be done by doping the core, typically with germanium or boron). In this way, the core index is higher than the cladding index, and the situation of Figure 2.1 is reproduced (the input medium will be the core, and the output medium will be cladding). Consequently, depending on the angle of incidence of light in the optical fibre (see Figure 2.3), there will be TIR or not. Those rays that satisfy the condition for TIR are transmitted with low losses (the ideal case of no losses is not possible in nature because all materials present intrinsic losses), and those rays that do not satisfy the condition for TIR are attenuated in a short distance and will not reach the detector.

From expression (2.1) it is possible to extract the critical angle. It is only necessary to set $\theta_2 = 90°$, and then the following expression is obtained:

$$\vartheta_1 = \arcsin\left(\frac{n_2}{n_1}\right) = \arcsin\left(\frac{n_{\text{clad}}}{n_{\text{core}}}\right) \tag{2.3}$$

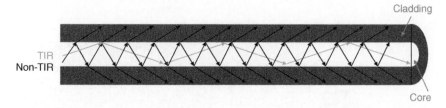

Figure 2.3   Light propagation in an optical fibre: rays transmitted through the core under TIR condition do not experience losses due to refraction, unlike rays under non-TIR condition.

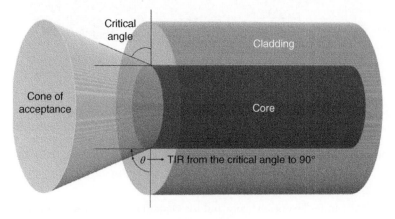

Figure 2.4   Cone of acceptance in an optical fibre.

where n_2 is the cladding index (henceforward n_{clad}) and n_1 is the core index (henceforward n_{core}).

Actually, since optical fibre presents a cylindrical geometry, the critical angle can be also interpreted as the maximum angle of acceptance (see Figure 2.4), because there is a cone of acceptance that includes those rays that can be propagated through the optical fibre.

In any case, may it be called critical angle or maximum angle of acceptance, the main conclusion that can be extracted from expression (2.3) is that this parameter depends on both n_{clad} and n_{core}.

However, it is generally preferred to use, instead of the critical angle, the numerical aperture (NA), which is as a parameter that ranges from 0 to 1 that is also related to the range of angles an optical fibre accepts:

$$NA = \sqrt{n_{core}^2 - n_{clad}^2} \qquad (2.4)$$

In expression (2.4) we can see again a dependence on n_{clad} and n_{core}, where a higher contrast between both values indicates a higher NA (a wider range of angles that are accepted by the optical fibre).

Finally, in order to calculate the transmitted optical power at each specific wavelength, the contribution of each ray is considered in the following equation [4–6]:

$$T(\lambda) = \frac{\int_{\theta_c}^{90°} p(\theta) R(\theta, \lambda)^{N(\theta)}(\theta, \lambda) d\theta}{\int_{\theta_c}^{90°} p(\theta)} \tag{2.5}$$

where λ is the wavelength, θ is the angle of incidence of each ray accepted by the optical fibre (ranging from the critical angle θ_c to 90°), $N(\theta)$ is the number of reflections for each angle along the fibre (in Figure 2.3 it is easy to observe that for angles approaching 90° the number of reflections decreases because $\tan \theta$ approaches infinite)

$$N(\theta) = \frac{L}{d \tan \theta} \tag{2.6}$$

and $R(\theta, \lambda)$ is the reflectivity at each point of the fibre where the corresponding ray with and angle θ is reflected. In the example of Figure 2.3, this occurs at the interface between the core and the cladding, while in cases where a thin film or a multilayer is deposited on the cladding, the problem turns into the calculation of the reflectivity in a stack of layers, which can be solved with different methods [5, 6]. Perhaps the most general and simplest one is the well-known plane wave method for a one-dimensional multilayer waveguide [7].

In any case, it must be pointed out that for the reflectivity can be calculated both for TE- or TM-polarized light. Consequently, when a set-up is used where polarization is not controlled, the best option is to calculate the average value between the reflectivity at TE polarization and the reflectivity at TM polarization [4, 6], while for set-ups where polarized light is used, such as SPR-based sensors, only one of the components is considered [8].

Finally, one of the main issues in solving expression (2.5) is the selection of an adequate model for the light source power distribution $p(\theta)$. For a collimated source, such a laser, the power distribution is expressed as [4, 5]

$$p(\theta) \propto \frac{n_{co}^2 \sin \theta \cos \theta}{\left(1 - n_{co}^2 \cos \theta^2\right)^2} \tag{2.7}$$

while for diffuse or Lambertian sources, such as a LED on an halogen lamp, the power distribution is expressed as [4, 6]

$$p(\theta) \propto n_{\text{co}}^2 \sin \theta \cos \theta \tag{2.8}$$

In this last case, it is also possible to use a Gaussian source [9]:

$$p(\theta) \propto \exp\left[-\frac{\left(\theta - \frac{\pi}{2}\right)^2}{2W^2} \right] \tag{2.9}$$

2.2 Wave Theory

It has been demonstrated that with geometric optics, it is possible to explain the guidance of light through the optical fibre and to calculate the transmitted optical power as a function of wavelength. However, a more thorough analysis is necessary to fully understand how light propagates through the optical fibre. For those readers who are not familiarized with the concepts explained below, it is highly recommended to read a good introduction on electromagnetics, such as the classic 'Fundamentals of electromagnetics' [10].

Light is an electromagnetic wave. Consequently, it is necessary to analyse the Maxwell equations in matter:

$$\nabla \mathbf{D} = \rho \tag{2.10}$$

$$\nabla \mathbf{B} = 0 \tag{2.11}$$

$$\nabla \times \mathbf{E} = -\frac{\partial \mathbf{B}}{\partial t} \tag{2.12}$$

$$\nabla \times \mathbf{H} = \mathbf{J} + \frac{\partial \mathbf{D}}{\partial t} \tag{2.13}$$

where $\mathbf{E}$ and $\mathbf{H}$ are, respectively, the electric and magnetic fields, $\mathbf{D}$ and $\mathbf{B}$ are the displacement and magnetic induction fields, and ρ and $\mathbf{J}$ are the free charges and currents. Assuming that there are no currents and charges, that losses are small, and that the refractive index is independent of the spatial coordinates both in the core and the cladding, the Helmholtz wave equation can be extracted [11]:

$$\nabla^2 \mathbf{E} + \frac{\omega^2 n^2}{c^2} \mathbf{E} = 0 \tag{2.14}$$

where ω is the frequency, n is the optical medium refractive index, and c is the speed of light in vacuum.

The solutions of the former equation are the modes – in other words, the paths through which light is transmitted.

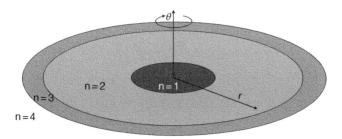

Figure 2.5 Multilayer cylindrical waveguide composed of $N = 3$ layers plus the external medium for a total of $N + 1 = 4$ layers.

There are multiple optical fibre geometries. For the sake of simplicity, we will analyse the most typical case: a multilayer cylindrical optical fibre. By this we will be able to understand not only the standard fibres with a core, a cladding, and the outer medium but also those optical fibres with two claddings, fibres deposited with a thin film, etc.

There are two basic ways of analysing this problem: with a scalar model and with a vectorial one. Initially we will start with the scalar model, and, once we know it, we will follow with the more complex vectorial analysis.

In Figure 2.5 the cross section of a multilayer cylindrical waveguide is shown. For this specific case the waveguide is composed of $N = 3$ layers plus the external medium for a total number of layers $N + 1 = 4$.

2.2.1 Scalar Analysis

Under weakly guiding condition (i.e. a low contrast between the refractive index of the core and the refractive index of the cladding), the wave equation can be simplified because the two polarizations in each mode travel with the same propagation constant and have the same spatial distribution [12]. Consequently, it is only necessary to analyse one of the two components or, in other words, a linear polarization analysis. Hence the term linearly polarized (LP) modes for the solutions obtained with this model.

It must be pointed out that, even though the scalar analysis is an approach to the vectorial analysis, most authors use and refer to this terminology because it is simpler and valid for many applications. Hence the importance of explaining the scalar analysis, which is the basic language all engineers and scientists that operate in the domain of optical fibre understand.

The transverse electric field component propagating along the 'z' axis in each layer of the cylindrical waveguide can be expressed for each mode $LP_{\nu j}$ as [13]

$$U_{\nu j,i}(r,\varphi,z) = \exp\left(-j\beta_{\nu j}z\right)\Psi_{\nu j,i}(r,\varphi) = \exp\left(-j\beta_{\nu j}z\right)\Phi_{\nu}(\varphi)R_{\nu j,i}(r)$$

$$= \exp\left(-j\beta_{\nu j}z\right)\left\{\begin{array}{c}\cos(\nu\varphi)\\\sin(\nu\varphi)\end{array}\right\}\times\left\{\begin{array}{c}A_{\nu j,i}J_{\nu}\left(r\gamma_{\nu j,i}\right)+B_{\nu j,i}Y_{\nu}\left(r\gamma_{\nu j,i}\right)\\A_{\nu j,i}I_{\nu}\left(r\gamma_{\nu j,i}\right)+B_{\nu j,i}K_{\nu}\left(r\gamma_{\nu j,i}\right)\end{array}\right\}$$

when $\quad\begin{array}{c}\beta_{\nu j}<k_0n_i\\\beta_{\nu j}>k_0n_i\end{array}$ (2.15)

where $R(r)_{\nu j,i}$ is the radial variation of the modal field of mode LP$_{\nu j}$ in a layer number i (the core is layer number 1, the cladding layer number 2, etc.), $\beta_{\nu j}$ is the propagation constant of the LP$_{\nu j}$ mode, $\gamma_{\nu j,i} = \sqrt{\left|k_0^2 n_i^2 - \beta_{\nu j}^2\right|}$ is the magnitude of the transverse wavenumber, ϕ is the azimuthal angle, and $A_{\nu j,i}$ and $B_{\nu j,i}$ are non-normalized field expansion coefficients determined by the boundary conditions within the cylindrical layer i. $J_{\nu}(r\gamma_{\nu j,\,i})$ and $Y_{\nu}(r\gamma_{\nu j,\,i})$ are the ordinary Bessel functions of first and second kind of order ν, while $I_{\nu}(r\gamma_{\nu j,\,i})$ and $K_{\nu}(r\gamma_{\nu j,\,i})$ are the modified Bessel functions of first and second kind of order ν, with ν being a non-negative integer number.

Now, the transfer matrix method (TMM) [13, 14] can be applied for the calculation of the propagation constants of the four-layer cylindrical waveguide problem. For the LP$_{\nu j}$ mode of azimuthal number ν and order j, the radial field of expression (2.15) can be written in the following form:

$$R_{\nu j,i}(r) = A_{\nu j,i}C_{\nu}\left(r\gamma_{\nu j,i}\right)+B_{\nu j,i}D_{\nu}\left(r\gamma_{\nu j,i}\right)$$

$$= \left\{\begin{array}{c}A_{\nu j,i}J_{\nu}\left(r\gamma_{\nu j,i}\right)+B_{\nu j,i}Y_{\nu}\left(r\gamma_{\nu j,i}\right)\\A_{\nu j,i}I_{\nu}\left(r\gamma_{\nu j,i}\right)+B_{\nu j,i}K_{\nu}\left(r\gamma_{\nu j,i}\right)\end{array}\right\}\text{ when }\begin{array}{c}\beta_{\nu j}<k_0n_i\\\beta_{\nu j}>k_0n_i\end{array}\quad(2.16)$$

By applying the continuity condition, R and dR/dr must be continuous along the interface between two cylindrical layers. At radius $r = r_i$, which coincides with the interface between i and $i+1$ layer, the following boundary conditions must be satisfied:

$$R_{\nu j,i}(r_i) = R_{\nu j,i+1}(r_{i+1}) \Rightarrow A_{\nu j,i}C_{\nu}\left(r_i\gamma_{\nu j,i}\right)+B_{\nu j,i}D_{\nu}\left(r_i\gamma_{\nu j,i}\right)$$

$$= A_{\nu j,i+1}C_{\nu}\left(r_{i+1}\gamma_{\nu j,i+1}\right)+B_{\nu j,i+1}D_{\nu}\left(r_{i+1}\gamma_{\nu j,i+1}\right) \quad (2.17)$$

$$\frac{dR_{\nu j,i}(r_i)}{dr} = \frac{dR_{\nu j,i+1}(r_{i+1})}{dr} \Rightarrow \gamma_{\nu j,i}\left[A_{\nu j,i}C_{\nu}'\left(r_i\gamma_{\nu j,i}\right)+B_{\nu j,i}D_{\nu}'\left(r_i\gamma_{\nu j,i}\right)\right]$$

$$= \gamma_{\nu j,i+1}\left[A_{\nu j,i+1}C_{\nu}'\left(r_{i+1}\gamma_{\nu j,i+1}\right)+B_{\nu j,i+1}D_{\nu}'\left(r_{i+1}\gamma_{\nu j,i+1}\right)\right]$$

$$(2.18)$$

where $C'_v(.)$ and $D'_v(.)$ denote the derivatives with respect to $r_i\gamma_{vj,\,i}$. For a single interface, the matrix equation that solves Eqs. (2.11) and (2.12) is

$$\begin{pmatrix} A_{vj,i} \\ B_{vj,i} \end{pmatrix} = \begin{pmatrix} \dfrac{m_{11}^{i,i+1}\left(\beta_{vj}\right)}{Q_i} & \dfrac{m_{12}^{i,i+1}\left(\beta_{vj}\right)}{Q_i} \\ \dfrac{m_{21}^{i,i+1}\left(\beta_{vj}\right)}{Q_i} & \dfrac{m_{22}^{i,i+1}\left(\beta_{vj}\right)}{Q_i} \end{pmatrix} \begin{pmatrix} A_{vj,i+1} \\ B_{vj,i+1} \end{pmatrix}$$

(2.19)

where the elements m_{11}, m_{12}, m_{21}, m_{22}, and Q_i are

$$m_{11}^{i,i+1}\left(\beta_{vj}\right) = \gamma_{vj,i}D'_v\left(r_i\gamma_{vj,i}\right)C_v\left(r_i\gamma_{vj,i+1}\right) - \gamma_{vj,i+1}D_v\left(r_i\gamma_{vj,i}\right)C'_v\left(r_i\gamma_{vj,i+1}\right)$$

$$m_{12}^{i,i+1}\left(\beta_{vj}\right) = \gamma_{vj,i}D'_v\left(r_i\gamma_{vj,i}\right)D_v\left(r_i\gamma_{vj,i+1}\right) - \gamma_{vj,i+1}D_v\left(r_i\gamma_{vj,i}\right)D'_v\left(r_i\gamma_{vj,i+1}\right)$$

$$m_{22}^{i,i+1}\left(\beta_{vj}\right) = \gamma_{vj,i}C_v\left(r_i\gamma_{vj,i}\right)D'_v\left(r_i\gamma_{vj,i+1}\right) - \gamma_{vj,i+1}C'_v\left(r_i\gamma_{vj,i}\right)D_v\left(r_i\gamma_{vj,i+1}\right)$$

$$Q_i = \gamma_{vj,i}\left[C_v\left(r_i\gamma_{vj,i}\right)D'_v\left(r_i\gamma_{vj,i+1}\right) - C'_v\left(r_i\gamma_{vj,i}\right)D_v\left(r_i\gamma_{vj,i}\right)\right]$$

$$= \begin{matrix} \dfrac{2}{\pi r_i} \\ \dfrac{-1}{r_i} \end{matrix} \quad \text{when} \quad \begin{matrix} \beta_{vj} < k_0 n_i \\ \beta_{vj} > k_0 n_i \end{matrix}$$

(2.20)

In the same way as for multilayer slab waveguides [15], the global matrix that relates the fields in the inner layer $i = 1$ to outer layer $i = N + 1$ can be expressed as

$$\begin{pmatrix} A_{vj,1} \\ B_{vj,1} \end{pmatrix} = M_{1,2}M_{2,3}\cdots M_{N,N+1}\begin{pmatrix} A_{vj,N+1} \\ B_{vj,N+1} \end{pmatrix}$$

(2.21)

For finite fields in the $i = 1$ layer and in the outer layer, coefficients A_{N+1} and B_1 must be zero. Otherwise, the field at the centre of the inner layer would be infinity, and the same would be true for the electric field in the infinity. Consequently, by satisfying the following condition, it will be possible to obtain the propagation constants of the modes of the waveguide for the particular azimuthal number v:

$$m_{22}^{1,N+1}\left(\beta_{vj}\right) = 0$$

(2.22)

A simple zero searching subroutine will serve for extracting the roots, which are the propagation constants or effective indices, since the propagation constant (β) is simply the effective index (n_{eff}) multiplied by a constant. This is the reason why these terms are interchanged quite often:

$$\beta = \frac{2\pi}{\lambda} n_{\text{eff}} \tag{2.23}$$

where λ is the free-space wavelength and the constant $2\pi/\lambda$ is the wavenumber, k_0.

Once the propagation constants are calculated, the transmitted power after some length for each mode will be

$$P_{\text{out}} = P_{\text{in}} \cdot e^{-j\beta L} \tag{2.24}$$

where P_{in} is the initial power (the optical power at the input of the optical fibre), L is the length of the optical fibre, and β is the propagation constant calculated from expression 2.22.

This indicates that in the ideal case that β is purely real, the mode propagates without losses, something that is not possible because, as we will see later, materials present absorption and there exist other phenomena that contribute to transmission losses.

The analysis for single-mode fibres is quite easier because only one mode is propagated and, once P_0, L, and β are known, the power at the end of the fibre can be easily calculated. However, for multimode fibres, this calculation must be performed for each of the multiple modes, and the contribution of each mode to the global power depends on the optical source, which is more difficult to model. These ideas apply as well for the vectorial analysed we will describe now.

2.2.2 Vectorial Analysis

Contrary to the scalar case, here we must consider both the magnetic and the electric field for the analysis:

$$E_z = \exp\left(j\beta_{\nu j}z\right)\sin(\nu\varphi + \delta) \times \begin{cases} A_{\nu j,i}J_{\nu}\left(r\gamma_{\nu j,i}\right) + B_{\nu j,i}Y_{\nu}\left(r\gamma_{\nu j,i}\right) \\ A_{\nu j,i}I_{\nu}\left(r\gamma_{\nu j,i}\right) + B_{\nu j,i}K_{\nu}\left(r\gamma_{\nu j,i}\right) \end{cases} \quad \text{when} \quad \begin{matrix} \beta_{\nu j} < k_0 n_i \\ \beta_{\nu j} > k_0 n_i \end{matrix}$$

$$H_z = \exp\left(j\beta_{\nu j}z\right)\cos(\nu\varphi + \delta) \times \begin{cases} C_{\nu j,i}J_{\nu}\left(r\gamma_{\nu j,i}\right) + D_{\nu j,i}Y_{\nu}\left(r\gamma_{\nu j,i}\right) \\ C_{\nu j,i}I_{\nu}\left(r\gamma_{\nu j,i}\right) + D_{\nu j,i}K_{\nu}\left(r\gamma_{\nu j,i}\right) \end{cases} \quad \text{when} \quad \begin{matrix} \beta_{\nu j} < k_0 n_i \\ \beta_{\nu j} > k_0 n_i \end{matrix}$$

$$\tag{2.25}$$

where $\gamma_{\nu j,i} = \sqrt{\left|k_0^2 n_i^2 - \beta_{\nu j}^2\right|}$ is the magnitude of the transverse wavenumber, n is the refractive index, $k_0 = 2\pi/\lambda$, λ is the free-space wavelength, and $A_{\nu j,i}$, $B_{\nu j,i}$, $C_{\nu j,i}$, and $D_{\nu j,i}$ are non-normalized electric field expansion coefficients determined by the boundary conditions within the cylindrical layer i. $J_\nu(r\gamma_{\nu j,i})$ and $Y_\nu(r\gamma_{\nu j,i})$ are the ordinary Bessel functions of first and second kind of order ν, while $I_\nu(r\gamma_{\nu j,i})$ and $K_\nu(r\gamma_{\nu j,i})$ are the modified Bessel functions of first and second kind of order ν, with ν being a non-negative integer number.

The transverse field components can be expressed as a function of E_z and H_z of expression (2.25):

$$
\begin{aligned}
E_r &= \frac{j\beta}{k_0^2 n^2 - \beta^2}\left(\frac{\partial}{\partial r}E_z + \frac{k_0 Z_0 u^2}{\beta}\frac{\partial}{r\partial\varphi}H_z\right) \\
E_\varphi &= \frac{j\beta}{k_0^2 n^2 - \beta^2}\left(-\frac{k_0 Z_0 u^2}{\beta}\frac{\partial}{\partial r}H_z + \frac{\partial}{r\partial\varphi}E_z\right) \\
H_r &= \frac{j\beta}{k_0^2 n^2 - \beta^2}\left(\frac{\partial}{\partial r}H_z - \frac{k_0 n^2}{\beta Z_0}\frac{\partial}{r\partial\varphi}E_z\right) \\
H_\varphi &= \frac{j\beta}{k_0^2 n^2 - \beta^2}\left(\frac{k_0 n^2}{\beta Z_0}\frac{\partial}{\partial r}E_z + \frac{\partial}{r\partial\varphi}H_z\right)
\end{aligned}
\tag{2.26}
$$

where u is the magnetic index and Z_0 is the vacuum impedance.

From expressions (2.25) and (2.26), the formulation for the TMM [16, 17] can be derived. In each uniform layer, the field components E_z, E_ϕ, H_z, H_ϕ can be expressed as a function of coefficients A, B, C, and D in the following form:

1) If $\beta_{\nu j} < k_0 n_i$:

$$
\begin{bmatrix} E_z \\ H_\varphi \\ H_z \\ E_\varphi \end{bmatrix} =
\begin{bmatrix}
J_\nu\left(\gamma_{\nu j,i}r\right) & Y_\nu\left(\gamma_{\nu j,i}r\right) & 0 & 0 \\
\frac{jk_0 n_i^2}{Z_0\gamma_{\nu j,i}}J'_\nu\left(\gamma_{\nu j,i}r\right) & \frac{jk_0 n_i^2}{Z_0\gamma_{\nu j,i}}Y'_\nu\left(\gamma_{\nu j,i}r\right) & -\frac{\nu\beta}{\gamma_{\nu j,i}^2 r}J_\nu\left(\gamma_{\nu j,i}r\right) & -\frac{\nu\beta}{\gamma_{\nu j,i}^2 r}Y_\nu\left(\gamma_{\nu j,i}r\right) \\
0 & 0 & J_\nu(\gamma_i r) & Y_\nu(\gamma_i r) \\
-\frac{\nu\beta}{\gamma_{\nu j,i}^2 r}J_\nu\left(\gamma_{\nu j,i}r\right) & -\frac{\nu\beta}{\gamma_{\nu j,i}^2 r}Y_\nu\left(\gamma_{\nu j,i}r\right) & -\frac{jk_0 Z_0 u_i^2}{\gamma_{\nu j,i}}J'_\nu\left(\gamma_{\nu j,i}r\right) & -\frac{jk_0 Z_0 u_i^2}{\gamma_{\nu j,i}}Y'_\nu\left(\gamma_{\nu j,i}r\right)
\end{bmatrix}
\begin{bmatrix} A_{\nu j,i} \\ B_{\nu j,i} \\ C_{\nu j,i} \\ D_{\nu j,i} \end{bmatrix}
$$

2) If $\beta_{\nu j} > k_0 n_i$:

$$
\begin{bmatrix} E_z \\ H_\varphi \\ H_z \\ E_\varphi \end{bmatrix} =
\begin{bmatrix}
I_\nu\left(\gamma_{\nu j,i}r\right) & K_\nu\left(\gamma_{\nu j,i}r\right) & 0 & 0 \\
-\frac{jk_0 n_i^2}{Z_0\gamma_{\nu j,i}}I'_\nu\left(\gamma_{\nu j,i}r\right) & -\frac{jk_0 n_i^2}{Z_0\gamma_{\nu j,i}}K'_\nu\left(\gamma_{\nu j,i}r\right) & \frac{\nu\beta}{\gamma_{\nu j,i}^2 r}I_\nu\left(\gamma_{\nu j,i}r\right) & \frac{\nu\beta}{\gamma_{\nu j,i}^2 r}K_\nu\left(\gamma_{\nu j,i}r\right) \\
0 & 0 & J_\nu(\gamma_i r) & Y_\nu(\gamma_i r) \\
\frac{\nu\beta}{\gamma_{\nu j,i}^2 r}I_\nu\left(\gamma_{\nu j,i}r\right) & \frac{\nu\beta}{\gamma_{\nu j,i}^2 r}K_\nu\left(\gamma_{\nu j,i}r\right) & \frac{jk_0 Z_0 u_i^2}{\gamma_{\nu j,i}}I'_\nu\left(\gamma_{\nu j,i}r\right) & \frac{jk_0 Z_0 u_i^2}{\gamma_{\nu j,i}}K'_\nu\left(\gamma_{\nu j,i}r\right)
\end{bmatrix}
\begin{bmatrix} A_{\nu j,i} \\ B_{\nu j,i} \\ C_{\nu j,i} \\ D_{\nu j,i} \end{bmatrix}
\tag{2.27}
$$

where i is the layer number starting from the core and the main matrix will be called T_i. The four field components are continuous across the interfaces. Using this boundary condition, the transfer matrix can be applied in the same way as for multilayer slab waveguides [15] and LP mode approximation in cylindrical multilayer waveguides [13]. The matrix that relates the fields of two adjacent layers can be expressed as

$$
\begin{pmatrix} A_{\nu j,i} \\ B_{\nu j,i} \\ C_{\nu j,i} \\ D_{\nu j,i} \end{pmatrix} = (T_i)^{-1}(T_{i+1}) \begin{pmatrix} A_{\nu j,i+1} \\ B_{\nu j,i+1} \\ C_{\nu j,i+1} \\ D_{\nu j,i+1} \end{pmatrix}
\tag{2.28}
$$

The global matrix that relates the fields in the inner layer $i = 1$ to outer layer $i = N + 1$ can be expressed as

$$
\begin{pmatrix} A_{\nu j,1} \\ B_{\nu j,1} \\ C_{\nu j,1} \\ D_{\nu j,1} \end{pmatrix} = (T_1)^{-1}(T_2)(T_2)^{-1}(T_3)\cdots(T_N)^{-1}(T_{N+1}) \begin{pmatrix} A_{\nu j,N+1} \\ B_{\nu j,N+1} \\ C_{\nu j,N+1} \\ D_{\nu j,N+1} \end{pmatrix}
\tag{2.29}
$$

$$
(T_1)^{-1}(T_2)(T_2)^{-1}(T_3)\cdots(T_N)^{-1}(T_{N+1}) = \begin{pmatrix} T_{11} & T_{12} & T_{13} & T_{14} \\ T_{21} & T_{22} & T_{23} & T_{24} \\ T_{31} & T_{32} & T_{33} & T_{34} \\ T_{41} & T_{42} & T_{43} & T_{44} \end{pmatrix}
\tag{2.30}
$$

For finite fields in the inner layer and in the outer layer, coefficients A_{N+1}, B_1, C_{N+1}, and D_1 must be zero for the same reason explained in Section 2.2.1.

Now that the initial conditions are established by satisfying the following expression, it will be possible to obtain the propagation constants of the modes of the waveguide for each particular azimuthal number ν:

$$
T_{22}^{1,N+1}\left(\beta_{\nu j}\right) T_{44}^{1,N+1}\left(\beta_{\nu j}\right) - T_{24}^{1,N+1}\left(\beta_{\nu j}\right) T_{42}^{1,N+1}\left(\beta_{\nu j}\right) = 0
\tag{2.31}
$$

Once the propagation constants are extracted, the transmitted power after some length for each mode will be calculated with expression (2.22).

After seeing the equation for calculating modes, an example of the modal profile will be presented. In Figure 2.6 the E_x field component of a set of different LP modes is calculated for S-SMF operating at wavelength 1300 nm. The modes present two subindices. The first one is the azimuthal order (related to the azimuthal axis of the fibre), and the second one is the radial number.

In the case of $LP_{0,x}$ modes (azimuthal order 0), there is no variation as a function of the angle, and the field changes as a function or radius with an increasing number of maxima and minima as a function of the radial order of the LP mode. For $LP_{1,x}$ modes, there is a variation as a function of the angle with two well-defined regions, whereas for $LP_{2,x}$ modes four different regions are observed, and the complexity of the field profile will increase as we continue increasing the azimuthal order.

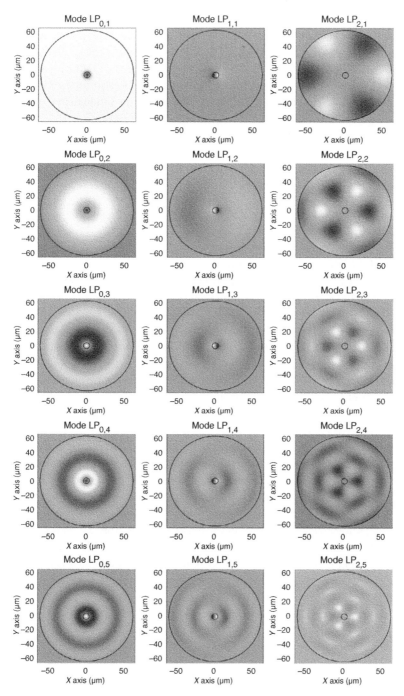

Figure 2.6 E_x field component of LP modes in a single-mode optical fibre at wavelength 1300 nm calculated with FIMMWAVE.

The results of Figure 2.6 also permit to understand the classification of fibres into single mode and multimode. There it is evident that the E_x field component is concentrated in the core for the $LP_{0,1}$ mode, while for the rest of modes the E_x field component is distributed between the cladding and the core. Light propagated through the cladding is more affected by bending, torsion, and losses induced by the surrounding medium. This explains why only the fundamental mode (i.e. the $LP_{0,1}$ mode) is transmitted through a single-mode fibre.

Similar conclusions can be obtained by analysing the electric field components (E_y, E_z) and the magnetic field components (H_x, H_y, H_z), as well as for the optical field intensity. For the sake of simplicity, just the optical field intensity will be analysed, which can be expressed as

$$I(x,y) = \frac{1}{4} \operatorname{Re}\left[\varepsilon(x,y)\right]\mathbf{E}(x,y)\cdot E(x,y)^* + \frac{\mu_0}{4}\mathbf{H}(x,y)\cdot\mathbf{H}(x,y)^* \tag{2.32}$$

where $\varepsilon(x,y)$ is the permittivity in the position x, y, μ_0 is the permeability of free space, and $\mathbf{E}$ and $\mathbf{H}$ are the field and magnetic vectors.

The results obtained in Figure 2.7 confirm the confinement of $LP_{0,1}$ mode in the core and the typical performance in a single-mode fibre.

The guidance of one or several modes in the core of the optical fibre has implications in terms of optical communications. When several modes are transmitted in the core of an optical fibre, there is a phenomenon called intermodal dispersion. Consequently, it is necessary to use repeaters at a shorter distance than in the case of a single-mode fibre. However, despite optical fibre sensors being a different field, the fact of being single mode or multimode will be important, as we will see soon. In general, it is easier to control the design and optimization of a single-mode optical fibre for sensing. However, all it will depend on the specific application the sensor is to be used for, exactly the same that happens in the domain of communications, where despite the better performance of single-mode fibres there are situations where multimode fibres operate adequately at a lower cost.

A parameter that is used to determine if an optical fibre is single mode or multimode is V parameter. This magnitude can be only used to describe the propagation of light modes in standard optical fibres as long as the light propagation is carried out under weak guidance conditions (i.e. low contrast between the refractive index of the core and of the cladding). Under this situation, the V parameter sets the condition to propagate several modes inside the fibre, according to the well-known equation

$$V = \frac{2\pi a}{\lambda}\sqrt{n_{core}^2 - n_{clad}^2} \tag{2.33}$$

where a is the core radius, λ is the wavelength, and n_{core} and n_{clad} are the refractive indices of the core and the cladding.

If the fibre is multimode, the number of guided modes asymptote to $\sim V^2/2$, whereas Figure 2.8 shows the effective index of each mode as a function V. It is

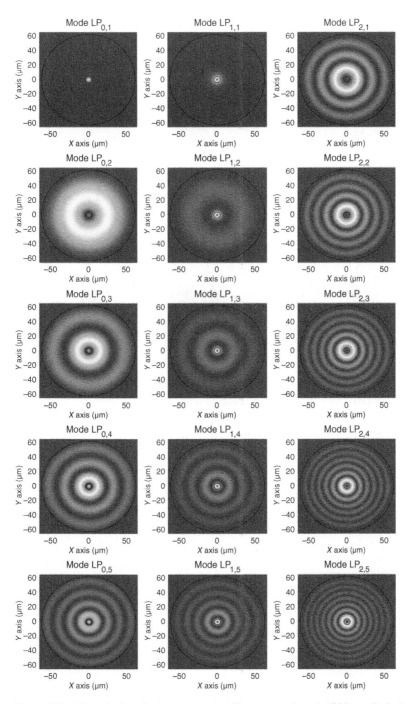

Figure 2.7 LP modes in a single-mode optical fibre at wavelength 1300 nm. Optical power intensity calculated with FIMMWAVE.

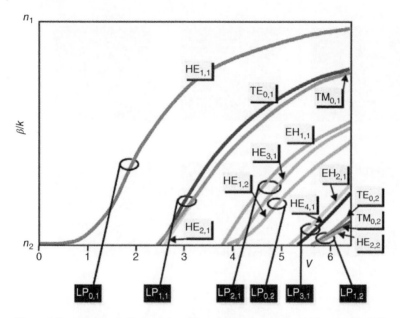

Figure 2.8 Effective indices of the different modes propagated in an SMF [18]. *Source:* Reproduced with permission of academica-e.unavarra.

important to indicate that below 2.405, only one mode is guided. This is the cut-off value for S-SMF (core diameter of 8.2 μm and diameter of 125 μm).

In the previous figure it is easy to observe the equivalence between the modes obtained with the LP approximation and with the vectorial analysis (Table 2.1):

Table 2.1 Equivalence between LP modes and HE, TE, and TM modes.

Linearly polarized	Exact solution
$LP_{0,m}$	$HE_{1,m}$
$LP_{1,m}$	$HE_{2,m}\ TE_{0,m}\ TM_{0,m}$
$LP_{1,m}$	$HE_{l+1,m}\ EH_{l-1,m}$

2.3 Fibre Losses and Dispersion

Taking into consideration expression (2.24), once P_{out} and P_{in} are known, power losses are considered typically in decibels dB via the following expression:

$$\text{Losses (dB)} = -10 \log_{10} \left(\frac{P_{\text{out}}}{P_{\text{in}}} \right) \tag{2.34}$$

The number 10 in the previous expression is replaced by 20 in case the amplitude of the signal is considered. However, here it is power, which is the square of the amplitude.

As an example, if the input power is 1 mW and the output power is 10 μW, it means the attenuation in the line was 20 dB, and many engineers and scientists express the average power losses in dB/km by dividing the power losses by the length of the fibre or the optical fibre-based set-up, which may consist of several connections. In the previous example, if the distance is 200 km, it would mean power losses of 0.1 dB/km. In distributed optical fibre sensing (Chapters 5 and 6), distances are quite long, and losses must not be neglected. However, in other set-ups such as an optical fibre endoscope, the length of the fibre is very small, and high losses are acceptable. This question must be considered before implementing a set-up based on an optic fibre sensor.

In any case, it is important to indicate that losses are due to multiple factors, though the most important ones are absorption, Rayleigh scattering, macrobending, and microbending.

Losses due to absorption are caused by conversion of light into heat by molecules in the glass. In this sense, it is important to remark the effect in the infrared region, typically called the infrared tail, which according to Figure 2.9 increases as a

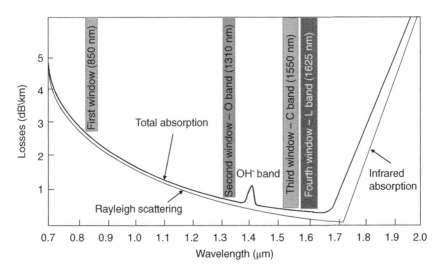

Figure 2.9 Losses in silica fibre and main bands used for communication.

function of wavelength in silica fibres. However, the technology for fabrication of optical fibre has been improved a lot during the last decades, and, as it will be shown in the chapter devoted to optical fibre sensors in harsh environments, there exist optical fibres, such as chalcogenide glass, silver halide polycrystalline, and sapphire-based fibres, that extend the operating wavelength range up to 20 μm. Therefore, the wavelength range where optical fibre sensors can be used ranges from 300–400 nm to 20 μm, being it possible that in the future the range is increased. Another important factor related to absorption is impurities in the optical fibre in the form of dopants used to increase the refractive index of the core, for instance, though the main factor is the presence of OH^-, related to the presence of water and typically observed at 1400 nm. This region is precisely located at the wavelength range where silica presents lower losses and it is what mainly caused the division between 1310 and 1550 nm telecommunication bands (see Figure 2.9), though nowadays manufacturers fabricate fibre without this problem. Another reason for the location of the telecommunication bands was dispersion (this concept will be explained below) and the availability of sources, amplifiers, and detectors at these wavelengths. This is what explains that the first telecommunication band, at 850 nm, is located at a wavelength where losses are much higher than at 1310 and 1550 nm. In Figure 2.9 also the L-band at 1625 nm is shown, because in that region low losses are obtained. Moreover, by suppressing OH^- impurities, it has been possible to enable the E-bands and the S-bands, which are located between the 1310 and the 1550 nm bands. Following the terminology of letters, the 1310 band is named O-band, and the 1550 nm band is called C-band.

The second effect that causes losses, also shown in Figure 2.9, is Rayleigh scattering, a phenomenon especially relevant at short wavelengths, which basically consists of the propagation of light in multiple directions due to inhomogeneities smaller than wavelength.

Finally, losses due to bending play also an important role in the global losses and must not be discarded. Microbending means small deformations of the fibre structure, for instance, a variation of the core diameter due to manufacturing errors. This causes a change in the angle of incidence at the core–cladding interface and can cause some rays to exceed the critical angle, and hence they do not reach the detector. Regarding macrobending, this other phenomenon is rather due to a bending that can be perceived by the human eye, and is typically generated when somebody curves an optical fibre by handling. This factor is quite relevant in many optical fibre sensors, such as fibre Bragg gratings (FBGs), which are sensitive to curvature. Consequently, if another variable like strain is to be measured, it is important to guarantee that the optical fibre is straight. Otherwise it will interfere with the measurement, and if this variation is not monitored by another method, it will not be possible to know what is the contribution of strain and of bending. There exist many optical fibre sensors for measuring two variables in order to solve this problem, but it is not a direct approach, and the designer must keep in mind this question.

A last question to consider in this section is dispersion, which is an effect that has been widely studied in communications in view of the implications it has. The main effect of dispersion is that pulses transmitted through optical fibre are broadened in time as they propagate through optical fibre to such a point that pulses transmitted through the optical fibre overlap each other and it is no longer possible for the detectors to interpret the information sent by the emitter.

This broadening is caused by several factors. The first one is chromatic dispersion, which is caused by the fact that the phase velocity, the velocity with which phase fronts propagate through optical fibre, depends on the medium refractive index and the geometry of the waveguide. Engineers and researchers have mitigated this problem by using wavelengths where this effect is minimum. In this sense, single-mode silica fibre presents typically a minimum dispersion at 1310 nm, and this was an important reason for selecting this band in communications, even though it presents higher losses than the 1550 band, while dispersion-shifted fibres were also developed in order to obtain zero dispersion at 1550 nm, i.e. the wavelength where minimum attenuation is obtained.

Also related to dispersion, another important phenomenon is intermodal dispersion, which is caused by the fact that different modes propagate with different velocities. This is the main reason why in communications it is better to use single-mode optical fibre, though at the same time the equipment required for the communication system is more expensive.

However, though dispersion is an important phenomenon in communications, it has been hardly exploited for sensors, mainly because optical fibre sensors, with the exception of distributed sensing, are not used for long-distance applications, where these effects are more evident. Nonetheless, it is difficult to predict if in the future the wavelength dependence will be used for improving the performance of optical fibre sensors. Regarding this research line, photonic crystal fibres (PCF), which will be explained in Section 2.4, seem to be the best candidate in view of the large amount of parameters that can be modified.

2.4 Propagation in Microstructured Optical Fibre

There are two main groups of microstructured optical fibre (MOF) if we attend to the propagation of light: hollow core fibre (HCF) and PCF. A couple of examples are shown in Figure 2.10, while the parameters of one of the two fibres are shown in Figure 2.11.

In both cases the optical fibre is composed of a single material where holes are introduced. However, the main difference between them is that in an PCF there is a silica core surrounded by holes, whereas in a HCF there is a central hole surrounded by

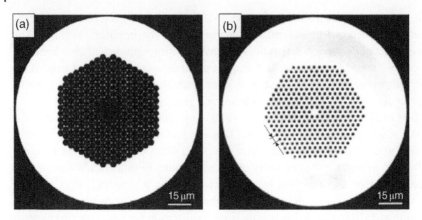

Figure 2.10 (a) Hollow core fibre. (b) Photonic crystal fibre. *Source:* Reprinted from [19] with permission from MDPI.

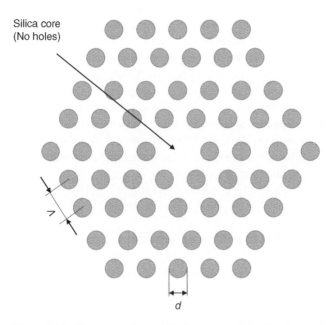

Figure 2.11 Parameters in an PCF: the centre of the fibre has no hole and it operates like the core in a standard single-mode fibre. The period of the holes is Λ and it is called pitch, and d is the diameter of the holes.

holes [20]. Consequently, in a PCF light is transmitted through silica and in a HCF light is transmitted through the air. PCF is more similar to the S-SMF because light is transmitted through a medium (silica) that presents a higher refractive index than the combination of holes and silica surrounding it (the average refractive index is lower). Hence, equation (2.33) can be transformed into this expression [21]:

$$V = \frac{2\pi\Lambda}{\lambda} \sqrt{n_{core}^2 - n_{PCFcladding}^2} \tag{2.35}$$

where $n_{PCFcladding}$ is the refractive index of the new cladding composed of silica with holes (the photonic crystal cladding) and Λ is the period of the holes or pitch.

If we compare Eq. (2.33) with Eq. (2.35), in the SMF, the ratio between the wavelength and the core radius determines the single-mode operation. However, here two modes of operation can be found. The first one is the short wavelength regime ($\lambda \sim \Lambda$), where the V parameter is virtually independent of the wavelength and the pitch, being more relevant the size of the air holes, which is related to $n_{PFCcladding}$. The second case is the long wavelength regime (i.e. longer wavelengths than the pitch). Consequently, a short V parameter is guaranteed. The result is that, by controlling the pitch, it is possible to obtain an endlessly single-mode fibre, one of the most interesting properties of PCF.

On the other hand, in HCF, light is transmitted through a hole surrounded by silica and air holes. Consequently, the TIR condition is not satisfied because the core (air) presents a lower refractive index than the cladding (a combination of silica and air). As a result, the only way to explain the guidance through the air hole is by the phenomenon of photonic bandgap. The periodic arrangement of holes is a photonic bandgap structure: a periodic modulation of a dielectric or metallic structure that inhibits the propagation of light in specific ranges of wavelengths [22]. In the HCF of Figure 2.10a, light transmitted through the hole in the middle is not propagated through the cladding composed of holes in a silica region because the cladding is itself a photonic bandgap structure.

2.5 Propagation in Specialty Optical Fibres Focused on Sensing

So far we have just focused on the conditions for guidance of light in an optical fibre. However, if the purpose is to design an optical fibre sensor, it is often interesting to obtain an interaction with the outer medium at the same time the light is transmitted. The interaction may take place during the complete path of the optical fibre, as it is the case in distributed sensors based on Rayleigh, Raman, or Brillouin scattering (they will be presented in Chapter 5) or by modifying specific sections of the optical fibre, as it is the case in FBGs, for instance.

In other words, an interaction with the surrounding medium must be obtained without compromising the guidance of light through the optical fibre. As we have seen in the previous sections, light is transmitted through the core by TIR. However, the wave theory allows observing that light is not completely transmitted through the core. The core mode transmits light partly confined in the core region (the guided field), and some part transmitted through the cladding (the evanescent field). The penetration depth (d_p) of the evanescent wave is a key parameter for sensing purposes. It is the distance from the interface at which the amplitude of the electric field is decreased by a factor equal to $1/e$ [23], and following the approximation of geometrical optics, this can be expressed with the following equation [24]:

$$d_p = \frac{\lambda}{2\pi\sqrt{n_{core}^2 \sin^2(\theta) - n_{clad}^2}} \tag{2.36}$$

where n_{core} is the core refractive index, n_{clad} the cladding refractive index, θ the angle of incidence at the core–cladding interface, and λ the incidence wavelength.

However, as it can be observed in Figure 2.12a, this field decays to zero before it reaches the outer medium (only the core and a small region around the core presents a significant optical intensity compared to the dark colour surrounding,

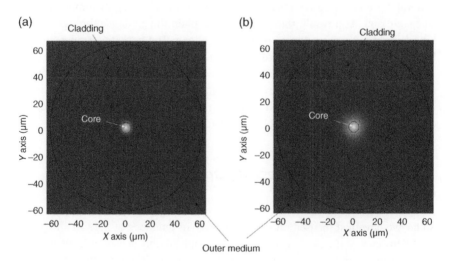

Figure 2.12 (a) $LP_{0,1}$ optical intensity in SMF (the refractive index of the core is 0.36% higher than the refractive index of the cladding). Some part of the optical intensity is located in the cladding. (b) $LP_{0,1}$ optical intensity in modified SMF (the refractive index of the core is 0.09% higher than the refractive index of the cladding). The evanescent field is increased compared with the case (a).

which means no optical intensity). In other words, there is no interaction with the outer medium. One way to solve this could be to decrease the ratio between n_{core} and n_{clad}, which according to expression (2.36) should lead to a higher penetration depth. As an example, in Figure 2.12b, a ratio 1.0009 instead of the 1.0036 of SMF is used. An improvement is visible. However, it is not enough to access the outer medium, and the ratio should be reduced a lot to achieve this, something technically very difficult.

A better approach is a no-core optical fibre. It is exactly like the S-SMF we have analysed so far, but instead of core and cladding, it has only a silica core, what is called no core fibre (NCF). Consequently, modes are propagated through the silica core, and there is a closer interaction with the outer medium. However, with this configuration, n_{core} is silica (refractive index approximately 1.45), and n_{clad} is air (refractive index 1). Consequently, according to expression (2.36), the penetration depth is reduced. In Figure 2.13a the evanescent field is not perceptible, but in Figure 2.14, by representing the field in another way, it is possible to observe a small evanescent field. This situation can be compensated but reducing the diameter of the core (in Figure 2.13b the evanescent field is perceptible without a zoom). However, this reduction of the fibre diameter (diameter of 1 µm) is also technically difficult to obtain.

In order to solve the previous issue, the deposition of the thin film allows enhancing the evanescent field without the need of drastically reducing the diameter of the optical fibre. In Figure 2.14 the optical intensity of the same NCF with and without coating is compared. The deposition of a 140 nm thick coating of refractive

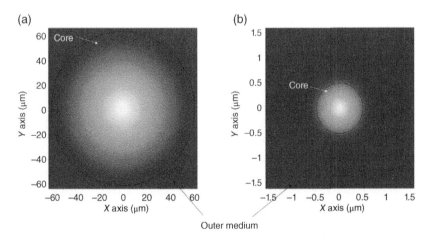

Figure 2.13 (a) $LP_{0,1}$ optical intensity in no-core fibre of diameter 125 µm. Apparently there is no evanescent field that penetrates in the outer medium. (b) $LP_{0,1}$ optical intensity in no-core fibre of diameter 1 µm. The evanescent field penetrates in the outer medium.

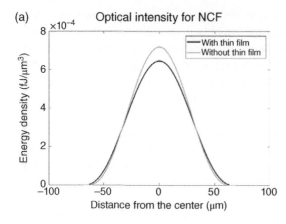

(a)

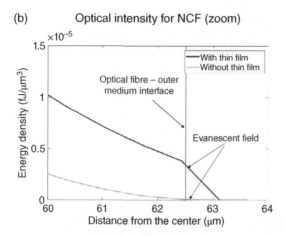

(b)

Figure 2.14 (a) Optical intensity for the LP$_{0,1}$ mode of a no-core fibre, with and without thin film. A zoom at the interface between the optical fibre and the outer medium allows seeing that the evanescent field can be enhanced.

index 1.8 leads to an enhancement of the interaction with the outer medium. However, this enhancement is not always related to a sensitivity increase. In intensity-based sensors the increase of the evanescent field leads to a sensitivity improvement, but for wavelength-based systems this is not always true. While in long-period fibre gratings deposited with a thin film the increase of the evanescent field leads to a sensitivity increase [25, 26], this is not true for D-shaped coated fibres, where the increase of the evanescent field only leads to an increase of the resonance depth and hence a sharp resonance can be created in the optical spectrum, but the wavelength shift as a function of the parameter to detect is not improved [27].

In addition to the no-core fibre case analysed in Figures 2.13 and 2.14, there are optical fibre architectures that can be used in combination with a thin film to interact with the outer medium. In Figure 2.15b the fibre is tapered, and, hence, the core dimensions are reduced to such a point that light cannot be propagated through the core; it is propagated through the cladding. Hence, again we are in a situation like that of Figure 2.15a; light is transmitted through a medium surrounded by the outer medium. So, provided the outer medium presents a lower refractive index than the cladding, guidance of light is satisfied at the same time the device is sensitive to the properties of the outer medium. Another structure is presented in Figure 2.15c. Due to the geometry of the cross section, it is called D-shaped fibre. The cladding has been polished to such a point that, even though light is transmitted through the fibre core by TIR, there is an interaction with the outer medium.

In the three cases analysed in Figure 2.15, light is transmitted through a medium that is so close to the outer medium that the outer medium plays a role in the transmission of light. Moreover, by addition of a thin film, as it is the case in the configurations proposed in Figure 2.15, it is possible to generate a guided mode resonance (also called lossy mode resonance), typically obtained with a polymer,

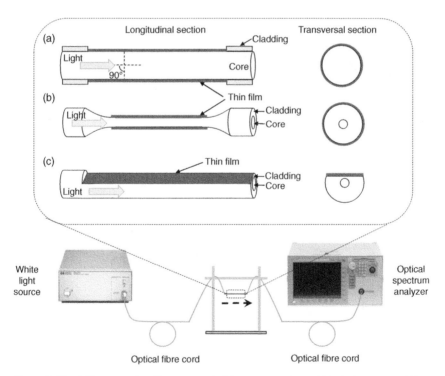

Figure 2.15 Different optical fibres coated with a nanocoating. (a) Uncladded optical fibre. (b) Tapered optical fibre. (c) D-shaped optical fibre.

semiconductor or metallic oxide, or an SPR on the basis of a deposition of a metallic thin film (a detailed explanation on the concept of SPR is given in Chapter 8, Section 8.2.4).

As an example of the generation of a resonance with a thin-film coated optical fibre, in Figure 2.16, the optical power transmitted by the core mode is shown for an indium tin oxide-nanocoated D-shaped fibre. The thin film soaks light towards the outer medium at the resonance wavelength, 1420 nm. Consequently, part of the light is partly transmitted through the core and partly through the thin film [27]. As a result, light transmission is inhibited in the wavelength range around 1420 nm, and the position of this resonance can be shifted if the thin-film properties are modified.

Another example of modification of light propagation and accessing the outer medium is, as mentioned before, the FBGs. This is one of the most well-known structures and has given rise to hundreds of publications and different designs since its first experimental demonstration [28]. Its basic principle is light coupling

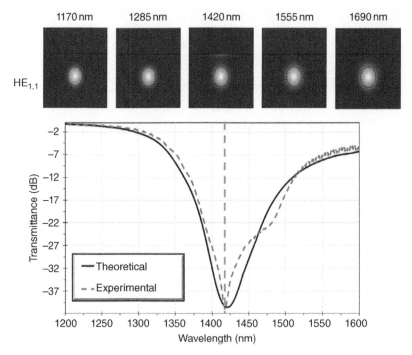

Figure 2.16 Optical spectrum (theoretical and experimental) for a thin-film coated D-shaped fibre. A resonance band corresponding to low transmission of light is created at 1420 nm. This is due to the guidance of light from the $HE_{1,1}$ mode ($LP_{0,1}$ with scalar analysis) in the thin film. *Source:* Reprinted from [27] with permission from Elsevier.

from the input mode to backward or forward propagating modes [29]. Similarly to the phenomenon observed in Figure 2.16, resonances can be created and shifted in the optical spectrum by applying strain, by immersing it in a liquid, or even by applying a thin film, like in Figure 2.15.

A summary of the typical configurations such as uniform standard FBGs, tilted fibre Bragg gratings (TFBGs) [30], and long-period fibre gratings [31] is presented in Figure 2.17 along with their typical transmission spectrum.

Among the three configurations, the most widely used one is the FBG, because it has the special property that a dip is created in reflection configuration, which permits to use it as a probe as well as for distributed sensing due to the narrow width of the resonance created, which allows for multiplexing resonances belonging to multiple FBGs.

In the case of the FBG sensor, the detection is based on the interaction along the grating region. However, it is even possible to create an interaction in the tip of an optical fibre. In this sense, multiple configurations can be obtained, from the simple deposition of a thin film on the tip [33] to complex structures that convert the

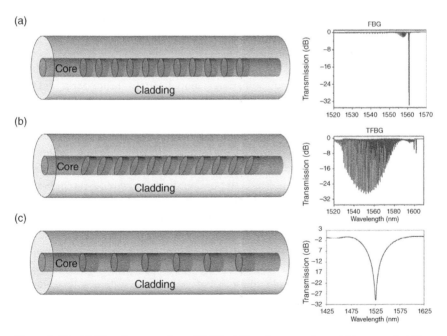

Figure 2.17 (a) Fibre Bragg gratings. (b) Tilted fibre Bragg gratings. (c) Long-period fibre gratings and their corresponding transmission spectra. *Source:* Reprinted from [32] with permission from Wiley.

fibre into a lab on tip [34]. These configurations and others will be described in more detail in Chapter 4.

Finally, we will conclude this section with the concept of polarization. There exist different types of fibres that exploit this property, with polarization-maintaining fibres and polarizing fibres being the most important groups.

In order to understand their operation, we will focus on single-mode fibre because multimode fibres are more complex in terms of polarization and the reality is that researchers and engineers focus mainly on single-mode fibres. We will consider first the standard single-mode optical fibre. In this structure light is propagated in the core through two independent modes with orthogonal polarization states. If LP light is introduced in the optical fibre, light transmitted through both modes will be depolarized in a short length. Oppositely, with polarization-maintaining fibres, the polarization is maintained even though the fibre is bent. This is possible because the structure presents a high birefringence (the effective indices of the two modes propagated in the core are quite different from each other). This difference prevents disturbances such as bending to affect both modes along the fibre, because this is only possible if there is a significant spatial Fourier component with a wavenumber that matches the difference of the effective index of the two modes and disturbances are typically too slowly varying to do an effective mode coupling. This property makes them suitable for applications where the polarization is not allowed to drift, for example, as function of temperature, such as gyroscopes or current sensors.

The most well-known types of polarization-maintaining fibre that are commercially available are bow-tie and panda fibres, though elliptical core fibres and PCF can also be designed for maintaining polarization.

In general, cylindrically symmetric structures are not affected by polarization. This is the case, for instance, in FBGs and long-period fibre gratings (Figure 2.16), which are polarization insensitive. However, TFBGs, where the plane of propagation is modified, are polarization sensitive. For this last case a polarization-maintaining fibre is a better option, though many researchers manage to work on TFBGs using a polarization-maintaining fibre [32, 35].

On the other hand, a clear case of a non-cylindrically symmetric optical fibre is D-shaped fibre. For this case it is again necessary to use a polarization-maintaining fibre or an in-line polarizer and a polarization controller that permits to control the polarization in an S-SMF [27, 36]. The latter option permits to use a less costly standard optical fibre and avoids the need of exciting the system with polarized light, but it is necessary to use two additional devices: the in-line polarizer and the polarization controller. In any case the control of the polarization, with either of these two options, makes it possible to optimize the quality of the resonances generated in the optical spectrum. These two options can be applied to TFBGs and to any other optical fibre sensor where controlling the polarization has some interest, such as many interferometers.

Regarding polarizing fibres, though they are applied in sensing domains that are similar to those of polarization-maintaining fibres, gyroscopes, interferometric sensors, and current sensors, their operation is completely different. Polarizing fibres guide one of the two modes, while the other is suppressed or strongly attenuated so that basically only one mode is propagated through the optical fibre. One successful commercial product is Zing^{TM}, where the polarization extinction ratio can be controlled by coiling the optical fibre, though it can lead to greater loss.

Finally, coming back to the concept of birefringence, it is important to conclude this section with an interesting application, which is the Faraday effect, a phenomenon that consists of the rotation of the polarization plane of light by a magnetic field that is enhanced when circular birefringence is potentiated [37], which can be obviously applied for magnetic field sensing.

2.6 Conclusion

In this chapter we have presented the basic concepts on propagation of light through optical fibre. The problem can be solved from a geometric optics approach in case the waveguide is thick (hundreds of micrometres), whereas in general the best option is to apply waveguide theory. As an introduction, here it has presented the well-known TMM for calculating the propagation constants. However, there exist special cases where subtler methods must be applied. One good example is the phase matching layer, which is used for simulating problems with open boundaries [38].

There exists also commercial simulation software such as FIMMWAVE, COMSOL, Optiwave, etc. These tools offer optimized finite difference and finite element methods. They are highly recommended, though all it will depend on the economical possibilities of the optical fibre sensor designer or developer. In this sense, the option of obtaining free software or to develop the one software must not be discarded. There are some problems that can be solved with very simple equations, for instance, those that involve multimode optical fibres of core diameter of 200 μm, where geometric optics can be applied with very good agreement with the experimental results [39].

In any case, before developing an application, it is important to analyse the performance of the device with a propagation analysis, which will probably permit to reduce costs and time spent in the laboratory, or oppositely, it will permit to understand the performance of the optical fibre sensor.

References

1 Johnston, W.K. (2004). The birth of fibreoptics from light guiding. *J. Endourol.* 18 (5): 425–426.

2 Dwivedi, Y.S., Sharma, A.K., and Gupta, B.D. (2007). Influence of skew rays on the sensitivity and signal-to-noise ratio of a fibre-optic surface-plasmon-resonance sensor: a theoretical study. *Appl. Opt.* 46 (21): 4563–4569.

3 Del Villar, I., Zamarreño, C.R., Hernaez, M. et al. (2010). Lossy mode resonance generation with indium-tin-oxide-coated optical fibres for sensing applications. *J. Light. Technol.* 28 (1): 111–117.

4 Xu, Y., Jones, N., Fothergill, J., and Hanning, C. (2000). Analytical estimates of the characteristics of surface plasmon resonance fibre-optic sensors. *J. Mod. Opt.* 47: 1099–1110.

5 Gupta, B.D. and Sharma, A.K. (2005). Sensitivity evaluation of a multi-layered surface plasmon resonance-based fibre optic sensor: a theoretical study. *Sens. Actuators B Chem.* 107 (1): 40–46. (Special issue).

6 Hernaez, M., Del Villar, I., Zamarreño, C.R. et al. (2010). Optical fibre refractometers based on lossy mode resonances supported by TiO_2 coatings. *Appl. Opt.* 49 (20): 3980–3985.

7 Yariv, A. and Yeh, P. (1977). Electromagnetic propagation in periodic stratified media. II. Birefringence, phase matching, and x-ray lasers. *J. Opt. Soc. Am.* 67 (4): 438.

8 Sharma, A.K. and Gupta, B.D. (2005). On the sensitivity and signal to noise ratio of a step-index fibre optic surface plasmon resonance sensor with bimetallic layers. *Opt. Commun.* 245 (1–6): 159–169.

9 Del Villar, I., Zamarreño, C.R., Sanchez, P. et al. (2010). Generation of lossy mode resonances by deposition of high-refractive-index coatings on uncladded multimode optical fibres. *J. Opt.* 12 (9): 095503.

10 Ulaby Fawwaz, T. and Ravaioli, U. (2019). *Fundamentals of Applied Electromagnetics*. Pearson.

11 Agrawal, G.P. (2001). *Nonlinear Fibre Optics*. Academic Press.

12 Bahaa, M.C.T. and Saleh, E.A. (1991). *Fundamentals of Photonics*. Hoboken, NJ: Wiley.

13 Anemogiannis, E., Glytsis, E.N., and Gaylord, T.K. (2003). Transmission characteristics of long-period fibre gratings having arbitrary azimuthal/radial refractive index variations. *J. Lightwave Technol.* 21 (1): 218–227.

14 Morishita, K. (1981). Numerical analysis of pulse broadening in graded index optical fibres. *IEEE Trans. Microwave Theory Tech.* 29 (4): 348–352.

15 Yeh, P. (1988). *Optical Waves in Layered Media*. New York, NY: Wiley.

16 Kawanishi, T. and Izutsu, M. (2000). Coaxial periodic optical waveguide. *Opt. Express* 7 (1): 10–22.

17 Guo, S., Albin, S., and Rogowski, R. (2004). Comparative analysis of Bragg fibres. *Opt. Express* 12 (1): 198–207.

18 Socorro, A.B. (2015). *Study and Design of Thin-Film-Deposited Optical Biosensing Devices Based on Wavelength Detection of Resonances*. Public University of Navarra.

19 Elosua, C., Arregui, F.J., Del Villar, I. et al. (2017). Micro and nanostructured materials for the development of optical fibre sensors. *Sensors (Switzerland)* 17 (10): 2312.

20 Russell, P. (2003). Photonic crystal fibres. *Science* 299 (5605): 358–362.

21 Rajan, G. (ed.) (2015). *Optical Fibre Sensors: Advanced Techniques and Applications*. CRC Press.

22 John, S.G.J., Joannopoulos, D., Winn, J.N., and Meade, R.D. (1995). *Photonic Crystals: Molding the Flow of Light*, 2e. Princeton University Press.

23 Baptista, J.M., Gouveia, C.A.J., and Jorge, P.A.S. (2013). Refractometric optical fibre platforms for label free sensing. In: *Current Developments in Optical Fibre Technology* (eds. S.W. Harun and H. Arof). Intech.

24 Angela Leung, R.M. and Mohana Shankar, P. (2007). Evanescent field tapered fibre optic biosensors (TFOBS): fabrication, antibody immobilization and detection. In: *Optical Fibres Research Advances* (ed. J.C. Schlesinger). Nova Publishers Inc.

25 Del Villar, I., Matias, I., Arregui, F., and Lalanne, P. (2005). Optimization of sensitivity in Long Period Fibre Gratings with overlay deposition. *Opt. Express* 13 (1): 56–69.

26 Cusano, A., Iadicicco, A., Pilla, P. et al. (2006). Mode transition in high refractive index coated long period gratings. *Opt. Express* 14 (1): 19–34.

27 Arregui, F.J., Del Villar, I., Zamarreño, C.R. et al. (2016). Giant sensitivity of optical fibre sensors by means of lossy mode resonance. *Sens. Actuators B Chem.* 232.

28 Hill, K.O., Fujii, Y., Johnson, D.C., and Kawasaki, B.S. (1978). Photosensitivity in optical fibre waveguides: application to reflection filter fabrication. *Appl. Phys. Lett.* 32 (10): 647–649.

29 Erdogan, T. (1997). Fibre grating spectra. *J. Lightwave Technol.* 15 (8): 1277–1294.

30 Shevchenko, Y.Y. and Albert, J. (2007). Plasmon resonances in gold-coated tilted fibre Bragg gratings. *Opt. Lett.* 32 (3): 211–213.

31 Bhatia, V. and Vengsarkar, A.M. (1996). Optical fibre long-period grating sensors. *Opt. Lett.* 21 (9): 692–694.

32 Urrutia, A., Del Villar, I., Zubiate, P., and Zamarreño, C.R. (2019). A comprehensive review of optical fibre refractometers: towards a standard comparative criterion. *Laser Photon. Rev.* 13: 1900094.

33 Arregui, F.J., Matias, I.R., Liu, Y. et al. (1999). Optical fibre nanometer-scale Fabry-Perot interferometer formed by the ionic self-assembly monolayer process. *Opt. Lett.* 24 (9): 596–598.

34 Vaiano, P. et al. (2016). Lab on Fibre Technology for biological sensing applications. *Laser Photonics Rev.* 10 (6): 922–961.

35 Albert, J., Shao, L.-Y.L.Y., and Caucheteur, C. (2013). Tilted fibre Bragg grating sensors. *Laser Photonics Rev.* 7 (1): 83–108.

36 Andreev, A. et al. (2005). A refractometric sensor using index-sensitive mode resonance between single-mode fibre and thin film amorphous silicon waveguide. *Sens. Actuators B Chem.* 106 (1): 484–488.

37 Rashleigh, S.C. and Ulrich, R. (1979). Magneto-optic current sensing with birefringent fibres. *Appl. Phys. Lett.* 34 (11): 768–770.

38 Berenger, J.P. (1994). A perfectly matched layer for the absorption of electromagnetic waves.pdf. *J. Comput. Phys.* 114: 185–200.

39 Zamarreño, C.R., Hernáez, M., Del Villar, I. et al. (2010). ITO coated optical fibre refractometers based on resonances in the infrared region. *IEEE Sens. J.* 10 (2): 365–366.

3

Optical Fibre Sensor Set-Up Elements

Minghong Yang and Dajuan Lyu

National Engineering Laboratory for Fibre Optic Sensing Technology (NEL-FOST), Wuhan University of Technology, Wuhan, China

3.1 Introduction

This chapter is devoted to describe the necessary elements to build an optical fibre-based sensing application system.

A fibre-optic sensing system is composed of a light source, a signal input optical fibre, a signal output optical fibre, and a detector (optionally the system may include other components such as an optical modulator and a demodulator). As shown in Figure 3.1, the basic principle is that the light of the source is sent to the sensor element through the incident optical fibre. The optical properties of light, such as intensity, wavelength, frequency, phase, and polarization state, are modulated in the sensor element and sent to the optical detector through the exiting fibre to obtain the measured parameters.

As mentioned in the Introduction, optical fibre sensors can be divided into intrinsic and extrinsic categories. The intrinsic optical fibre sensor makes use of the sensitivity of the optical fibre itself to the external conditions. The optical fibre is not only sensitive to the measured signal but also plays a role in transmitting the optical signal. The extrinsic optical fibre sensor is based on the sensitivity of non-optical fibre materials. The optical fibre is only used as the transmission medium of light.

Since the roles of the fibres in the two kinds of sensors are different, the requirements for the optical fibres are different. In the extrinsic optical fibre sensor, the optical fibre only plays the role of transmitting light, and the communication fibre or even the ordinary multimode fibre can meet the requirements, while the sensitive component can be flexibly selected by using high-quality materials.

Optical Fibre Sensors: Fundamentals for Development of Optimized Devices, First Edition.
Edited by Ignacio Del Villar and Ignacio R. Matias.
© 2021 The Institute of Electrical and Electronics Engineers, Inc.
Published 2021 by John Wiley & Sons, Inc.

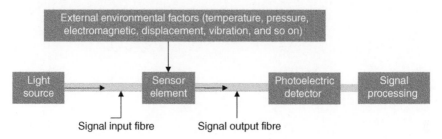

Figure 3.1 Detection principle of optical fibre sensor.

Consequently, the sensitivity of such a sensor can be very high, but it requires more optical coupling devices, and the structure is more complex. The structure of the intrinsic optical fibre sensor is relatively simple, and less coupling devices can be used. However, the requirements for the fibre are relatively high, and it is often necessary to use a special fibre that is sensitive to the signal and has good transmission characteristics. So far, the former is mostly used, but with the improvement of the optical fibre manufacturing process, intrinsic optical fibre sensors are being increasingly used.

In spite of this particular classification and of the many types of modulation techniques that can be used, we will try to present in the following sections the main elements that can be used in an optical fibre sensing system, building in this way the basis of a further complete understanding of any set-up we will use or we will want to design.

3.2 Light Sources

All radiation sources capable of generating light radiation are called light sources, including natural and artificial light sources. According to the phase characteristics of light waves in time and space, light sources are generally divided into coherent light sources and non-coherent light sources. The so-called coherent light sources emit infinite light wave trains from infinitely small point light sources, which are divided into two beams by optical methods, and then the same wave trains meet and overlap to obtain stable interference fringes. For example, laser light sources have good coherence, and the actions and steps of each light-emitting atom are orderly, regular, and coordinated with each other. There is a definite phase relationship between the wave trains emitted by the same atom in succession and between the wave trains emitted by different atoms in space. However, for non-coherent light sources, such as incandescent lamps and other common light sources, the same atom emits light with instantaneity, intermittence, contingency, and randomness, while different atoms emit light independently, that is, the light waves emitted by common light sources do not meet the coherence conditions, are not coherent light, and cannot generate interference phenomenon [1].

The interaction between light and matter is the physical basis of the light source. The interaction process between light and matter atoms includes three processes: spontaneous emission transition, stimulated emission transition, and stimulated absorption transition. Stimulated radiation is coherent, and spontaneous radiation is incoherent.

The characteristic parameters of the light source include the following points:

1) Radiation efficiency η_e and luminous efficiency η_v:

$$\eta_e = \frac{\Phi_e}{P} \tag{3.1}$$

$$\eta_v = \frac{\Phi_v}{P} \tag{3.2}$$

where in the range of wavelength $\lambda_1 \sim \lambda_2$, Φ_e is the radiant flux from a radiant source, Φ_v is the luminous flux from a radiant source, and P is the electrical power required to generate these radiant fluxes.

2) Spectral power distribution: The light source emits mostly multicolour light composed of monochromatic light, and the light power radiated at different frequencies is different. Spectral power distribution is commonly used to describe this relationship between optical power and frequency. The normalized optical power distribution is called relative spectral power distribution. Generally, the spectral power distribution of light sources has four situations: linear spectrum, band spectrum, continuous spectrum, and mixed spectrum. When selecting the light source, the spectral power distribution should be determined by the requirements of the measured object (see Figure 3.2).

3) Spatial light intensity distribution: The luminous intensity vector and the luminous intensity curve are commonly used to describe the spatial light intensity distribution of light source.

4) Colour of light source: It includes two meanings, namely, colour table (i.e. the daylight spectrum correlated at a specific temperature) and colour rendering (i.e. a scale from 0 to 100% that indicates how accurately a light source renders a colour when compared with a reference light source).

5) Colour temperature of light source: Colour temperature is used to express the relationship between the colour of the light source and the temperature. Generally, light source is often expressed by colour temperature, correlated colour temperature, and distributed colour temperature.

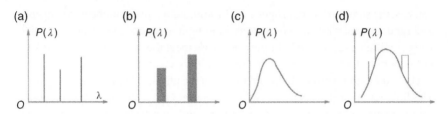

Figure 3.2 Typical spectral power distribution of light sources. (a) linear spectrum, (b) band spectrum, (c) continuous spectrum, (d) mixed spectrum.

When selecting a light source, the intensity, stability, and spectral characteristics of the light source should be considered comprehensively.

3.2.1 Light-Emitting Diodes

Light-emitting diodes (LEDs) are solid-state PN junction devices, and their electroluminescent mechanisms mainly include PN junction luminescence and heterojunction injection luminescence [2, 3]. Their work is based on spontaneous emission, and they belong to incoherent light sources.

Compared with semiconductor lasers, LEDs do not require thermal stabilization and light-stabilizing circuits, so the driving circuit is relatively simple, the manufacturing cost is low, and the output power of the LED is high. There are two main types of LED structures: surface-emitting diodes and edge-emitting diodes [4, 5].

3.2.1.1 Surface Light-Emitting Diode

The surface LED (see Figure 3.3a) emits light from a surface parallel to the plane of the junction and limits the light to a small area, which reduces the thermal resistance, can operate at a high current density, has high internal efficiency, and, as a result, obtains high radiation strength and high operating efficiency on the front side. The light it emits is in a large solid angle range and generally has an emission diameter of more than 50 μm. Consequently, the coupling efficiency with the optical fibre is low, often less than 10%. This type of LED is only suitable for multimode fibre systems.

3.2.1.2 Side Light-Emitting Diode

Side LED (see Figure 3.3b) emits light from the edge of the junction. Since the wave propagates perpendicular to the junction plane, the diverging beam of the edge-emitting diode is different from the surface-emitting diode, and its divergence angle is only 30° in the direction perpendicular to the junction plane. The divergence angle is reduced, and the radiation on the emitting side is eliminated. The light output is incoherent in time, but its spatial coherence is much higher than

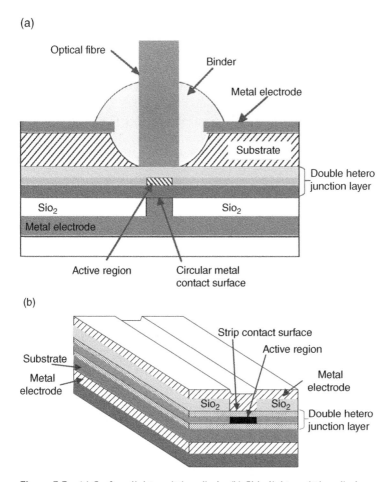

Figure 3.3 (a) Surface light-emitting diode. (b) Side light-emitting diode.

that of the surface-emitting diode. Therefore, the total power generated by the edge LED is smaller than that of the surface-emitting diode, but its brightness is very high. As a result, the LED is suitable not only for multimode fibre systems but also for single-mode fibre systems.

3.2.2 Laser Diode

The use of semiconductor lasers became practical after 1970, when continuous operation of such lasers at room temperature became possible [6]. Since then, semiconductor lasers have been developed extensively. They are also known as laser diodes (LD) or injection lasers, and their properties have been discussed in several books [7, 8].

The LD emits light by stimulated radiation, and the LD emits light of a single colour with good directivity, high intensity, and coherence. LD has high radiation power, narrow divergence angle, high coupling efficiency with single-mode fibre, narrow radiation spectrum, and high-speed direct modulation. Consequently, it is suitable for use as a light source for high-speed long-distance fibre-optic communication systems.

In standard semiconductor lasers, a Fabry–Pérot cavity is formed by two crystal cleavage planes acting as mirrors, which lock the light beam in the cavity reflecting back and forth.

In order to make the light oscillate stably in the resonator cavity, certain coherent conditions and threshold conditions must be satisfied. The coherent conditions are the conditions that make the forward and backward light waves in the resonator cavity coherent, and the threshold conditions are the conditions that make the light gain in the cavity exactly balance the losses of the cavity (above the threshold condition the light gain exceeds the losses of the cavity).

When the excited beam meets certain coherent conditions and threshold conditions, it will keep oscillating and will form a standing wave that penetrates through the two ends of the resonator cavity, obtaining an output coherent beam with very narrow spectral line.

3.2.2.1 Single-Mode Laser Diode Structure

Since the fibre dispersion causes the pulse to be widened, the side mode limits the transmission energy, and the transmitter used in a fibre communication system should be a single longitudinal mode transmission. However, a semiconductor LD having a Fabry–Pérot cavity generally exhibits multimode operation because the gain spectrum of the LD is wider than that of the single longitudinal mode. Consequently, a non-uniform widening mechanism is induced by the LD. For the sake of comparison, Figure 3.4 shows the spectral output characteristics and longitudinal modes for a LED, a multimode laser, and a single-mode LD.

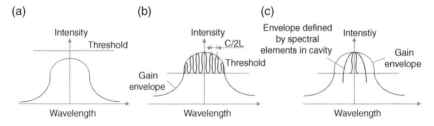

Figure 3.4 Spectral output characteristics of (a) light-emitting, (b) multimode laser, and (c) single-mode laser diodes and longitudinal modes.

In the case of high bit rate modulation, to achieve reliable longitudinal mode control, the techniques adopted for this purpose are mainly divided into the following two categories:

3.2.2.1.1 Short Cavity Laser Diode

The suppression of the edge in the cavity can be enhanced by reducing the length L of the cavity. If the pitch of the modes becomes comparable with the width of the gain curve, then only one mode will oscillate around the gain peak. In order to achieve stable single-mode operation, the LD must be extremely short. In order to overcome the high threshold current density caused by very high mirror loss, the refractive index at the end cleavage plane is required to be very good.

3.2.2.1.2 Frequency Selection Feedback

The second method of obtaining single-mode operation is to add a frequency selection element to the resonator structure. This can be achieved by a coupling cavity, an external grating or a Bragg grating:

- Coupling cavity: If one or more additional mirrors are introduced into the cavity, the increased boundary conditions due to reflection at the various interfaces severely limit the number of longitudinal modes. In order to achieve single-mode operation, the drive current or temperature must be constantly changed to adjust the cavity.
- External grating: The frequency selection can also be achieved by an external grating other than the resonator.

  Since the grating is not integrated on the wafer, the mechanical stability of such a laser is a key point. Lasers with external gratings are expensive devices and are not well suited for fibre-optic communication.
- Bragg grating: The most common method of not achieving single-mode emission is to add a Bragg grating that causes the complex refractive index and distribution feedback to periodically change throughout the cavity. If the threshold gain of the oscillating mode is significantly smaller than the threshold gain of the other modes, dynamic single-mode operation is possible.

  Devices using Bragg gratings can be roughly classified into two types: distributed Bragg reflection (DBR) and distributed feedback (DFB) LD (see Figure 3.5).

DFB semiconductor lasers were developed during the 1980s and are used routinely for WDM light wave systems [9–15]. The feedback in DFB lasers, as the name implies, is not localized at the facets but is distributed throughout the cavity length.

DFB lasers are lasers that are selected by a periodic grating and maintained at high-speed modulation and still have a single longitudinal mode and a transverse mode. The cavity loss of DFB laser has obvious wavelength dependence, so its coloration and stability are better than general Fabry–Pérot laser [16, 17].

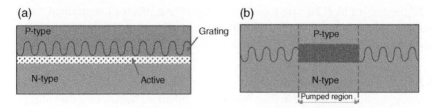

Figure 3.5 (a) Distributed feedback (DFB) laser diode. (b) Distributed Bragg reflection (DBR) laser diode.

3.2.2.2 Quantum Well Laser Diode

In addition to the limitation of carriers by double heterojunction semiconductor lasers, there is another completely different way of limiting carriers, which is the limitation of the state of energy allowed for electrons or holes. This LD is called quantum well (QW) LD. It has many advantages such as low threshold, narrow line width, high differential gain, temperature insensitivity, fast modulation speed, and easy control of the gain curve.

A typical QW device is shown in Figure 3.6, where Figure 3.6a is the QW structure schematic, and a very thin GaAs active layer is sandwiched between two very wide GaAlAs semiconductor materials, so it is a heterojunction device. In the LD, the thickness d of the active layer is very thin (typically about 10 nm), so the forbidden band potential in the conduction band can block electrons in the one-dimensional potential well in the x direction, but is free in the y and z directions.

This closure exhibits a quantum effect that causes the band to be quantized into discrete values $E1$, $E2$, $E3$, ..., which correspond to quantum numbers 1, 2, 3, respectively, as shown in Figure 3.6b. The active layer thickness of a QW semiconductor LD is only 10 nm, which is about 1/10 of that of a heterojunction device, so a small change in the injection current can cause a large change in the output laser.

The use of a plurality of thin-layer structure active regions having a thickness d of 5–10 nm improves the performance of the single QW device. This LD is a multiple quantum well (MQW) laser, which has the advantages of better modulation performance, narrower line width, and higher efficiency. Figure 3.7 shows a schematic and energy level diagram of a semiconductor LD with four QW separated by a three-layer GaAlAs barrier layer.

3.2.3 Superluminescent Diodes (SLD)

A superluminescent diode (SLD) is a semiconductor optoelectronic device interposed between an LD and an LED. SLD is a spontaneous emission single-pass optical amplifier component, which outputs not the laser, but the amplified spontaneous emission (ASE) light, i.e. super-radiative light, which belongs to

(a)

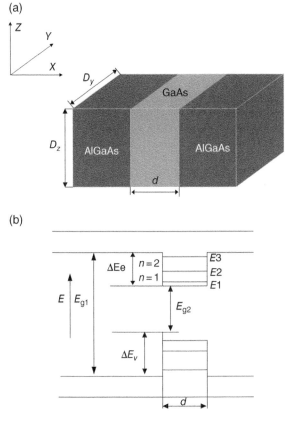

(b)

Figure 3.6 Typical quantum well device. (a) The QW structure schematic. (b) Energy band distribution.

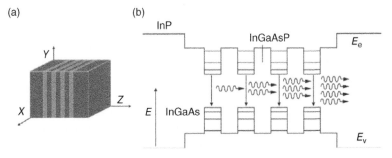

Figure 3.7 (a) Schematic. (b) Energy level diagram of a semiconductor laser diode with four quantum wells separated by a three-layer GaAlAs barrier layer.

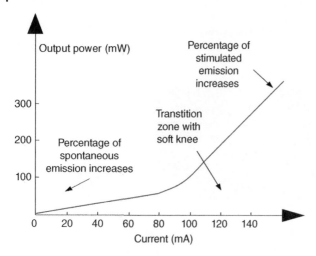

Figure 3.8 Superluminescent diode output versus current characteristics.

the low-coherence light source. Figure 3.8 shows the SLD output versus current characteristics.

The key to achieve super-radiation is to suppress Fabry–Pérot light oscillations and reduce light feedback as much as possible. The spectral output of an SLD is shown in Figure 3.9. In general, the ideal SLD would have output characteristics that are similar to LED at higher power levels. For superluminescent tubes, if the device does not completely suppress Fabry–Pérot oscillation, the emission spectrum is usually a superposition spectrum of multiple longitudinal modes and continuous spectrum, which results in a continuous spectrum with a certain modulation.

On the basis of the laser, lasing oscillation can be suppressed by introducing the light-absorbing region, reducing the reflectance of the end face of the device caused by the dielectric film, and adjusting the reflection angle of the end face of the device.

Figure 3.10 shows the optical power–current characteristic of SLD, LED, and LD. The light power of the LED varies linearly with the injection current from zero, and the slope of the curve is small. The LD optical power–current curve has an obvious threshold inflection point, and the linear part of the curve above the threshold increases rapidly. The SLD curve has no obvious inflection point and is in the middle between the LD and the LED.

Regarding the spectrum, the SLD, LED, and LD are compared in Figure 3.11. In terms of full width at half maximum (FWHM), the LED has the widest spectrum, and the LD has the narrowest spectrum. The spectrum of the SLD is in the middle. Finally, the main characteristics of the SLD, the LED, and the LD are compared in Table 3.1.

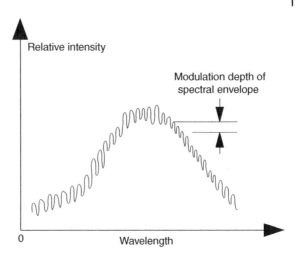

Figure 3.9 Spectral output of a superluminescent diode.

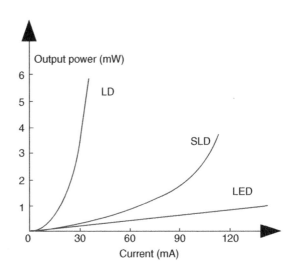

Figure 3.10 Optical power–current characteristic of SLD, LED, and LD.

3.2.4 Amplified Spontaneous Emission Sources

ASE light sources are also known as superfluorescent fibre sources. According to the different degree of excitation, the working substance (doped fibre) can be in three states, namely, weak excitation state, reverse excitation state, and over threshold excitation state. When the number of excited state level particles increases gradually, the number of spontaneous emission particles also increases

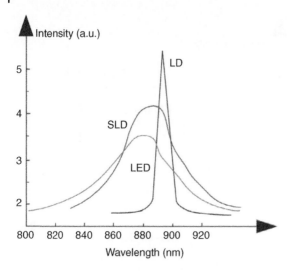

Figure 3.11 Comparison of SLD, LED, and LD spectra.

Table 3.1 Comparison of SLD, LED, and LD main characteristics.

Device characteristic parameters	SLD	LED	LD
Output power	High	Low	Highest
Coupling efficiency	High	Low	High
Coherence length	Short		Long
Temperature characteristics	Poor	Good	Poor
Modulation depth	Shallow		Deep
Response frequency	High	Low	High
Feedback noise	Low	Low	High
Device lifetime	Long	Longer	Short

gradually along with the interaction between them. When the number of excited state level particles exceeds the number of ground state level particles, the excited state is reversed, and the independent spontaneous radiation of a single particle gradually becomes the coordinated stimulated radiation of multiple particles, which is called ASE due to the amplification effect of doped fibre on spontaneous radiation. If the excitation is strong enough, ASE in a specific direction in the doped fibre will be greatly enhanced, and this enhanced radiation is called 'superfluorescence'.

In many fibre-optic sensors and fibre-optic detectors, broadband light sources with low time coherence are generally required. Currently, commercial broadband sources are mostly SLDs, but the maturity of rare earth doped fibre technology and the rapid development of pumping mechanisms have provided people with a convenient and reliable broadband fibre-optic light source. The ASE source is a high-stability, high-power output broadband source with a spectral range covering the C-band and the L-band. It can reduce the coherent noise of the system, the phase noise caused by fibre Rayleigh scattering, and the phase shift caused by the optical Kerr effect. By doping different optical rare earth elements in the fibre, such as Er^{3+}, Nd^{3+}, Yb^{3+} and Pr^{3+}, it is possible to obtain ultra-fluorescent output at many bands, meeting in this way the needs in various applications.

Depending on the direction of pump light and superfluorescence, the ASE source structure can be divided into two types: forward pump and backward pump [18, 19]. In addition, depending on whether there is reflection at both ends of the fibre, the ASE source can be divided into a single-pass structure and a two-way structure. Consequently, the superfluorescent fibre source is divided into several basic structures according to the similarities and differences between the direction of pump light and superfluorescence, and whether there are reflections at both ends of the fibre, as shown in Figure 3.12. If both ends of the fibre are non-reflective, it is called a one-way structure (Figure 3.12a, b). If one end of the fibre end is non-reflective and the other end is highly reflective, it is called a two-way structure (Figure 3.12c, d). In the single-pass structure, only one direction of spontaneous emission is utilized, and the two-way structure utilizes spontaneous radiation in two directions, so the two-way structure has higher conversion efficiency and lower pump power than the single-pass structure.

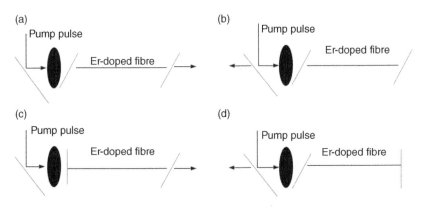

Figure 3.12 Structure of ASE light source.

3.2.5 Narrow Line Broadband Sweep Source

The narrow line wide-frequency sweep source refers to a light source with a good monochromaticity whose output wavelength can vary within a certain range and can be widely applied to a fibre grating demodulation system. It is mainly divided into two categories: swept-frequency laser sources based on tunable filters and the latest ultra-fast ultra-wideband swept sources.

The tunable filter-based swept laser source uses a tunable filter to filter and scan the broadband source. The filtering can be achieved by tuning the cavity length, temperature, energy gap, and refractive index of the filter. Regarding the fabrication of the sources, they can be manufactured with current control or with mechanical control. Current control technology is used to achieve wavelength tuning by changing the injection current, with nanosecond (ns) level tuning speed, wide tuning bandwidth, but low output power. The main control technology based on grating DBR (SG-DBR) laser and auxiliary gratings is directional coupled back-sampling scattering (GCSR) lasers. The mechanical control is mainly based on the microelectromechanical system (MEMS) technology, which permits to select the wavelength in a large adjustable bandwidth and with a high output power. Based on mechanical control technology, there are mainly DFB lasers, external cavity lasers (ECL), and vertical cavity surface-emitting lasers (VCSELs).

Finally, the ultra-fast ultra-wideband swept source uses the time-domain dispersion characteristics of the broadband pulse source to achieve time coding of the spectrum. That is, the wide-spectrum short-pulse light is stretched through the delay of the dispersive fibre, as shown in Figure 3.13.

3.2.6 Broadband Sources

Halogen light sources can generate a broadband spectrum (the model SLS202L from Thorlabs ranges from 450 to 5500 nm). These sources are composed of tungsten filament and quartz glass shell, which is introduced in the bulb with halogen gases

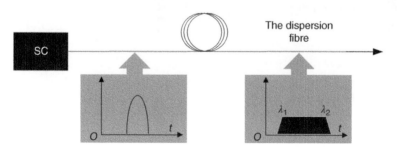

Figure 3.13 Ultra-fast ultra-wideband frequency sweeping light source.

such as iodine or bromine. The halogen lamp works as follow: when the filament heats up, the tungsten atoms evaporate and move towards the glass tube wall. When it approaches the glass tube wall, the tungsten vapour is cooled to about 800 °C and combines with the halogen atoms to form tungsten halide (tungsten iodide or tungsten bromide). Tungsten halide continues to move towards the centre of the glass tube, returning to the oxidized filament. Since tungsten halide is a very unstable compound, it regenerates into halogen vapour and tungsten when heated. Through this regenerative cycle, the service life of the filament is not only greatly extended (almost four times as long as that of an incandescent lamp), but also because the filament can work at higher temperatures, resulting in higher brightness, colour, and luminous efficiency.

With the continuous breakthrough of fibre laser technology, fibre light sources have been widely used in the market. Among them, the most successful one is the supercontinuous spectrum laser source based on photonic crystal. Supercontinuum light source refers to the light source whose spectrum is greatly broadened due to non-linear effect when light wave propagates in a transparent medium. Photonic crystal fibre (PCF) has advantages in non-linear optics because of its flexible dispersion characteristics. The generation of supercontinuum spectrum can be effectively controlled and adjusted by selecting fibre with different dispersion and non-linear characteristics and selecting appropriate input pulse parameters. The generation mechanism of supercontinuum involves many non-linear effects, including self-phase modulation (SPM), cross-phase modulation (XPM), four-wave mixing (FWM), and stimulated Raman scattering (SRS). These mechanisms depend on the parameters of the input pulse and the characteristics of the PCF, such as the width of the input pulse, the peak power, the dispersion curve of the fibre, the effective mode area, and the birefringence.

3.3 Optical Detectors

The most commonly used optical detectors for fibre-optic sensors are semiconductor photodiodes and avalanche photodiodes (APD). These types of detectors are generally used to monitor power returned from the sensor. Optical detectors should have high sensitivity, fast response, low noise, low cost, and high reliability, whereas the size should be compatible with the fibre core size.

According to the working principle, they can be divided into vacuum photoelectric devices and solid optical detectors, whereas according to their application, they can be divided into light intensity detectors, position detection devices, and imaging detection devices. This section mainly introduces the optical detectors commonly used in several fibre-optic sensor systems. To provide an overview of optical detection, we first discuss the basic principles of optical detectors.

3.3.1 Basic Principles of Optical Detectors

3.3.1.1 PN Photodetector

A PN junction is the simplest optical detector [20]. The basic principle is the photoelectric effect. When a reverse bias is applied to the PN junction, the direction of the applied electric field is the same as the direction of the electric field in the space charge region.

Since the photodiode operates in reverse bias mode, the carrier is substantially depleted in the space charge region. This region is called the depletion region when the photon energy incident on the detector is greater than the bandgap width of the semiconductor material. The electrons in the valence band absorb photon energy and transit to the conduction band, producing only electron–hole pairs. After the diffusion motion, the electrons and holes entering the depletion layer are accelerated by the electric field of the depletion layer, and each of them drifts in the opposite direction and is collected when reaching the edge of the depletion region so that the electromotive force is generated on both sides of the depletion layer of the PN junction. When the photodiode that receives the light is connected to the external circuit, there will be a current output in the circuit. The region where an incident photon is absorbed to produce an electron–hole pair is called the absorption region. The region that includes the depletion layer and a region whose width on both sides is the carrier diffusion length is called the active region.

Figure 3.14a shows the structure of a PN photodiode, whereas in Figure 3.14b the optical power decreases exponentially as the incident light is absorbed inside

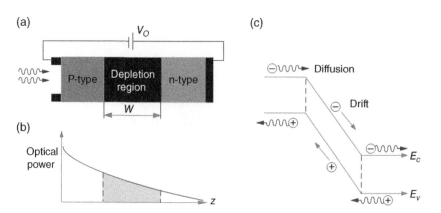

Figure 3.14 (a) A p-n photodiode under reverse bias. (b) Variation of optical power inside the photodiode. (c) Energy band diagram showing carrier movement through drift and diffusion.

Figure 3.15 Response of a p-n photodiode to a rectangular optical pulse when both drift and diffusion contribute to the detector current.

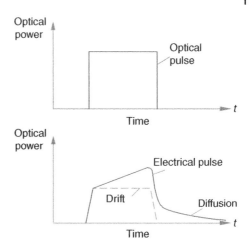

the depletion region. In Figure 3.14c, the electron–hole pairs generated inside the depletion region experience a large electric field and drift rapidly towards the P or N side, depending on the electric charge.

Figure 3.15 shows how the presence of a diffusive component can distort the temporal response of a photodiode. In order to absorb most of the incident optical power, the widths of the p and n regions can be decreased, and the depletion-region width can be increased. This way, the diffusion contribution can be reduced.

3.3.1.2 PIN Photodetector

The PIN detector is a special silicon photodiode. The silicon photodiode with conventional PN junction structure has a long response time, mainly because the diffusion time of the photo-generated carriers is long. In order to shorten the response time and improve the response speed, an intrinsic semiconductor material or a low-doped semiconductor material, called an I layer, is grown between the highly doped P-type and N-type semiconductors, and this structure is a PIN diode. The I layer not only shortens the response time of the detector but also broadens the effective working area of photoelectric conversion, improves quantum efficiency and sensitivity, and broadens the linear region.

Figure 3.16a shows the device structure together with the electric field distribution inside it under reverse bias operation. Figure 3.16b shows the design of an InGaAs p-i-n photodiode. The quantum efficiency can be made almost 100% by using an InGaAs layer 4–5 µm thick.

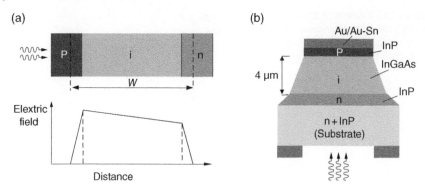

Figure 3.16 (a) A p-i-n photodiode together with the electric field distribution under reverse bias. (b) Design of an InGaAs p-i-n photodiode.

3.3.1.3 Avalanche Photodiode (APD)

The sensitivity of the PIN detector is low, and it is difficult to detect and receive weak optical signals. Therefore, APD detectors based on avalanche benefits have been developed. Under certain conditions, the accelerated electrons and holes get enough energy to collide with the crystal lattice to generate new electron–hole pairs, inducing a chain reaction. Consequently, APD has a high response capacity.

An alternative to the standard APD is another structure that enables uniformity of carrier multiplication and high quantum efficiency and high response speed. This structure is shown in Figure 3.17. This type of device is a P+IPN+ multilayer structure with one more P-type layer than the original PIN photodiode. Figure 3.17a shows the APD structure together with the variation of electric field in various layers. Figure 3.17b shows the design of a specific type of APD structure, the silicon reach-through APD.

3.3.2 Main Characteristics of Optical Detectors

3.3.2.1 Operating Wavelength Range and Cut-Off Wavelength

Semiconductor photodiodes fabricated from a particular semiconductor material can only detect optical signals in a certain wavelength range. In a photodiode that is exposed to light, the energy of the incident photon must be greater than or equal to the forbidden bandwidth of the semiconductor material to produce photo-generated carriers, $E = h\nu \geq E_g$.

From the previous equation, $\nu_c = E_g/h$ is called the cut-off frequency, and the corresponding wavelength $\lambda_c = h/E_g$ is called the cut-off wavelength, which is the upper limit wavelength that the semiconductor photodetector can detect. Figure 3.18 shows the working range of several common semiconductor materials [15].

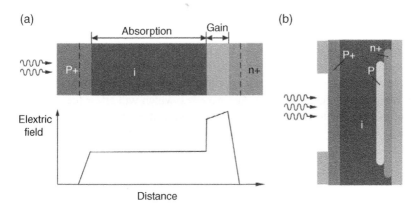

Figure 3.17 (a) An APD together with the electric field distribution inside various layers under reverse bias. (b) Design of silicon reach-through APD.

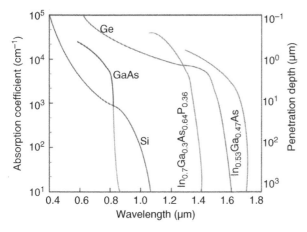

Figure 3.18 Wavelength dependence of the absorption coefficient for several semiconductor materials.

3.3.2.2 Quantum Efficiency and Responsiveness

The quantum efficiency of a photodetector represents the efficiency with which an incident photon is excited to be converted into photoelectrons. It is defined as the ratio of the number of photoelectrons generated per unit time to the number of incident photons:

$$\eta = \frac{I_p/e}{P/h\nu} \tag{3.3}$$

where P is the average optical power projected onto the semiconductor photodetector, I_p is the average photo-generated current, and e is the electron charge.

In order to achieve a high quantum efficiency, the photodiode depletion layer must be made so wide that a significant proportion of the photons fall in the depletion layer to generate electron–hole pairs, but if the depletion layer is too wide, the transit time of photo-generated carriers increases, which will affect the response speed of the photodetector. The responsiveness of a photodetector is defined as the ratio of the average output photocurrent to the average incident optical power:

$$R = \frac{\eta e}{h\nu} \tag{3.4}$$

A typical R is 0.5–0.7 A/W. Figure 3.19 indicates the responsiveness of the photodiodes of several materials.

3.3.2.3 Response Time

The response time of the photodetector to the light signal change is its response time. Usually, when the photodetector receives a step light pulse, the response time is the time interval between the 10 and 90% points of the output pulse front edge (the rise time). The fall time of the trailing edge of the pulse is the same as that

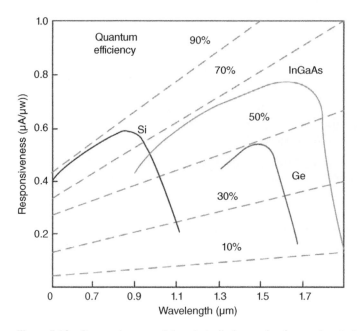

Figure 3.19 Responsiveness of the photodiodes made of several materials.

of the leading edge for a fully depleted photodiode but may be different at a low bias of non-depletion. In that case it is important to indicate both response times.

The response time is mainly determined by three factors: PN junction capacitance, diffusion time of photo-generated carriers in the active region, and transit time of photo-generated carriers in the depletion layer.

3.3.2.4 Materials and Structures of Semiconductor Photodiodes

In order to obtain the best conversion efficiency, quantum efficiency, and low dark current, which decreases exponentially with increasing bandgap energy, the bandgap energy E_g of an ideal photodiode material should be slightly smaller than the photon corresponding to the longest operating wavelength. In the short wavelength region around 0.85 µm, Si is the most desirable material, and its cut-off wavelength is 1.09 µm. In the long wavelength region, InP and InGaAs are ideal semiconductor materials (the response wavelength of the lattice-matched $In_{0.53}Ga_{0.47}As/InP$ reaches 1.68 µm). Figure 3.20 shows the structure of a typical InGaAs–InP photodiode.

Compared with the top incident structure, the bottom incident structure has a small capacitance (less than 0.1 pF), high quantum efficiency (75–100%), and dark current less than 1 nA. The depletion layer InGaAs layer thickness of these two types of devices is about 3 µm to obtain high quantum efficiency and bandwidth, and the low-doped InGaAs layer can be completely depleted at a low voltage of 5 V.

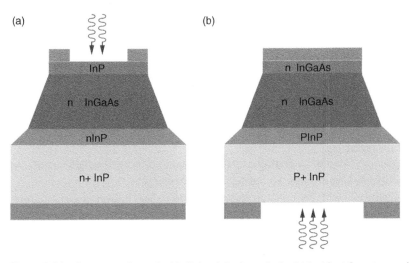

Figure 3.20 Structure of a typical InGaAs–InP photodiode: (a) incident from top and (b) incident from bottom.

The depletion layer is relatively narrow, which can shorten the transit time and make the theoretical bandwidth reach 15 GHz or even higher.

3.3.3 Optical Spectrometers

Optical spectrometers (often simply called 'spectrometers') show the intensity of light as a function of wavelength or of frequency. In the next chapters we will see that many optical fibre sensors are based on tracking the wavelength shift of a resonance in the optical spectrum. Therefore, this instrument is vital in optical fibre sensor domain.

The basic function of a spectrometer is to measure the spectral characteristics of the studied light (reflected, absorbed, scattered, or excited fluorescence), including wavelength, intensity, and other spectral line characteristics. Therefore, spectral instruments should have the following characteristics:

1) Light splitting: Separate the light according to a certain wavelength or wavenumber in a certain space.
2) Photosensitive: The light signal is converted into an electrical signal that is easy to be measured, and the intensity of each wavelength of light is measured accordingly.
3) Spectral diagram: Record or display the corresponding spectral diagram of the separated light waves and their intensities as a function of the wavelength or the wavenumber.

A typical spectrometer consists of an optical platform and a detection system. It includes the following major parts:

1) Incident slit: The object point of the spectrometer imaging system formed under the irradiation of incident light.
2) Collimating element: Makes the light from the slit become parallel. The collimating element may be an independent lens, a reflector, or directly integrated on a dispersion element, such as a concave grating in a concave grating spectrometer.
3) Dispersion element: A grating usually adopted to disperse the light signal into multiple beams in space according to the wavelength.
4) Focusing element: Focuses the scattered beam so that it forms a series of incident images of slits on the focal plane, where each image point corresponds to a specific wavelength.
5) Detector array: Placed on the focal plane to measure the light intensity of each wavelength image point. The detector array can be a CCD array or a different kind of light detector array.

There are many types and classification methods of spectrometers. According to the principle of spectrum decomposition adopted by spectrometers, they can be divided into two categories: classical spectrometers and new spectrometers. The classical spectrometer is an instrument based on the principle of spatial dispersion. The new spectrometer is an instrument based on the modulation principle, so it is also called modulation spectrometer.

According to its dispersion principle, the classical spectrometers can be divided into three categories: prism spectrometers, diffraction grating spectrometers, and interference spectrometers. According to the normal spectrum range of spectral instruments, spectrometers can be divided into the following categories and typical wavelength ranges, respectively: vacuum ultraviolet (far ultraviolet) spectrometer (6–200 nm), ultraviolet spectrometer (185–400 nm), visible spectrometer (380–780 nm), near-infrared spectrometer (780–2500 nm), infrared spectrometer (2.5–50 μm), and far-infrared spectrometer (50 μm to 1 mm).

3.4 Light Coupling Technology

Coupling technology is one of the key technologies of fibre-optic sensors. The coupling mainly includes the connection between the optical fibre and the light source, the optical fibre and the receiver, and optical fibre to optical fibre directly or indirectly (through various types of connecting devices such as lenses, couplers, etc.). The most common coupling techniques for fibre and light source (or detector) and for fibre and fibre are described below [21].

3.4.1 Coupling of Fibre and Light Source

3.4.1.1 Coupling of Semiconductor Lasers and Optical Fibres
In recent years, high-power semiconductor lasers have been increasingly used in production, such as direct material processing, fibre laser and amplifier pumping, free-space optical communication, printing, and medical applications. The semiconductor laser package enables the laser device to operate at high plug-in efficiency, improve stability, and save the user's cost of use, making it possible to automate large-volume machine packaging. In this sense, there exist two main coupling methods: direct coupling and lens coupling.

3.4.1.1.1 Direct Coupling
Direct coupling consists of directly opposing the processed fibre to the light-emitting surface of the laser. The main factors affecting the coupling efficiency at this time are the matching of the light-emitting area of the light source and

the total area of the fibre core and the matching of the light source divergence angle and the fibre numerical aperture angle.

3.4.1.1.2 Lens Coupling

The use of lens coupling can greatly improve the coupling efficiency and is therefore widely used. The types of lenses used are mainly end-face ball lenses (forming a fibre end face into a hemisphere), cylindrical lenses, convex lenses, conical lenses, and shaped lens couplings.

The main advantages of this optical planar package design are small size, vertical stacking, complete sealing, low thermal effect, low cost, and high output power (output power greater than 6 W).

All packaging processes are sealed in a fluid-free environment, making the laser highly reliable. The savings in materials and packaging procedures reduce the considerable cost of packaging and eliminate the need for all non-vertical assembly steps, making it possible to automate the packaging of semiconductor lasers with pigtail outputs. Other features include fixed passive connections and integrated fibre coupling.

Wedge-lens optical fibre is the common method for coupling a laser to optical fibre in the market. This type of moulding requires only a single step of fibre alignment. In addition, the fibre is fixed to a very robust, stable, and protective accessory with Au/Sn soldering. In this sense, such high reliability 6 W fibre output semiconductor lasers are currently available on the market. This type of coupling is known as the second type of technology for semiconductor packaging, and its products have been widely used in fibre laser, fibre remote sensing, fibre-optic ranging, and other fields.

3.4.1.2 Coupling Loss of Semiconductor Light-Emitting Diodes and Optical Fibres

From the perspective of coupling, the main difference between the semiconductor light-emitting tube and the semiconductor laser is that the former is spontaneous radiation and that the directivity of light emission is poor. Approximating a uniform surface-emitting device, its luminescence performance is similar to that of a cosine illuminator.

However, in spite of this difference, the method of coupling the LED to the lens for the optical fibre is similar to the method of coupling the laser and the optical fibre.

3.4.2 Multimode Fibre Coupled Through Lens

When the fibre and the lens are coupled, the matching of the numerical apertures and the aberration of the lens should be considered. The calculation of the coupling efficiency of three types of single lenses (ball lens, variable refractive index rod, cylindrical lens) considering aberrations is described in detail in the literature. With the continuous advancement of microelectronics technology, binary optics has developed rapidly. Therefore, some people have developed a micro-phase grating to change the spatial distribution of the output laser wave of the semiconductor laser, that is, from a long strip distribution to a circular symmetric distribution to improve the coupling efficiency of the semiconductor laser and the optical fibre. For the coupling of single-mode fibres, the fabrication, positioning, retention, and anti-interference of the micro-components are more difficult than the coupling of multimode fibres, because the core diameter of the fibre is much smaller.

3.4.3 Direct Coupling of Fibre and Fibre

Once the coupling from the source to the fibre has been explained, it is important to focus now on the couplings between fibres, because it is possible that between the source and the detector exists more than one fibre element (e.g. the sensor element is a PCF, and the fibre connected to the source and the detector is a standard fibre such as standard single-mode fibre).

The direct coupling of fibre to fibre has two ways: fixed connection and active connection. The fixed connection, that is, the fibre fusion, is realized by the optical fibre fusion splicer and has the advantages of small insertion loss and good stability; the disadvantage is that it is inflexible and inconvenient to debug. The active connection uses fibre-optic connectors and flanges to achieve direct fibre-to-fibre coupling. There are two types of commonly used connector types: bare fibre connectors and jumpers. The bare fibre connector is divided into a bare fibre adapter, a V-shaped slot, and a splicer and has subcategories such as FC/ST/SC/LC/SMA depending on the type of the connector. When calculating the loss, it is considered separately according to the different types of mutually coupled fibres. In addition, each connector may include a different polishing (PC, UPC, APC), which is related to the quality of the fibre-optic light wave transmission in terms of optical return loss and insertion loss. This is vital in the domain of communications and less important for sensing (typically the simplest PC polishing is enough). However, in some cases such as set-ups where laser sources are used, it is important to avoid return losses that may damage the source.

In addition, it is important to consider the influence of the axis and axial inclination on the coupling loss when the two fibres are directly connected and also the effect of the incompleteness (tilt or bend) of the fibre end face or the effect of connecting different fibre types.

3.5 Fibre-Optic Device

The unit that converts, connects, and controls the optical path in the optical fibre transmission system is called fibre-optic device, and it includes optical connectors, optical couplers, optical switches, optical attenuators, multiplexers, and demultiplexers. Fibre-optic devices have two basic parameters that must be considered: insertion loss and isolation. Regarding both parameters, as a general rule, an optical fibre transmission system requires low insertion loss and good isolation performance.

According to functional classification, optical fibre devices include fibre coupler, fibre circulator, fibre isolator, fibre attenuator, fibre polarizer, fibre switch, Faraday light source, fibre amplifier, fibre delay line, etc. In addition to optical fibre communication, optical fibre devices are widely used in optical fibre sensing, data processing, and computing technology, and this technology has evolved into a unique category of optoelectronic devices.

3.5.1 Fibre Coupler

The fibre coupler is a functional device intended for realizing the splitting / combining of optical signals. There are many classification methods for fibre couplers. One of them is according to the production method: polishing coupler, molten double-cone coupler, microbend coupler, and side compensation-type coupler. Couplers can also be classified into single-mode fibre couplers and multimode fibre couplers according to the transmission light mode. A schematic diagram of a 2×2 coupler is shown in Figure 3.21.

The main parameters characterizing the performance of a fibre coupler are insertion loss, additional loss, split ratio, uniformity, and directionality (isolation).

3.5.2 Optical Isolator

An optical isolator is an optical passive device that only allows light to pass in one direction, preventing adverse effects of the reverse transmitted light generated in the optical path for various reasons on the light source and the optical path system [22]. Its working principle is shown in Figure 3.22. Assuming that the incident light is vertically polarized, the direction of vibration transmission of the first polarizer is also in the vertical direction, and the direction of vibration transmission of the second polarizer is at an angle of 45° to the direction of vibration transmission of the first polarizer in order to accept all light whose polarization has been rotated in the Faraday rotator 45°. The reflected light passes again through the Faraday rotator, and, with the additional 45° rotation, it is 90° rotated

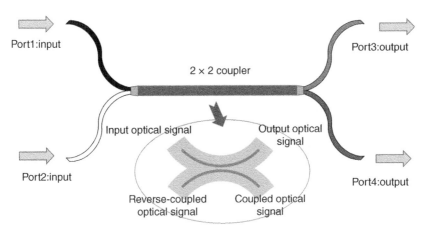

Figure 3.21 Schematic diagram of 2 × 2 coupler.

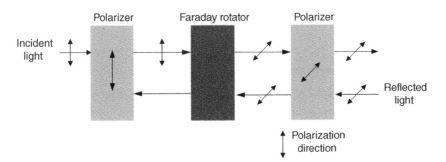

Figure 3.22 Schematic diagram of working principle of optical isolator.

compared with the input light (i.e. horizontally polarized). As a result, no reflected light passes through the first polarizer, which accepts only vertically polarized light.

The main characteristic parameters of the optical isolator include insertion loss, isolation, crosstalk, polarization mode dispersion, return loss, and so on. There are many factors that affect the isolation of the isolator, including:

1) Distance from the polarizer to the Faraday rotator
2) Optical component surface reflectivity
3) Polarizer wedge angle, spacing
4) Relative angle of the crystal axis

3.5.3 Optical Circulator

The optical circulator is a passive N-port non-reciprocal (forward sequential conduction, reverse transmission cut-off) optics [23]. A typical circulator generally has three to four ports, and its structure is similar to that of an optical isolator. The main difference is that the optical isolator is a two-port device whereas the circulator is an N-port device. As shown in Figure 3.23, the light received by port 1 is transferred almost without loss to port 2, and almost no light received by the other ports. After that, light received by port 2 is transferred almost without loss to port 3, while the rest of ports receive no light, and so on. In this way, there is a separation of optical signals transmitted in the forward and reverse directions. The main characteristics of optical circulators are insertion loss, isolation, crosstalk, polarization mode dispersion, and return loss.

3.5.4 Fibre Attenuator

A fibre attenuator is an optical passive device used to attenuate optical power to prevent signal saturation due to an excessively powerful input signal.

3.5.5 Fibre Polarizer

In coherent fibre communication and many functional fibre sensing, especially fibre-optic gyros, fibre polarizers are key components mainly used to retain one polarization mode and eliminate another polarization mode. They can be divided into two categories:

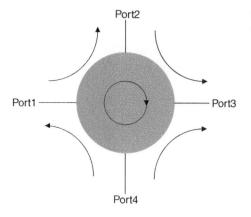

Figure 3.23 Schematic diagram of optical circulator.

1) Attenuating polarizer: It increases the attenuation difference of two polarization states.
2) Cut-off polarizer: It lets one of the polarization modes to be cut off, such as it is the case with birefringent crystal cladding fibre polarizer.

3.5.6 Optical Switch

It is a component that is similar to the coupler related to the ability to control several input signals, but it is much more versatile and sophisticated. It has one or more selectable transmission ports to control the on/off of the optical path (or the on and off of the signal) and can transfer the input optical signals in the optical transmission line to any of the outputs or even perform logical operations. Moreover, its mode of operation, unlike for couplers, can be modified continuously (i.e. in a 2×2 switch it is possible to connect the inputs to output number 1, but the operation mode can be modified and the same inputs can be connected to output number 2).

Its performance parameters mainly include insertion loss, switching time, extinction ratio, isolation, crosstalk, and so on.

3.6 Optical Modulation and Interrogation of Optical Fibre-Optic Sensors

The optical modulator is a key component in an optical fibre system. It has many functions such as amplitude, phase, frequency, polarization modulation, and so on [24]. Most optical modulators are solid-state devices. Light is modulated by changing the optical properties of the device material through an electrical control signal. The control signal is connected to the material characteristics by electro-optic, acousto-optic, or magneto-optic mechanisms. With the development of technology, more and more high-performance devices are used to demodulate optical signals of optical fibre sensors.

Three basic types of solid-state light modulators are bulk, integrated optics, and all-fibre devices (Figure 3.24). Signals in bulk modulators propagate through uniform material blocks. The bulk modulators lack waveguides and require high electric driving power and external optical devices to couple light into and out of the optical fibres. The integrated optical modulator (IOM) consists of waveguides manufactured directly in the modulator material. It requires less power and does not use external optical devices to couple optical fibres. In all-fibre modulators, optical signals are transmitted in the optical fibre, and the required modulation

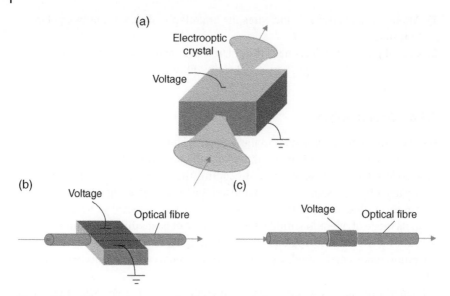

Figure 3.24 Three major types of solid-state optical modulators: (a) bulk, (b) integrated, and (c) all-fibre.

can be achieved by interfering with the optical fibre through the control signal. The advantage of this type of device is that it does not require optical fibre coupling devices or precise alignment. However, the modulation ability of ordinary optical fibre materials such as glass is relatively weak and requires relatively high driving power.

3.6.1 Intensity-Modulated Optical Fibre Sensing Technology

The principle of the intensity modulation in optical fibre sensing is to use the change of the measured object to cause changes in the refractive index, absorption, or reflection of the sensitive element, thereby causing a change in the light intensity to achieve sensitive measurement. Types of intensity-modulated fibre-optic sensors include reflective intensity modulation sensors, projected intensity modulation, optical mode (microbend) intensity modulation, refractive index intensity modulation, optical absorption coefficient intensity modulation, etc.

3.6.1.1 Reflective Intensity Modulation Sensor
The reflective intensity modulation sensor is a non-functional optical fibre sensor, and the optical fibre only serves to transmit light. The optical fibre is divided into two parts, i.e. an input fibre and an output fibre. The modulation mechanism is

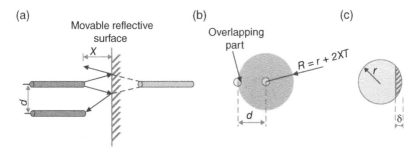

Figure 3.25 Schematic diagram of fibre reflection intensity modulation principle: (a) optical coupling diagram between fibres, (b) schematic diagram of coupled optical power, and (c) a simplified 'straight-edge' model for coupling processes.

based on the fact that the input fibre scatters the light from the light source towards the surface of the object to be measured and then it is reflected from the surface of the object to be measured to the other output fibre, which receives the intensity of the light. The size varies with the distance between the surface of the object being measured and the fibre.

As shown in Figure 3.25, a reflecting surface is placed at a position from the end face x of the fibre, which moves perpendicular to the input and output fibre axes, and a virtual image of the input fibre is formed on the reflecting surface. d is the distance between the output fibre and the input fibre, xT is the radius of the circle at the bottom of the emitting light cone, r is the radius of the fibre, and NA is the numerical aperture.

When $x < d/2T$, $T = \tan(\arcsin(NA))$, the optical power coupled into the fibre is zero; when $x > (d + 2r) \times 2T$, the light cones emitted from the images of the output and input optical fibres intersect at the bottom. The area where they intersect is always going to be πr^2, and the base area of the light cone is $\pi(xT)^2$. Therefore, the light transmission coefficient of the gap within this range is $(r/xT)^2$. When $d/2T \leq x \leq (d + 2r) \times 2T$, the flux coupled to the output fibre is determined by the area of the overlap between the bottom of the light cone emitted by the image of the input fibre and the output fibre. The area of the overlapped area can be calculated by linear approximation, that is, the edge of the intersection of the light cone bottom surface and the exit optical fibre end surface is approximated by a straight line.

The response of the modulator is equivalent to calculating the coupling between the virtual image fibre and the output fibre. If δ is the overlapping distance between the light cone edge and the output fibre, as shown in Figure 3.25c, under the premise of such approximation, the percentage of the surface exposed by the light cone on the end face of the output fibre can be given by geometric analysis.

3.6.1.2 Transmissive Intensity Modulation Sensor

The transmissive intensity-modulated fibre-optic sensor utilizes the spatial distribution characteristics of the optical fibre end face and modulates the light intensity by modulating the spatial position and direction of the optical fibre or the spatial refractive index. The schematic diagram is shown in Figure 3.26.

3.6.1.3 Light Mode (Microbend) Intensity Modulation Sensor

As shown in Figure 3.27, when the state between the fibres changes, mode coupling occurs in the fibre, and some of the waveguide modes become radiation modes, causing microbend losses. By combining the microbend losses with physical quantities such as the position and pressure of the device causing the microbend, various optical mode microbend intensity modulation sensors are constructed.

3.6.1.4 Refractive Index Intensity-Modulated Fibre-Optic Sensor

Generally, there are differences in the refractive index temperature coefficients of the core and the cladding in the optical fibre material, and the refractive index changes of the optical fibre and the environmental medium are different. The

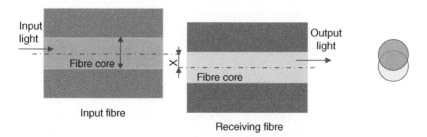

Figure 3.26 Structure diagram of transmission intensity-modulated fibre sensor.

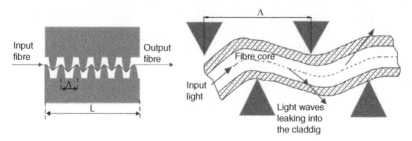

Figure 3.27 The schematic diagram of the optical mode microbend loss intensity-modulated fibre sensor.

variation of the light intensity signal caused by the change of the refractive index is detected by the photodetector and finally the physical quantity to be measured with a demodulation system.

The refractive index intensity modulation-type sensor includes three types: a fibre refractive index change type, an evanescent wave coupling type, and a reflection coefficient type. These techniques will be seen in the next chapters.

3.6.2 Wavelength Modulation Optical Fibre Sensing Technology

In some cases, the interaction between the parameters to be measured and the sensitive optical fibres results in the change of the wavelength of the transmitted light in the optical fibres. Consequently, the sensing method for determining the measured parameter according to the variation of the measured optical wavelength can be made into a wavelength modulation-type sensor, and the fibre grating sensor is a typical wavelength-modulated fibre-optic sensor. Demodulation of the wavelength-encoded signal of the fibre grating is the key to realize the fibre grating sensing. Generally, there are four kinds of modulation and demodulation schemes.

3.6.2.1 Direct Demodulation System

The direct method of fibre grating demodulation is to use spectrometer and spectral analysis as the most basic measurement method. The basic principle is to transmit the output light of the sensing probe to a spectrometer or an interrogator through the optical fibre. The detector monitors the light intensity distribution at different wavelengths. When an optical spectrum variation is induced, the corresponding wavelength shift is processed and evaluated in terms of the parameter that is being measured by the optical fibre sensors. This is what permits to monitor a specific parameter. A typical detection diagram is shown in Figure 3.28.

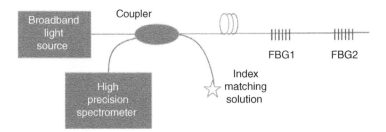

Figure 3.28 Schematic diagram of wavelength demodulation detection by spectrometer.

3.6.2.2 NarrowBand Laser Scanning System

The narrowband laser scanning system is based on the construction of a laser cavity with adjustable working wavelength. The sensing FBG is used as a mirror of the resonant cavity. When the operating wavelength is adjusted to the FBG reflection wavelength, the reflected signal light will excite a strong laser, and the peak wavelength information of fibre grating is obtained by tuning the corresponding relation.

Figure 3.29a shows a mode-locked modulation method. The laser cavity is composed of a broadband reflector or mirror, a sensing grating, an erbium-doped fibre segment, and an acousto-optic mode-locked modulator (MLM). After the pump light is coupled into the laser cavity, changing the frequency of the modulator can lock the laser output in different modes.

When the adjustment frequency meets the requirement of $f = c/(2L)$ (L is the length of the cavity formed by the reflector and a certain FBG), the laser mode is locked to form a strong laser pulse, depending on what FBG is addressed. However, the gain medium and MLM should be reasonably selected so that they can excite one wavelength and suppress other grating wavelengths.

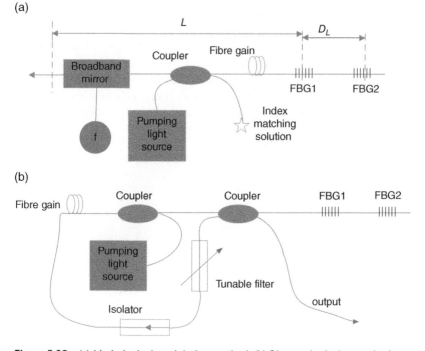

Figure 3.29 (a) Mode-locked modulation method. (b) Ring cavity lasing method.

Figure 3.29b shows a ring cavity lasing method. A tunable filter is used to adjust the working wavelength of an erbium-doped ring cavity. When the transmission wavelength of filter is tuned to the central reflection wavelength of an FBG, the reflected light obtained by the weak input light after passing through the fibre grating can pass through the tunable filter and can be amplified in the circular cavity in a one-way cycle to stimulate the strong laser. The wavelength of each fibre grating can be determined by synchronously detecting its output intensity. The output intensity of the system will be maximum when the wavelength of the tunable filter matches the wavelength of one of the FBGs.

3.6.2.3 Broadband Source Filter Scanning System

The typical broadband source filter scanning system uses a tunable Fabry–Pérot filter method that uses a closed-loop mode for a single raster and a scan mode for a raster of a multiplexed system. As shown in Figure 3.30, the light from the broadband source enters the sensing grating array through the isolator, and the reflected light signal passes through the coupler to reach the tunable filter. The filter operates in the scanning state, and the sawtooth scanning voltage is applied to the piezoelectric element. After that, it is important to adjust the length of the cavity to make the narrow passband scan within a certain range. When the waveform matches the Bragg wavelength of the sensing grating, the signal reflected by the sensing grating passes. Therefore, the position of the fibre grating reflection peak can be estimated from the relationship between the driving voltage of the Fabry–Pérot filter and the transmission wavelength.

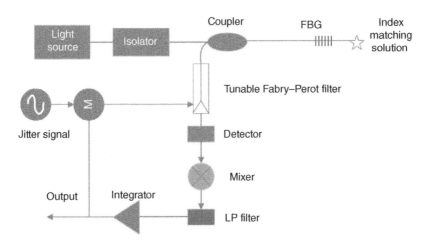

Figure 3.30 Principal diagram of tunable Fabry–Pérot filter.

3.6.2.4 Linear Sideband Filtering Method

The linear sideband filtering method converts the wavelength information into intensity information by using the wavelength range where the filter response function is linear. The reflected light of the sensing FBG is divided into two parts by a device such as a coupler: one part enters the detector after passing through the filter, while the other part directly enters the detector and is amplified to become a reference signal, as shown in Figure 3.31.

3.6.2.5 Interference Demodulation System

3.6.2.5.1 Non-equilibrium Mach–Zehnder (M–Z) Interferometer Method

As shown in Figure 3.32, the light emitted by the broadband light source is incident into the sensing fibre grating through the coupler, and the reflected light enters the M–Z interferometer with unequal arm length through another coupler. The

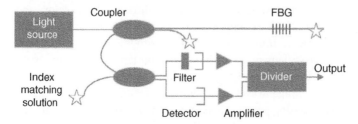

Figure 3.31 Schematic diagram of demodulation method for linear sideband filter.

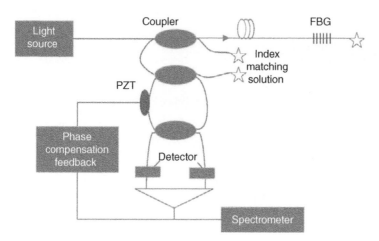

Figure 3.32 Structural diagram of demodulation system for non-equilibrium Mach–Zehnder (M–Z) interferometer.

interferometer transforms the offset of the central reflection wavelength of the sensing grating into the phase change to detect it. When the reflection wavelength of the FBG changes $\Delta\lambda$, the phase changes $\Delta\varphi$:

$$\Delta\varphi(\lambda) = \frac{-2\pi nd\Delta\lambda}{\lambda^2} \tag{3.5}$$

where n is the effective refractive index of the optical fibre, d is the length difference between the two arms of the interferometer, and λ is the central wavelength of the FBG reflected light. The change of FBG wavelength can be obtained from $\Delta\varphi$ measured by the detector, and the external measured physical quantities can be detected.

3.6.2.5.2 High Birefringence Fibre Environment Interference Demodulation Method

As shown in Figure 3.33, this demodulation system adopts a ring cavity structure, in which a section of the optical fibre in the ring cavity is a polarization-maintaining fibre, and the polarization controller is connected in series in the cavity. The signal light enters from one end, and the light propagating clockwise becomes a forward wave, and the light propagating counterclockwise is called a reverse wave (the forward wave comes out from the coupler port 3). Before entering the high birefringence fibre, a rotation transformation of the coordinate system is made with the polarization controller so that the directions of the X and Y axes of the coordinate axes are the fast and slow axes of the high birefringence fibre, respectively. Through propagation in the high birefringence optical fibre, phase difference is generated in the two polarization directions. The transmission route of the reverse wave is similar. Finally, the forward and reverse waves are extracted coherently from the coupler. By measuring the phase difference, the measured physical quantities can be detected.

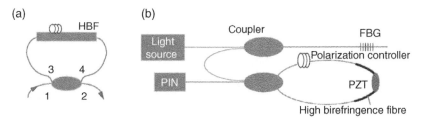

Figure 3.33 Schematic diagram of interference demodulation method for high birefringence optical fibre environment: (a) demodulation system architecture and (b) description of the complete demodulation system.

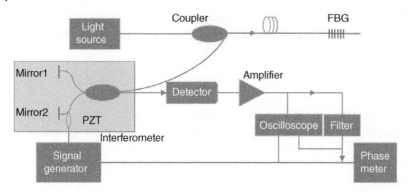

Figure 3.34 Schematic diagram of demodulation method for Michelson interferometer.

3.6.2.5.3 Demodulation of Unbalanced Michelson Interferometer

This method is similar to the non-equilibrium M–Z interferometer method. The system schematic diagram is shown in Figure 3.34. The mirror or reflector 1 and the mirror or reflector 2 are the reflective end faces of the fibre, respectively. Due to the difference of optical path, the two reflected beams will interfere at the coupler. After the interferometer is processed by the detector, the interferometer and the piezoelectric translator (PZT) actuator driving signal are input into the phase meter as the signal to be measured and the reference signal, respectively, and the phase meter is displayed. The value shown is related to the strain to be measured on the sensing grating. This method has high sensitivity but is susceptible to the influence of external environment.

3.6.3 Phase Modulation Optical Fibre Sensing Technology

The basic principle of the phase modulation optical fibre sensor is to use the action of the measured object on the sensitive component to change the refractive index or propagation constant of the sensitive component, resulting in a phase change of the light, leading to a variation in the interference fringes generated by the two monochromatic lights. The amount of change in the phase of the light is determined by detecting the amount of change in the interference fringe, thereby obtaining information of the object to be measured.

Phase modulation of optical fibres means that when the sensing fibres are subjected to external mechanical vibration or temperature field, the external signals change the geometric size and refractive index of the sensing fibres through the force-strain effect, thermal-strain effect, elastic-optic effect, and thermo-optic effect of the fibres, resulting in optical transmission delay and phase change in the fibres, exerting in this way a modification of the optical phase.

Since the photosensitive detector cannot respond to the high frequency of the laser, the phase modulation generated by the measured position cannot be directly detected. Usually, the phase modulation is first converted into amplitude modulation, and then the phase change is obtained by the change of the detected light intensity.

Phase-modulated optical fibre sensors are based on interference principle, which has the advantages of high sensitivity, high detection resolution, and wide range of sensing objects. The optical fibre interferometer is the core of the interferometric optical fibre sensor. The light emitted by the light source is usually divided into two or more beams. After travelling along different paths, the separated beams will converge again and reach the same detector. If the amplitudes of these two beams are A_1 and A_2, respectively, one of them is modulated by some factors (pressure, temperature, etc.), and it will interfere with the other. The light intensity after interference is

$$A_2 = A_1^2 + A_2^2 + 2A_1A_2 \cos(\Delta\phi). \tag{3.6}$$

In the formula, $\Delta\phi$ is is the phase difference between two coherent beams caused by phase modulation. The change of phase between two coherent beams can be determined by detecting the change of interference intensity, and the value of external parameters can be obtained.

There are four types of commonly used optical fibre interferometers: Michelson optical fibre interferometer, M–Z optical fibre interferometer, Sagnac optical fibre interferometer, and Fabry–Pérot optical fibre interferometer. They will be discussed in the Chapter 4.

References

1 Gall, D. (2001). Current topics in light source technology for lighting and radiation. *Advanced Engineering Materials* 3 (10): 775–780.

2 Krames, M.R., Shchekin, O.B., Regina, M. et al. (2007). Status and future of high-power light-emitting diodes for solid-state lighting. *Journal of Display Technology* 3 (2): 160–175.

3 Moiseyev, L.V. and Odnoblyudov, M.A. (2014). A review of modern light-emitting diode technologies in light sources for general illumination. *Light and Engineering* 22 (3): 11–19.

4 Trpper, A.C. and Hoogland, S. (2006). Extended cavity surface-emitting semiconductor lasers. *Progress in Quantum Electronics* 30 (1): 1–43.

5 Foutse, M., Kingni, S.T., Nana, B. et al. (2015). Edge-emitting semiconductor laser driven by a van der Pol oscillator: analytical and numerical analysis. *Optical and Quantum Electronics* 47 (3): 705–720.

6 Sherman, B. and Black, J.F. (1970). Scanned laser infrared microscope. *Applied Optics* 9 (4): 802–809.

7 Agrawal, G.P. and Dutta, N.K. (1993). *Semiconductor Lasers*, 2e. New York: Van Nostrand Reinhold.

8 Chuang, S.l., Liu, G., and Kondratko, T.K. (2008). High-speed low chirp semiconductor lasers. In: *Optical Fibre Telecommunications* (eds. I.P. Kaminow, T. Li and A.E. Willner), 53–80. Boston: Academic Press. Chapter 3.

9 Liu, D., Tang, K., Yang, Z. et al. (2011). A fibre Bragg grating sensor network using an improved differential evolution algorithm. *IEEE Photonics Technology Letters* 23 (19): 1385–1387.

10 Chung, W.H., Tam, H.Y., Wai, P.K.A. et al. (2005). Time- and wavelength-division multiplexing of FBG sensors using a semiconductor optical amplifier in ring cavity configuration. *IEEE Photonics Technology Letters* 17 (12): 2709–2711.

11 Yao, T., Zhu, D., Liu, S. et al. (2014). Wavelength-division multiplexed fibre-connected sensor network for source localization. *IEEE Photonics Technology Letters* 26 (18): 1874–1877.

12 Lopez-amo, M., Blair, L.T., and Urquhart, P. (1993). Wavelength-division-multiplexed distributed optical fibre amplifier bus network for data and sensors. *Optics Letters* 18 (14): 1159–1161.

13 Zhang, F., Wang, C., Jiang, S. et al. (2017). Dynamic fibre Bragg grating sensor array with increased wavelength-division multiplexing density and low crosstalk. *Optical Engineering* 56 (3): 037101.

14 Diaz, S., Cerrolaza, B., Lasheras, G. et al. (2007). Double Raman amplified bus networks for wavelength-division multiplexing of fibre-optic sensors. *Journal of Lightwave Technology* 25 (3): 733–739.

15 Rao, Y.J., Ran, Z.L., and Zhou, C.X. (2006). Fibre-optic Fabry-Perot sensors based on a combination of spatial-frequency division multiplexing and wavelength division multiplexing formed by chirped fibre Bragg grating pairs. *Applied Optics* A29 (2): 131–134.

16 Kubota, H., Oomi, S., Yoshioka, H. et al. (2012). Optical bending sensor using distributed feedback solid state dye lasers on optical fibre. *Optics Express* 20 (14): 14938–14944.

17 Foster, S.B., Cranch, G.A., Harrison, J. et al. (2017). Distributed feedback fibre laser strain sensor technology. *Journal of Lightwave Technology* 35 (16): 3514–3530.

18 Lara, H. and Eichenholz, J. (2004). Spectral processor and ASE source aid fibre sensing. *Laser Focus World* 40 (1): 99–102.

19 Liu, Y., Jia, Z., Qiao, X. et al. (2011). Superfluorescent fibre source achieving multisignal power equalization in distributed fibre Bragg grating sensing. *Optical Engineering* 50 (12): 125004.

20 Yamada, K., Tsuchizawa, T., Nishi, H. et al. (2014). High-performance silicon photonics technology for telecommunications applications. *Science and Technology of Advanced Materials* 15 (2): 024603.

21 Dutta, H.S., Goyal, A.K., Srivastava, V. et al. (2016). Coupling light in photonic crystal waveguides: a review. *Photonics and Nanostructures-Fundamentals and Applications* 20: 41–58.

22 Lutes, G. (1988). Optical isolator system for fibre-optic uses. *Applied Optics* 27 (7): 1326–1338.

23 Zizzo, C. (1987). Optical circulator for fibre optic transceivers. *Applied Optics* 26 (16): 3470–3473.

24 Zervas, M.N. and Giles, I.P. (1988). Optical fibre phase modulator with enhanced modulation efficiency. *Optics Letters* 13 (5): 404–406.

4

Basic Detection Techniques

Daniele Tosi[1,2] and Carlo Molardi[1]

[1] School of Engineering, Nazarbayev University, Astana, Kazakhstan
[2] Laboratory of Biosensors and Bioinstruments, National Laboratory Astana,
Astana, Kazakhstan

4.1 Introduction

Optical fibre sensors are strongly based on the waveguiding structure of the optical fibre, which allows propagating light from an optical source to an optical detector with excellent guiding capability [1]. Modern optical fibres are either based on silica, achieving a very low loss in the near-infrared region (down to <0.2 dB/km), or on polymers, which allow a high numerical aperture (NA) and therefore an easy coupling of light to/from the fibre. In optical fibre sensors, light propagates through the fibre and the sensing point(s) according to electromagnetic theory, carrying an optical power and a spectrum, having a polarization state and a phase, and being transported on specific mode(s). The role of sensors is to modulate one or more properties of incoming light, either travelling through the sensing point (transmission mode) or reflected by each sensing element (reflection mode).

The easiest sensing architecture is based on a modulation of the optical power, leading to the *intensity-based sensors* [2]. This type of sensing system is typically intended for low-cost sensing system, since it is possible to use an inexpensive light-emitting diode (LED) as a light source, a photodetector as a receiver, and a fibre link whose losses vary as a function of changes of the sensing parameter. Often, polymer optical fibres (POFs) are used for this architecture [3], operating in the visible wavelength range to simplify also the development of the sensing structure. Intensity-based sensors can work in *intrinsic* architecture, where the light is maintained in the fibre through the whole link, or in *extrinsic* format, where the

Optical Fibre Sensors: Fundamentals for Development of Optimized Devices, First Edition.
Edited by Ignacio Del Villar and Ignacio R. Matias.
© 2021 The Institute of Electrical and Electronics Engineers, Inc.
Published 2021 by John Wiley & Sons, Inc.

light travels through an external element in the sensing region. Intensity-based sensors have been demonstrated for detection of mechanical parameters [3] and for bending [4] or displacement measurement [5]. Recently, a growing trend is to port this type of fibre sensor on a smartphone device [6], removing the need for additional optical hardware. Intensity-based sensors can also work by transducing a polarization change along the fibre into a change of light intensity, by means of polarization filters [7, 8].

The other basic detection technique is to use reflective devices, which typically enables a wavelength selective behaviour [9]. In this case, the light reflected by the sensor has a different pattern for each wavelength, and a change of the reflection spectrum can be observed. This type of sensing is more robust than intensity-based sensing, because it is resilient to the fluctuation of the power of the optical source and the cable links. However, it requires a device capable to detect and resolve an optical spectrum such as a spectrometer or a scanning laser. Wavelength selective behaviour can also be used in transmission configuration by monitoring changes in the transmission spectrum.

Spectral detection is particularly effective when a resonant device is used as a sensing element in order to obtain a clear spectral pattern. A first approach is to use an interferometer, which consists of an optical cavity formed by multiple reflections along one or two arms. Classical interferometer-based sensors consist of two-arm interferometers such as the Mach–Zehnder interferometer (MZI) and the Michelson interferometer or the Sagnac loop based on a cavity made of a fibre loop [1]. However, in modern fibre-optic sensing approaches, the two main configurations are the Fabry–Pérot interferometer (FPI), which is made of a single cavity [10], or multimodal interference (MMI), typically obtained with the single-multi-single mode (SMS) interferometer, made by fibres having different core diameters [11]. These structures are simple to fabricate with a splicing tool, and they work 'in-line' without the need for external devices. All interferometers yield a reflection spectrum having a broadband periodical pattern, with an envelope that changes when the cavity changes its optical length.

The other popular method for sensing is based on the fibre Bragg grating (FBG), which is a periodical modulation of the refractive index of an optical fibre [12]. An FBG has a resonant spectrum that reflects a very narrow spectral line width, which shifts when the grating exhibits a change of temperature or strain. Using several types of FBGs (chirped, tilted, etched, long-period gratings [LPGs]), it is possible to enable advanced features such as refractive index sensing or detection of temperature patterns. FBGs also enable the fabrication of arrays of sensing elements, allowing a detection in multiple points.

In the following sections, we will outline the basics of fibre-optic sensors and sensing systems, with particular emphasis on modern sensing approaches with straightforward implementation. Section 4.2 describes the main interrogation methods, Section 4.3 describes sensors based on intensity methods, Section 4.4 describes sensors based on light polarization change, Section 4.5 describes simple fibre-optic interferometers, Section 4.6 describes FBG sensors, and Section 4.7 describes simple multiplexing techniques for fibre-optic sensors.

4.2 Overview of Interrogation Methods

Optical fibre sensing system can assemble several sensing unit and principles. However, an important distinction is to understand whether the system is working in *transmission* or in *reflection*. A schematic of the two architectures is shown in Figure 4.1. When a system is working in transmission (Figure 4.1a), it has a light source emitting light into the fibre, and the sensors (one or multiple sensing unit) are located along the fibre cable itself. The light detector is positioned at the end of the fibre link, collecting the output after each sensing point. A system working in reflection (Figure 4.1b) makes use of reflective elements as sensing units: in this case, the fibre is used both forward to deliver light to each sensor and backward to collect the backreflection. The system requires a three-port coupling element, such as a directional coupler (as the 1×2 coupler in the scheme) or a circulator in order to decouple the input light (from the source to the sensors) and the output component (from the sensors to the detector). A system working in transmission has the advantage of being simpler to implement, as no couplers are needed. In addition, some sensing elements such as LPGs or tilted fibre Bragg gratings (TFBGs) exhibit their spectral characteristics only in transmission mode. However, for several applications, it is preferable to use a reflection-based architecture. As the sensing fibres can be, for example, embedded into composites [3], or

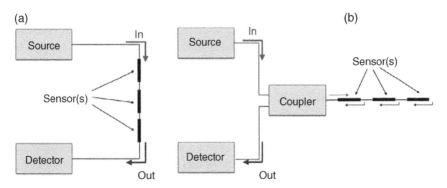

Figure 4.1 Arrangement of fibre-optic sensing system: (a) in transmission and (b) in reflection.

mounted on medical catheters [9], or installed on harsh environments [13], it is much easier to use a single fibre hosting the sensors rather than having to arrange a fibre that leads in and out of the sensing region.

An *interrogator* is a device, usually assembled as a benchtop device or a portable unit, that assembles all the optical elements (source, detectors, couplers, passive devices, switches) and electrical hardware (power supply, electrical circuitry, connection to a computer via USB, RS232, or GPIB protocol) in order to facilitate the use of the sensing system. Sensing cables are then connected to the interrogator via mating sleeves mounted on the interrogator box in order to facilitate interconnections.

Intensity-based sensors typically make use of a LED light source coupled into the sensing fibre and connected to a photodetector (usually a p-i-n photodiode [PD]). When the losses in the sensing fibre change, we observe a variation of the received power at the photodetector, which converts the optical power into a photocurrent that is subsequently amplified, usually through a transimpedance amplifier (TIA) [2]. This architecture is very simple but effective and used in displacement sensors [5] as well as in chemical sensors [6] where it is relatively easy to convert the measurand into a change of the fibre losses. The main weakness of this set-up is the long-term reliability, as fluctuations of the emitted power or of fibre performance (such as strain events or temperature variations) are also interpreted as a change of the losses in the sensing point. For this reason, it is common practice to design interrogation systems with multiple channels, including reference channels that measure the power in each part of the optical system, thus compensating power fluctuations and fibre extra losses. Another weakness, particularly when operating with fibres with large NA, is the presence of undesired ambient light or stray light that can also be photodetected in addition to the 'true' signal. This effect can be mitigated by modulating the LED source at a carrier frequency higher than the ambient disturbance (usually within 100 and 1500 Hz). Techniques such as lock-in amplification or a simple passband filter can be used to downconvert the signal at the photodetector [5].

When working with sensors having a wavelength selective behaviour (such as many interferometers and all gratings), the most common detection approach is to detect the reflectivity of the sensors throughout a bandwidth of several nanometres. This method allows extracting all the spectral features that depend on the type of sensors under interrogation. The two most popular set-ups are sketched in Figures 4.2 and 4.3.

The first schematic, shown in Figure 4.2, is based on a broadband source such as a LED, a superluminescent LED (SLED), or an Er-doped fibre amplifier (EDFA) without input. The detector is a spectrometer that ideally works on the same wavelength window of the light source. A coupler is used to route light to the sensing elements and to collect the reflection spectrum that is processed on the spectrometer, which converts it into an electrical signal proportional to the reflection

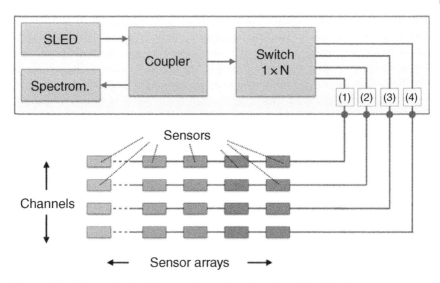

Figure 4.2 Schematic of an interrogation set-up based on a broadband source.

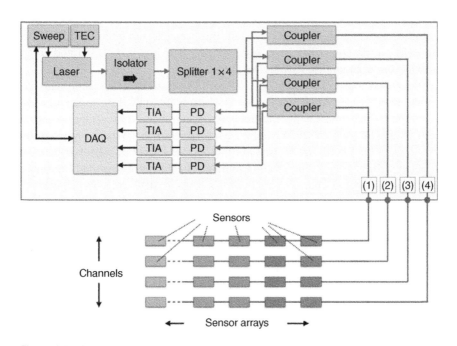

Figure 4.3 Schematic of an interrogation set-up based on a scanning laser.

spectrum. This set-up can interrogate a single sensor or an array of sensors. By using a $1 \times N$ switch, it is possible to generate multiple channels (for example, the schematic in Figure 4.1 is reported for a four-channel configuration), and the spectrometer detects the active channel. The approach in this scheme is popular for sensors operating at wavelength range centred at 1550 or 850 nm.

Broadband schemes have the advantage of being cheaper (about 4–6 times cheaper that a scanning laser), due to the lack of scanning elements, whereas the quality of the spectrometer and the stability of the SLED source determine the main performance figure. Modern systems based on this approach have an operative bandwidth of 50–80 nm and operate with fast spectrometers controlled via USB. The lack of bulky and expensive hardware can make this system portable, lightweight, and potentially powered by a battery-powered battery supply. The major limitation of this set-up is the poor spectral resolution that does not allow the detection of the narrower spectral features: typical infrared spectrometers have a resolution of 78–156 pm, corresponding to the difference between two adjacent wavelengths that can be discriminated. In order to obtain a fine detection of the spectral features, such as for FBG sensors, it is necessary to use signal processing and performing the so-called inter-pixel detection [9]. The other limitation is that the level of the detected signal is determined by the exposure time of the spectrometer, which cannot be rapidly changed. This provides problems to interrogate sensors having different reflectivity levels or connected to fibre links with different losses. Commercial instruments based on this principle and often used for interrogation are BaySpec FBG Analyzer (BaySpec, San Jose, CA, USA) and Ibsen I-MON (Ibsen Photonics, Farum, Denmark).

The second approach, sketched in Figure 4.3, is based on a fast scanning laser, which sweeps a wide wavelength range. The laser is controlled by a sweep function generator, which implements the frequency scanning, and by a thermoelectric controller (TEC), which sets the laser driving current and operative temperature. The laser is followed by an isolator that protects from the backreflections. A network formed by a splitter and multiple couplers decouples the laser signal into multiple output channels, each with a separate detector. Each detector is based on a PD, which converts the output power into a photocurrent, and a TIA, which converts the photocurrent into a voltage proportional to the reflectivity of the sensors at the laser wavelength. A data acquisition unit (DAQ) converts all the output into each spectrum.

The main advantage of this method is that the spectral resolution is determined by the wavelength grid used by the sweep function generator: in most systems, the spectral resolution is 8 pm over a wavelength range of 80–160 pm and with a stability <1 pm, making this method capable of capturing the features with great precision. Also, since the detector is a PD, it is possible to have a very large dynamic range, usually >40 dB, which means that the same interrogator is capable of simultaneously tracking high-reflectivity and low-reflectivity components. The drawback is that the method is more expensive and bulkier, since the optoelectronic hardware used to control the laser requires a power supply with a stabilization circuit. This scanning laser set-up has been integrated in many commercial devices, mainly featuring Micron Optics si255 (Micron Optics Atlanta, GA, USA), and HBM FS22 (HBM, Darmstadt, Germany) interrogators.

4.3 Intensity-Based Sensors

Intensity-based sensors have found their best application in low-cost sensing systems, whereas the simplicity of sensing approach make it a competitive and affordable technology [5]. Several approaches can be implemented for intensity-based sensors, which allow detecting the change of the optical losses through an optical path. We can distinguish between *intrinsic* sensors, where the optical path is closed from the source to the detector and the sensor is embedded in a specific region of the optical path, and *extrinsic* sensors, which make use of external elements to the optical path for sensing. Sensors that are based on intensity change can be based on standard single-mode glass fibres (SMF), which typically have core–cladding diameter of 10/125 μm and NA equal to 0.08, or on multimode fibres (MMF), which have larger core size and larger NA to facilitate the coupling of light into the sensor. However, the main low-cost systems are based on POFs, such as polymethyl methacrylate (PMMA) fibres. PMMA fibres have a perfluorinated cladding, with core–cladding diameter of 980/1000 μm and NA or 0.47, and operate at visible wavelengths [14]. This large value of NA allows an extremely favourable coupling between external light sources and the fibre without the need for external collimation, making the set-up simple to develop and package; the same simplicity occurs when coupling light from the output fibre to the photodetector. By operating at visible wavelengths, rather than at the infrared, operators have also a visual estimate of the optical power losses through the system. Conversely, systems based on low-NA fibres require the use of optical connectors in order to facilitate coupling and are therefore inherently more expensive and bulkier.

4.3.1 Macrobending

An important approach for intrinsic sensing is based on *macrobending*, where the sensing region of the fibre is a tightly bent optical fibre having an interdigitated bending shape that resembles the interdigitated strain gauges. When a strongly multimode fibre is tightly bent over its radius, extra losses are induced: this is a dual effect of the weakly guided modes converting into evanescent modes and the reduction of the confinement factor of the propagating modes. This effect

(a)   (b)

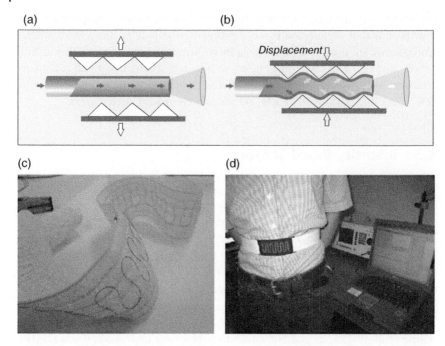

(c)   (d)

Figure 4.4 Working principle and application of a macrobending sensor (*Source:* Image taken and adapted from [15]). Working principle: when a pressure is exerted from the sensor in resting position (a), it transduces into an additional loss into the sensing fibre (b). The structure can be packaged into a textile (c) and used to detect breathing pattern (d). *Source:* Reproduced with permission of MDPI.

can be amplified by introducing multiple bending in a short space, thus increasing the path of fibre exposed to curvature. Figure 4.4, taken from [15], explains the working principle by showing how a tightly packaged fibre, bent over a tissue with approximately 2–4 distance between each bending side, can convert into an efficient sensor. The tighter the bending, the higher are the extra losses, depending also on the packaging. Macrobending POF sensors are an excellent tool for biomedical applications, since they are magnetic resonance imaging (MRI) compatible and they have an excellent sensitivity, while the POF fibre has a low Young's modulus that allows a tight bending, and its strongly multimode pattern (about four million modes propagating at visible wavelengths) allows having a substantial amount of light converting into evanescent field. Moreover, since the sensor is intrinsic, there are no open parts and therefore a lower stray-light effect (i.e. a low amount of light is lost and not used for sensing). The work described in [15] shows an application of macrobending to breathing rate detection, whereas the amplitude and the periodicity of the respiratory pattern are transduced into the optical intensity change.

4.3.2 In-Line Fibre Coupling

A major disadvantage of intrinsic sensors based on bending is that the sensor structure is not compact, even bending a POF fibre at its mechanical stability limit. Extrinsic sensors instead can have a more compact shape, since they can be proposed as an in-line structure using one or multiple fibres. The principle of operation of an in-line sensor made of two fibres, sketched in Figure 4.5a, is straightforward. When a transmitting (TX) fibre emits a power collected by a receiving (RX) fibre in absence of any displacement, virtually all the TX power is guided into the RX fibre. When a displacement is introduced within the TX/RX fibre pair, only a fraction of the TX power is collected by the RX fibre, corresponding to the amount of power that hits the fibre core, while the light that is transmitted into the cladding or out of the fibre is not collected.

Depending on the choice of fibre, it is possible to have different working ranges: SMF is rarely used for this application, while MMF achieves a working range up to hundreds of micrometres and POF fibres slightly over the millimetre [5, 17]. The

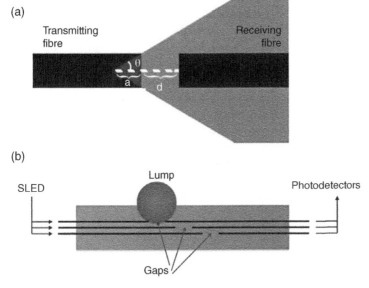

Figure 4.5 Working principle of in-line fibre coupling-based sensors. (a) Schematic of the working principle and (b) sketch of the lump detection proposed by Ahmadi et al. [16]. *Source:* Reproduced with permission of Springer.

response of the sensor, and its sensitivity, varies according to the relative position-ing of the RX fibre with respect to the TX fibre. In general, when both positive and negative displacement values are measured (for example, for detection of cracks [18]), the fibres are initially positioned at a reference value of displacement, allow-ing both compressions and extensions of the displacement. In-line fibre coupling configuration is usually performed to detect displacement. However, it is possible to convert this structure to detect any object that reduces the light transmission from TX to RX. The work reported in [16] by Ahmadi et al. and sketched in Figure 4.5b is used to detect and localize the presence of lumps or obstructions within a blood vessel and is based on three in-line structures with an intermediate gap. The fibres are mounted on V-grooves microfabricated on a silicon wafer in order to maintain their structure fixed and stable. In the presence of an obstruc-tion, the amount of light reflected and diffused by the inner lump induces a reduc-tion of the light coupled into the RX fibre. By using a splitter and multiple photodetectors, it is possible to obtain a plurality of channels, thus detecting and localizing the presence of lumps.

4.3.3 Bifurcated Fibre Bundle

The bifurcated fibre bundle (BFB) approach is similar to the previous method, but rather than measuring the displacement between a TX and a RX fibre, the set-up measures the displacement between a TX fibre and a target, with the support of one or multiple RX fibres. The theoretical analysis conducted by Faria [17] by means of imaging theory shows that a BFB simply behaves as an in-line fibre coupler by doubling the displacement value in order to account for both light transmitted into the reflector and the light backreflected into the fibres. The BFB configuration can be effectively used to measure the displacement from a moving target and can be used in industrial applications (for example, to monitor the thickness of components with a resolution of tens of micrometres). In a BFB configuration, the power received at the photodetector is proportional to the reflectivity of the target. However, it is possible to use a bundle with multiple RX fibres in order to compensate for the target reflectivity [5]. A two-fibre BFB can also be used in conjunction to a target constituted by a diaphragm, which bends when exposed to a pressure change. This architecture is at the base of pressure sensors, commercialized by Camino (now Integra LifeSciences, Plainsboro, NJ, USA), for intracranial pressure detection [19].

4.3.4 Smartphone Sensors

All the previously described architectures are based on a light source, usually a LED/SLED, and a silicon photodetector that detects the light intensity. Both ele-ments, as well as the elements to couple light into and out of the fibres, constitute the hardware of intensity-based sensors. A modern research trend is to replace the conventional hardware with smartphones, particularly exploiting the flashlight source and the photodetector [6] as the two building blocks. Since POF fibres have a large NA, the amount of light coupled into a POF fibre and detected on the smart-phone camera is sufficient to perform a detection. Aitkulov and Tosi reported very recently the main results on smartphone-based optical sensors, with a breathing sensor based on an in-line fibre coupler [20] and a chemical sensor based on a sil-ver-coated fibre [6]. The system proposed in these works is all-fibre, and the only external element is a connector socket that facilitates a stable coupling between the source and input fibre and from the detector to the camera. The connector socket has been 3D printed using commercial technology, and its shape can be adapted to the specific phone used for sensing, making the system low cost, with the total cost of the sensing elements external to the phone being about $2. The set-up of the system is shown in Figure 4.6: the principle of operation is similar to the in-line coupling sensors, implemented on a commercial POF; the main difference is that rather than using a PD, the camera of the smartphone is used for detection. The

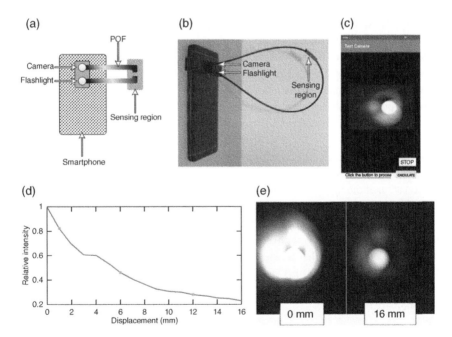

Figure 4.6 Sketch of a smartphone-based POF sensor reported in [20]. (a) Principle of operation; (b) photo of the set-up, (c) mobile app for camera detection, (d) calibration function, and (e) photos of the detected image with 0 and 16 mm displacement. *Source:* Reproduced with permission of IEEE.

intensity is obtained by estimating the amplitude of the correlation between the reference image (in the absence of displacement) and the detected image, which also allows to partially compensate for the movements of the fibre. The 3D printed connector is mounted on the back of the smartphone and hosts the sockets for both TX and RX fibres; the sensing region is covered in a black surface to prevent stray light to be coupled into the sensor. The camera shows that the intensity of the detected image decreases as the displacement increases, and the sensor behaves similarly to a POF bundle. The system control is implemented on a mobile Android app, developed in Java, which allows removing all the self-adjustment features for the camera, turn on/off the flashlight, collect and process the image, and store the data.

4.4 Polarization-Based Sensors

Polarization is one of the most important properties of light, and it can be exploited for several kind of sensing, ranging from pressure and temperature sensing to refractive index sensing. The operation principles of polarization-based fibre sensors are also quite variable since the information, embedded in the polarization state, can be collected and analysed in different ways. As an example, the modulation of polarization in an optical fibre-based sensor, usually induced by an external variation of pressure, can be easily transduced by a polarization analyser in a modulation of intensity, which can be detected by a common PD. The sensitive part of a polarization-based sensor system can be represented by an external birefringent element, in the case of extrinsic sensor [21], or can be embedded in-line in an all-fibre configuration [22, 23]. This kind of sensors, exploiting the information delivered by the polarization modulation through the intensity detection, offers a good compromise between accuracy and low cost. On the other hand, the polarization state of the light propagating in a fibre can be used in combination with the properties of thin-film coating to induce the lossy mode resonance (LMR) [8]. Typically, LMR is used to sense the refractive index of a solution coming in contact with the thin film. The resonance peak of the thin film is polarization dependent so that the variation of the external refractive index can be detected by a spectral analysis of the sensor output. This behaviour offers a double possibility for demodulation, i.e. the detection of the peak wavelength shift or the detection of transmitted intensity variation.

4.4.1 Pressure and Force Detection

Polarization-based fibre-optic sensors are, in broad sense, part of the larger family of phase sensors. These sensors are characterized by a phase shift between the two orthogonal light components accumulated during the passage through an element presenting high birefringence. The birefringent behaviour is responsible of the change of polarization state of the output light. The core operation principle of the sensor is related to the fact that the output polarization is affected by several external factors, such as stress, strain, pressure, and applied force [24]; with some restriction, also temperature, whose variation is able to modify the birefringence of the sensing element. Form this point of view, polarization-based fibre-optic sensors, characterized by a simple and cost-effective design, offer advantages in many fields of industrial applications and civil engineering.

The working principles of a polarization-based sensor for pressure and force detection are immediate to understand. The set-up, shown in Figure 4.7a, depicts an extrinsic configuration. A non-polarized source is delivered to the sensing element by an input fibre. The sensing element is made by a polarizer P, a birefringent material with photo-elastic properties, a quarter waveplate WP, and a polarization splitter A used as an analyser. The polarizer and the waveplate present an angle of 45° with respect to the axis of the polarization splitter. The output, transmitted through the splitter, is collected by a photodetector.

After the polarizer, the light, entering in the birefringent material, changes its polarization state according to the applied stress due the photo-elastic effect. The modulation of polarization is then filtered and transduced in an intensity modulation that can detected by a PD [21]. Some applications require a more compact design, and the extrinsic sensor can result too bulky. A polarized-based

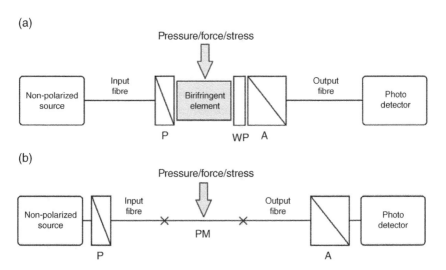

Figure 4.7 Polarization-based optical fibre sensor for pressure, force, or stress detection. (a) Extrinsic configuration and (b) all-fibre intrinsic configuration.

all-fibre sensor is preferable for such applications. This sensor design can be derived by the previous design by applying some simple modifications. The set-up is shown in Figure 4.7b. The signal in the input fibre is previously polarized by a polarizer P. The input fibre is then spliced to a polarization-maintaining (PM) optical fibre characterized by high birefringence. Typical PM fibres used for this application are: panda fibres, bow-tie-type fibres and side-hole fibres. The PM fibre, which represents the sensing element, is spliced to the input fibre with an angle of 45° with respect to the axis of the polarizer P. This is to allow the incoming light to equally excite both polarization modes in the sensing element.

Azimuthal alignment, characterized by high accuracy, is required to avoid reduction of sensor sensitivity. The other end of the PM sensing element is then spliced to a standard optical fibre that feeds a polarization analyser. The analyser transduces the polarization modulation in an intensity modulation that can be detected by a photodetector [22]. The all-fibre configuration, where the sensing element is a fibre, permits a compact design that can be easily embedded for civil engineering or industrial application to assess the structural integrity of foundations, pillars, or anchor.

4.4.2 Lossy Mode Resonance for Refractive Index Sensing

Optical fibres, coated by a thin film, have been demonstrated to be effective and accurate sensors. The presence of thin film affects the propagation of light inducing different types of resonances, strictly dependent on the properties of the thin film. Surface plasmon resonance (SPR) is a well-known and largely used phenomenon belonging to the thin-film resonance family [25] (see Section 8.2.4 of the book for a detailed explanation on SPR). Another type of resonance is the LMR [7]. This resonance appears when two conditions are satisfied. First, the real part of the thin-film permittivity is larger than its imaginary part. Second, the real part of the thin-film permittivity is larger than the permittivity of the silica used for the optical fibre and larger than the permittivity of the surrounded medium. Differently from SPR, which can be generated only by the TM mode of the light guided in the fibre, the LMRs are generated by both TM and TE modes. The polarization-dependent LMR can be exploited to obtain effective optical fibre sensors to detect the refractive index of the analyte coming in contact with the thin film. By coating a D-shaped polished single-mode optical fibre with a thin-film compound made of titanium oxide nanoparticles and polystyrene sulfonate, Zubiate et al. demonstrated the strong polarization dependence of the LMR [8]. According to their investigation, a polarized input shows a higher transmittance with respect to an unpolarized input. Moreover, the spectral separation between the TE and the TM mode wavelength peak is relevant. By controlling the TE or the TM input of the D-shaped sensing region, the authors demonstrate a clear shift of the resonance peak when the refractive index of the external solutions has changed. This resonance peak shift can be detected by a spectrum analyser or by a spectrometer. The measured sensitivity is relevant with values of 2893 nm/RIU for the TM mode and 2737 nm/RIU for the TE mode, obtained in the range of refractive index from 1.35 to 1.41. Furthermore, the transmitted intensity is also dependent on the refractive index, permitting a demodulation based on the intensity detection.

4.5 Fibre-Optic Interferometers

While intensity detection approach permits simple and cost-effective optical fibre sensor design, spectral detection largely improves the accuracy of the measurements. A way to further improve the spectral detection is to exploit the properties of resonant devices as a sensing element. An effective approach consists in taking in consideration the use of interferometers. A fibre-optic interferometer is made by creating an optical cavity by the use of one or two arms where the light can experience multiple reflections. Classical free-space interferometers can be built using optical fibre technology, as shown in Figure 4.8. In these devices light is forced to interfere with itself by following two different paths along two arms in case of Mach–Zehnder and Michelson interferometers or by entering in a closed loop in case of Sagnac interferometer [1].

In these designs one of the optical paths has to be put in contact with the external perturbation in order to create a spectral change of the interfering signal used to detect or measure the cause of the external perturbation. The spectral information gives a remarkable performance in terms of accuracy, sensitivity, and

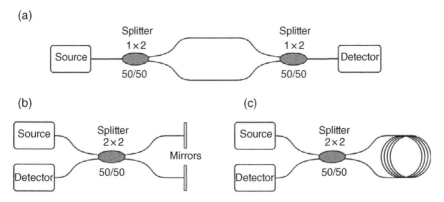

Figure 4.8 Schematic of optical fibre-based classical interferometers. (a) Mach–Zehnder, (b) Michelson, and (c) Sagnac.

range of measurement. The presented interferometers make use of discrete optical fibre components to properly operate, making them sometime too bulky for particular applications, like biomedical ones, that require both accuracy and miniaturization.

Nowadays the trend of fibre-optic interferometers is to substitute the traditional bulk optic components to operate on fibre scales. The best way to use in-line structures characterized by integrating the optical path in one single physical line. Moreover, in-line structures offer remarkable advantages including improvements in terms of stability and intrinsic alignment [26]. For such necessity of miniaturization, modern fibre-optic sensing area has focused the interest to other kinds of interferometers that make use of single in-line cavity. In particular research efforts are directed the study of FPI [27], the MZI, and the SMS interferometer [28]. There exist in-line Michelson and Sagnac interferometers, but they have attracted less interest.

4.5.1 Fabry–Pérot Interferometer (FPI)-Based Fibre Sensors

FPI-based sensors result particularly promising compared with other kinds of optical fibre sensors. They present high precision and responsivity with a design characterized by high versatility. The fabrication of Fabry–Pérot fibre sensors involves different strategies and techniques including design with air–glass reflectors, semi-reflective splices, and in-fibre Bragg gratings. Because the large variety of designs, FPI-based fibre sensors can be intrinsic, when the cavity is obtained in-line with the help of internal reflector, or extrinsic, when the cavity is externally fabricated.

The operation principles are simple. The FPI spectrum in both transmission and reflection configuration is wavelength dependent, presenting a modulation given by the interference between the forward and the backward light beam in the cavity. The maximum and the minimum peaks represent when the beams are in phase and out of phase. Therefore, the operation of an FPI-based fibre sensor is completely described by the phase difference:

$$\varphi_{\text{FPI}} = \frac{4\pi}{\lambda} nL \tag{4.1}$$

where the λ is the wavelength of the incident light, n is the refractive index inside the cavity, and L is the length of the cavity.

All the external factors such as pressure, strain, and temperature, which can influence the value of refractive index and the length of the cavity, change the pattern of the FPI spectrum. A typical FPI configuration, in this case for refractive index sensing, along with the spectral pattern, is shown in Figure 4.9.

(a)

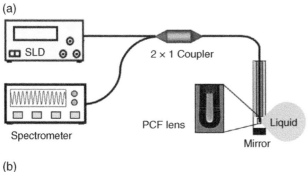

(b)

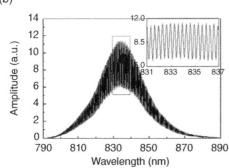

Figure 4.9 (a) Schematic of an extrinsic FPI fibre sensor for refractive index sensing in liquid. The system is based on a photonic crystal fibre (PCF) lens. (b) Spectral pattern of the FPI cavity [26]. *Source:* Reproduced with permission of MDPI.

4.5.1.1 Extrinsic FPI for Pressure Sensing

In this kind of optical sensors, the Fabry–Pérot cavity is usually located on the tip of an optical fibre so that the sensor operates in reflection. The cavity can be formed by a miniature glass diaphragm or can be a bulkier structure with a larger diaphragm [10]. In general, the sensitivity for pressure sensing is proportional to the area of the diaphragm. The light coming from a coherent source interferes with itself due to the reflection given by the cavity. For a low-finesse interferometer, only 96% of the light at the interface between the fibre and the cavity enters in the cavity, and only 4% of this light is reflected back by the second interface of the cavity. Therefore, the interference is given by the input beam and the roughly 4% of its reflection. The pressure variation of the free diaphragm, located at the tip of the fibre, is responsible of the cavity length variation, which can be easily detected as a spectral shift. Alternative solutions for a diaphragm include organic and plastic material to increase the sensitivity and ultra-thin metal or graphene diaphragms to improve the reflectivity. Integrated and compact solutions given by optical fibre pressure sensors based on FPI show a great potential for medical application, considering the good sensitivity [29]. FPI-based optical sensors find space also in industrial applications because of the large range of pressure detection and because of the good performance in terms of low measurement error [27].

4.5.1.2 In-Line FPI for Temperature Sensing

FPI-based fibre sensor represents a good and reliable solution for temperature measurement. In particular, in-line FPI presents high sensitivity and remarkable stability at high temperature. Operation with temperature around 1000 °C is reported in literature [30]. The design of in-line FPI usually involves complex and expensive microfabrication. By the use of advanced splicing technologies, special optical fibres like holey fibre or photonic crystal fibre (PCF) are fused to SMF in order to create effective cavities. Moreover, the use of femtosecond (fs) fibre laser, which permits precise microfabrication, is exploited to inscribe micrometre size cavity in single-mode fibres. An example of 80 μm microcavity FPI, fabricated directly in the core of an SMF by using a near-infrared fs laser, is shown in Figure 4.10a. The cavity is sensitive to both strain and temperature. Reported

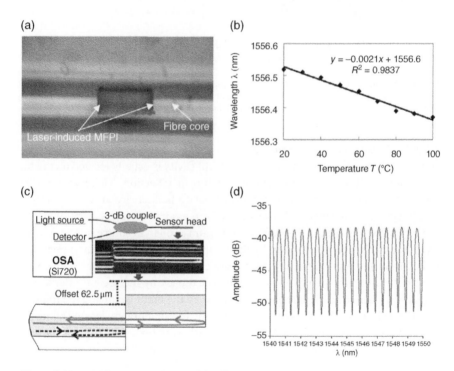

Figure 4.10 (a) Microscope picture of the FPI microcavity: the cavity is 80 μm cavity long, and it is inscribed in the core of an SMF. (b) Temperature sensitivity of the FPI microcavity. (c) Low-cost FPI-based fibre sensor obtained by SMF to SMF splice with lateral offset. (d) Reflection spectrum of the low-cost FPI [30]. *Source:* Reproduced with permission of MDPI.

results show a strain sensitivity of 0.006 nm/με and a temperature sensitivity of −0.0021 nm/°C, as depicted in Figure 4.10b. The temperature sensitivity is negative since the increase of temperature forces the interfaces of the cavity to expand inside the cavity itself so that the cavity length decreases [31]. However, not all the fabrication processes are expensive, and low-cost FPIs are possible. An example of low-cost FPI, created by splicing two cleaved segments of SMF together with a large lateral offset, is shown in Figure 4.10c.

In this sensor the cavity consists in the interface between the termination of second segment of SMF (1.65 mm long) and in the core–cladding interface between the first and the second segment of SMF. The interference fringe, reported in Figure 4.10d, has a visibility around 15 dB, which is sufficient for the sensing application. The temperature response over a large range of temperature is linear and stable, with a reported optical path difference (OPD) temperature sensitivity of 41 nm/°C [32].

4.5.2 Mach–Zehnder Interferometer (MZI)-Based Fibre Sensors

The original concept of MZIs with two independent arms, represented in Figure 4.8a, has been progressively replaced by designs where the two optical paths are contained in an in-line structure, with the aim of miniaturization, the same idea followed with the FPIs. With MZIs, interferometry is due to the phase difference between the two optical paths in the structure. To this purpose, there exist numerous configurations. Generally, SMF is the starting point of the MZI fabrication, and their inputs and outputs are SMF segments. In contrast, its configuration can vary substantially regarding the inner fibre fragments along the MZI structure. One example is shown in Figure 4.11a, where a structure composed of two LPGs separated from each other is presented (the first work on this structure was published in 2002 [33]). Part of the light transmitted through the first LPG is transferred to the cladding, generating a second optical path, alternative to the optical path of the core. The 5 cm separation between the LPGs permits to generate an interference pattern in the transmission optical spectrum in Figure 4.11b.

In addition to the structure presented in Figure 4.11, like for FPI-based sensors, there exist many other strategies for creating MZI-based sensors, such as in-line tapering [34], core-offset sections [35], in-line microcavities [36], etc.

4.5.3 Single-Multi-Single Mode (SMS) Interferometer-Based Fibre Sensors

SMS interferometer is a low-cost fabrication technique for creating intrinsic fibre sensors. The operation principle of this kind of interferometer is based on MMI. The technique of MMI along with SMS structures has been largely exploited for strain, temperature, refractive index, and displacement applications [37].

(a)

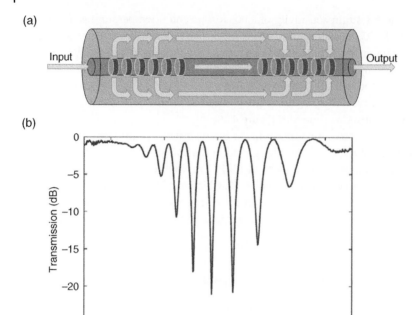

(b)

Figure 4.11 (a) In-line Mach–Zehnder interferometer: the first grating couples light to the cladding and the light guided through the core and through the cladding interferes and generates the fringes observed in (b) the optical transmission spectrum.

The properties of MMI are practically implemented in a SMS design by splicing two segments of standard SMF for telecommunication to a central segment of MMF. This fibre can be a standard MMF with a core of 50 µm or a custom made no-core fibre.

SMS structures present a spectrum characterized by a large number of maxima and minima generated by the interaction between the higher-order modes in the MMF section. Most of the investigation regarding the properties of these devices has focused in the special region of telecommunication windows. Recently the implementation of devices operating at wavelength shorter than 1 µm has been demonstrated [38].

The operation principle of SMS interferometer is based on the multimode propagation of the MMF section, stimulated by the SMF input. At the end of the MMF cut, the light is recoupled into the SMF. The performance of the SMS structure is given by the self-imaging effect. In a perfectly aligned structure, in the MMF segment, only the mode with circular symmetry belonging to the $LP_{0,x}$ family is excited. After a certain length, the phase difference between the higher-order modes vanish, thus presenting a prefect replica of the input field. The properties of the transmission spectrum are, therefore, controlled by the dimension of the MMF segment.

To locate the position of the transmission band at a specific wavelength λ, the MMF length Z is described by the following formula:

$$Z = \frac{pD^2 n_{co}}{\lambda} \tag{4.2}$$

where p is the order of the self-image and D is the diameter of the MMF section. The MMF section is, therefore, the sensing element, since the external perturbations influence and shift the transmitted spectrum. SMS-based sensors are particularly effective as refractive index sensors. In this case the diameter of the MMF section is fundamental to improving the sensitivity. Thinner MMF cladding increases the sensitivity to the external refractive index. For this reason, a further fabrication on SMS structures consists in a partial MMF cladding removal with HF acid.

Wu et al. have demonstrated that, by optimizing of the diameter D, a maximum sensitivity of 1815 nm/RIU for a refractive index range from 1.342 to 1.437 can be achieved [28].

4.6 Grating-Based Sensors

In this section we will refer to the most widely used grating-based structures, whose shape and typical spectra are described in Figure 2.17 in Chapter 2. Here more concepts on these structures will be given along with more results corresponding with them.

4.6.1 Fibre Bragg Grating (FBG)

The FBG is one of the most popular and straightforward architectures for fibre-optic sensing, and since the preliminary conception and results from late 1990s [12], a long path towards industrialization has been undertaken. Currently, FBGs find applications for measurement of strain, pressure, temperature, refractive index, and humidity in many applications in industry, structural engineering, aerospace, and medicine [39]. As interferometers, FBGs are also optical resonant structures that substantially implement the grating configuration inside an optical fibre. However, while interferometers are characterized by a broadband spectrum that alternates peaks and valleys of reflectivity, the FBG resonates at a specific wavelength, the so-called Bragg wavelength, and it is transparent to all the other wavelengths. To make a comparison with electronic circuits, the FBG behaves as a notch resonator with a quality factor ~10 000.

The FBG is defined as the periodical modulation of the refractive index of an optical fibre [12]. When incoming light travels through the modulation pattern, the presence of alternate high/low values of refractive indices creates a resonant pattern that resonates at the Bragg wavelength. If the refractive index modulation has an amplitude of Λ and the effective refractive index of the fibre is n_{eff}, the Bragg wavelength λ_B is equal to

$$\lambda_B = 2n_{eff}\Lambda \tag{4.3}$$

A theoretical description of the FBG spectrum through the coupled mode theory (CMT) has been proposed by Erdogan [40]. A uniform FBG is the most important configuration, having a periodical modulation of the fibre refractive index that is uniform both in amplitude and in period. Typically, FBGs are used in reflection, detecting the spectrum of the FBG with the set-ups described in Figures 4.2 and 4.3. For a uniform FBG, the CMT model provides a closed-form expression for the reflectivity of the grating, which can be measured with a spectrometer or with a scanning laser and PD. Figure 4.12 shows the spectrum of a uniform FBG having Bragg wavelength at 1550 nm, simulated with the CMT. The FBG has a resonant spectrum, with a main lobe centred at the Bragg wavelength that corresponds to the maximum reflectivity R_{max}.

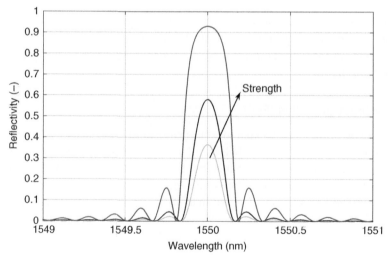

Figure 4.12 Reflectivity of a 1550 nm uniform FBG, simulated with coupled mode theory, for different values of grating strength.

Aside from the main lobe, we can observe a plurality of side lobes that are progressively attenuated and having reflectivity that is much smaller than R_{max}; side lobes can be further attenuated by using an apodization pattern [40]. The grating strength parameter kL is defined as the product between the grating length (L, usually 0.5–1.0 cm) and the so-called coupling coefficient (k), which depends on the amplitude of the refractive index modulation. The grating strength controls the reflectivity, as $R_{max} = a \tanh^2(kL)$. Weak gratings have a smaller bandwidth and a lower incidence of side lobes and achieve a reflectivity up to 10–30%. By increasing the grating strength, the reflectivity grows up to values close to 100%, enlarging the bandwidth of both the main mode and the side lobes. For the gratings in Figure 4.11, the full width at half maximum (FWHM) bandwidth ranges from 165 to 285 pm at a wavelength of 1550, which corresponds to an extremely narrow resonant spectrum.

The FBG structure can be turned into a sensor, as the Bragg wavelength has a dual dependence on both strain and temperature. This dependence origins from the dependence on both Λ and n_{eff} on temperature (ΔT) and strain ($\Delta \varepsilon$) variations, resulting in a linear variation of the Bragg wavelength. This relationship is typically expressed as

$$\Delta \lambda_B(\Delta T, \varepsilon) = k_T \Delta T + k_\varepsilon \Delta \varepsilon \tag{4.4}$$

where $\Delta \lambda_B$ is the wavelength shift of the Bragg wavelength from its reference position. Thus, when a temperature or strain change is experienced by the FBG, its entire spectrum shifts linearly, making the sensing principle robust. The sensitivity terms are k_T (pm/°C), the thermal sensitivity, and k_ε (pm/µε), the strain sensitivity. For a grating written in a silica fibre and operating at around 1550 nm, these values are approximately 10 pm/°C and 1 pm/µε, respectively [39]. Figure 4.13 shows the strain and temperature sensitivity of an FBG inscribed in a photosensitive glass fibre at 1560.5 nm central wavelength. The first chart shows the strain sensitivity, which has a linear trend and 1.02 pm/µε coefficient, up to values of 1500 µε (which correspond to an extension of the grating of 0.15% of its initial length). The second chart shows the temperature sensitivity, which also exhibits a linear pattern with a coefficient of 10.22 pm/°C. The linear relationship between the Bragg wavelength and the measurand is a key advantage for FBGs, as it allows a simple estimate of the strain or temperature from the Bragg wavelength shift [9].

4.6.2 FBG Arrays

A key advantage of FBG sensors is that the spectral occupancy of each grating is very narrow and the grating is transparent to the wavelengths outside of the Bragg wavelength. This gives the possibility to design arrays of FBGs, formed by a cascade of gratings located in different portions of the fibre, with each grating having a

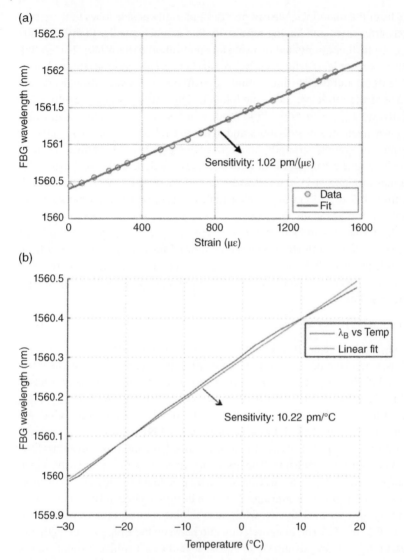

Figure 4.13 Variation of the Bragg wavelength of a 1560.5 nm FBG inscribed on a glass fibre for strain and temperature variations. (a) Strain calibration, obtained by applying strain to the FBG through a micropositioning system, and (b) temperature calibration, measured in a climatic chamber. A linear fit has been used to estimate the sensitivity from measured data.

different Bragg wavelength. In addition, all gratings can be interrogated in a single operation, since the detectors have a bandwidth sufficient to host many gratings.

The overall reflected spectrum of an FBG array is a comb-shape waveform, having multiple peaks, which allows clearly identifying each grating element. This

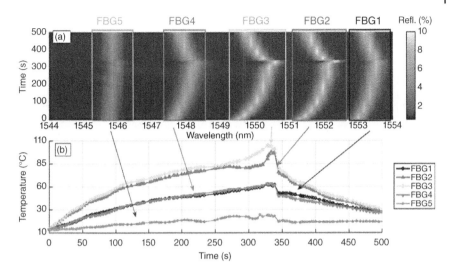

Figure 4.14 Application of a FBG array in WDM configuration. The chart shows the behaviour of an array of five FBGs exposed to a radio-frequency ablation into a phantom. (a) The chart shows the reflection spectrum measured at each time throughout heating and cooling processes. (b) By isolating each FBG and estimating the wavelength, it is possible to extract the five temperature traces.

process, often labelled wavelength division multiplexing (WDM), allows isolating each grating into a specific bandwidth. FBG arrays are often used in sensing, as they allow to detect a measurand in multiple locations, one for each grating.

An example of FBG array is shown in Figure 4.14. A five-element FBG array, with Bragg wavelengths ranging from 1545.7 to 1553.0 nm, has been inscribed on a glass fibre, achieving a maximum reflectivity of the order of 10%. Each grating has 0.5 cm length, with 1 cm distance between each FBG. The FBG array has been inserted in a phantom (a material mimicking the human body under test) and heated with a radio-frequency ablation module, such that each grating is exposed to a different temperature [41]. Through WDM, each spectral window can be isolated, and the Bragg wavelengths can be individually estimated, leading to a multiple temperature reading that is shown on the traces of Figure 4.14b.

4.6.3 Tilted and Chirped FBG

While in a uniform FBG the modulation of the refractive index within the fibre is uniform, it is possible to introduce other sensing features by changing the profile of the modulation of the refractive index. TFBG sensors are implemented by having a refractive index modulation within the fibre core that is tilted at an angle, usually 5–10° with respect to the fibre longitudinal axis [12]. The effect of the tilted

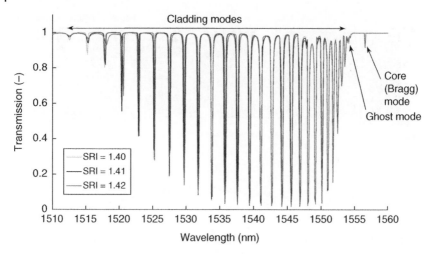

Figure 4.15 Transmission spectrum of a TFBG, with 5° tilt angle, for different values of SRI (surrounding refractive index).

modulation is to introduce, together with the standard Bragg grating resonance effect, a plurality of side modes that can be visualized within the transmission spectrum of the TFBG [42]. The transmission spectrum of a TFBG with 5° tilt angle is shown in Figure 4.15. As it can be observed, it is characterized by a core (Bragg) mode, which corresponds to the reflection of a uniform FBG; the Bragg mode is sensitive to both strain and temperature as a standard grating. However, moving towards the smaller wavelengths, we observe a plurality of cladding modes that are coupled in the fibre and are due to the excess of tilt angle introduced by the TFBG [42]. The spectral occupancy of cladding modes depends on the tilt angle, as higher tilt results in a wider transmission spectrum; a cut-off mode can be identified, for which all the modes preceding the cut-off are strongly attenuated and all the modes at wavelengths higher than the cut-off are visualized on the spectrum.

The key characteristic of the TFBG is that since cladding modes are propagated in the fibre cladding, at the boundary with the external medium, they are sensitive to the surrounding refractive index (SRI). An increase of the SRI results in a spectral shift of each mode, with sensitivity that is different for each mode and each SRI value, but in general is higher in the proximity of the cut-off condition.

TFBG sensors can be used either as refractometer or functionalized as biosensors for biological detection [42]. Moreover, it is possible to enhance its sensitivity by deposition of a metallic thin film that generates an SPR [43].

Another variation of the refractive index modulation is the chirped FBG (CFBG), in which the modulation period is not constant but changes in space [44]. The most popular configuration is the linearly chirped FBG, in which the Bragg wavelength

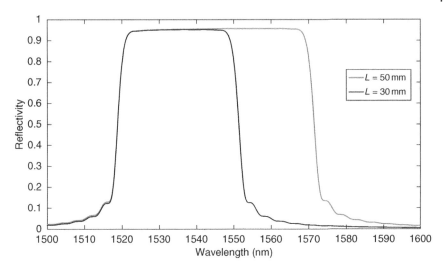

Figure 4.16 Reflection spectrum of a simulated CFBG, with linear chirp and 1 nm/mm chirp rate, for 50 and 30 mm grating length.

changes linearly along the grating. The rate of spatial change of the Bragg wavelength is defined as the chirp rate; for example, a chirp rate of 1 nm/mm in a 1550 nm CFBG of 1 cm length means that the Bragg wavelength reflected at the start of the grating is 1550 nm, the central wavelength is 1555 nm, and the end wavelength is 1560 nm. Since the CFBG reflects not a single Bragg wavelength, but rather a continuous of Bragg wavelengths, it is inherently a broadband device, with reflection spectrum of several nanometres width up to several tens of nanometres. Figure 4.16 shows the spectrum of two CMT-modelled CFBGs having 1 nm/mm chirp rate, one of length 30 mm and the other of length 50 mm. The grating has a wide bandwidth, causing a large portion of the spectrum to be reflected (30 and 50 nm respectively). CFBG sensors are gathering interest in modern applications, as they allow detecting not only physical parameters but rather their spatial distribution along the grating length [44]: for example, by comparing the left-side and right-side spectral shift, it is possible to measure temperature or strain gradients.

4.6.4 Long-Period Grating (LPG)

An LPG has a period longer than standard FBGs, of the order of hundreds of micrometres [12, 45]. Unlike FBGs, which are resonant structures affecting the mode propagating in the core, LPGs operate by coupling the fundamental modes into co-propagating higher-order modes, providing a transmission spectrum characterized by the presence of multiple transmission holes. As for a TFBG, the LPG

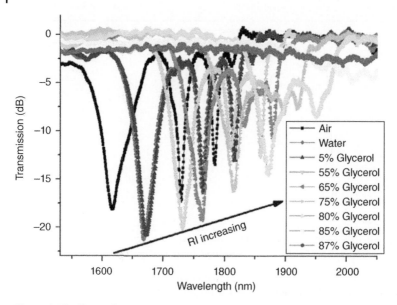

Figure 4.17 Transmission spectrum of an LPG with different values of surrounding refractive index. *Source:* Image taken from [46]. Reproduced with permission of IEEE.

finds its main application as an SRI sensor. Each transmission dip shifts when the SRI changes, due to the change of difference between cladding and outer refractive index. In Figure 4.17, taken from [46], we observe the spectrum of an LPG fabricated in a silica fibre. The spectrum is broader than in an FBG, as the bandwidth of the spectral mode analysed in this work is >10 nm. As the SRI increases, the spectrum translates towards the longer wavelengths, changing the amplitude of the transmission dip. The sensitivity of LPGs can be increased by reducing the cladding diameter, typically by etching, and with two other important phenomena: the dispersion turning point [46] and the mode transition [47, 48]. The latter phenomenon is induced by the deposition of a thin film of higher refractive index than the fibre, and it causes a reorganization of the effective indices of the cladding modes when a mode starts to be guided in the thin film.

4.6.5 FBG Fabrication

The most extended method for FBG inscription is *photosensitivity*: when exposed to intense and focused ultraviolet (UV) light, the glass compound of the optical fibre exhibits a permanent change of the refractive index, due to the permanent damages introduced within the silica [12]. The photosensitivity effect can be recorded on silica fibres without any treatment in a weak form, usually sufficient to inscribe weak gratings.

In order to increase the photosensitivity, and therefore fabricate gratings with high reflectivity, two methods are commonly used. The first method is to introduce in the fibre a strong germanium doping, as the presence of Ge within the fibre core enhances the photosensitive effects. Currently, photosensitive fibres having the same effective refractive index and size as SMF are available commercially for FBG inscription.

The second method is based on hydrogenation, or hydrogen loading, which allows instead fabricating FBGs on standard SMF fibre: the fibres are soaked in a high-pressure (~150 atm) hydrogen gas chamber at a temperature of 20–75 °C, and subsequently the FBG is fabricated as soon as the fibre is removed from the chamber.

While photosensitivity allows the permanent change of refractive index, several methods are used to generate the periodic modulation of the refractive index, which gives rise to the resonance effect in the grating. The most popular method, first demonstrated in the 1990s, is based on the *phase mask* [47]: the phase mask is a diffractive element that, when exposed to normally incident UV light, cancels the 0th order diffraction and maximizes the −1st and +1st order diffractions. The interference pattern created by the phase mask allows having regions of constructive and destructive interference on the incoming UV light, therefore generating a periodical pattern of high/low refractive index: with a phase mask having periodicity Λ_{pm}, it is possible to inscribe FBGs with periodicity $\Lambda = \Lambda_{pm}/2$. The phase mask is a very robust and repeatable method, as the FBG wavelength, chirp, tilt, or apodization patterns are all encoded into the phase mask itself (or, as the tilt angle, can be easily adjusted on the inscription set-up), thus being able to inscribe always identical gratings. The phase mask method has been also industrialized in the NORIA FBG inscription device (Northlab Photonics, Nacka, Sweden), equipped with multiple phase masks. The main drawbacks of this method are the impossibility to design a custom-made grating (as each phase mask inscribes only one type of grating), the need to strip the fibre before inscription, which may reduce its mechanical stability, and the impossibility to easily inscribe FBG arrays (each grating has to be inscribed separately).

In order to solve these drawbacks, a method based on drawing tower and labelled *draw-tower* inscription has been proposed [48] and industrialized by FBGS International (Geel, Belgium). In this method, the gratings are inscribed in the fibre during its drawing process, thus drawing the preform and inscribing the gratings altogether as a single process. This method enables writing arrays of gratings in a single fabrication process, with gratings having a uniform distance between each element of the array as well as a uniform spacing between each Bragg wavelength. Up to 80 FBGs per fibre can be inscribed with the draw-tower method, and it can be extended to all-grating fibres. Also, since the fibre drawing is completed after the FBG inscription, a draw-tower grating maintains the fibre protective buffer. To date, this method has been implemented only for uniform gratings, and the reflectivity of the FBGs is lower than phase mask inscription.

The *interfering beams* method [49] for FBG inscription also makes use of an interference pattern applied to a UV or KrF laser. In this case, however, the input light is split into the two arms of an interferometer and recombined in order to impress an interference pattern. The most popular method uses a low-coherency UV or KrF source, split into a 50/50 beam splitter and separately propagating each beam in air. Each beam is then steered with a mirror at the incidence angle θ_i and focused by means of a pair of cylindrical lenses into the fibre at the FBG inscription point. This structure produces an interference fringe in the fibre with period $\Lambda = \lambda_L/(2\sin\theta_i)$, where λ_L is the excimer laser wavelength. The interferometric set-up is harder to implement and control, as each element of the air path requires a precise alignment and the accuracy of the translational stages is crucial. In addition, in general, it requires monitoring the incident UV/KrF light power focused onto the FBG fibre core. The great advantage of the interfering beams method is that it allows the inscription of custom-made gratings: the Bragg wavelength and tilt angle, chirp, or apodization pattern can all be selected by adjusting the air path. The precise control of this method, usually implemented with software programmed in LabVIEW, allows optimizing the preparation of the set-up for an efficient FBG inscription. While versatility is the main advantage, the repeatability is lower as focusing errors lead to changes of the Bragg wavelength.

The direct inscription, often referred to as *point by point* (PbP) [50], is the most recent method for FBG inscription, based on a fs laser. A fs source delivering pulses with strong energy is focused into the fibre core with high precision (focusing can be supported by index matching oils). Then, inscription is performed by rapidly scanning the laser across the fibre core, hitting each section of the fibre, PbP, with a periodical pattern (such as a sawtooth or a sine wave). The narrow duration of the pulse induces that a permanent refractive index change in the core remains confined, whereas the scanning speed determines the Bragg wavelength. The PbP method is well suited to inscribe arrays of FBGs, particularly with narrow spacing between each grating.

Finally, it is important to mention that LPGs, with a longer period than FBGs, are easier to manufacture. Consequently, all the techniques used for FBGs can be applied to LPG inscription, along with other methods like microtapering, CO_2 laser inscription. But perhaps the most popular and cost-effective alternative to fs or UV laser inscription is arc-induced inscription, which can be achieved with a simple splicing machine [51].

4.7 Conclusions

In this chapter we have reported the basic methods and techniques to design fibre-optic sensors, which lead to the most popular architectures and commercial devices. The low-cost methods are based on intensity detection or transducing measurands into a change of optical losses. However, most commercial systems are based on spectral detection and are mainly based on interferometers and gratings. Among interferometers, the most widely used architectures are Fabry–Pérot and multimode interferometry, whereas among grating-based sensors the most popular architecture is uniform FBGs, though CFBGs, TFBGs, and LPGs add some sensing properties that may lead to a more extended use in the future of these technologies.

References

1 Udd, E. and Spillman, W.B. Jr. (2011). *Fibre Optic Sensors: An Introduction for Engineers and Scientists*. Wiley.
2 Vallan, A., Carullo, A., Casalicchio, M.L., and Perrone, G. (2014). Static characterization of curvature sensors based on plastic optical fibres. *IEEE Transactions on Instrumentation and Measurement* 63 (5): 1293–1300.
3 Peters, K. (2010). Polymer optical fibre sensors – a review. *Smart Materials and Structures* 20 (1): 013002.
4 Sai, V.V.R., Kundu, T., and Mukherji, S. (2009). Novel U-bent fibre optic probe for localized surface plasmon resonance based biosensor. *Biosensors and Bioelectronics* 24 (9): 2804–2809.
5 Tosi, D., Perrone, G., and Vallan, A. (2013). Performance analysis of a noncontact plastic fibre optical fibre displacement sensor with compensation of target reflectivity. *Journal of Sensors* 2013: 781548.
6 Sultangazin, A., Kusmangaliyev, J., Aitkulov, A. et al. (2017). Design of a smartphone plastic optical fibre chemical sensor for hydrogen sulfide detection. *IEEE Sensors Journal* 17 (21): 6935–6940.
7 Wang, Q. and Zhao, W.M. (2018). A comprehensive review of lossy mode resonance-based fibre optic sensors. *Optics and Lasers in Engineering* 100: 47–60.
8 Zubiate, P., Zamarreño, C.R., Del Villar, I. et al. (2015). Experimental study and sensing applications of polarization-dependent lossy mode resonances generated by D-shape coated optical fibres. *Journal of Lightwave Technology* 33 (12): 2412–2418.
9 Tosi, D. and Perrone, G. (2017). *Fibre-Optic Sensors for Biomedical Applications*. Artech House.

10 Rao, Y.J. (2006). Recent progress in fibre-optic extrinsic Fabry–Perot interferometric sensors. *Optical Fibre Technology* 12 (3): 227–237.

11 Shao, M., Qiao, X., Fu, H. et al. (2014). Refractive index sensing of SMS fibre structure based Mach-Zehnder interferometer. *IEEE Photonics Technology Letters* 26 (5): 437–439.

12 Othonos, A. and Kalli, K. (1999). *Fibre Bragg Gratings: Fundamentals and Applications in Tele-communications and Sensing*. Norwood, MA: Artech House.

13 Mihailov, S.J. (2012). Fibre Bragg grating sensors for harsh environments. *MDPI Sensors* 12 (2): 1898–1918.

14 Yang, H.Z., Qiao, X.G., Luo, D. et al. (2014). A review of recent developed and applications of plastic fibre optic displacement sensors. *Measurement* 48 (2): 333–345.

15 Massaroni, C., Saccomandi, P., and Schena, E. (2015). Medical smart textiles based on fibre optic technology: an overview. *Journal of Functional Biomaterials* 6: 204–221.

16 Ahmadi, R., Arbatani, S., Packirisamy, M., and Dargahi, J. (2015). Micro-optical force distribution sensing suitable for lump/artery detection. *Biomedical Microdevices* 17 (1): 10.

17 Faria, J.B. (1998). A theoretical analysis of the bifurcated fibre bundle displacement sensor. *IEEE Transactions on Instrumentation and Measurement* 47 (3): 742–747.

18 Casalicchio, M.L., Penna, A., Perrone, G., and Vallan, A. (2009). Optical fibre sensors for long- and short-term crack monitoring. In: *2009 IEEE Workshop on Environmental, Energy, and Structural Monitoring Systems*, 87–92. Crema: IEEE.

19 Roriz, P., Frazão, O., Lobo-Ribeiro, A.B. et al. (2013). Review of fibre-optic pressure sensors for biomedical and biomechanical applications. *Journal of Biomedical Optics* 18 (5): 1–19.

20 Aitkulov, A. and Tosi, D. (2019). Optical fibre sensor based on plastic optical fibre and smartphone for measurement of the breathing rate. *IEEE Sensors Journal* 19 (9): 3282–3287.

21 Wang, A., He, S., Fang, X. et al. (1992). Optical fibre pressure sensor based on photoelasticity and its application. *Journal of Lightwave Technology* 10 (10): 1466–1472.

22 Domanski, A.W., Wolinski, T.R., and Bock, W.J. (1994). Polarimetric fibre optic sensors: state of the art and future. In: *Proceedings of SPIE International Conference on Interferometry '94*, vol. 2341. Warsaw: SPIE.

23 Hegde, G. and Asundi, A. (2006). Performance analysis of all-fibre polarimetric strain sensor for composites structural health monitoring. *NDT & E International* 39 (4): 320–327.

24 Nierenberger, M., Lecler, S., Pfeiffer, P. et al. (2015). Additive manufacturing of a monolithic optical force sensor based on polarization modulation. *Applied Optics* 54 (22): 6912–6918.

25 Slavík, R., Homola, J., and Čtyroký, J. (1999). Single-mode optical fibre surface plasmon resonance sensor. *Sensors and Actuators B: Chemical* 54 (1): 74–79.

26 Lee, B.H., Kim, Y.H., Park, K.S. et al. (2012). Interferometric fibre optic sensors. *MDPI Sensors* 12 (3): 2467–2486.

27 Islam, M.R., Ali, M.M., Lai, M.H. et al. (2014). Chronology of Fabry-Perot interferometer fibre-optic sensors and their applications: a review. *MDPI Sensors* 14 (4): 7451–7488.

28 Wu, Q., Semenova, Y., Wang, P., and Farrell, G. (2011). High sensitivity SMS fibre structure based refractometer – analysis and experiment. *Optics Express* 19 (9): 7937–7944.

29 Poeggel, S., Tosi, D., Duraibabu, D.B. et al. (2015). Optical fibre pressure sensors in medical applications. *MDPI Sensors* 15 (7): 17115–17148.

30 Zhu, T., Wu, D., Liu, M. et al. (2012). In-line fibre optic interferometric sensors in single-mode fibres. *MDPI Sensors* 12 (8): 10430–10449.

31 Rao, Y.J., Deng, M., Duan, D.W. et al. (2007). Micro Fabry-Perot interferometers in silica fibres machined by femtosecond laser. *Optics Express* 15 (21): 14123–14128.

32 Duan, D.W., Rao, Y.J., Wen, W.P. et al. (2011). In-line all-fibre Fabry-Perot interferometer high temperature sensor formed by large lateral offset splicing. *Electronics Letters* 47 (6): 1702–1703.

33 Allsop, T., Reeves, R., Webb, D.J., and Bennion, I. (2002). A high sensitivity refractometer based upon a long period grating Mach–Zehnder interferometer. *Review of Scientific Instruments* 73 (4): 1702–1705.

34 Wang, Q., Wei, W., Guo, M., and Zhao, Y. (2016). Optimization of cascaded fibre tapered Mach–Zehnder interferometerand refractive index sensing technology. *Sensors and Actuators B: Chemical* 222: 159–165.

35 Yao, Q., Meng, H., Wang, W. et al. (2014). Simultaneous measurement of refractive index and temperature based on a core-offset Mach–Zehnder interferometer combined with a fibre Bragg grating. *Sensors and Actuators A: Physical* 209 (1): 73–77.

36 Jiang, L., Zhao, L., Wang, S. et al. (2011). Femtosecond laser fabricated all-optical fibre sensors with ultrahigh refractive index sensitivity: modeling and experiment. *Optics Express* 19 (18): 17591–17598.

37 Tripathi, S.M., Kumar, A., Varshney, R.K. et al. (2009). Strain and temperature sensing characteristics of single-mode–multimode–single-mode structures. *Journal of Lightwave Technology* 27 (13): 2348–2356.

38 Rodríguez-Rodríguez, W.E., Del Villar, I., Zamarreño, C.R. et al. (2018). Sensitivity enhancement experimental demonstration using a low cutoff wavelength SMS modified structure coated with a pH sensitive film. *Sensors and Actuators B: Chemical* 262: 696–702.

39 Kersey, A.D., Davis, M.A., Patrick, H.J. et al. (1997). Fibre grating sensors. *Journal of Lightwave Technology* 15 (8): 1442–1463.

40 Erdogan, T. (1997). Fibre grating spectra. *Journal of Lightwave Technology* 15 (8): 1277–1294.

41 Jelbuldina, M., Korobeinyk, A.V., Korganbayev, S. et al. (2018). Fibre Bragg grating based temperature profiling in ferromagnetic nanoparticles-enhanced radiofrequency ablation. *Optical Fibre Technology* 43: 145–152.

42 Guo, T., Liu, F., Guan, B.O., and Albert, J. (2016). Tilted fibre grating mechanical and biochemical sensors. *Optics & Laser Technology* 78: 19–33.

43 Shevchenko, Y.Y. and Albert, J. (2007). Plasmon resonances in gold-coated tilted fibre Bragg gratings. *Optics Letters* 32 (3): 211–213.

44 Korganbayev, S., Orazayev, Y., Sovetov, S. et al. (2018). Detection of thermal gradients through fibre-optic Chirped Fibre Bragg Grating (CFBG): medical thermal ablation scenario. *Optical Fibre Technology* 41: 48–55.

45 Esposito, F., Srivastava, A., Iadicicco, A., and Campopiano, S. (2019). Multi-parameter sensor based on single long period grating in Panda fibre for the simultaneous measurement of SRI, temperature and strain. *Optics & Laser Technology* 113: 198–203.

46 Shu, X., Zhang, L., and Bennion, I. (2002). Sensitivity characteristics of long-period fibre grating. *Journal of Lightwave Technology* 20 (2): 255–266.

47 Del Villar, I., Matias, I., Arregui, F., and Lalanne, P. (2005). Optimization of sensitivity in long period fibre gratings with overlay deposition. *Optics Express* 13 (1): 56–69.

48 Cusano, A., Iadicicco, A., Pilla, P. et al. (2006). Mode transition in high refractive index coated long period gratings. *Optics Express* 14 (1): 19–34.

49 Li, Q.S., Zhang, X.L., Shi, J.G. et al. (2016). An ultrasensitive long-period fibre grating-based refractive index sensor with long wavelengths. *MDPI Sensors* 16 (12): 2205.

50 Hill, K.O., Malo, B., Bilodeau, F. et al. (1993). Bragg gratings fabricated in monomode photosensitive optical fibre by UV exposure through a phase mask. *Applied Physics Letters* 62 (10): 1035–1037.

51 Lindner, E., Mörbitz, J., Chojetzki, C. et al. (2011). Draw tower fibre Bragg gratings and their use in sensing technology. In: *Proceedings of SPIE Fibre Optic Sensors and Applications VIII*, vol. 8028. Orlando, FL: SPIE.

52 Meltz, G., Morey, W.W., and Glenn, W.H. (1989). Formation of Bragg gratings in optical fibres by a transverse holographic method. *Optics Letters* 14 (15): 823–825.

53 Lacraz, A., Polis, M., Theodosiou, A. et al. (2015). Femtosecond laser inscribed Bragg gratings in low loss CYTOP polymer optical fibre. *IEEE Photonics Technology Letters* 27 (7): 693–696.

54 Rego, G. Arc-induced long period fibre gratings. *Journal of Sensors* 14: 3598634, 20.

5

Structural Health Monitoring Using Distributed Fibre-Optic Sensors

Alayn Loayssa

Department of Electrical, Electronic and Communications Engineering, Public University of Navarre, Pamplona, Spain

5.1   Introduction

Structural health monitoring (SHM) can be defined as the set of techniques, methods, and systems that aim to assess the integrity of a structure non-destructively using permanently attached transducers. It is closely related to the non-destructive evaluation (NDE) field, with the difference that NDE typically uses removable transducers that are temporally fixed to assess the integrity of the structure in measurements that are usually infrequent. By contrast, SHM is based on frequent measurements that enable to interpret the results by comparing them with previous baseline measurements in a process that can be automated. SHM can be also defined to include the closely related area of condition monitoring, which aims to assess the status of a machine to optimize its maintenance cycles so that they are carried out at the optimum time [1].

The SHM concept has been widely developed during the last decades using a large number of different and competing technologies such as vibration sensing using accelerometers, acoustic emission sensors, guided wave transducers, and electrical strain gauges. In the last two decades, fibre-optic sensors have also started to be deployed for SHM mainly as strain sensors alternative to conventional electrical strain gauges. They are particularly interesting for deployment in two subcategories within SHM [1]: the global monitoring of large structures and the monitoring for localized damage. Here is where fibre-optic sensors excel by

Optical Fibre Sensors: Fundamentals for Development of Optimized Devices, First Edition.
Edited by Ignacio Del Villar and Ignacio R. Matias.
© 2021 The Institute of Electrical and Electronics Engineers, Inc.
Published 2021 by John Wiley & Sons, Inc.

providing a large number of measurement positions. Fibre-optic point sensors such as fibre Bragg gratings (FBGs) have been used for this purpose. In the context of SHM, FBGs provide measurements similar to electrical sensors such as strain gauges. However, the installation of a large number of strain gauges in a structure is cumbersome and costly due to the required electrical wiring and reading equipment. By contrast, FBG sensors can be easily multiplexed along a low-loss optical fibre deploying a purely passive optical network that can be interrogated from kilometres distance. Furthermore, this network is passive (not requiring electric power supply), dielectric, and immune to electromagnetic interference.

Fibre-optic point sensors can be certainly regarded as a significant advancement for SHM over the use of conventional electrical sensors. However, the truly remarkable step is to deploy distributed fibre-optic sensors (DFOS) for SHM, which is the focus of this chapter. DFOS are unique in the sense that they use standard optical fibre, typically identical to the one used for communications, for sensing. They turn the fibre into a spatially continuous transducer that provides a distribution of the measurand, usually temperature or strain, along its length. The spatial resolution of this measurement is such that thousands of simultaneous measurement locations are provided along the fibre. Therefore, these sensing fibres can be fixed or installed in a structure with a dense layout so that very precise quantification of its structural health is gained.

The chapter starts with a brief introduction to the fundamentals of DFOS in Section 5.2. Then, three major application areas for DFOS-based SHM are described: civil and geotechnical engineering (Section 5.3), hydraulic structures (Section 5.4), and electric power transmission and distribution (Section 5.5). These are the areas where the application of DFOS technology has been more successful or where it is regarded to have a brighter future. The oil and gas industry is another field that has seen very successful use of DFOS for applications that can be loosely defined as SHM. These are described in a specific chapter of this book. Finally, other application areas such as aerospace and composite materials monitoring have been left outside the scope of this chapter because there are still few demonstrations of DFOS technology, although it is expected that this will change in the future, particularly if DFOS fully capable of dynamic measurements are developed.

5.2 Fundamentals of Distributed Fibre-Optic Sensors

In this section, we briefly introduce the fundamentals of DFOS from the point of view of the engineer or scientist that is to make use of them for SHM applications. Hence, advanced details concerning the underlying physics or the techniques and methods that are required to conceive, design, and/or build a sensor are omitted. We will focus on the main DFOS types that are in the market today and that are deemed suitable for the SHM of the relatively large structures that are of interest in this chapter. These are distributed temperature sensors (DTS) based on the Raman scattering effect and distributed temperature and strain sensors (DTSS) based on the Brillouin scattering effect [2].

There are other types of DFOS that are based on Rayleigh scattering effect either working in the time or frequency domains. An optical frequency-domain reflectometer (OFDR) Rayleigh DTSS is commercially available, and its use has been widely demonstrated in the literature [3]. However, this instrument has a reduced range (less than 70 m) for simultaneous distributed measurements. This can be enough for laboratory or proof-of-concept demonstrations, but it is not deemed sufficient to perform the field monitoring required by SHM applications described in this chapter, which concern mainly large structures.

In addition, there is another family of DFOS based on Rayleigh optical time-domain reflectometry: the so-called distributed vibration sensors or distributed acoustic sensors [2]. These sensors are based on the interference of optical waves backreflected from Rayleigh scattering centres that are illuminated by a propagating optical pulse. This interference makes these sensors very sensitive to small vibrations or acoustic waves that induce strains variations in the fibre in the nano-strain range. For this reason, they are widely used in perimeter security applications and for the detection of acoustic signals in oil and gas wells during production and for other applications in which the spectral and temporal characteristics of the nano-strain signal detected are more important than the very precise quantification of the induced strain. Furthermore, these sensors have not yet demonstrated the capability to perform the quantified high-precision large-dynamic-range strain measurements that are required in the applications described in this chapter. Hence, their discussion is omitted here.

Figure 5.1 schematically depicts the basic elements that a DFOS system comprises. The optical fibre that is used for sensing is fixed, installed, or embedded in the structure under test in such a way that the magnitude of the physical quantity of interest experienced by the fibre is as close as possible to that experienced by the structure. The characteristics of this fibre, as well as the cabling type that is used, depend on the specific application under consideration. The sensing fibre is connected to the interrogator, which is the electronic equipment in charge of generating the optical signals that are propagated along the sensing fibre as well as detecting and processing the optical signals associated with the measurement process so as to obtain the results. These measurement results, as represented in the figure, give the longitudinal distribution or profile of the measurand, typically temperature or strain, along the sensing fibre.

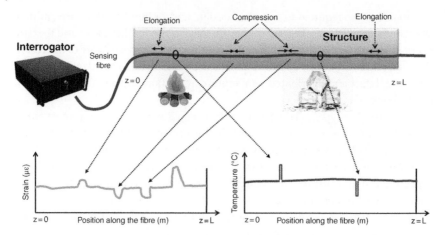

Figure 5.1 Configuration of a distributed fibre-optic sensor measurement system.

5.2.1 Raman DTS

DTSs based on Raman scattering were the first DFOS to be widely deployed in industrial applications. Raman scattering is a non-linear effect that takes place in optical fibres due to the interaction of a pump light wave with molecular vibrations of the material. The effect of this interaction is that part of the pump light that travels along the fibre is backscattered and frequency shifted in two main spectral components: a Stokes waves (lower photon energy that the pump, higher wavelength, λ_s) and an anti-Stokes wave (higher photon energy, lower wavelength, λ_a). For commonly used pump wavelength, the shift between the pump wave and the backscattered Stokes and anti-Stokes components is of the order of tens of nanometres. The use of Raman scattering for sensing takes advantage of the fact that the ratio of the scattered intensity in the Stokes and anti-Stokes bands, I_s and I_a, is temperature dependent:

$$\frac{I_a}{I_s} = \frac{K_a}{K_s} \exp\left(-\frac{h\Delta\nu}{k_B T}\right) \tag{5.1}$$

where K_a and K_s are constants, h and k_B are the Planck and Boltzmann constants, respectively, and $\Delta\nu$ is the Raman frequency shift. Therefore, it is possible to determine the temperature of an optical fibre by performing a simple measurement of the power detected in a pair of photodiodes preceded by optical filters that select either the Stokes or anti-Stokes component.

However, an issue that needs to be addressed is the localization of measurements. In order to perform a distributed measurement that provides a profile of the temperature distribution along an optical fibre, it is necessary to deploy some method to obtain the particular Raman interaction from each individual position in the fibre. This is done by applying conventional reflectometric techniques either in the time or frequency domains. The Raman OTDR is more popular in commercial systems than the Raman OFDR. In Raman OTDRs, a pulse of pump light is launched into the fibre, and the time-dependent Stokes and anti-Stokes power backscattered as the pulse travels along the fibre are recorded. This is easily translated to power of the Stokes and anti-Stokes components versus position by using the velocity of light in the fibre. Moreover, the application of expression (5.1) provides temperature versus position measurements with a spatial resolution given by the duration of the pump pulse. A previous calibration procedure is required to eliminate all dependencies of the measured temperature on the propagation characteristics of the backscattered waves and on the material properties of the fibre.

5.2.2 Brillouin DTSS

The Brillouin scattering effect is another inelastic scattering phenomenon that has been harnessed to provide distributed temperature and strain measurements. It can be used either as spontaneous Brillouin scattering (SpBS) or stimulated Brillouin scattering (SBS). SpBS occurs when a pump wave propagates along an optical fibre and is backscattered due to interaction with hypersonic acoustic phonons originated via thermal agitation of the material. The optical frequency of the optical signal (Stokes wave) backscattered due to the SpBS effect is shifted from that of the pump by the so-called Brillouin frequency shift, which is around 10.8 GHz at 1550 nm for standard single-mode fibre. Sensing applications of this effect take advantage of the fact that the Brillouin frequency shift depends on the temperature and strain experienced by the optical fibre.

For the localization of Brillouin scattering measurements, there are methods that operate in the time, frequency, or correlation domains, although just the first two have been used in commercial Brillouin DTSS to date. Brillouin OTDRs operate with a similar measurement principle to Raman OTDRs, although the optical signals used and the detection and measurement process are very different. In Brillouin OTDRs, a pump pulse of a narrowband single-mode laser is launched into the fibre, and the position-dependent spontaneous Brillouin backscattered signal is detected and processed to determine the Brillouin frequency shift of the radiation for each location.

The other main commercial Brillouin DTSS is based on the Brillouin optical time-domain analysis (BOTDA) principle. In this scheme, a pump pulse that propagates along the fibre generates a gain, via the SBS effect, for probe waves that are launched from the opposite end and counter-propagate with it. This Brillouin gain spectrum is quite narrowband, a few tens of MHz for standard single-mode fibre, and has its peak at a frequency separated from the pump by the Brillouin frequency shift. In BOTDA, the probe wave is detected to determine the gain that it has experienced at each location. Then, the probe and pump optical frequency separation is tuned so as to scan the Brillouin gain spectrum and extract the position-dependent Brillouin frequency shift.

In both cases, Brillouin OTDR and BOTDA, the temperature and strain in the fibre are derived from previous calibration data of the dependence of the Brillouin frequency shift on these measurands. In fact, one factor that needs to be taken into account in practical measurement scenarios is that the Brillouin frequency shift depends on *both* temperature and strain simultaneously. Hence, some technique or experimental arrangement is needed to compensate for this cross-sensitivity. The most common method used to solve this issue is to co-locate two sensing fibre cables at every location of the structure under test. One of the fibre cables has a special construction that ensures a good transfer of the strain experienced by the structure to the sensing fibre core inside the cable. The other fibre uses a loose-tube construction in which the sensing fibre is introduced in a tube filled with thixotropic gel so that it remains isolated from the possible strain experienced by the cable outer sheath. Hence, this fibre can be used to independently measure the temperature and use this information to remove this dependence on the other fibre so that it can measure just the strain of the structure. Alternatively, sensing cables that incorporate both the loose-tube fibres and the strain measuring fibres in a single unit have been developed and are available on the market.

5.3 DFOS in Civil and Geotechnical Engineering

In this section, we discuss the various applications of DFOS for SHM in civil and geotechnical engineering. The focus is on the applications of DFOS that require the measurement of strain, typically using a Brillouin DTSS. Nevertheless, as explained above, the monitoring systems also require the measurement of temperature to compensate for cross-sensitivity of DTSS to temperature and strain.

The fundamental advantage that DFOS bring to the monitoring of civil and geotechnical structures is the possibility to deploy a multitude of measurement locations along the structure. As explained before, Brillouin DTSS provide thousands of measurement locations along a single-mode optical fibre that can be fixed or embedded in the structure in a very dense layout. This increased availability of simultaneous measurement locations can be used for two main proposes in the SHM of civil and geotechnical engineering structures: integrity monitoring and structural assessment.

Integrity monitoring aims at detecting any damage to a structure without previous detailed knowledge of where this failure may develop. This damage can manifest as local changes of strain (or temperature). The multitude of measurement locations provided by DFOS is a fundamental advantage over the use of point sensors such as FBGs for this application because the point sensors need to be located at the precise location where the damage occurs to be able to detect it. In principle, a large number of point sensors could be deployed to cover the whole structure, but this is rarely either feasible or cost effective.

A research area in integrity monitoring using DFOS that has been especially active in the last years is that of crack detection in concrete, metallic, or composite structures. Particularly, the detection of cracks in reinforced concrete is paramount in civil engineering. The use of DFOS for this purpose is complicated by the fact that the spatial resolution of commercial Brillouin DTSS is larger than the typical crack size. Therefore, as the DTSS measurement gives the mean strain in a section of length equal to the spatial resolution, the crack can be overlooked because the strain it generates, although it can be rather large, is confined to a very small length within the spatial resolution [4]. However, methods have been developed to detect cracks even when the detailed strain mapping of the crack is not possible. For instance, one proposal is the use a fibre-optic sensing cable that delaminates in the event of a crack [4]. Delamination is useful, firstly, to avoid the breakage of the fibre due to the very large local strain that would develop along the crack if perfect bonding is maintained. Furthermore, another side benefit of this delamination is the spread of the high local strain at the crack to a larger delaminated area so that it shows in the measured Brillouin gain spectra as a secondary gain peak separated from the main peak associated with the mean strain experienced within the larger spatial resolution of the Brillouin DTSS. Other approaches that have been proposed in the literature to detect cracks are based on performing dynamic measurements in short lengths of sensing fibres attached to beams [5] or in the detection of crack-induced changes to the curvature of beams [6].

Apart from integrity monitoring, the other area where the large number of measurements positions provided by DFOS can be leveraged is structural assessment. The objective of these structural analyses is to check the ultimate limit states and serviceability limits of a given structure. In DFOS-based structural analysis, useful engineering parameters such as moments, stresses, curvature, displacement, frictions, etc. are derived from the strain measurements provided by the DFOS.

One of the most basic structural devices that can be analysed is a beam. In fact, beams are a very important example because many complex structures can be regarded as a combination of beams. A beam can be instrumented with optical fibre as schematically depicted in Figure 5.2 and then use these fibres to measure strain and calculate several derived parameters. From the classical theories of Euler–Bernoulli and Saint-Venant, the strain experienced by an optical fibre fixed to or embedded in a beam can be derived to be

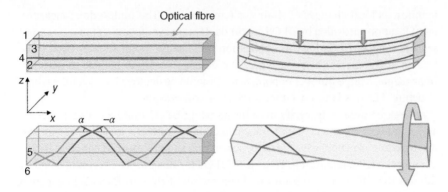

Figure 5.2 Instrumentation of a beam to measure bending and torsion.

$$\varepsilon_r = \left(\frac{\sin^2(\alpha) - \gamma \cos^2(\alpha)}{E}\right)\left(-\frac{M_z}{I_z}y + \frac{M_y}{I_y}z\right) + \left(\frac{\sin(\alpha)\cos(\alpha)}{G}\right)\left(\frac{M_T}{W_T}\right)$$

(5.2)

where y and z are the location coordinates of the fibre, α is the angle of the fibre with the longitudinal axis, γ is Poisson's ratio, E is Young's modulus, G is the shear modulus, W_T is the torsional constant, and M_z, M_y, and M_T are the bending moments in the two axes and the torsional moment, respectively. Therefore, the distribution of bending and torsional moments can be calculated from the strain measured with the fibres fixed to the beam as follows:

$$M_z = \frac{\varepsilon_1 - \varepsilon_2}{d}\frac{EI_z}{\gamma}$$

(5.3)

$$M_y = \frac{\varepsilon_3 - \varepsilon_4}{l}\frac{EI_y}{\gamma}$$

(5.4)

$$M_T = (\varepsilon_5 - \varepsilon_6)W_T G$$

(5.5)

where ε_1, ε_2, ε_3, and ε_4 are the strains measured with the longitudinal fibres and ε_5 and ε_6 the strains measured with two fibres that are helically fixed to the beam with a 45° angle in opposite directions; d and l are the distances between opposing fibres in vertical and horizontal directions, respectively.

Furthermore, the curvature, κ, and the vertical displacement (shape change), u, are given by

$$\kappa(x) = \frac{\varepsilon_1(x) - \varepsilon_2(x)}{d}$$

(5.6)

$$u(x) = \iint \kappa(x)\,\mathrm{d}x + K_1 x + K_2$$

(5.7)

where K_1 and K_2 are integration constants that depend on the boundary conditions. Moreover, other quantities such as axial and shear forces, F_a and V_s, can be obtained from beam theory as

$$V(x) = \frac{d}{dx}[EI\kappa(x)] \tag{5.8}$$

$$F_a(x) = EA\varepsilon_a(x) \tag{5.9}$$

where I and A are the moment of inertia and cross-sectional area, respectively, and $\varepsilon_a(x) = 1/2(\varepsilon_1 + \varepsilon_2)$ is the average strain. These expressions are for vertical bending, but similar expressions can be derived for lateral bending by using the strain measured by fibres fixed in the two other faces of the beam. Notice that all the described derived parameters can be also calculated using strain measured with point sensors, but the higher density of measurement positions for the distributed sensing case leads to smaller errors in their determination [7].

The use of DFOS has been demonstrated for the SHM of bridges, tunnels, piles, retaining walls, geotechnical structures, etc. In the following, we review the most significant applications with the aim of illustrating the possibilities of the technology and the various trade-offs involved in its use.

5.3.1 Bridges

Bridges are one of the most important civil infrastructures in our society, playing a fundamental role in saving natural obstacles for transportation networks. The failure of these structures can be very costly in terms of human lives and economic impact for the areas in which they are located. However, the assessment of the condition of bridges is not simple, and many countries need to cope with the need to ensure the safety of an ageing bridge fleet. In this context, the application of DFOS can play a very significant role in implementing SHM for bridges.

The first documented application of DFOS to bridge monitoring was the Götaälv Bridge in Gothenburg (Sweden) [8]. This is a 1 km long bridge that was originally built in 1930 with a design based on a concrete deck on top of nine continuous steel girders supported by multiple columns. The bridge underwent renovation in the 2000s after fatigue cracks were found on the girders. During the renovation works a system for integrity monitoring based on Brillouin DTSS was installed and commissioned. The objective of the monitoring system was to be able to detect additional cracks or any other unusual structural behaviour during the remaining lifespan of the bridge. With this purpose, around 5 km of strain sensing fibre-optic cable was glued to five of the steel girders. In addition, temperature measurement cable was also installed to remove the cross-sensitivity effects on the strain measurements. The sensing cable deployed presented a high attenuation loss once glued; hence a multichannel measuring system using two DTSS systems was necessary in order to have enough dynamic range to accommodate the cable loss. The system includes the automatic measurement system together with alarms coupled to the detection of damage in the bridge.

Another example of the application of DFOS in bridges is described in [9]. In this case, the DTSS was used for structural assessment of the bridge, instead of just pure integrity monitoring. A total of 1.16 km of sensing fibre was glued to the girders of the bridge. First, the system was tested during a load test in which the deflection calculated by integrating the strain-derived curvature (Eqs. 5.6 and 5.7) was compared with geodetic measurements using a total station. Excellent agreement was obtained between these two measurement methods. Furthermore, the measured strains were compared with those of a finite element (FE) model to check that they were in good agreement.

There are other applications on the use of DFOS in bridges reported in the technical literature. One of the most thorough and long-term deployments has been the study of the Streicker Bridge. This is a bridge at Princeton University campus that was instrumented with embedded sensing fibre for Brillouin DTSS and long-gauge point sensors at the time of its construction [10]. The SHM system deployed on this bridge performs integrity monitoring, which includes measurements of the evolution of strain from the early age of the bridge and detection of cracks and reduced joint stiffness, and also structural identification analysis for the determination of curvatures, neutral axis position, stiffness, deformed shape, natural frequencies, etc. The static measurements were performed using a Brillouin DTSS. However, all the dynamic measurements were performed in this study using long-gauge point sensors due to the inability of Brillouin DTSS at the time to provide dynamic strain measurements over medium distances. Nevertheless, it is expected that the advancement of commercial Brillouin DTSS technology will eventually incorporate techniques for dynamic sensing over long distances that are being researched in labs worldwide. This will open a whole new measurement mode for integrity and structural assessment of civil engineering structures. For instance, incorporation of these dynamic measurement capabilities to Brillouin DTSS will serve to implement other methods for condition assessment of Bridges that have been demonstrated with scaled models [11] or with point sensors [12].

5.3.2 Tunnels

Tunnels are another civil engineering structure that is fundamental for our society. Tunnels are widely used for transportation networks, the electric power grid, sewage systems, water supply, public utility, canals, etc. Tunnels are extremely expensive and complex to build, especially in areas with difficult geological conditions. Therefore, the deployment of an SHM system for these infrastructures is very important.

DFOS can play a role in the SHM of tunnels in a variety of scenarios such as:

- Monitoring of the tunnel while it is being built. This can be very important when the most popular method for tunnel construction, the so-called New Austrian Method, is deployed. A key component of this method is monitoring and measuring in order to adapt the tunnel construction to the observed ground conditions. This is achieved by instrumenting the different tunnel elements (lining, ground, boreholes, etc.) during construction and installing support elements to stabilize the tunnel only when needed, hence reducing costs.
- Monitoring of the structural effects on a tunnel of nearby construction. For instance, controlling the deformations induced by the construction of a tunnel on a pre-existing tunnel.
- Monitoring of the structural effects on a tunnel of a landslide, ground fault movement, or other geological events.
- Monitoring the effects of tunnelling on another structure or construction. For instance, on a nearby pipeline system.

Several types of measurements can be made using DFOS in tunnels. One of the most important measurements is tunnel distortion. This measurement can be performed using DFOS and applying similar principles than for shape sensing of beams: optical fibre cables for sensing can be installed on the intrados (inner face) and extrados (outer face) of the tunnel lining as schematically depicted in Figure 5.3a. Then the curvature can be calculated subtracting the measured strain in both fibre lines as in Eq. (5.6) and then obtaining the lining shape by integration [13]. However, there are situations in which access to both intrados and extrados is not possible, for instance, because an already built tunnel is to be instrumented. In these cases, fibre can be fixed just in the inner part of the tunnel. This approach has been demonstrated, for instance, in the construction of a subway tunnel in Singapore [13]. In this case, the tunnel was instrumented with DFOS to assess its deformation as a parallel tunnel was bored. Hooks were attached at multiple sections of the lining intrados at approx. 1 m intervals (gauge length) and optical fibre glued to them. The fibre was pre-tensioned so as to be able to measure compression as well as elongation. A total of 12 rings were monitored in a difficult area for tunnel construction. The strain in these rings was monitored during the advancement of the tunnel boring machine on the other tunnel. These strain measurement results displayed qualitative agreement with the expected mode of deformation of the tunnel lining. However, the integration of the measured strain to obtain the displacement or change in the shape of the lining was constrained by the fact that strain sensing fibre was available just in the intrados so that axial strain due to adjacent tunnel construction could not be evaluated. Therefore, the calculated distortions could not be precisely matched to those measured using tape extensometers. A similar

(a)

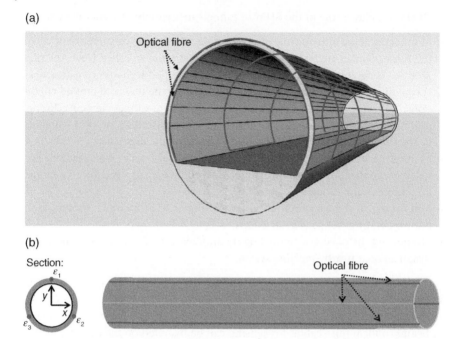

(b)

Figure 5.3 (a) Instrumentation of a tunnel with optical fibre to measure deformation. (b) Shape sensing using a tube instrumented with optical fibre for axial strain measurement.

tunnel ring monitoring method has been demonstrated in other tunnels, for instance, in the long-term monitoring of tunnels at CERN [14].

Another important measurement related to tunnel construction that can be performed using DFOS is surface ground settlement due to underground boring activity. This is especially needed in urban environments where tunnel construction can seriously impair other structures. Two different methods for settlement monitoring using DFOS have been demonstrated. One is to bury a sensing fibre directly in a trench above the tunnel and derive fundamental greenfield parameters associated with the tunnelling process, e.g. volume loss and inflection point, from measurements of the fibre longitudinal strain [15]. This method uses a detailed ground displacement model and takes advantage of the fact that the fibre tends to follow the soil deformation due to its low rigidity. This method has been deployed using Brillouin DTSS interrogation to detect cross-border smuggling tunnel construction [16] as well as to monitor tunnelling in a water system [15]. Brillouin DTSS has also been used to monitor the settlement of a pipeline due to tunnel construction underneath [17]. For this purpose, an optical sensing fibre was fixed to the crown of the pipeline, and its settlement was calculated by integrating the measured strain.

Another method for settlement monitoring is based on measuring the shape change of a tube inserted on the ground. The tube is instrumented with optical fibre, for instance, as depicted in Figure 5.3b with several fibres glued, embedded, or fixed to the tube with a 120-degree separation. This arrangement allows to calculate the two-axis curvature and, from that, by double integration, the shape change of the tube, which follows the soil settlement. This device has been demonstrated, for instance, in the monitoring of a metro tunnel in Cairo [18].

This same concept of instrumented tube and shape calculation from axial strain measurement at various angles has been also deployed to create a deformation sensor based on a PVC tube that is instrumented with optical fibre and that can be attached to the sidewalls and ceiling of the tunnel [19]. In this way, the deformation of the tunnel is translated to the instrumented tube and calculated from the axial strain measurements along that tube.

5.3.3 Geotechnical Structures

Soil instabilities constitute a severe risk to human life, transportation networks, and infrastructure and buildings integrity. Slope monitoring is particularly needed in areas where there are geological instabilities, extreme weather events, and mining or construction activities or in the vicinity of dams. DFOS can play a significant role to develop effective monitoring methods and systems that can be used to assess and evaluate the risk related to slopes. In addition, DFOS can be instrumental in the study of the soil dynamics associated with ground movements, for instance, in the study of the soil behaviour just before and during landslides. These studies can lead to the identifications of the factors that are indicative of the triggering of a landslide, which can be used in turn to develop early warning systems.

Inclinometers are the most widely used instrument to evaluate displacement in the ground and also in geotechnical engineering structures such as retaining walls. Inclinometers comprise a tube or inclinometer casing, typically made of plastic or metallic material, that has guides or grooves in its inert part. These grooves serve to lower the inclinometer probe, which is a small container with electronic components that include inclination sensors, typically based on MEMS technology. For their installation, a borehole is drilled, and multiple inclinometer casing segments are joint and introduced down to the desired depth. Then, the measurements are performed by lowering the inclinometer probe, which records the inclination of the casing at fixed intervals of a few tens of centimetres. From this inclination data, the relative displacement among the different ground layers can be calculated, and ground movement detected and quantified.

Inclinometer measurements are a firmly established slope monitoring method, but their drawback is that they are labour intensive and relatively cumbersome to perform. DFOS-based inclinometers have been proposed in order to overcome these constraints [20, 21]. These fibre-optic inclinometers follow the same principles outlined in Section 5.3.2 for the tubes instrumented with optical fibre that are used to measure settlement or tunnel distortion. Optical fibres are fixed to the tube so that curvature and shape change measurement can be derived from a combination of the measured axial strains. Deployment of these novel inclinometers can lead to the vision outlined in Figure 5.4, in which a network of fibre-optic inclinometers is simultaneously and automatically monitored. The inclinometers are connected to a single Brillouin DTSS interrogator using low-loss fibre links. This, together with the long-range capabilities of Brillouin DTSS, means that the inclinometer network can be scattered in a fairly large geographical area. For instance, a problematic landslide area can be fully covered by the network.

However, fibre-optic inclinometers face some constraints due to the method used to calculate displacement, which relies on the double integration of curvature. As explained above, Brillouin DTSS have a limited spatial resolution that translates to limitations when measuring close curvature changes, such as the ones that take place, for instance, at a surface of rupture between soil layers that are relatively sliding. At those locations, curvature changes take place at scales of a few centimetres that cannot be followed with precision by current Brillouin DTSS measurements. Therefore, large errors can accumulate when calculating displacement from those measurements. On the other hand, fibre-optic inclinometers are very suitable for use when smooth curvature variations are expected.

Another method to evaluate slope displacement and landslides is the direct insertion of fibre-optic cables on the ground. This approach has been demonstrated, for instance, in the monitoring of a landslide area in Switzerland and also

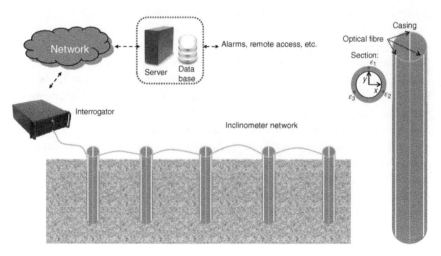

Figure 5.4 Vision for a network of fibre-optic inclinometers.

in the monitoring of slope displacements that could affect a gas pipeline in the Andes [22]. For the landslide, a sensing cable was buried at shallow depth along an 80 m long trench crossing the area where the creeping zone boundary was expected to lie. A measurement campaign was then capable of detecting the qualitative behaviour of the landslide. As for the pipeline, a 60 km section in the most geohazard-prone section along its route was instrumented with optical fibre cable for landslide monitoring. This cable was installed during construction in the same trench but with some separation from the pipeline. Other examples of the use of DFOS with direct insertion of the fibre on the soil exist, but most of the examples in the literature refer to proof-of-concept test in laboratory environments where model slopes with embedded fibres are constructed and tested.

It must be pointed out that systems demonstrated in the field to date have been capable at most of qualitatively detecting soil displacement and identifying landslide boundaries, but the ambitious objective of correlating the complex landslide dynamics to the measurements obtained using a DTSS has proved elusive for the time being. This objective is fundamentally hindered by the somewhat deficient transfer of strain from the soil to the sensing fibres [23]. Solutions to increase the coupling between the soil and the sensor fibre have been proposed, such as the use of small anchors attached to the fibre cable [22]. Nevertheless, its effectiveness for different soils and different soil conditions is yet to be rigorously analysed. An added advantage of the use of this anchoring system is that the fibre cable can be pre-strained during installation so that compression, as well as elongation, can be measured. Altogether, even if DFOS-based methods may not be capable of adequately capturing the detailed soil dynamics of landslides, they can be good enough for early warning and safety assessment systems.

Another important application of DFOS is for the monitoring of geotechnical engineering structures such as retaining walls, piles, or soil nails. Retaining walls are used to prevent slope failures by laterally containing soil in a slope. They are used as permanent walls in high terrain areas to support soil masses and allow roadway overpasses. Moreover, retaining walls are also extensively used on a temporary basis in excavation sites within construction areas to contain soil and water. Many times, for the safety of the construction environment, it is necessary to continuously monitor these walls, typically using geodetic measurements or conventional inclinometers. An alternative method is the use of DFOS by introducing in the wall at least two fibres in both the soil and excavation side. Then, the lateral displacement of the wall can be calculated using equations similar to Expressions (5.6)–(5.7). In addition, the total vertical displacement, w, can be obtained integrating the average axial strain [24]:

$$w(y) = \int_0^y \varepsilon_a(y) \, \mathrm{d}y + C \tag{5.10}$$

This method has been used in the monitoring of a secant piled wall used during the construction of a building with parking lot [24].

Soil nails are used to stabilize slopes. They are slender elements, typically reinforcing bars, that are inserted in the slope either by direct drilling or by drilling holes and grouting. Soil nailing is an increasingly popular technique of stabilizing the soil, but detailed knowledge of its structural behaviour is required in order to evaluate its performance and effectiveness and in the process obtain valuable information related to soil behaviour. The structural performance of soil nails can be evaluated by instrumenting them with strain sensing optical fibre. For instance, the fibre can be glued or anchored to the rebar, or ad hoc soil nails with embedded sensing fibre can be directly fabricated. This technique has been demonstrated, for example, in the monitoring of multiple GFRP bar soil nails [25]. The use of Brillouin DTSS provides measurements of axial strain along the nail and its evolution in time. Then, the distributed axial force can be calculated with knowledge of Young's modulus and the cross-sectional area (Eq. 5.9). Instrumented soil nails also serve to monitor the instability of a slope.

Another structure closely linked to the geotechnical characteristics of the soil is piles. Piles are the elements within a foundation that are designed to transfer the load of a building or other surface structures to deeper strata with hard soil. The geotechnical performance of a pile is derived from skin friction and base resistance. Several methods can be used in the pile design phase to estimate the performance that is going to finally be achieved. However, the use of instrumented piles is recommended to finally qualify the pile's performance. Options for this instrumentation are the use of conventional sensors such as vibrating wire strain gauges (VWSG) and linear voltage distance transducers (LVDT). Compared with these classical sensors, the use of DFOS provides the possibility of a larger number of measurement positions and hence enhanced spatial resolution and precision of the measurements.

The installation of the sensing fibre takes place at the time of the pile construction [26]. Strain and temperature sensing cables are fixed to the steel reinforcement cage segments as they are lowered into the pile's borehole. Then, concrete is poured, and the sensing cables are left safely embedded within the pile.

After construction, load tests are performed to verify that the pile and soil responses are consistent with the initial design. This is necessary because of the uncertainties in the soil behaviour and the variability of the construction method. These tests are performed in difficult soil scenarios that justify their required time and costs. A proper test can actually save a lot of money that would be otherwise required to solve problems that may arise afterwards due to insufficient capacity of a pile within a structure. The test's main objectives are to determine the allowable bearing capacity of the pile and its adequacy to the design parameters and to determine its settlement under working load.

The use of DFOS during the load test provides measurements of the axial strain distribution along the pile for the various loads. Moreover, the axial force can be calculated from the axial strain using the pile axial rigidity (Eq. 5.9). Also, the vertical displacement of the pile under load can be calculated integrating the axial strain distribution (Eq. 5.10). Additionally, the use of an FE model for extracting soil characteristics from the measurements has been demonstrated [26]. The FE model soil parameters are optimized so that the calculated strain and displacement match the measurements. This allows calculating the shaft friction profiles and the ultimate shaft capacity.

5.4 DFOS in Hydraulic Structures

Leakage detection in dams is a very successful SHM application of DFOS. Seepage can lead to internal erosion that is one of the most frequent causes of failure or degradation of embankment dams. Therefore the availability of methods to detect those effects is paramount. DTS based on either Raman or Brillouin provide an efficient means to detect and measure water leakage. For this purpose, fibre-optic cables are introduced within the body of the dam in the areas of interest.

There are two approaches for the detection of water leakage: passive and active methods [27]. The passive method relies on the detection of seepage by the measurement of the temperature, i.e. a leakage is detected by the onset of a temperature change in a given section of the dam due to the seepage of water that is at a different temperature than the surrounding material. The problem with the passive method is that its use is limited to situations in which there is indeed a gradient between the temperature of the water and the dam material. However, that is not the case, for instance, when the fibre-optic cable is installed close to sealing devices whose temperature is close to that of the retained water. For these cases, the so-called heat pulse method was devised. In this method, a hybrid cable is used containing optical fibres for measuring temperature with a DTS as well as electrical wire. Current is passed through the electrical wires so as to generate heating of the cable. Then, from the thermal response of the medium to this heating, as recorded by the DTS, the onset of leakage can be detected. One approach is to measure the total temperature increment after a given time. Areas of higher water content or flow can then be identified because the temperature change brought by heating is reduced by the increased heat transport. Another approach is to determine the apparent thermal conductivity of the surroundings by analysing the temporal evolution of the heating response. The conductivity can be obtained from the inverse of the slope of the asymptotic temperature increment in a logarithmic timescale [28]. As the water saturation of the medium increases, the measured thermal conductivity also increases. Moreover, once convective effects start, either natural or due to the flow of water, the constantly incrementing temperature of the pure conductive case starts to evolve to an asymptotic value. In the measurement, this appears as an abrupt increase in the apparent conductivity.

Therefore, the heat pulse method serves to efficiently detect seepage. This technique has been widely deployed worldwide since its first use in 1996 with more than 180 km of linear heating and temperature measurement cable installed as of 2013 [27]. Moreover, with this method, it is possible to even estimate pore velocities of seeping water in dams with the development of a detailed model for the heat transport of specific hybrid cable designs to its surroundings. Different pore velocities lead to different asymptotic values of the final temperature increase after a heating cycle.

Figure 5.5 depicts an example of a leakage detection system based on the heat pulse method installed by the author's research group in an asphalt-core rock-fill dam in the north of Spain in a project with the Confederación Hidrográfica del Ebro organization. A hybrid cable was laid along the axis of this dam at an elevation of 575.2 m above sea level. The interrogation system comprised a Raman DTS, an electrical current generator, and a PC to automate the measurement process. The figure also depicts an example measurement of thermal conductivity along the dam in two consecutive days highlighting the detection of water flowing

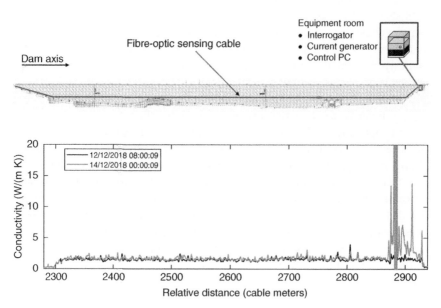

Figure 5.5 Installation of a fibre-optic leakage detection system in a dam and example of the measurement of conductivity along the dam axis.

downslope along the right abutment of the dam after a rain episode. The detection of water flow manifests as a great increase in the apparent thermal conductivity of the soil. The heating power in these measurements was 7.5 W/m for a 120 minute heating cycle.

DFOS can also be used to detect seepage in other hydraulic structures such as dikes or water channels [29]. Nevertheless, a complication of the heat pulse method for application to the monitoring of the long lengths of fibre cable required in these cases is the need to provide sufficient heating power. These are constrained by the maximum voltage ratings of the hybrid cables and by the difficulties associated with the generation of high voltages and currents.

5.5 DFOS in the Electric Grid

The electric power distribution network is currently in an evolutionary process in which it is being transformed into a smart grid where the transmission and distribution of electricity are enhanced under greater control and improved management. High- and medium-voltage transmission lines are the key elements of the transmission and distribution of energy in the smart grid. The current boom in the development of renewable energies, particularly wind electric generation, together with the growth in electrical load, is making it necessary to construct new transport lines and, at the same time, is straining the capacity demands placed on existing lines.

The fundamental measure of the capacity of a line is its current carrying capacity or ampacity. This is limited by mainly two factors: the network stability and the thermal limits of the cables or conductors deployed, with the latter being commonly the most restrictive. In overhead lines, the heating induced by the Joule effect associated with current can generate the annealing of the metal that makes the conductor material, which reduces its strength and increases the resistance of the line. Furthermore, the temperature increase leads to the thermal expansion of the conductor, thus increasing the sag of the line and reducing the clearance to ground, which is a fundamental safety concern. Maximum core temperature is determined depending on the maximum admissible sag. The temperature of the line depends on the heat balance relationship that takes into account the heating due to current flowing and the incidence of solar radiation on the cable and the cooling due to radiation and convection to the surrounding air, which greatly depends on the wind speed and direction.

Once temperature limits are imposed, current rating methods are used to determine the ampacity of the line at any given time. Two types of rating methods are used: static line rating (SLR) and dynamic thermal line rating (DTLR). SLR methods are based on using historic seasonal weather data such as air temperature and wind speed to estimate the maximum current that a line can carry. The problem with this method is that the resultant estimates tend to be rather conservative and the capacity of a line can be largely wasted. At the same time, it is possible, in extreme weather situations, that the ground clearance limits are overcome. In DTLR methods, real-time environmental and line conditions are used to continuously determine the ampacity of the line instead of relying on approximated estimates. The use of DTLR can greatly increase the capacity of a line. For instance, in the very important case of wind energy integration, the use of DTLR can take advantage of the fact that the peak outputs of wind farms are accompanied of a greater cooling effect of wind on transmission lines so that their ampacity increases. In addition, the use of DTLR allows the operator to deal with infrequent short-time peak demand situations facilitating access to extra transmission capacity. Altogether, the main benefits that DTLR can bring to the power grid have been identified [30]: mitigate transmission line congestion, facilitate wind energy integration, enable economic benefits, and improve reliability performance of power systems.

There are two types of DTLR [30]. Indirect methods of DTLR estimate the line rating from weather data either measured or forecasted for the transmission line area, whereas direct methods are based on measurement of conductor characteristics such as temperature, line tension, and ground clearance of the line or sag. The use of DTS for DTLR constitutes a direct method that provides the distribution of temperatures along every span of the transmission line [31, 32]. Optical fibres have been routinely deployed in the ground wire in the so-called optical ground wire (OPGW) cables. These combine conductors to connect adjacent towers to earth ground and optical fibres that are used for communications. However, to implement DTS-based DTLR, it is necessary to deploy optical fibres in the current-carrying conductors. Fortunately, this is not a huge problem because the electrical cable industry has developed the so-called optical phase conductor (OPPC) cables that incorporate optical fibres within its structure. These cables were developed for communication applications in cases in which there was no ground wire, and hence OPGW could not be deployed. However, they can be reused for temperature sensing applications. In OPPC cables one or more of the conducting wires are substituted by metallic tubes containing a number of fibres.

Optical fibres are also been deployed in high- and medium-voltage underground and subsea cables [33, 34]. Underground power distribution systems are gaining popularity because they are considered safe, robust, and optimized for space compared with overhead lines. The ampacity of an underground cable is also limited by the maximum allowable temperature of the conductor, cable insulation, and sheath. For typical, cross-linked polyethylene (XLPE) insulated cables, the maximum temperature is around 90°. The temperature of the cable at a given current load depends on the soil type, porosity, water content, and possible water flow in its vicinity. These characteristics may vary on a spatial scale much smaller than the comparable environmental variations of the air surrounding overhead lines. Therefore, the use of DTS to provide a detailed high-spatial-resolution thermal profiling along the length of the cable is necessary to detect possible hot spots. This need is increased by the fact that buried or subsea cables cannot be visually inspected. Furthermore, subsea cables are subjected to additional risk due to anchor drags, fishing nets, etc. that may damage the cable. Subsea cables may be subjected to particularly fast environment changes. For instance, soil migration can lead to the cable being cover by mud, radically changing the dynamics of heat transfer at that spot.

Raman-based and Brillouin-based DTSS can be used to monitor the temperature along cables. However, when deploying a Brillouin DTSS, it is necessary that the fibre measuring temperature is in a loose-tube configuration, typically using fibre in metal tube (FIMT) to ensure that it is decoupled from the strain experienced by the fibre. However, loose-tube fibres have a margin of strain isolation, and for strains larger than a given threshold, the measurement would also respond to strain. This can be useful, for instance, in the monitoring of excess strain in subsea cables due to damage or deficient installations in which they may be experiencing unwanted mechanical tensions. One issue that needs to be taken into account in overhead lines, as well as underground or subsea cables, is the radial location of the optical fibre used for sensing within the cable. It has been found that depending on this location, the difference between the measured DTS temperature and the actual conductor temperature can be significant [35].

5.6 Conclusions

In this chapter, we have reviewed the main applications of DFOS in the field of SHM. As it should have become apparent during the explanations, most of the deployed systems to date have been limited or proof-of-concept demonstrations. This is particularly true for the application of DFOS to the SHM in civil engineering. Much work is still needed to counteract the natural scepticism of potential end users. These users will not adopt the technology unless they see clear evidence that these systems can deliver their potential and help them in the design, maintenance, and management of their structures. However, this is not an easy task. For instance, a thorough demonstration of the damage detection capabilities of the technology would require its deployment in a large number of real-world test structures in which controlled damage is introduced.

Furthermore, in these initial stages of development, a very careful application of the technology by experts should be ensured because end users will be discouraged if the technology capabilities are oversold and fail to deliver its promise or if it is applied incorrectly. In addition, it is very important that the rationale behind the SHM system deployed is clearly identified and disclosed. The mere installation of instrumentation without a requirement to interpret the measurements and act on the findings should be avoided.

Another factor that needs to be taken into account to facilitate the adoption of DFOS technology in SHM is the clear justification of the business case for monitoring. It is necessary to convince the infrastructure owners and operators of the real value of SHM. One needs to bear in mind that the key to successful application lies not purely on technology but on awareness of its value to the end users. The main economic motivations for the use of SHM, in general, have been identified [1] and can be equally applied to DFOS-based SHM: reduce inspection time and cost; improve repair planning; increase, optimize, or customize inspection intervals; extend the economic life of the structure; and enable new design principles and maintenance concepts. Another fundamental economic motivation is that monitoring can reduce the costs that may be associated with the catastrophic failure of a structure.

Another big issue to be sorted out in the deployment of DFOS for SHM is what to do with the measurement data. A DFOS permanently installed and programmed for frequent measurement generates a large amount of data that can drown the end user in information if not correctly processed. Measurement data needs to be automatically processed to a digestible form before being passed to decision-making levels. The emerging application of machine learning algorithms to SHM can play a significant role in this.

Finally, regarding the application of DFOS itself, the main area for improvement is the development of standardized application guidelines. The procedures for installation of the sensors in the different types of structures should be clearly defined as well as the methods to collect and process the measurement data to obtain meaningful engineering parameters.

References

1 Cawley, P. (2018). Structural health monitoring: closing the gap between research and industrial deployment. *Structural Health Monitoring* 170 (5): 1225–1244. https://doi.org/10.1177/1475921717750047.
2 Hartog, A.H. (2017). *An Introduction to Distributed Optical Fibre Sensors*. CRC Press.

3 Froggatt, M. and Moore, J. (1998). High-spatial-resolution distributed strain measurement in optical fibre with Rayleigh scatter. *Applied Optics* 370 (10): 1735–1740. https://doi.org/10.1364/AO.37.001735.

4 Glisic, B. and Inaudi, D. (2012). Development of method for in-service crack detection based on distributed fibre optic sensors. *Structural Health Monitoring* 110 (2): 161–171. https://doi.org/10.1177/1475921711414233.

5 Babanajad, S.K. and Ansari, F. (2017). Mechanistic quantification of microcracks from dynamic distributed sensing of strains. *Journal of Engineering Mechanics* 1430 (8): 04017041. https://doi.org/10.1061/(ASCE)EM.1943-7889.0001230.

6 Goldfeld, Y. and Klar, A. (2013). Damage identification in reinforced concrete beams using spatially distributed strain measurements. *Journal of Structural Engineering* 1390 (12): 04013013. https://doi.org/10.1061/(ASCE)ST.1943-541X.0000795.

7 Sigurdardottir, D.H., Stearns, J., and Glisic, B. (2017). Error in the determination of the deformed shape of prismatic beams using the double integration of curvature. *Smart Materials and Structures* 260 (7): 075002. https://doi.org/10.1088/1361-665x/aa73ec.

8 Enckell, M., Glisic, B., Myrvoll, F., and Bergstrand, B. (2011). Evaluation of a large-scale bridge strain, temperature and crack monitoring with distributed fibre optic sensors. *Journal of Civil Structural Health Monitoring* 10 (1): 37–46. ISSN 2190-5479. doi:https://doi.org/10.1007/s13349-011-0004-x.

9 Matta, F., Bastianini, F., Galati, N. et al. (2008). Distributed strain measurement in steel bridge with fibre optic sensors: validation through diagnostic load test. *Journal of Performance of Constructed Facilities* 220 (4): 264–273. https://doi.org/10.1061/(ASCE)0887-3828(2008)22:4(264).

10 Sigurdardottir, D.H. and Glisic, B. (2015). On-site validation of fibre-optic methods for structural health monitoring: Streicker bridge. *Journal of Civil Structural Health Monitoring* 50 (4): 529–549. ISSN 2190-5479. doi:https://doi.org/10.1007/s13349-015-0123-x.

11 Scarella, A., Salamone, G., Babanajad, S.K. et al. (2017). Dynamic brillouin scattering-based condition assessment of cables in cable-stayed bridges. *Journal of Bridge Engineering* 220 (3): 04016130. https://doi.org/10.1061/(ASCE)BE.1943-5592.0001010.

12 Domaneschi, M., Sigurdardottir, D., and Glisic, B. (2017). Damage detection on output-only monitoring of dynamic curvature in composite decks. *Structural Monitoring and Maintenance* 40 (1): 1–15.

13 Mohamad, H., Soga, K., Bennett, P.J. et al. (2012). Monitoring twin tunnel interaction using distributed optical fibre strain measurements. *Journal of Geotechnical and Geoenvironmental Engineering* 1380 (8): 957–967. https://doi.org/10.1061/(ASCE)GT.1943-5606.0000656.

14 Di Murro, V., Pelecanos, L., Soga, K., et al. (2016). Distributed fibre optic long-term monitoring of concrete-lined tunnel section TT10 at CERN. *Paper presented at International Conference on Smart Infrastructure and Construction*, Cambridge, UK (25 June 2016). doi: https://doi.org/10.1680/tfitsi.61279.027.

15 Klar, A., Dromy, I., and Linker, R. (2014). Monitoring tunneling induced ground displacements using distributed fibre-optic sensing. *Tunnelling and Underground Space Technology* 40: 141–150. ISSN 0886-7798. doi:https://doi.org/10.1016/j.tust.2013.09.011.

16 Klar, A. and Linker, R. (2010). Feasibility study of automated detection of tunnel excavation by brillouin optical time domain reflectometry. *Tunnelling and Underground Space Technology* 250 (5): 575–586. ISSN 0886-7798. doi:https://doi.org/10.1016/j.tust.2010.04.003.

17 Vorster, T.E.B., Soga, K., Mair, R.J. et al. (2006). The use of fibre optic sensors to monitor pipeline response to tunnelling. *GeoCongress 2006: Geotechnical Engineering in the Information Technology Age Volume 2006*, Atlanta, USA (26 February 2006 – 1 March 2006), 33pp.

18 Dewynter, V., Rougeault, S., and Magne, S. (2009). Brillouin optical fibre distributed sensor for settlement monitoring while tunneling the metro line 3 in Cairo, Egypt. In: *20th International Conference on Optical Fibre Sensors*, vol. 7503 (ed. et al.), 75035M. International Society for Optics and Photonics.

19 Moffat, R.A., Beltran, J.F., and Herrera, R. (2015). Applications of botdr fibre optics to the monitoring of underground structures. *Geomechanics and Engineering* 90 (3): 397–414.

20 Lenke, P., Wendt, M., Krebber, K., and Glätzl, R. (2011). Highly sensitive fibre optic inclinometer – easy to transport and easy to install. *Proceedings of SPIE – The International Society for Optical Engineering*, vol. 7753, *21st International Conference on Optical Fibre Sensors*, Ottawa, ON, Canada (15 May 2011 through 19 May 2011), Article number 775352.

21 Minardo, A., Picarelli, L., Avolio, B. et al. (July 2014). Fibre optic based inclinometer for remote monitoring of landslides: on site comparison with traditional inclinometers. In: *2014 IEEE Geoscience and Remote Sensing Symposium*, 4078–4081. IEEE https://doi.org/10.1109/IGARSS.2014.6947382.

22 Iten, M., Ravet, F., Niklès, M. et al. (2009). Soil-embedded fibre optic strain sensors for detection of differential soil displacements. *Proceedings of 4th International Conference on Structural Health Monitoring on Intelligent Infrastructure (SHMII-4)*, Zurich, Switzerland (22 July 2009 through 24 July 2009), pp. 22–24.

23 Zhang, C.-C., Zhu, H.-H., and Shi, B. (2016). Role of the interface between distributed fibre optic strain sensor and soil in ground deformation measurement. *Scientific Reports* 6: 36469 EP. https://doi.org/10.1038/srep36469.

24 Mohamad, H., Soga, K., Pellew, A., and Bennett, P.J. (2011). Performance monitoring of a secant-piled wall using distributed fibre optic strain sensing.

Journal of Geotechnical and Geoenvironmental Engineering 1370 (12): 1236–1243. https://doi.org/10.1061/(ASCE)GT.1943-5606.0000543.

25 Hong, C.-Y., Yin, J.-H., and Zhang, Y.-F. (2016). Deformation monitoring of long GFRP bar soil nails using distributed optical fibre sensing technology. *Smart Materials and Structures* 250 (8): 085044. https://doi.org/10.1088/0964-1726/25/8/085044.

26 Pelecanos, L., Soga, K., Elshafie, M.Z.E.B. et al. (2018). Distributed fibre optic sensing of axially loaded bored piles. *Journal of Geotechnical and Geoenvironmental Engineering* 1440 (3): 04017122. https://doi.org/10.1061/(ASCE)GT.1943-5606.0001843.

27 Dornstädter, J. (2013). Leakage detection in dams – state of the art. *20th SLOCOLD*, Ljubljana (16 October 2013),77–86111.

28 Weiss, J.D. (2003). Using fibre optics to detect moisture intrusion into a landfill cap consisting of a vegetative soil barrier. *Journal of the Air & Waste Management Association* 530 (9): 1130–1148. https://doi.org/10.1080/10473289.2003.10466268.

29 Perzlmaier, S., Straßer, K.H., Strobl, T., and Augleger, M. (2006). Integral seepage monitoring on open channel embankment dams by the dfot heat pulse method. *Transactions of the International Congress on Large Dams* 22: 145.

30 Karimi, S., Musilek, P., and Knight, A.M. (2018). Dynamic thermal rating of transmission lines: a review. *Renewable and Sustainable Energy Reviews* 91: 600–612. ISSN 1364-0321. doi:https://doi.org/10.1016/j.rser.2018.04.001.

31 Fernández De Sevilla, S., Gonzalez, G., Juberias, G. et al. (2014). Dynamic assessment of overhead line capacity for integrating renewable energy into the transmission grid. *International Conference on Large High Voltage Electric Systems 2014*, Paris, France (24 August 2014 through 30 August 2014).

32 Martínez, R., Useros, A., Castro, P., Arroyo, A., and Manana, M. (2019). Distributed vs. spot temperature measurements in dynamic rating of overhead power lines. *Electric Power Systems Research* 170: 273–276. ISSN 0378-7796. doi:https://doi.org/10.1016/j.epsr.2019.01.038.

33 IEEE *Xplore* (2012). 1718–2012 – IEEE Guide for Temperature Monitoring of Cable Systems. IEEE Std, 1–35. https://ieeexplore.ieee.org/document/6214562 (accessed 8 June 2012).

34 Ukil, A., Braendle, H., and Krippner, P. (2012) Distributed temperature sensing: review of technology and applications. *IEEE Sensors Journal* 120 (5): 885–892. ISSN 1530-437X. doi:https://doi.org/10.1109/JSEN.2011.2162060.

35 Yang, L., Qiu, W., Huang, J. et al. (2018). Comparison of conductor-temperature calculations based on different radial-position-temperature detections for high-voltage power cable. *Energies* 11 (1): 117. https://doi.org/10.3390/en11010117.

6

Distributed Sensors in the Oil and Gas Industry

Arthur H. Hartog

Worthy Photonics Ltd, Winchester, UK

The hydrocarbon extraction industry has seen an active, although not ubiquitous, adoption of distributed optical fibre sensors (DOFS).

To illustrate this point, thousands of articles[1] on the use of DOFS in the industry may be found in the OnePetro (OnePetro is a database operated by the Society of Petroleum Engineers that provides scientific articles from journals and conferences for the hydrocarbon exploration and production industry). In addition, specific workshops on DOFS in the oil and gas industry are organized on a regular basis, such as the Society of Petroleum Engineers Annual Workshop on Fibre-Optic Sensing Applications for Well, Reservoir, and Asset Management. The major geophysical societies (Society of Exploration Geophysicists, European Association of Geoscientists and Engineers) also organize workshops on DFOS immediately before or after their main annual meetings. The subject of DFOS is thus becoming an important part of technological advances in the oilfield. Finally, an industry forum, SEAFOM,[2] is driving standards in the characterization and technology for fibre-optic sensing in the upstream oil and gas industry.

In this chapter, a DOFS will be taken to mean an optical fibre sensor that determines the spatial distribution of the measurand along a section of sensing fibre that is often many kilometres long.

The oil and gas industry is particularly appropriate to the deployment of DOFS for reasons including:

---

1 The number of DOFS references in OnePetro was more than 9000 in 2019.
2 https://seafom.com/.

Optical Fibre Sensors: Fundamentals for Development of Optimized Devices, First Edition.
Edited by Ignacio Del Villar and Ignacio R. Matias.
© 2021 The Institute of Electrical and Electronics Engineers, Inc.
Published 2021 by John Wiley & Sons, Inc.

1) The fact that production wells are long linear objects that match the one-dimensional nature of an optical fibre.
2) The environment is very hostile to electronics. In addition, any form of point sensor, including optical sensors, is vulnerable owing to the potential for single-point failures at connections. An optical fibre installed in a single straight deployment is thus the simplest conceivable type of sensor to be installed in a well.
3) The distributed nature of the sensor, reporting the measurand along the entire length of the fibre, suits the need to collect data along the well or at least along the production interval.

However, the industry has needed to overcome many challenges in order to use DOFS effectively, such as protecting the fibre against the very harsh environment and finding deployment methods that are compatible with commonly accepted oilfield practice.

While the industry was progressively adopting fibre-optic sensing technology (and specifically DFOS), it also faced the issue of managing the massive streams of data created by DFOS that are orders of magnitude larger than previously known sensing methods. Conventionally, permanently installed downhole gauges were restricted to one or perhaps a few points per well, each delivering at most one temperature and one pressure reading per second. The data storage and software techniques that were available when distributed temperature sensing (DTS) was introduced to the industry were simply not appropriate for DTS data sets that are densely sampled on the distance axis and that deliver readings for an entire well at, say, one profile per minute. Such profiles had previously only been available on surveys conducted by intervening in a well by lowering a measuring tool on a wire, in which case only one profile is obtained for the whole well from each survey.

The introduction of DTS in the oilfield therefore forced a redesign of the data storage and transmission systems. An advanced multiplexed electrical gauge system might be capable of logging tens of points once per second (although the full capability was rarely installed on grounds of cost). At 10 points installed, with 2 measurands (temperature and pressure) per point and 1 reading per second, this represents a data rate of $\sim 1.7 \times 10^6$ points/day. A DTS system installed along a 5 km well, sampled at 0.5 m intervals, and updated at 1 profile/min generates about 14×10^6 points/day. Apart from the larger data volumes, distributed sensors add a new dimension (distance) that was not present in the design of prior databases for recording well data from permanently installed gauges. The availability of these new data sets, updated several times per hour during the entire life of the well, resulted in new interpretation methods and workflows to derive value from the newly available data. The data must be organized as a two-dimensional array of measurand vs. distance and time, which was not how data was previously recorded and stored.

More recently, the development of distributed vibration sensing (DVS) or distributed acoustic sensing (DAS) systems forced yet another step change in data rates: with DAS, the profiles are updated on a millisecond timescale, so at equal spatial resolution, the incoming data rates are four to six orders of magnitude larger than even for DTS. The sampling rates differ between suppliers and are dictated by how the data are acquired and processed to extract the information rather than the spatial sampling requirements for the ultimate application. However, as an example, if we assume a 0.25 m sampling interval, the same 5000 m well length, and a 2 kHz pulse repetition frequency, the raw data rate is then 40 Mpoints/s, i.e. at 2 bytes/point, almost 7 TB/day. Although these data rates are usually reduced after a processing phase has extracted the useful information, this is still a massive flow of data to be handled, stored, interpreted, and used by the service company or operator.

In this chapter, we start by providing a primer on the relevant stages in the exploration, development, production, and use of hydrocarbon resources. We then discuss the challenges in the deployment of optical fibres in the oilfield environment and describe common methods of installing sensing fibres in this applications domain. We then move on to the main successful applications of DFOS in the industry.

6.1 The Late Life Cycle of a Hydrocarbon Molecule

The hydrocarbons that provide much of the energy used in industrial societies, particularly for transportation and electricity generation, originated from the decomposition, under high-temperature and high-pressure conditions, of organic material deposited in earlier geological eras in lakes, lagoons, and seabeds. Over millions of years, this material is transformed into kerogen and then into the oil and gas that are used in modern societies.

We will now describe the processes that the hydrocarbons go through from discovery to consumption.

The oil and gas industry may be separated into:

- Upstream operations, involving the discovery and production of the hydrocarbons from the subsurface.
- Midstream operations where the material may be separated to remove undesired components, such as water and hydrogen sulphide (H_2S), and then transported to a refinery site.

- Downstream operations, in which the hydrocarbons are converted from the input material (crude oil, natural gas, etc.) into the refined products (gasoline, diesel, kerosene, propane, methane, etc.) and distributed to the consumers.

6.1.1 Upstream

Upstream operations may be divided into a number of distinct phases, each of which is preceded by extensive planning and approval processes. Apart from the regulatory approvals, each step of the process is evaluated for its likely return and risk, and the project only proceeds to the next step if the likely financial reward is determined to meet the criteria for investment applicable in the oil company. The investment decision balances the expected costs and risks with the anticipated return based on the volume and quality of the reserves and the selling price of the hydrocarbon at the anticipated production time.

6.1.1.1 Exploration

The exploration phase involves the discovery of a potential hydrocarbon reservoir, followed by the drilling of exploration well and appraisal wells. Surface methods such as seismic and gravity surveys and the studying of the local geology from nearby known formations are used to define probable locations of reservoirs. Surface seismic surveying, in particular, provides two- or three-dimensional (3D) images of the subsurface showing the underlying geological structures based on the reflections of low-frequency acoustic (seismic) waves. Analysis of the seismic images, often complemented by other information (such as gravity maps and local geology), allows regions that could meet the necessary requirements for the presence of a hydrocarbon reservoir to be identified. The prerequisites include the presence of a trapping structure of impermeable rock to capture hydrocarbons produced below and the existence of a source rock resulting from earlier deposits of organic materials, e.g. in ancient lagoons or estuaries.

The seismic surveying process requires the transmission of powerful seismic waves into the subsoil and the collection of the reflected waves [1]. A seismic data set consists of combinations of data acquired with a range of source and receiver positions. Typically, large numbers of receivers (tens of thousands) are arrayed on the soil or in cables towed behind acquisition vessels at sea, and the sources are moved in predefined patterns to cover the required area with sufficient density of source positions.

Seismic exploration requires a wide combination of technologies and competencies, including the design of the survey, the ability to deploy highly sensitive receivers and collect the data that they generate, the design and control of seismic sources, controlling the timing of source and receivers, managing the massive data sets that are generated, and the task of processing the data sets to provide a seismic image and then interpreting that image in terms of the geology that it reveals. Seismic exploration is also a large exercise in logistics involving custom-designed acquisition and source vessels (in marine surveys) and large arrays of sources and receivers on lands with crews to deploy and reposition the sources and receivers. Apart from the technical challenge, it is also a challenge in optimizing the efficiency of the acquisition process.

6.1.1.2 Well Construction

The construction of a hydrocarbon well involves the drilling of the well, protecting it from collapse using a steel casing, the annulus between the rock formations and the casing being filled with cement to prevent any leakage of hydrocarbons from the reservoir to the surface. Figure 6.1 illustrates schematically the structure of a few examples of well construction.

A single-producing zone vertical well is shown in the centre of Figure 6.1. In general, the drilling process involves drilling a large diameter hole near the surface and lowering into that hole a first casing, a strong tube usually made from steel (labelled 'surface casing' in the figure). The gap between the drilled rock and the casing is filled with cement that is pumped into the casing and forced to flow back up the gap. The cement forms a barrier to prevent fluids moving up to surface from lower geological levels, and it is essential to avoid pollution at the surface or in subterranean aquifers. After the cement has cured, a further section is drilled, through the upper casing, and the process of cementing is repeated often with multiple reducing casing sizes. Changes in casing size allow the pressure to be controlled by selecting a drilling fluid that is dense enough to avoid fluids flowing from the geological formation during the construction process and yet is not so dense that the hydrostatic pressure in the well leads to fracturing the formations through which the drilling is proceeding. Sometimes a narrow window of appropriate pressure forces a frequent change of casing size.

Once the packers and safety valves are installed, the casing can be perforated, usually by means of clusters of shaped charges that, beyond opening a hole in the casing, can form a channel in the rock over distances of order 1 m, which allow the fluid to drain from the reservoir through a high-permeability path into the wellbore. The completion equips the well to allow the hydrocarbon to be produced safely and controllably. In practice, this usually includes adding a second internal tubing (known as the production tubing) inside the casing through which the produced fluids flow, together with seals (packers) that prevent fluid from travelling in the annulus between production tubing and casing, and valves including one just above the production level to block production in the event that the well needs to be shut in (formation isolation valve). Where the well is drilled through multiple reservoir layers (illustrated in right-hand schematic of Figure 6.1 with two such

Figure 6.1 Schematic illustration of a few types of hydrocarbon wells. Centre: a vertical well, cased and perforated to produce from a single reservoir layer. Right: a vertical well, also cased and perforated but producing from two separate formations. Left: an example of a horizontal well producing from a thin layer of hydrocarbon through an uncased hole that includes a gravel-packed sand screen. *Source:* Background image is courtesy of Christine Maltby.

layers), it is often desirable to separate the production intervals by providing a packer (seal) to separate them and separate perforations to allow the fluids to flow in from the location of interest. Often, inflow control devices are used to balance the flow from each reservoir layer to avoid producing only from that under the highest pressure. In the right-hand well illustrated in Figure 6.1, the production tubing has apertures between the two packers to allow the fluid to move from the upper production interval into the production tubing.

As illustrated in the left-hand schematic of Figure 6.1, modern wells are commonly drilled with steerable tools that allow the trajectory of the well to be controlled in real time. Thus, a well starting from a platform might be deviated from the vertical to separate it from other wells starting from the same platform (or, on land, the same drilling pad). Further along the well, the trajectory might be selected to remain within a particular geological formation for reasons of rate of penetration or wellbore stability. In the reservoir section, it is common to steer the well to remain within a selected hydrocarbon formation or to weave through multiple portions of a meandering formation. In Figure 6.1, the well trajectory illustrated is placed towards the bottom of the reservoir layer of interest. Directional drilling is thus used to maximize the contact between the wellbore and the rock formation. The permeability of reservoirs varies over a wide range of values, from a few hundred millidarcys in poorly consolidated sands to nanodarcys in tight formations such as shale. Reducing the distance through which the fluids need to flow within the rocks improves the productivity of the well. In some cases, satellite wells are spurred off the main wellbore to allow a single main bore to drain a larger volume of the reservoirs from a single well head. This arrangement is known as a multilateral well.

Some of the science and technologies in well construction involve,

- Rock physics and how optimally to cut through the several kilometres of formation needed to place the well as intended.
- The drilling process including the cutters (usually industrial synthetic diamond).
- Drill tool design, i.e. how the cutters are arranged to optimize the cutting process.
- Drill collar including the actuators needed to steer the well.
- The drilling mud.
- Instrumentation to measure the position of the drill tool and many important parameters of its essential elements.
- Telemetry to return data to surface and to receive commands from surface regarding steering.

The drilling mud fulfils many functions, such as well control (provide a counterbalancing pressure in the wellbore to prevent fluid from the formation entering the wellbore until the well is completed), returning rock cuttings to surface, cooling and powering downhole electronics, and providing a communications channel for up- and down-telemetry; the mud flow also lubricates the interface between the already drilled section and the drill string.

6.1.1.3 Formation and Reservoir Evaluation

A host of petrophysical measurements are performed downhole to assess electrical resistivity, acoustic velocity, rock density, permeability, porosity, and the properties of the fluids in each section of the reservoir. In the case of resistivity and acoustic velocity, the output amounts to an image of the formation often over a full 360° angular range and with fine axial resolution that shows details of the geological layers and their inclination relative to the wellbore axis. Analytical methods including nuclear magnetic resonance (NMR) [2], optical spectroscopy [3, 4], and chemical sensing (e.g. H_2S and CO_2) are carried out *in situ*, and fluid samples are captured and carried back to surface in containers that preserve the downhole pressure and so allow more detailed laboratory analysis.

Traditionally, formation evaluation (known as well logging) is carried out on tools lowered into the well after it is drilled but before casing. The logging tool, often comprising many interconnected measurement modules, is carried on a wireline cable that carries the weight (often several tonnes) of the downhole equipment as well as delivering power and providing telemetry to and from the downhole logging tools. Thus, the measurements are relayed to surface in real time. Some wireline cables incorporate optical fibres. The optical cables were initially intended for telemetry, but, as improved electrical telemetry rates over copper cables improved (for example, using techniques derived from the ADSL signalling used in the delivery of broadband Internet to the home), the fibres are now more likely included in the cables for the purpose of distributed optical fibre sensing.

It is also possible to carry out these petrophysical measurements while drilling on special modules conveyed just above the drill bit. However, at present, there are no drilling techniques that allow optical fibres to be deployed from surface to the drilling tool.

6.1.1.4 Production

The design of the fluid entry into the wellbore varies widely depending on the geological conditions. For example, in unconsolidated formations, sand management is essential to prevent the production of fine solids and so the design of the section in the well where fluids are produced (the production interval) requires a sand screen – essentially a filter that passes the fluids but blocks as much of the fine solids as possible. Sand screen designs include filling the gap between the wellbore and the formation with gravel (gravel pack, illustrated in the horizontal well of Figure 6.1) and metal mesh structures, often formed by wrapping wires around a central pierced tubular structure.

Hydrocarbon production requires careful management of the reservoir. For example, producing the fluids as fast as their natural pressure allows usually results in early unwanted production of water or gas rather than oil. The produced water needs to be handled carefully and disposed of, for example, by reinjection. The early production of gas reduces the pressure drive that brings the oil to the surface, and in some cases, there are no surface facilities for handling the gas. In earlier times, unwanted gas was frequently flared (burned), which is environmentally highly undesirable (and wasteful) and is now banned in many jurisdictions. Therefore, for maximum recovery of the hydrocarbons in place, careful monitoring and management of the production rate all along the well, and between the various wells in the field, is essential.

In addition, the permeability of the reservoir usually varies between geological layers, and so it is important to balance the flow from these layers to avoid producing merely from the most permeable. To this end, inflow control devices are frequently placed in completions to adjust the inflow profile from the various reservoir layers. Some of these devices are fixed at installation, but some are controlled hydraulically or electrically from surface or in some cases by lowering a tool in the well to alter their setting.

After initial production under natural pressure, it is usually necessary to continue producing by secondary production, which could involve artificial lift (pumps or gas injection to lower the density of the fluid column). However, pressure support is also used to restore some of the pressure that initially drove the oil from the formation into the wellbore. This is usually achieved by injecting fluids (water, CO_2, specific chemicals) via injection wells into the same formation at some distance from existing production wells.

Each of the processes in hydrocarbon production benefits from real-time monitoring, and fundamentally, this means determining the fraction of oil, gas, and water at each location in the well. It also means monitoring the status of artificial lift devices and the injection profile in injector wells. These are the measurements that are made with the help of optical fibre sensors in increasing numbers of cases.

Heavy oil formations are produced by injecting heat (usually as steam) into the formation to lower the viscosity and so allow the oil to flow. A number of processes such as cyclic steam injection or steam-assisted gravity drainage (SAGD) are used for this purpose. Steam injection is also used towards the end of the life of a reservoir when the produced fluids have left heavy fractions behind; this is notably the case in the Bakersfield area of California and in some Indonesian fields.

Other issues affecting the ability to continue to produce as planned include the deposition of solids (e.g. waxes, asphaltenes, scales) in the well, the production of sand that needs to be separated at surface and also erodes the completion, and the stability of the wellbore, for example, due to subsidence (possibly due to depletion of shallow reserves), heave (perhaps due to heating in some recovery processes), or shearing in seismically active regions.

Thus, monitoring of the well while it is producing is a vital tool for optimizing the production process, by maximizing the fraction of the fluids in place that are recovered, minimizing the production of unwanted fluids, ensuring the flow can proceed uninterrupted by deposits and the integrity of the well itself. It is in the area of production that optical fibre sensors have found their largest role in the hydrocarbon industry.

6.1.1.5 Production of Methane Hydrates

The production of methane hydrates is an emerging topic in hydrocarbon exploration and production although it is still in the phase of research-oriented field trials. Methane hydrates are combinations of water and methane that are stable under certain conditions of temperature and pressure. They may be found in offshore continental shelves in unconsolidated sands and also under frozen arctic tundra. In spite of the drive to avoid the use of fossil fuels, in all likelihood the artic methane reserves will evaporate under worsening climate change conditions and will contribute to further acceleration of global warming, whether or not they are used. In fact, burning these reserves is likely, in the short term, to mitigate their contribution to driving climate change owing to the much stronger greenhouse effect of methane in comparison with carbon dioxide. In any event, the estimated reserves of natural gas in the form of hydrates are massive and equivalent to several tens of years of hydrocarbon production at current rates [5, 6].

Producing methane hydrates requires disturbing the conditions under which methane is stable, for example, by reducing the pressure [6]. Monitoring the temperature and pressure of the formations holding the hydrates is critical to the safe operation of their extraction, and DTS is one of the techniques that is used for this purpose [7–9].

6.1.1.6 Well Abandonment

After reaching the end of their useful economic lives, hydrocarbon wells are decommissioned with care: it is vital to ensure that residual fluids are blocked from rising and polluting the surface environment or subsurface aquifers. The process involves evaluating the quality of the cement behind the casing and, if necessary, intervening to rectify the integrity of the fluid barriers. Detecting very low rate leaks behind casing is a crucial part of this process. In addition, how the well is eventually plugged and abandoned depends on jurisdiction, some regulators requiring that the wells are monitored periodically for long periods after the plugging process. The technologies for long-term monitoring of plugged and abandoned wells are still being developed and evaluated. However, DFOS with its fully passive sensors and its ability to monitor the entire length of the well are likely to be applied to this final part of the life of the field.

6.1.2 Midstream: Transportation

The midstream segment of oil and gas activity involves preparing the product for transportation and transferring it to refineries for separation into the fluids of interest (petrol, diesel, fuel oil, kerosene, tar, lubricants, etc.). There are strict requirements on the composition of the fluids transported, for example, to avoid corrosion in pipelines, to avoid scouring from entrained solids, separation of gas from oil, elimination of oils (in gas pipelines), and many others. As an example, at the time of writing, a major problem has arisen [10] in the East European pipeline network owing to contamination of the product with organochlorides well in excess of the refineries' ability to handle the oil together with health and safety hazards associated with the contaminants. As a result, movement of oil in this part of the network is on hold, and shipment of crude oil from Russia is down by about one million barrels per day.

More generally, the specification of products in a pipeline system includes limits on the H_2S content, gas content, and water content as well as the fractions of specified hydrocarbon molecular weights. An additional factor restricting the composition of products carried in a pipeline is the fact that each refinery is optimized for particular types of crude oil. Thus, although the processes can be adjusted, the operators prefer a consistent type of feedstock and often blend different crudes to maintain the input within their operating parameters.

6.1.3 Downstream: Refinery and Distribution

A hydrocarbon refinery is in essence a large-scale chemical processing plant. Some of the processes involve high temperatures (>350 °C) and pressure, often in oxygen-rich environments. Although the size of the process vessels is limited, there are opportunities for fibre-optic monitoring because, in many cases, it is required to monitor the entire surface area of the vessel. For example, some refractory-lined vessels are monitored by applying finely spaced grids of DTS sensing fibre in order to detect damage to the lining that would lead to a failure of the pressure vessel.

Some of the pipework interconnecting the storage and process vessels is also monitored. In this case, the motivation for the monitoring is to do with flow assurance. For example, heavy oil fractions need to be heated to ensure that they flow freely, and monitoring the pipes ensures that failures of the trace heating system are detected before they cause the fluid to cool and block the pipe network.

The key benefits of distributed fibre sensors in the refinery context are their intrinsic safety and the reduced wiring. In contrast, with conventional single-point electrical sensors (e.g. platinum resistance thermometers) connected point by point to a control room, the cost of the wiring easily exceeds that of the sensor and signal conditioning/acquisition circuitry.

6.2 Challenges in the Application of Optical Fibres to the Hydrocarbon

6.2.1 Conditions

Hydrocarbon production wells operate over a range of temperature and pressure conditions. Naturally producing wells reach about 225 °C and pressures of order 30 000 psi (about 2 000 bar) although more commonly the temperature is in the range of 100–150 °C and the pressure below 15 000 psi. In any case, although electronic gauges are available for the lower temperature range, they certainly require specific designs and manufacturing techniques to ensure long-term functioning given a requirement for permanently installed sensors to demonstrate operational lifetimes of 10 years or more. At the higher end of the temperature range, electronic sensors function only for short durations, suitable for use during logging and appraisal, but not for permanent installation.

The high pressures bring the need for heavier-duty housings involving thicker walls and special metals.

However, further issues arise from a corrosive chemical environment, caused by the presence of hydrogen disulphide (H_2S). At sufficiently high temperatures (>300 °C), even water is corrosive. The corrosion risk is specific to each reservoir, but in the worst cases it requires the use of corrosion-resistant alloys that are expensive and can be difficult to machine and weld.

The presence of hydrogen is almost inevitable in a hydrocarbon well, if only as a corrosion product. Hydrogen, particularly at high temperature, permeates through essentially any barrier: above 200 °C a steel barrier only retards the penetration of hydrogen.

In addition to the chemical and pressure/temperature issues facing the installation of sensors downhole, the sheer mechanical environment is hostile. The sensors must be installed in small spaces between large mechanical objects, such as the tubulars, and components such as valves and packers that are heavy and that are manipulated from surface at distances of hundreds or thousands of metres. The packaging must be sufficiently sturdy to avoid the sensors and their telemetry connections being crushed or cut.

It should be appreciated that the risks faced by optical sensors in downhole applications also apply to electrical sensors – the issues of confined spaces, high temperature, high pressure, and corrosive environments apply to all types of sensor. Optical fibres, and DOFS in particular, require few connectors (which are a key point of failure and a significant cost for electrical sensors) and move the complexity and delicacy of the electronics to surface, where the environment is much more lenient. With the right fibre design and cabling methods, the long-term survival of optical fibres downhole is now established, and so the acceptance of the technology is growing. Nonetheless, the adoption of DOFS in the downhole environment is not ubiquitous, partly for reasons of natural conservatism in an industry that presents many safety and environmental hazards and perceived risk with any new technology, but also because it requires some changes to industry practice, in the planning and installation of the sensing fibres and also in the exploitation of the data.

6.2.2 Conveyance Methods

Given the harsh environment, several approaches to place a fibre in a well have been devised. They all involve protecting the fibre with one or more metallic barriers. The technology of laser-welding a thin stainless steel tube around a fibre [11], developed originally for subsea telecommunications cables, is critical to many of the approaches for protecting fibres downhole. Thus, a fibre-in-metal-tube (FIMT) provides a hermetic package with wall thickness of 100–200 μm and an outer diameter of 0.9–2.0 mm as a first barrier to the ingress of fluids; it also forms a structure that is sufficiently robust to withstand further cabling and encapsulation processes.

6.2.2.1 Temporary Installations (Intervention Services)

For temporary placements, the fibre is inserted and then retrieved after completing some short-term measurements usually in conjunction with other planned operations on the well.

Wireline cables with integrated fibres have been developed [12] that are fully compatible with conventional wireline-conveyed tools, in terms of load-bearing, power conveying, and electrical telemetry. These cables can therefore be used exactly like a conventional wireline cable for the usual measurement (e.g. resistivity, sonic logs, dynamic testing, vertical seismic profiling [VSP], etc.), and the fibre component within the cable is available for distributed measurements.

Slickline is a smooth-surfaced cable that in the simplest form is simply a steel wire, similar to a piano wire and having a typical diameter of 1/8″ or 3/16″. It is used for operations in completed wells, for example, adjusting the settings of flow control devices or replacing gas-lift valves. Optical versions of this type of cable have been produced in a FIMT that is reinforced with a carbon fibre composite and then embedded in a steel wire similar in outer dimensions and properties to a conventional slickline. The optical slickline can be used for most normal operations that are conducted with slickline, but with the benefit of real-time distributed measurements that can validate the intervention carried out with the slickline.

Coil tubing is a semiflexible steel tube (typically 2–3″ in outer diameter) used for intervening in wells, for example, to pump liquids into a particular zone, to clear sands from the bottom of the well, and many other activities including conveying fluids and proppants for hydraulic fracturing. Bottom-hole assemblies (BHAs) are frequently fitted to the remote end of the tubing, for example, for milling or cutting, and some BHAs also contain sensors for *in situ* measurements. An interconnecting cable from the surface to the BHA can be provided inside the coil tubing for telemetry to recover the sensing data in real time. Some companies [13] incorporate optical fibres within the internal cable for telemetry but increasingly for distributed sensing. This enabled distributed measurements to be carried out during coil tubing operations.

6.2.2.2 Permanent Fibre Installations

Permanently installed fibres have generally followed the industry practice of connecting to downhole gauges and controlling hardware from the surface [14]. The cables for permanent installations are based around hydraulic control lines (initially used for hydraulic actuation of downhole valves). These are typically 1/4″ steel tubes, with wall thicknesses designed for the pressure anticipated, and using, if necessary, a corrosion-resistant alloy. The control lines are normally encapsulated in a plastic material to an 11 mm square profile and strapped to the production tubing as the completion is lowered into the well using steel bands. Cable protectors are generally placed over the cable at joints between sections (stands) of tubing because the diameter of the tubing increases at the joints to allow for the threads that link successive tubing sections. A cable passing over a tubing joint without a protector would be at risk from wear and crushing damage as the tubing is lowered into position.

Optical downhole cable designs build on the established practice by inserting an FIMT (see Figure 6.2) containing one or more fibres inside a control line and its encapsulation. The fibres are thus protected by a double steel barrier, internal separators placed between the FIMT and the control line, and finally by the plastic encapsulation. Apart from providing a protection against wear, the encapsulation also limits the degree to which corrosion can take place by separating dissimilar metals in the well and thus slowing any electrochemical reaction. In the example of Figure 6.2, the cable is a hybrid cable containing both optical fibres for distributed sensing and electrical conductors to connect to downhole electrical gauges (typically a single-point pressure and temperature gauge).

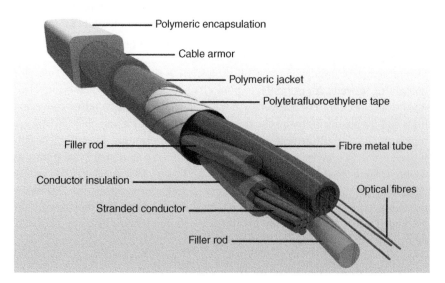

Figure 6.2 Example of a hybrid electrical and optical cable structure designed for permanent installation. *Source:* Reproduced with permission from [14].

Optical fibres are also installed (pumped) in wells into existing hydraulic control lines by fluid drag [15, 16]. This practice is similar to the air-blown fibre techniques for installing network fibres in buildings. In wells, however, a liquid is used to entrain the fibre in the control line that can either be looped back to surface or terminated downhole with a non-return valve. Lengths of 10 km or more can be installed in 1/4″ lines, and 15 km lengths have been installed in 3/16″ lines. The limiting factor on the installation length is the need to maintain a sufficient flow rate to entrain the fibre. As the control line length is increased, the pressure required to maintain a given flow increases, and eventually the pressure required exceeds the burst pressure rating of the control line.

Installing fibres by fluid drag is attractive for a number of reasons. Firstly, it is minimally disruptive of the well completion process: the control line is installed using normal industry practice, and when all the completion work is finished, the fibre pumping can take place, when the rig is less busy and the main crews are no longer on-site. Secondly, there are no optical connectors involved: if a downhole connection is required, a well-established hydraulic connection method is used [17], and so no time is lost for making splices and rebuilding the protection around the fibre on the rig floor. Thirdly, where the control line is looped back to surface, the fibre can be replaced if it degrades or (less commonly) if a different type of fibre is required. If a new fibre is required, the old fibre can be pumped out of the control line, and the new one pumped in to replace it without any interference in the operation of the well. In spite of the very large pressures involved, the installation process is relatively gentle on the fibre because the force that propels the fibre is distributed over the entire length of the fibre in the control line; it is quite different from attaching a drag device at the far end of the fibre and using that to pull the fibre in place. The presence of flow around the fibre tends to keep the fibre away from the walls of the control line and so minimize friction.

The ability to replace a fibre while the well is still operating is particularly important in conditions where the fibre lifetime is limited, such as at very high temperature (>200 °C). The pumping approach to fibre installation is also particularly valuable in multistage completions, where the bottom part of the well is fitted with its completion (e.g. sand screen, packer) and an upper completion is then brought in, sometimes months later, to connect to the lower completion and thus make the well ready for production. The issue of debris is a major challenge for the design of downhole wet mate connectors, and the problem is somewhat less severe, but by no means trivial, for hydraulic connectors.

6.2.3 Fibre Reliability

When fibres were first installed in high-temperature wells, their useful life was measured in days [18]. The need for high-temperature coatings was well understood, and the acrylate coatings used in conventional telecommunication applications (specified to typically 70 °C) were replaced with more robust coatings, such as silicone and ultimately polyimide (which is rated for operation above 300 °C).

However, survival of the coating is only part of the problem. At high temperature, hydrogen (in molecular form or as H^+ ions) is present, and it penetrates hermetic steel barriers increasingly rapidly above 150 °C and then diffuses into the glass [19, 20]. Once in the glass, molecular hydrogen gives rise to absorption bands at 1080, 1170, and 1240 nm [20]. Worse still is the chemical reaction with the glass to form OH^- radicals that have much stronger absorption bands than molecular hydrogen, for example, at around 1385 nm, but also at 1240 and 945 nm [21]. The reaction of the glass with hydrogen also causes dimensional and optical path length changes to the fibre, causing sensors relying on the stability of the optical path length (e.g. fibre Bragg gratings or interferometric sensors) to drift, sometimes dramatically [22].

The fibre reliability is addressed partly with hermetic barriers to diffusion (graphitic carbon deposited onto the fibre as it is drawn and below the polymer coating and the steel barriers in the cable construction). These barriers are effective at moderate temperatures (<150 °C), but at high temperatures, they only extend the fibre life by delaying the arrival of the hydrogen. The same applies to materials that absorb the hydrogen (known as getters) but that eventually are saturated and thus become ineffective.

Given that hydrogen barriers have little long-term value at high temperatures, for reliable long-term operation, the design of the fibre itself is optimized to reduce the propensity of the glass to react to hydrogen. Whereas conventional fibres are doped with germania (GeO_2) and sometimes with traces of P_2O_5, these additives react with the hydrogen, and fibres having pure silica cores and fluorine-doped cladding are preferred for high-temperature operations. The design of the glass and how it is processed is also optimized to minimize defects that can become colour centres and thus induce losses. Thus, the drawing conditions and stoichiometry of the glass are carefully selected. Although these optimized fibres do not eliminate the absorption caused by interstitial hydrogen, that particular form of absorption is weaker than that caused by OH^- ions or defects, and it is reversible if the hydrogen diffuses back out of the fibre.

As a result of the active research and development on improving fibre reliability, optical fibres optimized for downhole use are now available and suitable for reliable operation over many years.

6.2.4 Fibre Types

The signals returned by Raman DTS have very low coherence (the spontaneous Raman bandwidth is of order 6 THz), and therefore coherent detection techniques and optical amplification are ineffective, meaning that direct detection is used to convert them to electrical signals. They are also very weak (a 50/125 graded-index fibre [23] operating at a wavelength of 1064 nm returns only ~1 nW for each W of probe power at a spatial resolution of 1 m). For DTS applications, therefore, multimode graded-index fibres are strongly preferred because they maximize the strength of the available signal.

However, other distributed measurements including Brillouin OTDR and DAS involve interferometry in the physics of the measurement or in the design of the interrogators. For these measurements, a single-mode fibre is therefore much more appropriate.

The dilemma in downhole applications is, therefore, to choose the fibre type to install in an intervention cable or in a permanent installation.

Raman DTS can operate on a single-mode fibre, but this choice comes at the cost of a degraded signal-to-noise ratio: the smaller core and numerical aperture of a single-mode fibre reduces both the power that can be launched (limited by non-linear optical effects) and the fraction of the scattered light that is returned owing to the lower numerical aperture of typical single-mode fibres. Furthermore, if the single-mode fibres under consideration are designed for operation at or beyond 1300 nm, the performance of Raman DTS systems is further degraded by the lower scattering coefficient at long wavelength and the worse performance of detectors relative to that of silicon avalanche photodiodes.

Conversely, whereas it is possible to operate Brillouin [24] and DAS [25] systems on multimode fibres using spatial filters, mode conversion (e.g. at splices) can lead to spurious signals if the mode conversion changes on a timescale commensurate with the measurement.

Where possible, therefore, fibres of both types are installed, particularly in permanent installations, where the need to connect to downhole point sensors (e.g. pressure gauges) is also considered. Thus downhole cables usually contain at least one single-mode fibre for connecting to a pressure and for interrogation by DAS and one or two multimode fibres for DTS. However, this is not an easy decision because the fibre designed for downhole use is two to three orders of magnitude more expensive than single-mode fibre used in surface telecommunication applications. Moreover, each additional fibre requires splicing, pressure-blocking feedthroughs (to block the migration of fluids to the surface), and other provisions that all add to the cost of installation and in particular to the time taken for installation. In some cases, the space available (e.g. in wireline cables) is limited, and so the number of fibres is restricted.

As a result of these constraints, DFOS measurements are often carried out on non-ideal fibres, and the optical engineers are therefore required to use the fibres that are available and mitigate the effects of the use of non-optimal fibres.

6.3 Applications and Take-Up

The main applications for which DOFS has been used downhole are discussed below. By and large, the descriptions start with DTS applications, but it should be recognized that in some cases DTS and DAS are used together so the separation is not always clear-cut. A monograph on the interpretation methods used for applying DTS in the oilfield is recommended for further reading [26].

6.3.1 Steam-Assisted Recovery; SAGD

DOFS were first used in the oilfield to monitor steam-assisted oil recovery. An early field trial of DTS and an optical pressure sensor was reported in 1996 [27]. The relatively short wells and low-cost environment of the particular field (Coalinga, California), together with a moderate temperature, provided an ideal first test bed for the technology. This field test also demonstrated the concept of installing the sensing fibre using fluid drag.

In steam-flood fields such as Coalinga, steam is injected into vertical wells, and the intent is to sweep the oil into a producer well by lowering the viscosity of the oil in place. These wells are relatively shallow (~500 m) and cheap to drill. The interest in DTS relates to the cost of steam; at the time this was purely a financial consideration, but it is now also a carbon-budget issue. It is vital to optimize the recovery for a given input of energy in the form of steam.

DTS monitoring has been widely adopted in a steam recovery process known as steam-assisted gravity drainage [28]. This process was developed for the oil sands of Western Canada, where formations of bitumen extend over a wide area at a depth beyond that at which open-cast mining is economically viable. Bitumen formations are typically a few tens of metres thick and at a depth of order 200–600 m. The SAGD process involves drilling two wells that are horizontal in the bitumen formation (Figure 6.3). The wells are placed one above the other and separated vertically by typically 5 m with the lower well positioned near the lower boundary of the oil-bearing formation.

The upper well is used to inject steam into the formation; it is designed with concentric tubing so that steam can be circulated to heat the entire horizontal section of the well. After an initial period in which the reservoir adjacent to the well is heated by the circulating steam, the steam is injected into the formation itself. After a sufficient heating time, a so-called cavern of hot bitumen forms around, but mainly above, the steam injector well. The viscosity of this part of the bitumen is reduced sufficiently to allow it to flow into the lower producer well from where it can be pumped to surface. Typically, the horizontal section is about 1000 m long, giving a typical measured depth (linear distance from well head to the most remote part of the well) of order 1500 m.

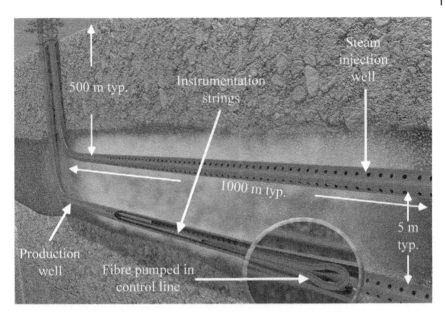

Figure 6.3 Illustration of the SAGD process and thermal monitoring with fibres deployed by fluid drag in a control line. *Source:* Adapted with permission from Duey [79].

The SAGD process is remarkably efficient – some operators claim a recovery of some 70% of the oil in the heated part of the formation. From a single pad, many well pairs are drilled in parallel arrays, with spacing between pairs of typically 200 m, in order to cover the entire area of the formation as efficiently as possible. The co-location of the well heads allows a single steam-generating plant to heat tens of wells in its vicinity.

The SAGD process is also quite delicate. The formation must be heated evenly, and if a hot spot is allowed to develop in the region between the two wells, the steam will bypass all other regions and simply flow straight into the producer well at this location of highest permeability. As a result, the steam that is injected is almost completely wasted. If this situation occurs, it is necessary to switch the steam to recirculate until the normal conditions are restored. This process can take weeks, and it is purely remedial, so little of the energy expended during the recirculation contributes to future production. This is why DTS is used extensively in SAGD production; it allows the well to be controlled to achieve the optimum temperature, known as subcool, to all portions of the well by adjusting the steam temperature and the injection locations. DTS monitoring has allowed the process to be monitored to maintain optimal conditions [29, 30], ultimately with closed-loop control involving the DTS as an input to the control process.

Conditions in SAGD wells are quite challenging – the temperature is typically in the range of 225–275 °C and the low-cost environment frequently means that carbon steel (rather than more expensive grades) is used, which enhances the rate of corrosion and so leads to the presence of hydrogen in the well. As a result, the fibres initially used in this application failed very rapidly. The problem of fibre degradation in SAGD is made more severe by the small space available that in many cases precludes a turnaround in the optical cable and so the DTS is operated in single-ended mode (i.e. it is probed only from one end). For a conventional anti-Stokes/Stokes Raman DTS, the increasing fibre loss initially causes a measurement error because the differential loss between anti-Stokes and Stokes wavelengths is affected (a change in the differential loss of a 0.03 dB over the length of the sensing fibre is sufficient to induce a calibration error of 1 K). Consequently, it was commonly found in early installations that the measurement error rose to tens of K in a few days. As the loss continues to grow as a result of hydrogen ingress and reaction with the glass, the reduction in signal level eventually increases the noise on the temperature measurement, but, in the early stages of fibre degradation, the total loss effect tends to be secondary to the change in the differential loss between the wavelengths (anti-Stokes and Stokes Raman bands) used for the measurement.

The problem of fibre degradation has been addressed in the 20 or so years since DTS has been applied to SAGD firstly by replacing the fibre at regular intervals in those wells where the fibre had been installed by fluid drag. However, this is an expensive and unsatisfactory solution if the fibre lifetime is only a few days or weeks.

The fundamental problem with a Raman DTS that probes the fibre from one end only of the fibre is that the data (in the form of the anti-Stokes/Stokes ratio as a function of temperature) depends on both the local temperature at a location z and the cumulative differential loss between the interrogator and z: there is simply not enough information in the single-ended technique for distinguishing, in the estimated temperature, a change due to a genuine variation of the temperature profile and one due to a variation of the differential loss between anti-Stokes and Stokes bands [31].

A more robust solution, implemented even in the early SAGD installations, was to use the double-ended technique [26, 31–35], which assesses the differential loss as well as the temperature and corrects changes in the differential loss in real time. The double-ended measurement is not always employed, partly because of the additional fibre (and associated cost) that is involved but also because of space constraints in the completion that prevent the control line being looped back. Nonetheless, those installations made with the double-ended technique proved much more resilient than single-ended installations. However, even with the double-ended measurement, increasing losses eventually degraded the performance below an acceptable level.

As a result of the fibre reliability issues, research and development in fibre technology was initiated in the early to mid-2000s. This led over the following 5–10 years to more robust solutions based on fibre that is specifically designed to resist the effects of hydrogen at high temperature through changes in composition and processing conditions of the glass forming the fibre. The improvement has been dramatic [20] to the point that excess loss below about 1080 nm due to hydrogen reactions is almost unnoticeable and much of the induced loss is due to interstitial hydrogen rather than the result of chemical reactions with the glass to form O—H bonds.

The second line of attack has been the design of DTS systems that are intrinsically accurate in single-ended operation and that do not rely on prior knowledge of the differential attenuation profile of the fibre (which will change during ageing at high temperature) [36]. A summary of the DTS design techniques for accurate single-ended operation may be found in [31].

6.3.2 Flow Allocation: Conventional Wells

As mentioned previously, optimizing the production from an oil well requires the understanding of which reservoir layer is producing and being able to estimate the composition of the fluid entering at each level. DOFS and DTS specifically have been used [26] in conventional wells (i.e. those not relying on steam-assisted production) to allocate the flow between different locations along the well, and, in a few cases, some inference can also be drawn on composition.

The principles of using temperature profiles for flow allocation were established by Ramey [37] for injector wells and adapted for producing wells by Curtis and Witterholt [38]. However, these methods were applicable, at the time of their development, only to the interpretation of occasional temperature logs, obtained by lowering a sensor in the well and measuring the temperature as a function of depth. The availability of DTS in wells created a profound change because temperature profiles acquired frequently and at negligible incremental cost were suddenly available and also these profiles were acquired under the normal flowing conditions of the well when it is producing – there is no disturbance whatsoever in the well due to the data being acquired.

The approach used to interpret the temperature profile obtained from DTS in terms of inflow rate was developed by G. A. Brown [26, 39] and is illustrated in Figure 6.4. The geothermal profile, i.e. the naturally occurring temperature variation that increases with depth from surface, is crucial to how DTS data is used for flow allocation. The fluid enters the wellbore at a temperature that is close[3] to that

---

3 There is a small temperature change as the fluid enters the wellbore owing to the pressure drop between the formation and the reservoir (the Joule–Thomson effect). For gas, a reduction in pressure causes cooling of the fluid, whereas for water and oil, a pressure drop warms the liquid slightly (0.06 K/bar for oil and 0.024 K/bar for water).

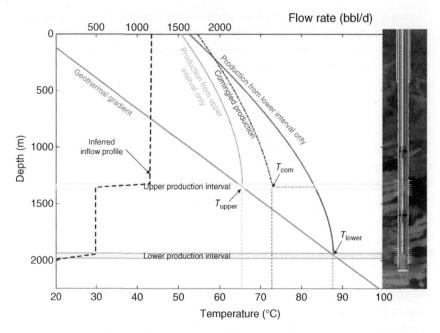

Figure 6.4 Simplified model of wellbore temperature (abscissa) vs. depth (ordinates) with fluid production from a well. The solid lines indicate the geothermal gradient (green in online edition), the modelled temperature profile when producing from the lower reservoir layer (orange in online edition) and the modelled temperature profile when producing from the upper reservoir layer (mustard in online edition). The dot-dash curve shows the modelled temperature profile when commingled production occurs from both layers at the same rate as the single-layer case (purple in electronic version). Black broken line: inferred inflow profile (plotted against the upper horizontal axis).

determined by the local geothermal profile. As the fluid moves up the well, it cools somewhat through heat exchange with the lower-temperature formations near the surface. In a steady-state flow, the temperature profile approaches asymptotically a line parallel to the geothermal gradient but displaced to a higher temperature by an offset that is proportional to flow rate. The rate of cooling depends on the fluid velocity, and so the temperature profile can be interpreted as a measure of flow rate. Since the flow rate can be measured at surface, the temperature profile is not particularly useful in the case of flow originating from a single entry point. However, a second entry point further up the well will contribute a new component to the fluid, which adds to the flow rate but mixes the fluid already moving up the well with cooler oil. The result is a step-down in temperature at the second entry and a steeper temperature gradient in the wellbore above it. In Figure 6.4, we illustrate fluid entering from the lower production interval at a depth of about

1900 m, at temperature T_{lower}, and at a rate of 500 barrels/day (bbl/d). As the fluid moves up the well, it cools (orange curve) gradually. If this were the sole fluid entry, the temperature would simply follow the orange curve. Likewise, for a single fluid, entry from the upper reservoir at T_{upper} and at 700 bbl/d would follow the mustard-coloured curve. The temperature for commingled production, where both layers produce at the previously stated rates, follows the dot-dash curve, which is the same as for the first fluid entry below the upper reservoir layer. At the upper entry point, it drops to T_{com} as the fluids mix, and the warmer fluid from the lower entry point is diluted by that entering at the local geothermal temperature. The increased flow rate reduces the cooling of the well above the upper entry point. In the simple case of two separate entry points, the proportion of the inflow profile at the upper production interval can be deduced simply from the temperature drop at that location.

Generalizing this concept, a measured temperature profile in a well with a distributed inflow will have a complex temperature shape that depends on the precise contribution from each portion of the reservoir [39]. In practice, measured temperature profiles are inverted using a model that simulates the effect of fluid entry and adjusts the inflow profile until the modelled temperature matches the measured value. Given that the total flow can usually be measured at surface, this allocation process only needs to establish the relative contribution from each reservoir layer.

Of course, in general, the geothermal profile is not perfectly linear, and, in any case, its slope varies according to the local geology. One would, therefore, normally start from a geothermal curve measured before the well has produced (and after the thermal disturbance from the well construction has thermalized back to the natural profile). It should also be stated that producing from multiple layers will affect the flow from each layer, given that the pressure within the wellbore is altered by the production from the other layers.

The discussion above is applicable only to single-phase liquid flow and in wells that have sufficient geothermal gradient to allow the temperature data to be processed as described. As the deviation of the well from the vertical increases (i.e. as the well trajectory becomes more horizontal), the geothermal gradient is reduced, and eventually, the resolution of the temperature profile is insufficient for a reliable flow allocation. The Joule–Thomson effect could in principle be used to infer flow rates even in perfectly horizontal wells. However, in practice, the temperature resolution of DTS systems is not sufficient to provide a reliable inversion to flow using purely that effect.

There is no general solution to inverting DTS data in the case of multiphase flow. However, in some specific cases, it has been possible to infer the main regions of water or gas entry. For example, gas entering the wellbore and coming out of solution (due to the pressure drop) results in a strong cooling and often generates an erratic temperature profile.

Contextual information is also used to support the interpretation, so, for example, if a change in the water cut (fraction of water in the total liquid produced) is found to increase from surface measurements, it is often possible to detect where the entry point for the additional water might have occurred from a study of the total inflow profile and relating changes in the profile as a function of time to changes in the water cut. Although these methods are not exact, they provide a low-cost means of identifying problems in the well performance, and this allows the preparation of remedial work (for example, sealing off water-producing zones) with greater confidence.

The new capability that DTS brings to production management is to be able to observe changes in the inflow profile throughout the life of the well without intervention. In spite of the fact that the DTS interpretation is sometimes ambiguous and relative to the total flow rate measured at surface, it still represents a step change in the ability to understand how the well is performing and thus manage the production more effectively. In turn, this knowledge helps improve hydrocarbon recovery rates and minimize undesirable effects, such as the need to handle unwanted fluids, such as water or in some cases gas.

6.3.3 Injector Monitoring

Injector wells are used to provide pressure support to a producing reservoir and to drive the remaining oil to a producer well. The basic ideas of using the temperature profile to infer the injection rate at each level could in principle be adapted from Ramey [37]. However, in practice, when an injector well has been operating for some time, the entire wellbore and the near formation have all been cooled by the injected fluid. There is therefore very little sensitivity in the measured steady-state DTS traces to the injectivity profile.

However, other techniques, in which the injection is interrupted, have been devised [26]. In the first approach, the injection is stopped, and the temperature profile is recorded as the wellbore warms back towards its natural geothermal condition. Low-permeability layers, which have not accepted much of the injected fluids, warm much faster than high-permeability regions where a substantial amount of low-temperature fluid has been injected into the formation. Thus, by studying the warm-back rate for each portion of the well, one can infer the relative degree to which the different layers have accepted the injected fluids. The limitation of the warm-back technique is that it is not very sensitive for wells that have been injected for a long time (months or years), and it is necessary to stop the injection process for a very long time (which could in some cases stretch to weeks or months). Such a long interruption is unacceptable given that during this time, the well is not fulfilling its main function of supporting the production in the reservoir.

An alternative approach is to interrupt the injection (shut-in) for a much shorter time (of the order one day), and the injection is then restarted [40]. In this case, the portion of the well above the reservoir (which is isolated from the surrounding formation by the casing) warms up quickly given that there has been no injection into the formation at this level in the well. The fluid in the upper part of the well therefore also warms, and, when the injection process is restarted, a distinct temperature feature (referred to as a *hot slug*) may be observed travelling down the well and displacing the fluid in the reservoir section (which will not warm up substantially over the shut-in period).

The hot slug method involves measuring the DTS profiles with a fast update rate (a few seconds). In this case, the emphasis is on tracking the hot slug, not on a fine temperature measurement; so the coarser resolution that is achieved with a fast update rate is acceptable. The principal information that is gleaned from the DTS data is the location of the boundary between the hot slug and the fluid that it displaces, and, from that, the speed of the fluid is inferred for each location in the well. Knowing the diameter of the wellbore, the flow rate can be determined. At the start of the reservoir section, the flow rate is, of course, the same as the injection rate. However, as fluid penetrates into some of injection zones, the flow rate in the remainder of the wellbore is reduced, and so, by tracking the speed of the hot slug as a function of depth, the injectivity profile can be deduced.

Of course, it would be possible to heat a tank of the injection fluid at surface and divert it into the well at a selected time. However, the volume involved in filling a reasonable length of the wellbore is quite large, and so in general the additional infrastructure (including tanks, valves, heating devices) is not justified when a similar effect can be achieved by using the natural temperature features that develop in an injector well after it has been operational for some time.

6.3.4 Thermal Tracer Techniques

The hot slug technique is an example of a thermal tracer, where a temperature change is induced in the well at a particular location and used to infer the fluid velocity. In slow-flowing wells, this is a particularly effective method [41] because the amount of heat required to be added or removed is small. One implementation of [41] involves pumping pressurized nitrogen, which is often available in the oilfield for other reasons, into a heat exchanger, where the diameter is increased to allow the pressure (and thus the temperature) to drop; the drop in temperature is transferred to the flowing oil via the heat exchanger. A downhole valve allows the nitrogen flow to be controlled and turned on abruptly to form a transient temperature feature that is tracked with DTS to assess the velocity of the oil.

Permanently installed DTS systems are uniquely suitable for thermal tracing because of their ability to measure the temperature profile along the sensing fibre.

They are naturally suited to tracking temperature features that move up or down the well, an approach that is simply not possible with single-point sensors.

6.3.5 Water Flow Between Wells

The injection process, used for supporting production, causes water to flow into the formation. This water is usually at a much lower temperature than the formation into which it is injected, and, consequently, it cools the formation. As a result, injected water flowing into a neighbouring producing well will be colder than the geothermal temperature and can therefore be identified. However, there is a time lag between the arrival of the injected fluid and a temperature response because the water first arriving at the well is warmed by the formation. If the producing well is shut in, regions that have received injected water can be identified because they are colder than the geothermal temperature.

Long-term monitoring with DTS is therefore capable of providing some information about flows from injectors to producers.

6.3.6 Gas-Lift Valves

We introduced the subject of gas lift for assisting the production from reservoirs where the pressure has depleted. Once the downhole pressure has fallen to a point where it is insufficient to drive the column of fluid up the well, natural production ceases. The cause of the insufficient pressure is usually a combination of a decline in the reservoir pressure and an increase in the water cut. Water, being denser than oil, requires a higher pressure to maintain production.

Insufficient pressure is addressed either through pumps (for deep wells, these are electrical submersible pumps) or gas lift. In gas lift, a gas, usually nitrogen, is injected into the well near the bottom of the liquid column in order to reduce the average density of the liquid and so allow the remaining reservoir pressure to drive the produced liquids to surface. The gas is injected through gas-lift valves that are designed to be open (allow gas to flow) in a defined range of pressure that is set by a mechanical arrangement in the device. In general, there is a string of gas-lift valves, and the system is designed so that initially the upper valve opens; this relieves the pressure on the next lower valve that then opens and creates the conditions for the upper valve to close. This process continues until only the lowest valve is injecting gas into the well.

Gas-lift systems require maintenance because the valves occasionally fail and they also need adjustment as a result of changes in the well, such as reducing reservoir pressure or increased water cut. Keeping the system operating as designed is important for the economics of the field: for example, gas injected too high in the well contributes little to production but still requires the gas to be compressed and injected. Valves are retrieved using a slickline intervention. With a DTS-enabled slickline survey, the service company can rapidly assess which valves are functioning properly and which, for example, are working intermittently. With that information, the valves that need attention can be retrieved, then adjusted or replaced, and repositioned. A new DTS survey can then be run with the same slickline cable to confirm the proper operation of the gas-lift system before the service team leaves the site, making the entire process efficient and verifiable.

DTS is particularly effective at monitoring gas-lift valves because they generate a clear cooling signal when operating. However, small gas leaks at valves can be difficult to detect, and, in this case, DAS/DVS systems have been proven to be able to locate suspected leaks. DAS/DVS sensors, although designed for sensing dynamic strain, also respond to temperature (and are far more sensitive than a Raman DTS).

6.3.7 Vertical Seismic Profiling (VSP)

Surface seismic acquisition is widely used in oil and gas exploration. The technique consists in launching low-frequency acoustic energy into the ground and detecting the reflected waves returning from subterranean formations, or more precisely from changes in acoustic impedance between geological layers. In general, large arrays of receivers are placed over areas that cover many km in one dimension at least. The source(s) is(are) moved to collect reflected energy from multiple source locations.

Signal processing is a major part of the science of seismic acquisition, and its output is primarily a 3D map of the subsurface showing the geological layers. From such maps, the presence of potential hydrocarbon reservoirs can be deduced. In the production phase, repeated seismic surveys (4-D seismic imaging) are sometimes used to understand how the subsurface has changed since the start of production, and this identifies regions of hydrocarbon that have not been produced (i.e. have been bypassed) perhaps because of non-uniformities in the geological formation that favour the flow from one part rather than another. It is then possible to drill another well specifically to target the bypassed reserves.

Seismic acquisition is one of the few physical techniques that can reach far into the earth (the others being gravity and very-low-frequency electromagnetics that offer very poor spatial resolution and of course measure different quantities). Seismic surveying relies on very-low-frequency sounds (up to about 120 Hz) because the earth attenuates the high frequencies and its spatial resolution is therefore limited to a few tens of m (depending on depth and the composition of the surface).

The data provided by surface seismic acquisition is measured, in the depth direction, in elapsed time from the emission of the seismic signal. In order to relate the time to actual physical depth, it is necessary to know the speed of sound in each geological layer. VSP is one of the techniques that is used for calibrating the speed of sound in the local geology (a measurement known as checkshot).

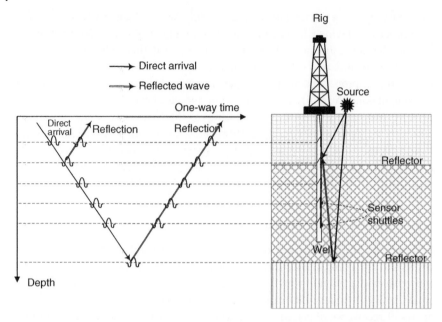

Figure 6.5 Schematic illustration of borehole seismic acquisition with a conventional array of electrical seismic sensors. Right: simplified sketch of the well, the seismic source, an array of sensor shuttles, and a simple model of the geological formation containing three distinct densities. Left: illustrative seismic waves and the signals recorded at each level for an impulsive source.

In VSP (or borehole seismic acquisition), a string of acoustic sensors is lowered into a well and used to record seismic signals emitted from surface and their reflections (Figure 6.5). The most fundamental output is simply the first arrival, i.e. the time at which the direct seismic wave travelling from the source reaches each seismic receiver. The first arrival, in the form of a time–depth curve, allows the velocity model for the subsurface formations to be calibrated and thus redefined in terms of physical depth. Therefore, it is critical for drilling wells to follow precisely the desired trajectory in order to place the production section of the well in the best part of the hydrocarbon-bearing formation. A single source location, near the well head, is sufficient for calibrating the velocity model in the case of a vertical well (this is known as a zero-offset VSP).

Another benefit of vertical seismic acquisition is that the distance travelled by the seismic waves is much reduced and this allows a higher-frequency content to reach the sensors, which improves the spatial resolution. Thus, borehole seismic acquisition is also used to provide high-resolution images of the subsurface, in which case the source must be placed in multiple locations or offsets. A grid of source positions might be selected for a 3D imaging, but other configurations are used such as a circular locus for the source positions to estimate the tilt (departure from horizontal) of the subterranean formation. In the case of near-horizontal wells, a walk-above survey where the source is moved in succession to a set of positions above the well is used to calibrate the average speed of sound from the surface to the well.

Although a 3D VSP provides high-resolution imaging of the subsurface, it is effective only relatively near to the well because the source–receiver geometry limits the range of angles that are available for locations away from the well. However, as the technology matures, the use of DAS simultaneously in multiple wells will allow the useful volume covered by DAS-based VSP to expand. This concept is being evaluated, for example, in fields in Oman where there is already a dense pattern of completed wells [42].

Borehole seismic surveys are frequently conducted with three-component sensors that discriminate according to the angle from which the wave returns. They are usually conducted during the logging phase of the well, i.e. shortly after drilling. With conventional VSP tools, the surveys can be quite time consuming because the string of sensors is assembled as it is lowered in the well and, for each level, the section of cable already in the well needs to be secured, the next sensor package is then connected to the logging cable, which is tensioned to allow the cable to carry the weight of the string of sensors, and the whole process is repeated many times as the sensor string is assembled and lowered into the well. Another factor that contributes to the time taken for a VSP is the usually limited number levels in the sensor string: this can force the operator to move the string and repeat the source activation to acquire further sets of data with different receiver positions. The need to acquire data at multiple source positions compounds the time taken for the survey.

A typical 3D VSP survey can thus require 12 hours to complete. In some environments, such as deepwater offshore, this represents a considerable cost even if one considers only the day rate charged for the rig. As a result, the number of VSPs that are actually conducted is suboptimal. DAS changes this picture somewhat.

A borehole seismic survey performed with DAS can occur much faster than with a conventional tool: the cable can be installed at the normal deployment rate of a logging cable (typically 2000 ft/min), and the entire length of the cable is seismically sensitive; as a result no repositioning of the sensors is required to fill in the coverage [43]. More importantly, in the case of borehole seismic using optical wireline cable, the VSP data can be acquired at the same time as other (e.g. petrophysical) measurements, and thus for relatively few source positions, no additional rig time is required for the VSP acquisition [44].

However, VSPs can also be acquired using other optical cables, for example, the fibres used to connect downhole optical sensors (e.g. pressure gauges) to the surface [45]. When using a permanently installed optical cable, the seismic survey can take place after the rig, and its crew have departed: this is particularly appropriate for large surveys, with many source positions. Given that there are many permanent fibre installations in wells for connecting to pressure gauges and/or for DTS acquisition, existing cables are frequently repurposed for DAS [45], including in cases where the well is producing [46].

In some cases, the optical cable is placed in the annulus between the formation and the casing and cemented in place, especially for permanent seismic monitoring. The earliest examples of DAS used in borehole seismic applications were based on observation wells [47] where the coupling can be optimized without any concern for the function of the well as a producer or injector.

Some of the players in the petroleum industry see permanently installed DAS as a tool for managing their production by conducting frequently repeated borehole seismic surveys to understand the movement of fluids in their reservoirs [42, 48–52]. Achieving this level of maturity has required a systematic evaluation of the technology through increasingly large-scale field trials and the development of efficient signal processing treatment.

One milestone along this road was a large-scale field trial in the Gulf of Mexico involving 2 deepwater wells and 50 000 shots of the seismic sources using fibres installed originally for connecting pressure gauges to surface instrumentation. The source positions covered an area of 13×9 km. A parallel survey with ocean-bottom nodes was recorded at the same time; this was one of the established technologies at the time, and it provided a reference for the optical borehole seismic work. This number of shots, combined with the large number of resolvable points along the sensing fibres, resulted in ~70 million individual seismic records (each representing the vibration signal vs. time at a specific location for a particular source shot).

Although this large-scale trial was essential for establishing the technology, research has then been directed at reducing its cost by using more sparse source positions [50]. Redundancy in the source positions simplifies the signal processing, and it also improves the signal-to-noise ratio because the additional information is integrated in the processing whereas the noise contributions will be averaged out. Thus, improved signal-to-noise performance on the acquisition (and the certainty of good coupling) allows the source effort to be minimized.

The converse of this argument is illustrated by an example [53] where multiple interrogators were connected to separate fibres in a single sensing cable in order to guarantee adequate signal quality; in this case, the usual approach for improving the signal-to-noise ratio (by repeating the source shots and combining the results) was not available because the source was a dynamite explosion that could not easily be repeated.

A key issue in DAS borehole seismic acquisition is the coupling of the seismic waves from the formation to the fibre: the cable must be in contact with the wall of the well in order for the small dimensional distortion caused by the seismic wave to be transmitted to the fibre (the coupling from the well to the cable through a fluid in the well when the cable is not in contact with the wall is minimal [54]). In general, the dynamic strain is small (of order a few tens of nm, over the gauge length, which itself is usually in the range of 10–50 m), and therefore the forces involved are also quite small. Nonetheless, some force holding the cable to the wall of the well is essential to transmit the strain and also to prevent the vibration of the cable between points of contact, which can result in a spurious ringing signal [54, 55]. In conventional VSP tools, this problem is addressed using deployable arms that act against the borehole to push the sensor package against the opposite side of the well. However, this solution is not available with a continuous small dimensional cable. For very shallow wells, a deployable liner has been used [56] to achieve a clamping effect. The cable is deployed, followed by a flexible liner. The latter is then filled with water that expands the liner and pushes the cable against the borehole wall.

Brillouin OTDR has also been used to measure the strain on the wireline cable and so to infer the locus of its contact to the borehole wall; the tension and slack on the cable can then be adjusted to optimize the cable-borehole wall coupling [57].

From a geophysical point of view, cementing the sensing cable behind the casing provides the best coupling, and this is the configuration that is preferred when the objective is a long-term seismic surveillance, and this requirement is determined before the well is constructed. The use of cables permanently installed on production tubing is very common because there is an installed base of such cables for connecting to pressure gauges or for DTS. The coupling from the formation to the sensing fibre is not quite as good in this configuration as is the case for cables cemented behind casing, partly because of the different barriers (steel casing, annulus from casing to production tubing) but also because the coupling is transmitted via the contact of the production tubing to the casing at the tubing joints. In a perfectly vertical well, it is conceivable that there is no contact between tubing and casing. In practice, this is seldom a serious issue because a minor deviation from the vertical (e.g. 0.01°) is typically sufficient to ensure that the tubing is in contact with the casing over the majority of its length.

The most challenging approach for VSP is the use of a deployable cable, for the reasons outlined above. However, where is it essential to collect the geophysical information quickly (perhaps to inform the continued drilling of the well on which the measurement is conducted or the next wells in the same field), then the immediacy of the wireline-based VSP dictates this deployment method, in spite of its drawbacks. It is also common to combine a DAS-based VSP with a conventional survey by adding one or a few levels of conventional VSP tools below the optical cable. In this way, the best aspects of both techniques can be exploited [58].

DAS is not a direct equivalent to modern multilevel borehole seismic tool [59]. Firstly, it measures a different quantity. The electrical tools measure acceleration or particle velocity (geophones); in contrast, phase-measuring DAS outputs the dynamic strain across a gauge length derived from a differential phase measurement [60, 61]. Although the resulting seismic images appear similar, the sign of the differential phase change in DAS in response to an acoustic wave travelling parallel to the fibre axis is insensitive to the direction of that wave (i.e. travelling towards or away from the interrogator). The sign of the wavelets recorded with the electrical sensors in contrast does depend on the direction of the waves.

The sensitivity of DAS as a function of the angle of arrival is also different from that of conventional accelerometers or geophones ($\cos^2$ vs. cos). As a result, a DAS is insensitive to compressive waves arriving perpendicular to the fibre axis (broadside) [62]. This effect can be understood by considering that a compressive acoustic wave displaces the medium in the same direction as that in which the wave travels; an acoustic wave reaching the fibre broadside will displace the fibre, but not strain it to first order. In contrast, the fibre responds strongly to an acoustic shear wave arriving broadside and polarized parallel to the fibre axis.

In some applications of DAS, the directional response to compressive waves is a serious issue. For example, a potential application for DAS is surface seismic where cables are laid on the ground, or trenched, and a source fires acoustic energy into the ground that is expected to be reflected by a subterranean layer. In this case, most of the returned signal arrives near the broadside to the sensing cables, which are therefore insensitive to the main signal of interest. In order to solve this problem, one option is to design the path of the fibre within the cable so that its direction varies and some part of the fibre is sensitive to the arriving energy. One approach is to arrange the fibre to occupy a helical path within the cable [63–65], i.e. to form a helically wound cable (HWC).

By an appropriate choice of the pitch of the helix, the response is aimed to be approximately omnidirectional, and test results certainly show [65, 66] a much lower degree of extinction of the broadside waves than it is the case for a straight fibre in a cable. According to one analysis, the optimum cable design depends on the elastic properties of the soil [64], and so in general no cable will be perfectly omnidirectional.

The idea of shaping the path of the fibre can be extended to provide some discrimination of the signals' direction of arrival. Thus, a straight fibre is preferentially sensitive to on-axis compressive waves and an HWC approaches an omnidirectional response. Now, a fibre path that meanders in one plane only can be designed to respond to compressive waves arriving in that plane, but will have no sensitivity to signals arriving in a direction orthogonal to that plane [67].

The cable can also be designed so that it responds asymmetrically to waves arriving from various azimuths, for example, by making the compliance of the materials greater in one axis than in another [68]. Thus, by combining several cables with differential azimuthal response and interrogating them separately, in principle, a multicomponent sensitivity can be achieved. In practice, some field tests have been conducted at surface or in shallow horizontal boreholes [63, 65, 66] but so far not in typical oilfield boreholes, probably because the special cables are made of low-temperature materials and the prototypes manufactured so far have been bulky.

The gauge length is the distance over which the dynamic strain is measured in a differential phase DAS [31, 60]. In some interrogators, it is defined by the physical length of a fibre in a phase-recovery interferometer [69]; in dual pulse-interrogation DAS systems, it is defined by the time separation between the pulses in a pulse pair [48], whereas in the heterodyne DVS (hDVS) technique, it is defined by software after the acquisition [31, 70, 71].

The gauge length is one of the factors limiting the spatial resolution in a DAS. However, it turns out that it has a profound influence on the signal properties [72]. The DAS response is proportional to the change in length of the sensing fibre over the gauge length. Thus, at very short gauge lengths (much shorter than the acoustic wavelength), the measured differential phase caused by the passage of the wave is proportional to the gauge length. However, as the gauge is increased in relation to the acoustic wavelength and reaches half the value of the acoustic wavelength, the signal response begins to reduce. Indeed, at a gauge length equal to precisely one acoustic wavelength, there is no response. In this case, the strain in the first half of the wave within the gauge length is exactly compensated by the strain in the second half. Consequently, there is a notch in the wavenumber response of the DAS, which therefore acts as an acoustic wavelength filter. Thus, the response to an impulsive signal (which covers a range of frequencies) is distorted, and, in the frequency domain, some frequencies are absent in the case of a very long gauge length. Even in the case of a gauge length shorter than a half-wavelength (for the highest frequencies present in the signal), this gauge-length dependence still colours the spectrum by enhancing the higher frequencies.

It turns out that the gauge length can be chosen to maximize the signal-to-noise ratio while minimizing the spectral distortion [73, 74]. However, this optimum is a function of the spectral content of the acoustic energy and also of the acoustic velocity in the particular geological formation that is studied. It should also be noted that the relevant acoustic velocity is the apparent velocity (i.e. the velocity divided by the cosine of the angle of incidence), and so, strictly, the optimum gauge length is also a function of the angle of arrival of the seismic signals on the sensing fibre. This means that (i) the optimum gauge length cannot be determined before the survey (although a good guess can be made) and (ii) the optimum gauge length is a function of depth because the acoustic velocity varies according to the type of rock and (iii) in some cases, for very distant source positions where the energy arrives close to broadside, the gauge length can be selected, in addition, depending on the geometry of the source and well trajectory.

6.3.8 Hydraulic Fracturing Monitoring (HFM)

Hydraulic fracturing is a technique to enhance the productivity of a well by fracturing the formation using pressurized fluid and solid particles (known as proppant) to prevent the fractures from re-sealing after the treatment is completed.

Although hydraulic fracturing has been used for many years, it is now associated with the production from shale formations that have extremely low permeability (of the order of nanodarcy). It is the combination of directional drilling, fracturing, and improved methods of designing the fracturing for particular locations that have enabled the sudden development (since c. 2010) of oil and gas from shale formations, particularly in the United States.

DFOS have a number of applications in HFM. The imaging of the fractures as they develop would be very informative in showing where the rock is failing under pressure, the degree to which this fracturing is uniform and preferential planes for the rock to open in.

Conventional electrical tools are commonly used to map (relative to the well trajectory) the location of the many fractures that form during hydraulic fracturing. Given earlier section on borehole seismic applications of DAS, microseismic detection would appear to be ideal extension of DAS to HFM. The detection with DAS of microseismic events reported in the form of seismic waves has been reported [75, 76].

However, the translation of DAS technology to microseismic detection and location is not straightforward. Microseismic events are generally quite weak (of order of magnitude −2 to 0). The acoustic signals are therefore weak and also at quite high frequency (containing energy up to 500–1000 Hz). The high-frequency content implies a need for fast sampling and for a short gauge length, given the short acoustic wavelengths. The microseismic events can therefore be difficult to detect (equivalently, equipment of a given sensitivity will miss the weaker events). In [75], where a conventional electrical tool was used at the same time as a DAS system, it was reported that the DAS system detected only 3.5% of the events observed with the electrical system. The relative performance of DAS systems will improve as the optical technology advances and also as the signal processing improves.

A second disadvantage of DAS systems is the fact that they record mainly in the axial direction. This means that locating the fracture in 3D is not possible from a single well. The distance of the fracture to the sensing fibre can be estimated from the delay between the arrival of the compressive and shear waves, and, where the well has vertical and highly deviated sections, some triangulation is possible, but the azimuth of the fracture around the well is still difficult to quantify. Data collected from a nearby well would help define the location of the fracture causing a particular event, but the weakness and high-frequency content of the events limit the distance at which they can be detected and often this distance is less than the well separation in typical fields. It is expected that the problem of locating the fractures will also improve with refinement of the interrogation and signal processing of DAS.

DAS, in combination with DTS, is increasingly applied to understanding the progress of an HFM job from the fracturing and the injection of proppants and of various fluids that attempt to make the flow through the fissures uniform and finally to assess the distribution of production from each of the fracturing stages (wells are generally fractured in multiple stages in turn, by separating the stages with packers so the hydraulic energy can be applied to each stage optimally).

The DAS data is generally translated into acoustic energy in separate frequency bands by applying a Fourier transform to the data at each location over a processing time window (of order of 1–30 seconds) and integrating the acoustic energy found in each of a set of frequency bands during that window. The particular frequency band is selected to give best discrimination of events during the operation [77]. An extension of the frequency band energy approach has allowed a number of phenomena to be detected and quantified, such as the initiation of pressure waves and estimation of bulk flow rate, and has provided some discrimination between fluid types [78].

6.3.9 Sand Production

The production of oil in unconsolidated sands brings with it the very serious issue of sand breaking through the sand screen, which causes erosion of the tubing and valves in the well and requires separation at the surface, and the removal of the sand around the well can cause instability around the well.

One of the earliest applications of DAS in the oilfield was the detection of sand in an unconsolidated formation in order to understand where the sand was breaking into the well and to allow remedial treatment [17]. In this particular case, multimode fibres that had been installed for DTS flow allocation were used, so this is an early example of the use of multimode fibre for DAS; it is also an example of a two-stage completion in which a downhole wet mate hydraulic connection was used to form a complete hydraulic circuit so that a fibre could be pumped after the installation of the upper completion of the well [17].

6.4 Summary

DFOS have been applied in the upstream oil and gas industry since about 1996. In that time the technology has grown from the research into the installation and use of the technology in a few conceptually relatively simple applications (such as monitoring steam injection) to a vast range of techniques for extracting valuable information from the data using ingenious ways of inferring the behaviour of the well and the fluids. The sensing technologies have broadened from DTS to include DAS and on occasion Brillouin static temperature and strain measurements.

This chapter covers only a few of the many ways in which DFOS technology is applied to the oilfield; the reader is referred to [26, 31] for further examples.

With the basic aspects of fibre installation and fibre reliability now well understood, the use of the technology is broadening to encompass much of the life of the field, from exploration, reservoir evaluation, production, and well abandonment.

There is still scope for much more adoption of the DFOS techniques, for example, in designing the downhole assemblies (e.g. sand screens, valves, pumps) to include fibres as they are built so that the placement and protection of the fibre is a given in fibre-enabled cables and completion. In addition, the signal processing and the physics of how the measured profiles are translated into useful, actionable, and timely information are still rapidly evolving. It is therefore reasonable to expect an even wider adoption in coming years.

References

1 Kearey, P., Brooks, M., and Hill, I. (2013). *An Introduction to Geophysical Exploration*. Wiley.

2 Kenyon, B., Kleinberg, R., Straley, C. et al. (1995). Nuclear magnetic resonance imaging – technology for the 21st century. *Oilfield Review* 7 (3): 19–33.

3 Fujisawa, G. and Yamate, T. (2013). Development and applications of ruggedized VIS/NIR spectrometer system for oilfield wellbores. *Photonic Sensors* 3 (4): 289–294.

4 Fujisawa, G. et al. (2003). Analyzing reservoir fluid composition in-situ in real time: case study in a carbonate reservoir. In: *SPE Annual Technical Conference*. Denver, CO: SPE.

5 Bouffaron, P. and Perrigault, T. (2013). Methane hydrates, truths and perspectives. *International Journal of Energy, Information & Communications* 4 (4): 23–31.

6 Moridis, G.J. (2008). *Toward Production from Gas Hydrates: Current Status, Assessment of Resources, and Simulation-Based Evaluation of Technology and Potential*. Lawrence Berkeley National Laboratory.

7 Yamamoto, K. et al. (2017). Thermal responses of a gas hydrate-bearing sediment to a depressurization operation. *RSC Advances* 7 (10): 5554–5577.

8 K. Fujii, Yasuda, Masato; Cho, Brian et al. 2008. Development of a monitoring system for the JOGMEC/NRCan/Aurora Mallik gas hydrate production test program. *6th International Conference on Gas Hydrates (ICGH 2008)*, Vancouver, BC (6–10 July 2008).

9 Kanno, T. et al. (2014). In-situ temperature measurement of gas hydrate dissociation during the world-first offshore production test. *Offshore Technology Conference*, Houston, USA (5 May 2014 through 8 May 2014).

10 Cohen, A. (2019). *Russia Loses Billions in Druzhba Oil Pipeline Contamination Crisis*. https://www.forbes.com/sites/arielcohen/2019/05/10/russia-loses-billions-in-druzhba-oil-pipeline-contamination-crisis/ (accessed 19 May 2019).

11 Karlinski, H.E. (1989). Apparatus and method for continuous manufacture of armored optical fibre cable. US Patent 4,852,790 A.

12 Varkey, J. et al. (2008). Optical fibre cables for wellbore applications. US Patent 7,324,730 B2.

13 Lovell, J.R. et al. (2009). System and methods using fibre optics in coiled tubing. US Patent 7,617,787 B2.

14 Algeroy, J. et al. (2010). Permanent monitoring: taking it to the reservoir. *Oilfield Review* 22 (1): 34–41.

15 Kluth, E.L.E. and Varnham, M.P. (2001). Apparatus for the remote measurement of physical parameters. GB Patent 2,311,546 A.

16 Kluth, E.L.E. et al. (2000). Advanced sensor infrastructure for real time reservoir monitoring. *SPE European Petroleum Conference*, Paris (24–25 October 2000).

17 Mullens, S.J., Lees, G.P., and Duvivier, G. (2010). Fibre-optic distributed vibration sensing provides technique for detecting sand production. *Offshore Technology Conference*, Houston, TX (3–6 May 2010), p. OTC-20429.

18 Smithpeter, C. et al. (1999). Evaluation of a distributed fibre-optic temperature sensor for logging wellbore temperature at the Beowave and Dixie Valley geothermal fields. In: *24th Workshop on Geothermal Reservoir Engineering*. Stanford, CA: Stanford University.

19 Semjonov, S.L. et al. (2006). Fibre performance in hydrogen atmosphere at high temperature. In: *Reliability of Optical Fibre Components, Devices, Systems, and Networks III* (eds. H.G. Limberger and M.J. Matthewson). SPIE.

20 Ramos, R.T. and Hawthorne, W.D. (2008). Survivability of optical fibre for harsh environments. In: *SPE Annual Technical Conference and Exhibition*. Denver, CO: SPE Paper SPE 116075-MS (21–24 September 2008).

21 Stone, J. (1987). Interactions of hydrogen and deuterium with silica optical fibres: a review. *Journal of Lightwave Technology* 5 (5): 712–733.

22 Clowes, J.R. et al. (1997). Effects of high temperature and pressure on silica optical fibre sensors. In: *12th Conference on Optical Fibre Sensors*. Williamsburg, VA: Optical Society of America.

23 ITU (2008). G.651. 1 (2007)/Amd.1 (12/2008) Characteristics of a 50/125 μm Multimode Graded Index Optical Fibre Cable for the Optical Access Network.

24 Davies, D. et al. (2011). Measuring a characteristic of a multimode optical fibre. US Patent 8,077,314 B2.

25 Davies, D., Hartog, A.H., and Kader, K. (2010). Distributed vibration sensing system using multimode fibre. US Patent 7,668,411 B2.

26 Brown, G.A. (2016). *The Essentials of Fibre-Optic Distributed Temperature Analysis*, 2e, vol. 1. Sugarland: Schlumberger Educational Services https://www.slb.com/resources/publications/books/fibre_optic_distributed_temperature_analysis_book.aspx.

27 Karaman, O.S., Kutlik, R.L., and Kluth, E.L. (1996). A field trial to test fibre optic sensors for downhole temperature and pressure measurements, West Coalinga Field, California *SPE Western Regional Meeting*, Anchorage, AK.

28 Butler, R.M. (1998). SAGD comes of AGE! *Journal of Canadian Petroleum Technology* 37 (7): 9–12.

29 Mohajer, M.M. et al. (2010). An integrated framework for SAGD real-time optimization. In: *SPE Intelligent Energy Conference and Exhibition*. Society of Petroleum Engineers.

30 Bailey, W., Güyagüler, B., Stone, T., and Law, D. (2014). Practical control of SAGD wells with dual-tubing strings. *Journal of Canadian Petroleum Technology* 53: 32–47.

31 Hartog, A.H. (2017). *An Introduction to Distributed Optical Fibre Sensors*. CRC Press/Taylor and Francis.

32 Bolognini, G. and Hartog, A.H. (2013). Raman-based fibre sensors: trends and applications. *Optical Fibre Technology* 19 (6B): 678–688.

33 Hartog, A.H., Gold, M.P., and Leach, A.P. (1987). Optical time domain reflectometry. EP Patent 0,213,872 B2.

34 Fernandez Fernandez, A. et al. (2005). Radiation-tolerant Raman distributed temperature monitoring system for large nuclear infrastructures. *IEEE Transactions on Nuclear Science* 52 (6): 2689–2694.

35 Hartog, A. and Leach, A. (1986). A practical optical-fibre distributed temperature sensor. In: *Colloquium on Distributed Optical Fibre Sensors*. Savoy Place, London: IEE.

36 Chen, Y. et al. (2008). Accurate single-ended distributed temperature sensing. *SPE Annual Technical Conference and Exhibition*, Denver, CO (21–24 September 2008), p. SPE 116655.

37 Ramey, H.J. (1962). Wellbore heat transmission. *Journal of Petroleum Technology* 14 (04): 427–435.

38 Curtis, M.R. and Witterholt, E.J. (1973). Use of the temperature log for determining flow rates in producing wells. In: *Fall Meeting of the Society of Petroleum Engineers of AIME*, 12. Las Vegas, NV: Society of Petroleum Engineers.

39 Brown, G.A. (2003). Method and apparatus for determining flow rates. EP Patent 1,196,743 B1.

40 Brown, G.A. (2012). *Method to measure injector inflow profiles*, US 8,146,656 B2.

41 Brown, G.A. (2004). Method and apparatus for flow measurement. US Patent 6,826,954 B2.

42 Kiyashchenko, D. et al. (2013). Steam-injection monitoring in South Oman from single-pattern to field-scale surveillance. *The Leading Edge* 32 (10): 1246–1256.

43 Hartog, A. et al. (2014). Vertical seismic optical profiling on wireline logging cable. *Geophysical Prospecting* 62: 693–701.

44 Kimura, T. et al. (2017). Borehole seismic acquisition using fibre-optic technology: zero rig-time operation. *SEG Technical Program Expanded Abstracts,* Houston, TX (24–29 September), pp. 936–940.

45 Barberan, C. et al. (2012). Multi-offset seismic acquisition using optical fibre behind tubing. *74th EAGE Annual Conference and Exhibition*, Copenhagen (4 June through 7 June 2012).

46 Madsen, K.N. et al. (2013). A VSP field trial using distributed acoustic sensing in a producing well in the North Sea. *First Break* 31 (11): 51–55.

47 Mestayer, J. et al. (2011). Field trials of distributed acoustic sensing for geophysical monitoring. *2011 SEG Annual Meeting*, San Antonio, TX, 4253–4257.

48 Mateeva, A. et al. (2014). Distributed acoustic sensing for reservoir monitoring with vertical seismic profiling. *Geophysical Prospecting* 62 (4): 679–692.

49 Mateeva, A. et al. (2015). *Frequent Seismic Monitoring for Pro-Active Reservoir Management.* Society of Exploration Geophysicists.

50 Chalenski, D. et al. (2016). Climbing the staircase of ultralow-cost 4D monitoring of deepwater fields using DAS-VSP. *SEG Technical Program Expanded Abstracts,* Dallas, TX (16–21 October 2016), pp. 5441–5445.

51 Mateeva, A. et al. (2013). Distributed acoustic sensing (DAS) for reservoir monitoring with VSP. *2nd EAGE Borehole Geophysics Workshop*, St Julian, Malta, 15685.

52 Hatchell, P.J. et al. (2013). Instantaneous 4D Seismic (i4D) for Water Injection Monitoring. *2nd EAGE Workshop on Permanent Reservoir Monitoring: Current and Future Trends,* Stavanger, Norway (2 July 2013 through 5 July 2013).

53 Zwartjes, P. and Mateeva, A. (2015). High resolution Walkaway VSP at Kapuni recorded on multi-fibre DAS and geophones *77th EAGE Conference and Exhibition 2015,* IFEMA, Madrid, Spain (1–4 June 2015).

54 Schilke, S. (2017). Importance du couplage des capteurs distribués à fibre optique dans le cadre des VSP. *Paris Sciences et Lettres.* Thèse de doctorat en Géosciences et géoingénierie

55 Schilke, S. et al. (2016). Numerical evaluation of sensor coupling of distributed acoustic sensing systems in vertical seismic profiling. *SEG*, Dallas, TX (16–21 October 2016).

56 Munn, J.D., Coleman, T.I., Parker, B.L. et al. (2017). Novel cable coupling technique for improved shallow distributed acoustic sensor VSPs. *Journal of Applied Geophysics* 138: 72–79.

57 Constantinou, A. et al. (2016). Improving DAS acquisition by real-time monitoring of wireline cable coupling. In: *SEG Technical Program Expanded Abstracts 2016*, 5603–5607. Society of Exploration Geophysicists.

58 Dean, T. et al. (2015). Vertical seismic profiles: now just another log? In: *SEG technical Program Expanded Abstracts 2015*, 5544–5548. Society of Exploration Geophysicists.

59 Dean, T. et al. (2016). Distributed vibration sensing for seismic acquisition. *The Leading Edge* 35 (7): 600–604.

60 Hartog, A.H., Liokumovich, L.B., and Kotov, O.I. (2013). The optics of distributed vibration sensing. In: *Second EAGE Workshop on Permanent Reservoir Monitoring*. Stavanger: EAGE.

61 Frignet, B.G. and Hartog, A.H. (2014). Optical vertical seismic profile on wireline cable. *SPWLA 55th Annual Logging Symposium*, Abu Dhabi, UAE (18–22 May 2014).

62 Papp, B. et al. (2016). A study of the geophysical response of distributed fibre optic acoustic sensors through laboratory-scale experiments. *Geophysical Prospecting* 65 (5): 1186–1204.

63 Hornman, K. et al. (2013). Field trial of a broadside-sensitive distributed acoustic sensing cable for surface seismic. *75th EAGE Conference & Exhibition Incorporating SPE EUROPEC 2013*, London (10–13 June 2013).

64 Kuvshinov, B.N. (2016). Interaction of helically wound fibre-optic cables with plane seismic waves. *Geophysical Prospecting* 64 (3): 671–688.

65 Hornman, J.C. (2016). Field trial of seismic recording using distributed acoustic sensing with broadside sensitive fibre-optic cables. *Geophysical Prospecting* 65 (1): 35–46.

66 Lumens, P. et al. (2013). Cable development for distributed geophysical sensing, with a field trial in surface seismic. In: *Fifth European Workshop on Optical Fibre Sensors*. International Society for Optics and Photonics.

67 Den Boer, J.J. et al. (2013). Detecting broadside acoustic signals with a fibre optical distributed acoustic sensing (DAS) assembly. WO Patent 2,013,090,544.

68 Martin, J.E. et al. (2018) Fibre optic distributed vibration sensing with directional sensitivity. GB Patent 2,529,780 B.

69 Posey, R.J., Johnson, G.A., and Vohra, S.T. (2000). Strain sensing based on coherent Rayleigh scattering in an optical fibre. *Electronics Letters* 36 (20): 1688–1689.

70 Hartog, A.H. and Kader, K. (2012). *Distributed fibre optic sensor system with improved linearity*. US Patent 9,170,149 B2.

71 Hartog, A.H. et al. (2018). The use of multi-frequency acquisition to significantly improve the quality of fibre-optic distributed vibration sensing. *Geophysical Prospecting* 66: 192–202.

72 Dean, T., Papp, B., and Hartog, A. (2015) Wavenumber response of data recorded using distributed fibre-optic systems. *3rd EAGE Workshop on Borehole Geophysics* (19–22 April 2015).

73 Dickenson, P. et al. (2015). Gauge length optimisation for distributed acoustic sensing. *SPE Distributed Fibre-Optic Sensing Workshop*, London.

74 Dean, T., Cuny, T., and Hartog, A.H. (2017). The effect of gauge length on axially incident P-waves measured using fibre-optic distributed vibration sensing. *Geophysical Prospecting* 65: 184–193.

75 Webster, P. et al. (2013). Micro-seismic detection using distributed acoustic sensing. *SEG 2013 Annual Meeting*, Houston, TX, 2459.

76 Molteni, D., Williams, M.J., and Wilson, C. (2017). Detecting microseismicity using distributed vibration. *First Break* 35 (4): 51–55.

77 In't panhuis, P. et al. (2014) *Flow Monitoring and Production Profiling using DAS*, Society of Petroleum Engineers.

78 Paleja, R. et al. (2015). *Velocity Tracking for Flow Monitoring and Production Profiling Using Distributed Acoustic Sensing*. Society of Petroleum Engineers.

79 Duey, R. (September 2008). *Unfogging the Glass*. E&P.

7

Biomechanical Sensors

Cicero Martelli, Jean Carlos Cardozo da Silva, Alessandra Kalinowski, José Rodolfo Galvão, and Talita Paes

Graduate Program in Electrical and Computer Engineering, Federal University of Technology –Paraná, Brazil

In this chapter optical fibre sensing concepts applied to biological systems are presented under the perspective of mechanics, particularly that of living tissues: biomechanics. The main focus of this chapter is the review of technologies and techniques that are most used and published in the literature. The objective is to present optical fibre sensors as well as their forms of integration that have obtained greater applicability in the biomechanics area. The chapter is divided into four main sections. First, a historical review is presented, where statistical analysis shows the most common technologies and application areas. The second section presents several application examples with focus on body parts and functions spanning from the angular movement of joints to the pressure in blood vessels and the use of prostheses. In the third section encapsulation of the optical fibre sensors as well as the integration of the optical fibres with other materials is presented. The fourth section presents the future perspective for the application of optical fibre sensors in biomechanics.

7.1 Optical Fibre Sensors in Biomechanics: Introduction and Review

The daily life of people in this postmodern age is closely linked to optical fibres, especially in telecommunications, allowing ever faster levels of information transport and having deep participation in the cultural construction of the society. Its application in health-related areas was made possible through the knowledge

Optical Fibre Sensors: Fundamentals for Development of Optimized Devices, First Edition.
Edited by Ignacio Del Villar and Ignacio R. Matias.

developed and consolidated in monitoring technologies using optical fibre sensors in several industrial sectors. Thus, the sectors such as civil construction, aeronautics, and oil, gas, and structural geology also enabled the application of such sensors to biological systems.

The work published by Charles K. Kao and George Hockham in 1966 [1] on low-loss optical fibre led to the technology level as it is today. In the mid-1970s, research began in fields related to health and the biomechanics with applications to neurology in the monitoring of intracranial pressures during the post-operative period [2, 3]. Even in the first papers about optical fibres, biomedical applications were already being sought using the concepts of biomechanics [4–6].

Biomechanics is understood as the mechanics of living bodies [7]. From the early stages of classical mechanics, scientists such as Aristotle, Galileo Galilei, and da Vinci sought to transpose the relations of force, displacement, pressure, and angular momentum for living bodies [7, 8]. Hatze in 1974 defined that biomechanics studied the structures and functions of biological systems through the methods of mechanics [9]. It is believed that biomechanics plays an important role in the study of all biological systems, both at the macroscopic level, recognizing the individual as a system, or going deeper into molecular-size scales in microsystems [10].

In this context, questions about the human body performance arise. For example, how to know if an athlete has more or less power in their jump? How to tell if a patient is evolving in gaining muscle strength? How to know where the forces are concentrated in a lower-limb prosthesis? With technological advances, there are several electrical or mechanical equipment to assess each issue.

Optical fibre enters this scenario with its intrinsic advantages to solve problems for application in biomechanics, for example, sensor lightness, small size, and ease of integration into other media – primarily polymers such as polymethyl methacrylate (PMMA) and glass-fibre-reinforced polymers, electromagnetic immunity, and the ability to integrate multiple sensors into a single strand are extremely attractive features in optical fibre sensors. Such a set of characteristics, when combined into a single device, provide a simple constructive feature sensor provided that there are no multiple mechanical parts and also because it is a passive sensor.

The study of biomechanics gets the attention of several health professionals, from orthopaedic physicians and physiotherapists who need to know, for example, the range of motion of a joint to diagnose and treat it, to orthodontists who need to know the force transferred from an orthodontic appliance to the bone.

The biomechanical evaluation of humans and animals within the optical fibre universe began in the 1980s and has been growing continuously presenting a more significant increase since the 2000s. The analysed parameters cover the musculoskeletal system (bones, muscles, tendons, ligaments, cartilages, joint capsules, and intervertebral discs, among others), the cardiovascular, respiratory, and digestive systems, dentistry, and the mechanical characterization of objects used close to the body. As for the kind of experiment, it is noticeable that *in situ* experiments have been reported in the largest number of publications, as shown by Figure 7.1, surpassing the other types of tests considered here: *in vivo*, *ex vivo*, *in vitro*, and external phantom (physical or computational archetypes). The term *in situ* represents the integration at the place, being used by the patient/user themselves in a non-definitive and non-invasive way.

Some terminologies are not unanimous, with some being more consecrated than others. Divergences in the medical literature occur, as is the case of the terms *in vivo*, *ex vivo*, *in vitro*, and *in situ*. For comprehension purposes, we use *in vivo* when the device is internal to the body, i.e. it is an invasive or minimally invasive procedure; *in situ* when the device is external to the body, i.e. it is a non-invasive procedure; and *ex vivo* is when it comes to a human body segment or a recently dead animal. *In vitro* experiments are performed on tissue samples or isolated parts that do not constitute a body part, while the term 'phantom' is used to designate tests with results obtained through physical (non-biological) or computational models. In Section 7.2 we will deal with the experiments according to their biomechanical investigations.

In general, trials are initiated with phantoms, undergoing *in vitro* and *ex vivo* development until finally reaching *in vivo* applications. However, research does not always follow this order. A sign of this is the prevalence of studies that have their applications *in situ*, as shown in Figure 7.1. This may be explained by the fact that the studies use non-invasive approaches, offering no risks to the individual. Figure 7.1 also shows the evolution over time of the different types of optical fibre assays employed as biomechanical sensors. It should also be noted that interest in

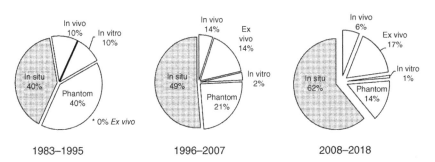

Figure 7.1 Estimative evolution by subdivision of experiments performed in biomechanics: *in vivo*, *ex vivo*, *in vitro*, *in situ*, and phantom [11–169].

in vitro studies has shown an important reduction the same way as for phantoms and *in vivo* experiments.

Within biomechanics, we have the classical subdivisions of kinematics and kinetics. Kinematics aims to analyse the characteristics of movements without investigating the forces that cause them. For example, with the analysis of distance and time, we have the speed and acceleration obtained by a given body segment when performing a movement as well as its trajectory.

The kinematic parameter analysis, based on the acquisition of images during the execution of the movement, is called cinemetry. The first analyses were made using two-dimensional (2D) images, with a frame-by-frame analysis, and, later, through the composition of multiple cameras, capturing composite images of three dimensions (3D). Nowadays, such studies may be performed by cameras reading the infrared spectrum or even by systems that use light detection and ranging (LIDAR) lasers [11–13]. The kinetic study may be performed by investigating the electromyographic signals and by dynamometric analysis, for example.

According to the literature, there are nine principles of biomechanics: balance, inertia, coordination continuum, range of motion, segmental interaction, force–motion, force–time (impulse), spin, and optimal projection [14]. In this chapter, we will not delve into such approaches, but it is interesting to know their definitions since biomechanics is interdisciplinary.

Concern about the integrity of body structures and their behaviour over time, as well as their wear, is of great interest in areas such as healthcare, sports medicine, performance, and rehabilitation and also for monitoring vital signs in particular situations. These are the areas in which the monitoring techniques applying optical fibre sensors are concentrated and, especially, disseminated.

Regarding their type, as stated in other chapters of this book, optical fibre sensors are commonly subdivided into two broad categories: extrinsic and intrinsic sensors. They are classified as extrinsic when the phenomenon of light interaction occurs outside the fibre, with the fibre only conducting the light. Intrinsic is, therefore, when the sensor element is also in the fibre itself. Intrinsic sensors are constituted to be optically sensitive and responsive to external stimuli [15].

Figure 7.2a shows the proportion of intrinsic and extrinsic sensors from articles published between 1983 and 2018.

The most researched technology during this time was the fibre Bragg gratings (FBG) (Figure 7.2b), with 47% of the articles in the leading scientific platforms involving optical fibre sensors applied to biomechanics. Optical fibres themselves as sensors, in interferometric systems, for instance, are the second most used sensors with 39% of publications. Polymer optical fibres (POFs) are the third, and long-period gratings (LPG) the fourth with 11 and 3%, respectively. Figure 7.3 shows the chronological order of the evolution and the number of applications with the different types of sensors. One may notice that, in the last decade, the

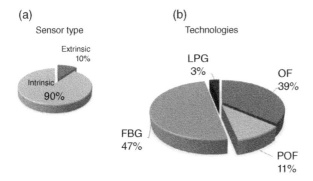

Figure 7.2 Data of publications from 1983 to 2018. (a) Intrinsic versus extrinsic sensors. (b) Large technological groups involving optical fibre. OF, optical fibre; POF, polymeric optical fibre; FBG, fibre Bragg gratings; LPG, long-period fibre gratings [11–169].

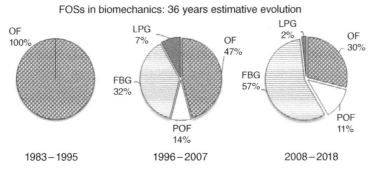

Figure 7.3 Estimative evolution over the decades of, POF, FBG, and LPG sensors with applications in biomechanics. Publications from the period of 1983 to 2018 [11–169].

work with FBG sensors became evolved into the most studied sensing element. These sensors will be discussed in Section 7.3 of this chapter, describing the various techniques for coupling and integrating optical fibres into other systems, such as carbon fibre prostheses and fabrics.

As for optical fibre, we consider fibres with core and cladding made of silica, and the PMMA coats the cladding, giving the glass fibre mechanical protection. With smaller application than silica fibres, the POFs are mostly constructed of PMMA. They have the advantage of presenting greater fracture toughness, flexibility, and high numerical aperture. Their disadvantages are related to (i) the large dimensions, (ii) high propagation losses, (iii) impossibility to operate at temperatures above 60 °C typically, and (iv) low performance regarding single modeness.

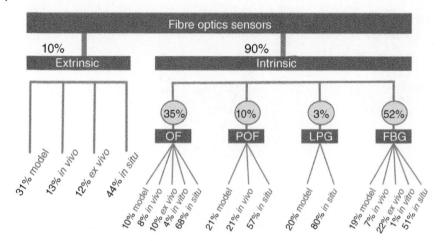

Figure 7.4 Summary of the 36 years of applications of optical fibres in biomechanics. Percentage of publications by type of sensor technology and type of experiment [11–169].

The general research panorama carried out over the last 36 years is summarized in Figure 7.4, which shows the publications involving each type of technology and experiments in percentage numbers.

7.2 Optical Fibre Sensors: From Experimental Phantoms to *In Vivo* Applications

This section covers the areas of application of optical fibre sensors for the study of biomechanical parameters such as force, pressure, angulation, and velocity. Studies exploring each of the following five experimental forms – phantoms, *in vitro*, *ex vivo*, *in vivo*, and *in situ* – are presented.

The following sections are subdivided and organized in application areas as follows: joints, bones and muscles, teeth, mandible and maxilla, prostheses and extracorporeal devices, soles and insoles, smart fabrics, blood vessels, mechanical properties of tissues, cardiac monitoring, and respiratory monitoring.

7.2.1 Experimental Phantoms and Models

Experimental phantom and models are a good option to start tests when it is intended to reach an invasive application. Tools that simulate the practical test response without the sensor being applied to humans or animals can be used.

The experimental model may use computer software, employing the finite element method (FEM), for instance. This method is generally used in the mechanical analysis of materials but can also be employed in the analysis of heat conduction, fluid dynamics, infiltration flow, and electric and magnetic fields [170]. In biomechanical studies, the FEM allows simulating in software the responses of biological materials to mechanical stimuli.

Another way of conducting research without using a living body is to use or construct a prototype, a phantom, with the approximate mechanical properties of the structure that is intended to be measured in the future.

7.2.1.1   Joints

Joints are structures that bind the bones and make possible the movement of body segments. The structures that make up the joints are formed by connective tissue, which acts in the connection and sustentation of other tissues and whose great differential is the extracellular matrix, which has a series of fibrous proteins (predominantly collagen) and macromolecules that form the ground substance.

The articular components are the following: (i) articular cartilage, which is located near the bones, participates in joints, and has the function of protecting the bony ends from wear as well as of providing a better distribution of the imposed loads; (ii) articular fibrocartilage, which is disc shaped, improves load distribution and the attachment of joint surfaces, limits certain movements, protects, lubricates, and absorbs shocks; and (iii) ligaments and joint capsules, which connect the bones and are not contractile but slightly distensible [171].

Through experimental models, optical fibres may be used to measure joint angulation, which assesses the degree of freedom of movement of a patient's joint and is essential for a specialist to prescribe the best treatment [16]. The sensor may also be used to measure the degree of tremor in patients with Parkinson's disease [17]. Patil and Prohaska [18] used optical fibre as an intensity sensor to measure the angle formed between two phalanges of the hand. Zawawi et al. [19] presented a calibration model as well as a study about the optimization of the signal captured by plastic optical fibres to measure spine angulation.

With a different goal from the measuring of angles, FBG sensors have also been encapsulated and tested under loads with a prosthetic femoral component for validation [20] so to subsequently measure intra-articular pressures.

7.2.1.2   Bones and Muscles

Biomechanical parameters of the musculoskeletal system are investigated to improve diagnoses and treatments, as well as assist the development of medical devices.

Bone is a living tissue that is rigid and resistant and has the mechanical function of supporting and protecting the body. Bone tissue adapts to the imposed loads, modifying its structure according to mathematical laws [172].

In their microstructure, bones are composed of connective tissue, which is constituted by an organic matrix, a mineral matrix, and osteogenic cells. The organic matrix is composed primarily of type I collagen, which provides tensile strength. The mineral matrix is composed mainly of calcium and phosphate ions in the form of hydroxyapatite crystals, which provide strength under compressive loads. Finally, osteogenic cells, namely, osteoblasts, osteoclasts, and osteocytes, are the primary cells that act on bone growth, adaptation, and remodelling [173]. Given the complexity of the bone composition, it is important to develop systems that help understand their mechanical behaviour under different conditions.

Talaia et al. [21] considered in their studies synthetic femurs, one pristine and one fractured with a fixation plate joining the fracture. The authors used FBG sensors to characterize the deformations of the femur and the fixation plate when subjected to various loads.

Muscles, unlike bones, are contractile structures that are extensible, elastic, and responsible for the execution of movements. Muscle cells contain filaments composed of proteins that can transform chemical energy into mechanical energy, thus producing muscle contractile force. Koozekanani and Makouei [22] aimed to study how to improve the sensitivity of an FBG to measure muscle deformations through tests changing fibre parameters.

7.2.1.3 Teeth, Lower Jaw (Mandible), and Upper Jaw (Maxilla)

In a more specific area of bone research, orthodontists have used optical fibres to study the mechanical behaviour of teeth and adjacent bones when subjected to specific forces.

The effectiveness of mouth guards was studied by Tiwari et al. [23] to investigate the force transmitted by teeth protectors. Initially, a study of the mouth protector was done through the FEM analysis, and, later, impact tests using a pendulum were performed with a synthetic jaw and the mouth protector. FBG sensors were glued to the regions of the mandible and teeth where objects could collide. The FEM has also been used to analyse deformations in a mandible, with FBG sensors glued to a synthetic mandible being used to validate the results obtained with the simulation [24].

The objective of the study by Milczewski et al. [25, 174] was to show the possibility of using FBG sensors as tools to measure the internal tension of an artificial jaw, transmitted by the teeth as a consequence of the use of orthodontic appliances.

7.2.1.4 Prosthesis and Extracorporeal Devices

Devices used near the human body aiding in some function or having the purpose of protecting some structure must have their mechanical behaviours properly defined, given that their efficiency depends on their mechanical properties and good connections with the corporeal segments.

Regarding the study of prosthesis, in the work of Al-Fakih et al. [26], an integrated sensor measured the pressure that the stump of an amputee would impose to the prosthesis. A mechanical device that uses an air compressor was installed inside the socket to simulate pressure. The distribution of stresses at the interface of the limb and the prosthesis may generate discomfort. High pressures applied to the skin result in ulcers, skin irritation, and partial or total vascular occlusions [175]. The continuation of this study by Al-Fakih et al. [27] developed a machine that simulates the gait of an individual to analyse the pressures between the stump and the prosthesis more completely. The result of these studies assists in the development of more comfortable prostheses.

In another research, the efficiency of a helmet was tested by incorporating FBG sensors to an aluminium structure that was subjected to impact tests performed by a machine [28]. Another example is the work of Leal-Junior et al. [29], where POFs as intensity sensors were calibrated and used to measure the strength that a person would perform when using a lower-limb exoskeleton, a device that aids in the treatment of limb disorders.

7.2.1.5 Sole and Insoles

Measuring sole pressure and its ground reaction forces is important in two situations: (i) in order to diagnose deviations in the individual's footing that may involve complications in joints, muscles, tendons, and posture or (ii) prevent ulcerative lesions on diabetic patients (who have reduced sensitivity in the hands and especially feet) or individuals who have reduced sensitivity in the feet for some other reason.

Wang et al. [30, 162] constructed a mesh of optical fibres arranged in rows and columns and immersed in rubber material to detect plantar pressures. Detection occurred by the curvature caused in the fibre when subjected to external pressure. This mesh was calibrated with the help of a mechanical loading machine.

7.2.1.6 Smart Fabrics

Sensors incorporated into fabrics may be used in film animations, robotics, and collision tests, among others [31]. Curvature optical fibre sensors were fixed to a non-elastic band applied around body segments to find the object in space through a computer program [31]. Cowie et al. [32] developed and characterized a unidirectional and 2D tactile monitoring system using FBG sensors, which could be used in the future for gait analysis.

7.2.1.7 Blood Vessels

Blood travels through several vessels from the moment it is ejected from the heart. Arteries and arterioles carry oxygenated blood, while gas exchanges with the tissues occur in the capillaries, and venules and veins carry the carbon-rich blood back into the heart. Numerous diseases may alter the biomechanical behaviour of this system. For example, systemic arterial hypertension alters the blood pressure in the arteries and may change heart muscle tropism as well as the heartbeat pattern, among other factors. Therefore, there is a need to develop tools to measure the parameters of cardiovascular mechanics in a non-invasive or minimally invasive way.

An optical sensor intended to measure blood pressure has been developed and tested [33]. In more recent work [34, 35], a model of a blood vessel with a cerebral aneurysm was fabricated with hydrogel and is illustrated in Figure 7.5. This model was equipped with interferometric pressure sensors made of optical fibres. Shi et al. [36] used FBG sensors on the tip of a cardiac catheter to measure the forces that interact with it. With the intent of measuring pressure in blood vessels, the FEM analysis was performed along with an experimental testing [37].

Gurkan et al. [38] glued FBG sensors to a subwoofer that reproduces cardiac sounds for the purpose of using this technique in ballistocardiography, which

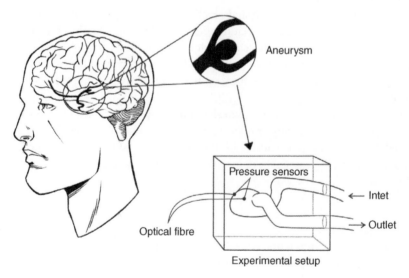

Figure 7.5 Representative model of an aneurysm made of hydrogel (PVA-H) equipped with interferometric pressure sensors. *Source:* Based on [34]. Reproduced with permission of IEEE.

records body movements produced and transmitted by the heartbeat or the passage of blood in the vessels.

7.2.1.8 Respiratory Monitoring

The primary function of the respiratory system involves the exchange of oxygen and carbon dioxide from the alveoli into the blood and from the blood to the alveoli. However, the respiratory system is more complicated than just performing gas exchanges. Magnitudes such as volume and air pressure are involved in its performance. This system depends on the correct mechanical functioning of the lungs, thoracic cavity, respiratory tract, pleura, and respiratory muscles, among other neurovascular structures. Measuring the respiratory rate, volumes, and pressures that vary during breathing is important for understanding and monitoring the system.

The main advantage of monitoring respiratory movements with optical fibres is the immunity to the electromagnetic interference of the fibre. Respiratory movements are usually measured in hospital environments, which present a large number of electronic equipment.

Allsop et al. [39] created a device using LPG sensors to monitor respiratory movements. For the validation of the sensor, a phantom of the human thorax used in resuscitation training was employed. An airbag between the structure and the skin of the manikin was inflated, simulating the respiratory movements in the same volumetric proportions as in a human being. Jonckheere et al. [40] used FBG sensors integrated into an elastic band, and, for validation, a machine that simulated respiratory movements was constructed.

7.2.2 *In Vitro*

Here are considered those *in vitro* studies that remove some material from a living being and manipulate this material in the laboratory, changing its properties, usually with chemicals. Regarding the application of optical fibres for biomechanical analysis, there are very few papers in the literature. In 1994, Moore used optical fibres to investigate the mechanical properties of embryonic tissues [41]; through the study of biomechanical characteristics during morphogenesis, such as viscoelasticity and active mobility, normal and abnormal tissues can be differentiated. In 2002, Erdemir et al. [42] employed optical fibres as force transducers in *in vitro* tests in calcaneal tendons. This study investigated the error rate caused by the cable migration of the transducer according to the loading rates imposed to the tendon. This study highlights the complexity of this same application in *in vivo* tests. Mishra et al. employed FBG sensors to measure tensions in a decalcified goat bone and compared it with a healthy bone, which represents an important research for the study of osteoporosis [43].

7.2.3 *Ex Vivo*

The use of body parts extracted from animals or humans is an alternative when the performance evaluation of invasive tests is necessary given that it consists of a model that is as close as possible to the living being. Body parts extracted from pigs, for example, have appeared frequently in this methodology of studies, notably for their tissue characteristics being very similar to the human being.

7.2.3.1 Joints

Inserting sensors into the joint space or intervertebral discs can generate information about pressure distribution and joint alignment, which is important for the study of spinal pathologies such as herniated discs and arthrosis. Mohanty et al. used FBG sensors on the knees of human cadavers [44]. A research group from Canada developed different encapsulations for FBG sensors with the purpose of exploring joint pressures using pig intervertebral discs [45–47], human intervertebral discs [48], and human hips [49]. Figure 7.6 shows the schematic drawing of a dorsal functional unit (two vertebrae and an intervertebral disc), demonstrating the application of the FBG for intradiscal pressure measurement.

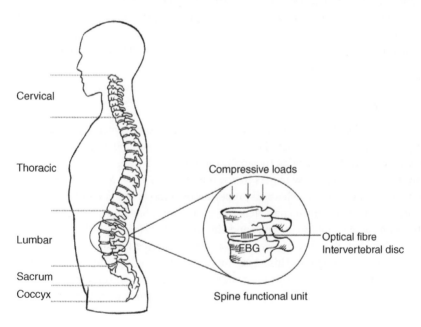

Figure 7.6 Schematic drawing demonstrating the application of FBG sensors on an intervertebral disc for pressure measurement. *Source:* Based on [48]. Reproduced with permission of Elsevier.

Esmonde-White et al. [50] used the Raman spectra to detect the cartilage and subchondral bone quality in the human arm and knee after performing the FEM and phantom tests. To generate data regarding healthy and pathological cartilages, the mechanical properties of bovine cartilage have been studied with the support of FBG sensors [51].

7.2.3.2 Bones and Muscles

In a study on bone properties, FBG sensors associated with strain gauges were used in the external and internal cortical wall of a femur, between the bone and a hip prosthesis, to characterize the relationship of the prosthesis and the bone when submitted to mechanical loading [52].

Tendons are muscle extensions composed of connective tissue. They are noncontractile structures with high collagen concentrations, and their function is to connect muscles to bones. As important as studying the biomechanics of muscles is investigating that of tendons. Komi et al. used a rabbit calcaneal tendon and inserted an optical fibre, which worked as an intensity sensor to measure forces acting on the tendon when it was submitted to different loading conditions [53]. Using the same method, forces present on the human calcaneal tendon have been studied [54].

In the work of Ren et al., FBG sensors were used to measure forces in a human calcaneal tendon and then in medial and lateral collateral ligaments of a human knee [55]. In turn, Behrmann et al. used FBG sensors in deer tendons [56].

7.2.3.3 Teeth, Lower Jaw (Mandible), and Upper Jaw (Maxilla)

Human mandible with a dental implant was studied using FBG sensors positioned in the mandible to measure deformations caused by impacts on the implanted tooth, and the results assist in developing new dental implants [57]. Karam et al. and Romanyk et al. used an FBG sensor inserted between the alveolar bone and the periodontal ligament in pig mandibles to study the mechanical properties of the ligament when submitted to loading [58, 59]. The positioning of the sensor is demonstrated in Figure 7.7.

Wosniak et al. attached an FBG to a lower jaw in a goat skull to measure local deformations generated by the chewing at distinct feeding conditions with results being relevant for agribusiness [61]. Karam et al. used the FBG sensors for the same purpose of attaching them to the lower jaw in an ox skull, describing a surgical technique for the implantation of the sensor [62].

7.2.3.4 Blood Vessels

Agah et al. studied aortic arteries using optical fibres as pressure sensors, aiming to understand the mechanisms of traumatic rupture of the aorta [63].

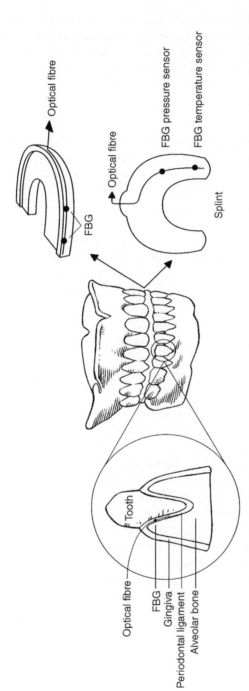

Figure 7.7 Left: schematic drawing showing the dental structure as well as sensor location at the periodontal ligament [59]. Right: schematic drawing showing the dental splint views of different angles emphasizing of the positioning of the FBG sensors. *Source:* Based on [60]. Reproduced with permission of Elsevier.

7.2.3.5 Mechanical Properties of Tissues

To measure the resistance of different tissues, Mo et al. used, inside a needle, an optical fibre with a Fabry–Pérot sensor at its end as a force sensor. In this way, it is possible to access a specific tissue layer with only the information of resistance that the different tissues offer against the needle. After the calibration of this system, it was tested on pig tendons/ligaments and on a pig spine [64]. In a more recent study from the same research group, the same measurement system was used in the fabrication process of artificial tissues with mechanical properties similar to those of humans and also in different pig belly tissues. The advantage of the proposed system for use in robotic surgery is also highlighted [65].

7.2.4 *In Vivo*

The application of sensors *in vivo* is a great step towards the validation of its application and the possibility of reaching the population. All the work cited in the following paragraphs was tested on humans or living animals. Among the applications, there were sensors tested in invasive and minimally invasive ways. Minimally invasive sensors are those inserted into a body cavity without compromising the skin or mucosa of the living being, whereas sensors that are invasively tested are those that compromise the integrity of the skin or the mucosa, breaking down the safety barriers of the human body and reaching structures inside. Such studies present risks of infection and other side effects and require more complex methodologies.

7.2.4.1 Joints

Fabry–Pérot cavity-based optical fibre sensors have been used to measure pressure in intervertebral discs, with the optical fibre being inserted with a needle and the loading done with a device fixed in the vertebrae of pigs [66].

7.2.4.2 Bones and Muscles

Ferreira et al. designed a probe with FBG sensors inside it so that vaginal muscle forces were measured. Such a device may aid in the diagnosis and evolution of the treatment of pelvic disorders [67].

Arkwright et al. introduced a catheter with FBG sensors into the oesophagus of patients to measure the peristaltic pressures produced by swallowing [68]. In another publication, FBG sensors were used in the construction of a colonoscopy catheter, and using 72 multiplexed sensors, it was possible to monitor the peristaltic movements of the colon 24 hours' measurement [69]. According to the authors, catheters produced with optical fibres are sensitive, smaller, and more flexible for this type of measurement.

Finni et al. presented a technique using optical fibres as an intensity sensor inserted into the calcaneal tendons of healthy individuals. The purpose of this technique is to measure forces that act on the tendon in real time [70]. Using the same measurement technique, optical fibres have also been inserted into patellar tendons to study the kinematics of jumping in amateur athletes [71].

7.2.4.3 Teeth, Lower Jaw (Mandible) and Upper Jaw (Maxilla)

Tjin et al. incorporated FBG sensors into a dental splint so as to monitor the strength and temperature to which the splint was exposed while patients with sleep apnoea slept [60]. The schematic design of the proposed system is shown in Figure 7.7. Similar works used FBG sensors in occlusal splints to assess the bruxism conditions of patients [72] and inserted them into a device to evaluate the bite force of different volunteers [73].

In another approach, FBG sensors have been encapsulated with metallic meshes and screwed to the mandible of a bovine, with results being published by the same research group in various documents with different focuses [74–76]. The purpose of such studies was to develop an alternative and minimally invasive tool to analyse the food consumed by the animal and the period of consumption of each type of food. Such data are of interest to the agroindustry. The recognition of the type of food was done by software, only identifying bone deformations through the encapsulated FBG sensor fixed to the ramus of the mandible.

7.2.4.4 Blood Vessels

Hansen, in 1983 [77], introduced the optical fibre into a blood vessel during a catheterization. Similar sensors were used to measure the internal pressure of the bladder, urethra, and rectum of patients, such as an optical fibre Fabry–Pérot cavity to measure blood pressure being inserted with the aid of a needle into the right atrium, left atrium, left ventricle, and aorta of a goat [78] or a fibre sensor with a specific encapsulation that has been developed to measure the blood pressure in the aortic arch and right coronary artery of a pig [79].

7.2.4.5 Respiratory Monitoring

Cavaiola et al. used an optical fibre with six FBG sensors inserted into a needle to access some sites in a pig liver, employing the FBG sensors to monitor the temperature. Since FBG is sensitive to deformations, the noise generated by respiratory movements was also measured [80].

7.2.5 *In Situ*

Sensors classified as *in situ* are non-invasive and used outside the human body without compromising the integrity of the skin or mucous membranes. Such sensors may also be integrated into structures that are used in contact with the body. This modality of application represents the majority of the publications optical fibre sensors for measuring biomechanical parameters.

7.2.5.1 Joints

Several research groups have studied hand, finger, and wrist movements. Back in 1990, a glove was constructed using several optical fibre sensors operating with attenuation caused by curvatures in the fibre, with the goal of creating a device that would aid in determining the angulation of all fingers at the same time, recording this data on a computer [81]. Later on, other researchers have developed gloves using hetero-core fibres [82], FBG sensors [83], and fibre as a curvature (attenuation) sensor [84] for the purpose of measuring hand movements that could be applied in physiotherapy, the study of gestures in sports, and virtual reality, among others.

Optical fibres have also been used as goniometers to measure lumbar spine curvatures and to relate sedentary work with low back pain [85, 86]. Researchers have used optical sensors to analyse the posture of wheelchair users while they performed wheelchair propulsion movements [87, 88]. Sportswear with hetero-core optical fibre sensors (based on curvature attenuation) was used to measure simple trunk movements; later, the clothing was used on a golf player's trunk to describe the gesture characteristic of the sport [89, 90]. Sensors incorporated into clothing also caught the attention of Nishiyama et al. [91], who used hetero-core optical fibres to measure the positioning of the human elbow and knee.

To measure human knee movements, optical sensors have been incorporated into elastic bands [92, 93], articulated structures [94, 95], and polymeric tissues [96] and then adapted to the knee with the future objective of monitoring sports movement or assisting in physiotherapeutic treatment.

7.2.5.2 Bones and Muscles

Optical fibre sensors as myographs (a device that measures muscle contraction graphically) in the forearm region were also developed [97–99]. Since the muscles responsible for the hand movements are located in the forearm, it was possible to recognize the posture of the hand through the data measurement. Besides measuring movement in hand joints, Guo et al. used a highly flexible optical fibre sensor to detect speech and respiratory movements [100].

In addition, FBG sensors have been glued to structures in contact with the calf for researchers to observe walking frequency [101] as well as attached directly to the skin of volunteers in the calf region while they performed exercises to avoid deep vein thrombosis (DVT), which is common in long flights [102]. The results of this last research were compared with ultrasound measurement of blood velocity to define the most efficient exercises.

7.2.5.3 Prostheses and Extracorporeal Devices

A recent and innovative application of FBG sensors has been in prostheses. Al-Fakih et al. [103] presented an interface that measures the force exerted by a transtibial prosthesis on the stump of the patient. FBG sensors have also been incorporated into composite materials in the form of a transtibial and foot prostheses, so as to study the behaviour of the prostheses [104, 105].

In [104], volunteers walked on a treadmill. The results obtained allow monitoring the distribution of force along the prosthesis during the gait cycle, which may be visualized through a colour map or the resultant force vectors. Also, the authors were able to identify the main phases of gait cycle support. Optical fibre instrumentation for applications involving amputee prostheses contributes to medical evaluations, during the amputee rehabilitation process, during a competition, or even to customize a new prosthesis design. Figure 7.8 shows a real test performed by the authors on an ergonomic treadmill.

Another interesting field of research is the incorporation of different optical fibre sensors in beds [106–109], chairs [110–112], and pillows [113] with the purpose of monitoring the movements, heart rates, and breathing of individual. Intelligent structures for the basic monitoring of a bedridden patient may be used in hospitals and nursing homes, aiding in the prevention of pressure ulcers, patient falls, and abnormal cardiac and respiratory behaviours without the inconvenience of a wire attached to the body, and may also be used during examinations such as magnetic resonance [114, 115].

7.2.5.4 Soles and Insoles

A subject that has attracted the attention of some groups recently is the study of the distribution of foot forces and forces of reaction to the ground to assist in diagnosing

Figure 7.8 Illustrative picture of a volunteer walking on a treadmill and a portable computer processing the data and informing on the deformation points in real time [104]. *Source:* Reproduced with permission from IEEE.

inadequate posture [116] and incorrect gait [117–119, 176], as well as helping diabetic patients who lose sensation in their feet [120]. FBG sensors [116, 118, 120, 176], hetero-core fibres [117], and plastic fibres [119] have been incorporated into devices, insoles, and shoes in different studies, with the possibility of remote monitoring real-time data made by a mobile device [121–123]. Also, for the purpose of gait analysis in horses, FBG sensors have been incorporated into composite horseshoes, with tests carried out in different stages: gait, trot, and gallop [124]. The results obtained are promising for the clinical and biomechanical study and medical evaluations of horses, even during dynamic training and competitions. Figure 7.9 shows a schematic drawing of the application of instrumented horseshoes, in which information on horse gait, temperature, and pressure can be viewed in real time.

7.2.5.5 Cardiac Monitoring

In the field of vital sign monitoring, several studies have been published with the aim of measuring arterial pressure [125–128], arterial pulse shape [129, 130], cardiac movements [131], and/or heart rate [132] with the optical fibres encapsulated into extracorporeal devices or directly glued to the skin.

7.2.5.6 Respiratory Monitoring

Besides cardiovascular measurements, the use of optical fibres for the measurement of respiratory parameters stands out. It is the most studied area among

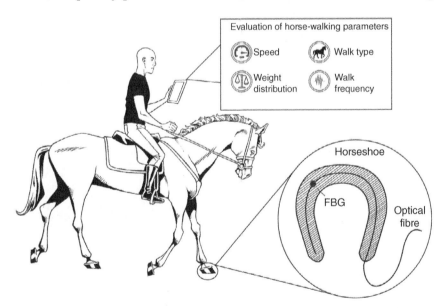

Figure 7.9 Schematic representation of an instrumented horse. The FBG sensors are promising to evaluate animal hull-related pathologies. *Source:* Based on [124]. Reproduced with permission of IEEE.

biomechanical applications with optical fibres. Curvature attenuation-based sensors[133–137], POFs [138, 139], LPG sensors [140, 141], and FBG sensors [142, 143] attached to elastic bands and to the thorax and/or abdomen of patients have been demonstrated under different respiratory conditions and for different purposes, such as measuring respiratory volume and rate as well as chest movements. Moreover, sensors have also been incorporated into T-shirts [144, 145]. In this sense, there are studies that performed cardiac and respiratory measurements simultaneously using interferometric optical fibre sensors [146], FBG sensors [147, 148], POFs [149], and curvature sensors [177]. Figure 7.10 shows the positioning of sensors on the thorax and abdomen in one of the studies that predicts

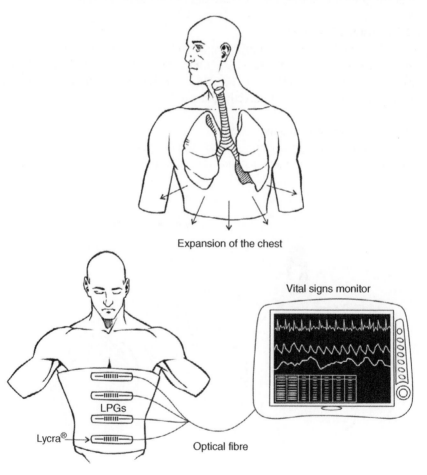

Figure 7.10 Schematic drawing demonstrating the position of LPG sensors on the thorax and abdomen. *Source:* Based on [150]. Reproduced with permission of Springer.

respiratory volumes [150]. The values recorded by LPG sensors were compared with those recorded by a spirometer for validation of the proposed system.

7.3 FBG Sensors Integrated into Mechanical Systems

In the domain of biomechanics, optical fibre sensors allow real-time monitoring of several physical parameters such as temperature, deformation, and pressure as presented in the preceding sections. For the correct and safe sensing, the optical fibre sensor must be installed properly and mechanically protected by using adequate integration or encapsulation methodologies.

Figure 7.11 shows the percentage of each methodology adopted for optical instrumentation with FBG sensors. Results indicate that in 44% of FBG sensor applications in biomechanics, the authors used a polymer-based glue to attach the FBG sensors to the investigated region. Next, the methodology with the sensors integrated into polymers is present in 32% of the applications, while the other methods appear in lower percentages. The reason for the difference among the methodologies adopted by the authors is very likely related to the maturity and ease of use of the optical FBG technology.

A review of such methodologies is presented in this section with some examples of applications where the sensors are glued with polymers, integrated into polymers, and with smart fibre reinforced polymers. Table 7.1 shows the classifications of the methodologies described in articles published in journal and conference papers from 2000 to 2018 involving the application of FBG sensors in biomechanics.

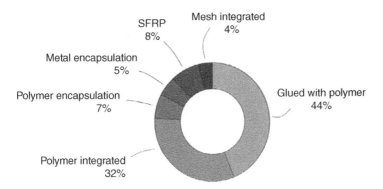

Figure 7.11 Percentage of the encapsulation methodologies using FBG sensors in different applications in the field of biomechanics from 2000 to 2018, according to the reference list.

Table 7.1 Classification of the methodologies adopted for fixing optical fibre sensors in the field of biomechanics according to the articles published from 2000 to 2018.

Methodology	Description
Glued with polymer	Applications in which FBG sensors are glued directly to the region where a physical quantity is to be monitored. The adhesives used are polymer based. Application example: characterization of mechanical stresses in *ex vivo* bones (mandible, femur) [21, 24]
Polymer integrated	Applications in which the FBG sensor is integrated into a polymeric material, i.e. the optical fibre is inserted into the material during manufacturing. Usually, a mould is used, in which the optical fibre is positioned. The mould is filled with polymer, for example, an epoxy resin. After the polymerization process, the sensor is characterized for a given application. Application example: development of an insole for the study of human gait [118]
Mesh integrated	The FBG sensors are integrated into elastic fabrics. During the fabric-making process, an optical fibre is sewn among the yarns of the fabric. Usually, this technique is used for *in vivo* applications to monitor respiratory or muscle movements [40, 148]
Smart fibre reinforced polymer (SFRP)	Applications in intelligent materials using fibre- reinforced polymer. The fibres may be glass, carbon, aramid, or a combination thereof. The optical fibre with FBG sensors is inserted between the reinforcing layers during the manufacturing process of the composite material. Examples of applications: prosthetic fittings for amputees or mechanical protection devices for optical fibres [104, 105]
Metal encapsulation	Applications in which the optical fibre is inserted inside a metallic capillary or embedded in a metal body. The first is a commonly used methodology to isolate variations of mechanical deformation and monitor only temperature variations [46, 48, 51]
Polymer encapsulation	Methodology similar to the metal encapsulation procedure. It involves applications that require the use of nonconductive, biocompatible, or flexible materials, such as catheters and probes [56, 68]

7.3.1 FBG Sensors Glued with Polymer

Among the large diversity of polymers that are commonly used to fix optical fibres, cyanoacrylate is the preferred one in biomechanical applications given that it is already of common use in medicine and it is also very efficient to attach in glass.

A cyanoacrylate adhesive is a liquid substance (monomers) with low viscosity and colourless, deriving from cyanoacrylic acid and presenting the general chemical structure of $CH_2=CH-COOR$, where R may represent a methyl, ethyl, butyl, or isobutyl radical or octyl cyanoacrylate [178]. The polymerization process takes place in a few seconds when it comes into contact with air. Because it is a liquid-form product, it penetrates uneven surfaces, promoting good adhesion. However, for a satisfactory result, it is important to sanitize the surface before the glue application.

Zhang et al. [179] used gauze with cyanoacrylate-based adhesive aiming at stopping bleedings, and tests were performed with volunteers. Another example of an application in medicine is described by Bugden et al. [180], who used glue to attach intravenous catheters. The study was carried out with several patients and showed to be effective in controlling the adhesion of the catheter to the surface on which it is applied. The study reports that this technique decreases the risk of failure during treatment.

Several papers present laboratory research studies. da Silva et al. [24] investigated the mechanical deformations generated in a phantom jaw bone. The FBG sensor was attached to the external surface of the bone, and the results of the FBG sensors are promising when compared with electrical strain gauge sensors, mainly because of their geometry allowing less intrusive sensing. An example of optical instrumentation applied to a mandible is shown in Figure 7.12, whereas Talaia et al. [21] instrumented a femur to evaluate the distribution of forces along the bone and used cyanoacrylate-based glue to attach sensors to the bone. Figure 7.13 illustrates a femur instrumented with the FBG sensors.

7.3.2 Polymer-Integrated FBG Sensor

In biomechanics it is often not possible to fix the optical fibre directly onto the region of interest, primarily because of the optical fibre mechanical fragility against shear forces. One of the alternatives to protect the optical fibre is the integration of FBG sensors into a polymeric matrix. Besides the mechanical properties, the advantage of polymers is the ease of preparation of the device: using different moulds and polymerization processes at controlled temperatures, it is possible to make sensing devices of various geometric shapes and with good thermal and mechanical properties. The commonly used polymers are those classified as thermoplastic or thermoset [181].

Thermoplastic polymers have the ability to soften and flow with increasing temperature and pressure. Subsequently, when removed from the heat, they solidify into an object with a defined shape. New applications of temperature and pressure produce the same softening and flowing effects. This change is a physical and reversible transformation. In addition, when the polymer is semi-crystalline, the

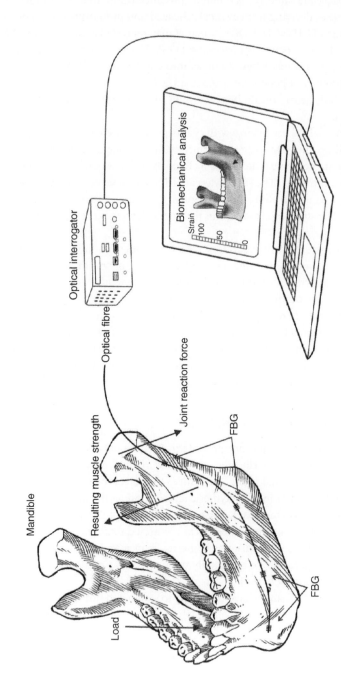

Figure 7.12 Schematic representation of a mandible instrumented with FBG sensors. *Source:* Based on [24]. Reproduced with permission of Springer.

Figure 7.13 Schematic representation of a femur instrumented with FBG sensors. *Source:* Based on [21]. Reproduced with permission of Springer.

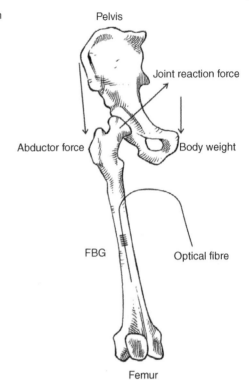

softening occurs with the melting of the crystalline phase. They are also fusible, soluble, and recyclable. Typical examples of thermoplastics are polyethylene, polyvinyl chloride (PVC), polypropylene, and polystyrene [181].

Thermoset polymers are formed by an irreversible chemical reaction between the resin and a curing agent (hardener or catalyst) that leads to a rigid state of the material. When later heated, they do not alter their physical state, i.e. they do not soften. After polymerization, they are infusible and insoluble. Thermosetting polymers are not usually recyclable. Their structure is amorphous, that is, it does not have a defined atomic structure, being the opposite of crystalline structures. Typical examples are the epoxy resins [181].

The integration of FBG sensors into polymers adds advantages for both materials given that the optical fibre is protected when integrated into the polymer and the FBG sensor assists in the polymerization characterization process: the polymer must be polymerized at controlled temperature and time so as to avoid the maximum residual stresses after curing, which may result in fractures in the material. An example of an application is in the dental area, where polymer resins are used for restorative procedures. The resins are polymerized with visible light, and, for a satisfactory result, it is necessary to control the time of exposure to light. Otherwise, if the time is shorter/longer than necessary for the bonding of the chemical chains, the restoration may present problems such as cracks, cavities, and, consequently, bacterial contamination. Based on this concept, Milczewski et al. [182, 183] integrated FBG sensors into two distinct commercial resins to investigate the polymerization process. With the use of an optical interrogator, the authors recorded the central wavelength of the Bragg peak before and during the curing process. Hence, it is possible to analyse the difference between the variation of the final and the initial wavelengths and convert this value into a deformation ($\mu\varepsilon$).

In the field of rehabilitation, the use of polymer-integrated FBG sensors is described by da Silva et al. [83], who integrated the sensors into a glove made of polyvinyl polychloride (PVC) polymer. During the glove manufacturing process, optical fibres with FBG sensors were incorporated into the PVC material. For this application, the ease of multiplexing of the FBG sensors made it possible to instrument several points of interest in a single hand.

In addition, Domingues et al. [118] instrumented an insole, integrating FBG sensors into small cubes made of epoxy resin. To integrate the sensors, the authors built a mould, inserted the FBG sensor into it and filled it with epoxy resin. To characterize the sensitivity of the FBG sensors, the authors used a universal testing machine that applied a constant force on the epoxy resin cubes. The sensitivity of the sensors was of 11.06 pm/N (wavelength shift in picometres per Newton). Finally, trials with volunteers walking on a commercial force platform were performed. The generated data may be used to analyse foot structure and correlated anomalies, being promising for biomechanical applications. Figure 7.14 shows a schematic drawing illustrating an example of the instrumentation of a glove and an insole.

7.3.3 Smart Fibre Reinforced Polymer (SFRP)

Integrating optical fibre sensors into composite materials is a promising technique to mechanically protect the optical fibre in *in vivo* applications of biomechanics. Composite materials are designed to combine the properties of two or more materials. Generally, the composites are composed of two phases, denominated matrix and reinforcement. In the composite, the matrix phase has a structural function, completing the voids in the fabric. This phase also assists in positioning and protecting the reinforcement phase, which guarantees the mechanical, electromagnetic, or chemical resistance of the material. Thermal and mechanical properties depend upon several factors such as the chemical composition, the geometric arrangement of reinforcements, the type of reinforcement used, and the methods for making the composite [181].

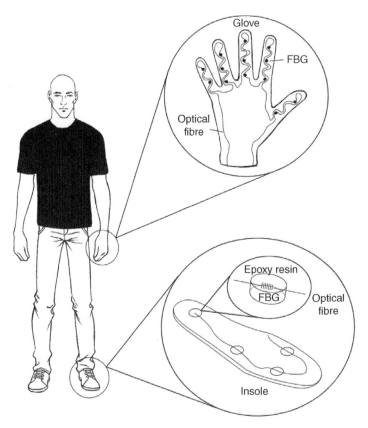

Figure 7.14 Schematic representation of a glove and an insole instrumented with FBG sensors. *Source:* Based on [83, 118].

With the use of composite materials, it is possible to develop intelligent composite materials [184] formed by a sensor and/or an actuator. The system receives a stimulus that the sensor detects and can respond to it through an external actuator. Sensors used in intelligent composite materials may be of optical fibres, electromechanical devices, and piezoelectric materials [104, 184]. The actuators may be piezoelectric ceramics, electromagnetic/magnetorheological fluids and memory alloys [181].

The most commonly used composite materials in biomechanics are carbon fibre fabrics in conjunction with epoxy resins. Figure 7.15 shows an example of the methodology for integrating FBG sensors among the layers of carbon fibre fabric. The optical fibre is positioned between the reinforcement layers during the material manufacturing process, which usually follows a polymerization process with controlled temperature cycles and with the aid of vacuum and pressure [181].

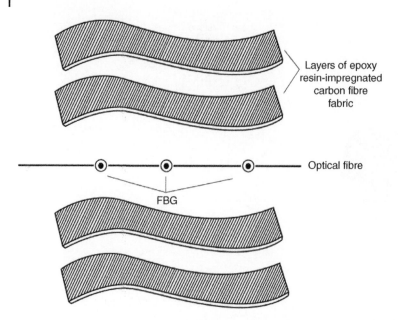

Figure 7.15 Schematic representation of the integration process of FBG sensors in a composite material. *Source:* Based on [104]. Reproduced with permission of IEEE.

In the field of amputee rehabilitation, composite materials are used for the development of prosthetic components because of their mechanical characteristics such as elasticity modulus and low weight. The integration of the FBG sensors into these materials allows monitoring physical quantities that are significant for the rehabilitation process of a lower-limb amputee. Al-Fakih et al. [26] made a socket, which is a prosthetic component that supports the stump (residual limb) of a lower-limb amputee, as described in Section 7.2. During the process of making the socket, FBG sensors were integrated into the polymer material. The purpose of the authors was to investigate the pressure distribution along the socket. The advantage of optical instrumentation for this application is the possibility of monitoring several points along the socket. In the case of using an electric sensor, the larger volume of electrical wires would limit the number of sensors.

Galvão et al. [104, 185] instrumented a prosthesis for a lower-limb amputee with FBG strain and temperature sensors. The prosthesis was made of carbon fibre with FBG sensors integrated into the composite material. Two optical fibres with strain sensors were used, both incorporated into the layers of the carbon fibre prosthesis and one of the FBG sensors encapsulated into a metallic capillary to monitor only temperature variations. Figure 7.16 shows an example where the prosthesis can be used by a transtibial amputee, and the gait cycle information can be analysed in real time.

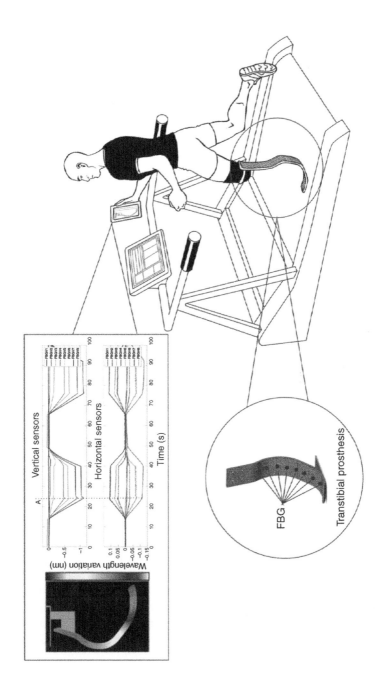

Figure 7.16 Schematic representation of an amputee using a FBG instrument carbon fibre transtibial prosthesis. *Source:* Based on [104]. Reproduced with permission of IEEE.

(a)   (b)

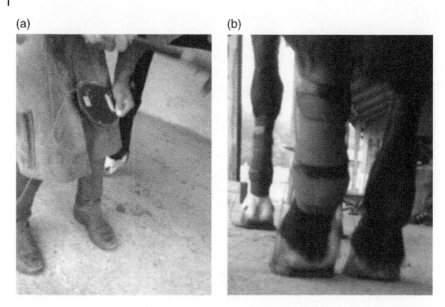

Figure 7.17 Photographic images of the optical instrumentation of a horse's hoof: (a) method of fixing the horseshoe using a hammer, showing the robustness of the developed system, and (b) detail of the horseshoe installed on the horse's limb and the positioning of the output of the optical cord on the side [124]. *Source:* Reproduced with permission of IEEE.

Another application using the same encapsulation technique allowed the design, fabrication, and application of horseshoes in *in vivo* experiments. Galvão et al. [124] installed four of such sensors in the hooves of a horse. During the manufacturing process, the FBG sensors were integrated into the composite material, and the horseshoes were nailed to each member of the animal the same way it is done with standard horseshoes. Figure 7.17 illustrates the process of installing the horseshoes on the horse's limb by fixing it using blackheads.

7.4 Future Perspective

In spite of the numerous technical possibilities and solutions that have been investigated using optical fibre sensors in biomechanics, it was not possible to find a real commercially available product opportunity based on this technology. Market sales are a very good measure of a technology success, if not the best. It also shows, conversely, where one can focus when thinking about opportunities for investment and business development.

However, there are several reasons why we believe optical fibre sensors and bio-mechanics get along well and can surely become a case of success. Firstly, most of the clothing technology is already based on fibres, both synthetic and organic. The numerous forms of fabrics use very fine fibres with diameters ranging from 20 to 60 μm, typically. It is well known that tapered fibres can reach such dimensions while keeping low attenuation losses and good propagation and confinement properties. Therefore, producing fabrics of optical fibres similarly to the way it is made with other conventional fabrics is possible. Secondly, optical fibres can measure pretty much anything, any variable, being it physical or chemical. For example, active or sensitive layers of special coatings can be added to the fibres providing them with enhanced performance to measure things like O_2, CO_2, H_2O, and other chemicals like hormones, exalted by the skin. No other technology can do this in real time and distributed over the entire body. Thirdly, a smart skin of optical fibres can measure motion and body reactions and provide data to prevent actions or movements that can harm the human health, all these when considering solely the development of smart fabrics. But there are many more opportunities.

The integration of the optical fibre sensors in composite materials, used to fabricate several types of body prosthesis, has the potential to revolutionize this field. The first obvious applications are those related to external prosthesis of legs, feet, arms, and so on, including exoskeletons. However, they gain an unprecedented level of importance when the optical fibre sensors are integrated to internal body prosthesis such as artificial bones, joints, veins, etc. Their miniaturization, allied with their capacity to conform to pretty much any surface or shape and being able to have tens to hundreds of multiplexed measurement points, is unique and should definitely be more explored.

Finally, one should mention that these possibilities will rely upon the development of novel interrogation and communication methods, which are more cost effective than the technology that is available in the market today. Hence, the next success will likely depend upon the capacity of connecting the sensors to advanced and compact electronic systems that can be wearable and will explore some sort of artificial intelligence methods to process the data and generate information.

Acknowledgment

The authors acknowledge CAPES (Coordination of Superior Level Staff Improvement), CNPq (National Council for Scientific and Technological Development), FINEP (Funding Authority for Studies and Projects), Araucaria Foundation, and State Secretariat of Science, Technology and Higher Education of Paraná for the financial support. This study was financed in part by the Coordenação de Aperfeiçoamento de Pessoal de Nível Superior – Brasil (CAPES) Finance Code 001.

References

1 Kao, K. and Hockham, G.A. (1966). Dielectric-fibre surface waveguides for optical frequencies. *Proceedings of the Institution of Electrical Engineers* 113 (7): 1151–1158.

2 Levin, A.B. (1977). The use of a fibreoptic intracranial pressure monitor in clinical practice. *Neurosurgery* 1 (3): 266–271.

3 Numoto, M., Wallman, J.K., and Donaghy, R.P. (1973). Pressure indicating bag for monitoring intracranial pressure. *Journal of Neurosurgery* 39 (6): 784–787.

4 Ramirez, A., Hood, W. Jr., Polanyi, M. et al. (1969). Registration of intravascular pressure and sound by a fibreoptic catheter. *Journal of Applied Physiology* 26 (5): 679–683.

5 Matsumoto, H., Saegusa, M., Saito, K., and Mizoi, K. (1978). The development of a fibre optic catheter tip pressure transducer. *Journal of Medical Engineering & Technology* 2 (5): 239–242.

6 Cole, J.S. and Conn, R.D. (1971). Assessment of cardiac apex impulse using fibre optics. *British Heart Journal* 33 (4): 463.

7 Fung, Y.-C. (2013). *Biomechanics: Mechanical Properties of Living Tissues.* Berlin: Springer Science & Business Media.

8 Ross, W.D. and Smith, J.A. (1912). *The works of Aristotle*, vol. 5. Clarendon Press.

9 Hatze, H. (1974). The meaning of the term "biomechanics". *Journal of Biomechanics* 7 (2): 189.

10 International Society of Biomechanics. (2018). *General information: History and purpose", International Society of Biomechanics*, Auckland Bioengineering Institute, University of Auckland, New Zealand. [Online]. https://isbweb.org/about-us/general-information (accessed 19 May 2020).

11 Taipalus, T. and Ahtiainen, J. (2011). Human detection and tracking with knee-high mobile 2D LIDAR. In: *2011 IEEE International Conference on Robotics and Biomimetics*, 1672–1677. IEEE.

12 Noguchi, H., Ogawa, Y., Nakagami, G. et al. (2013). Measurement system of body shift during head of bed elevation based on robust tracking of mattress edges using LIDAR. In: *2013 35th Annual International Conference of the IEEE Engineering in Medicine and Biology Society (EMBC)*, 4686–4689. IEEE.

13 Yokota, S., Hashimoto, H., Chugo, D., and Matsumoto, A. (2018). Modeling of human body movement on personal mobility interface using LIDAR. In: *2018 IEEE Industrial Cyber-Physical Systems (ICPS)*, 585–590. IEEE.

14 Knudson, D. (2007). *Fundamentals of Biomechanics.* Springer Science & Business Media.

15 Silva, A., Rocha, R.P., Carmo, J.P.P., and Correia, J.H. (2013, Chapter 6). Photonic sensors based on flexible materials with FBGs for use on biomedical applications. In: *Current Trends in Short-and Long-period Fibre Gratings* (ed. C. Cuadrado-Laborde), 105–132. Intech.

16 Jung, G.-I., Kim, J.-S., Lee, T.-H. et al. (2014). Fibre-optic goniometer for measuring joint angles. *Journal of Mechanics in Medicine and Biology* 14 (06): 1440014.

17 Jensen, J., Li, J.-J., and Sigel, G. (1989). A fibre optic angular sensor for biomedical applications. In: *Images of the Twenty-First Century. Proceedings of the Annual International Engineering in Medicine and Biology Society*, 1118–1119. IEEE.

18 Patil, M.M. and Prohaska, O. (1988). Fibre optic sensor for joint angle measurement. In: *Proceedings of the Annual International Conference of the IEEE Engineering in Medicine and Biology Society*, 803–804. IEEE.

19 Zawawi, M., O'Keeffe, S., and Lewis, E. (2013). Plastic optical fibre sensor for spine bending monitoring with power fluctuation compensation. *Sensors* 13 (11): 14466–14483.

20 Mohanty, L. and Tjin, S.C. (2006). Pressure mapping at orthopaedic joint interfaces with fibre Bragg gratings. *Applied Physics Letters* 88 (8): 083901.

21 Talaia, P.M., Ramos, A., Abe, I. et al. (2007). Plated and intact femur strains in fracture fixation using fibre Bragg gratings and strain gauges. *Experimental Mechanics* 47 (3): 355–363.

22 Koozekanani, R. and Makouei, S. (2017). High sensitive FBG based muscular strain sensor. *Advanced Electromagnetics* 6 (2): 71–76.

23 Tiwari, U., Mishra, V., Bhalla, A. et al. (2011). Fibre Bragg grating sensor for measurement of impact absorption capability of mouthguards. *Dental Traumatology* 27 (4): 263–268.

24 da Silva, J.C.C., Ramos, A., Carvalho, L.M.R. et al. (2005). Fibre Bragg grating sensing and finite element analysis of the biomechanics of the mandible. In: *17th International Conference on Optical Fibre Sensors, SPIE Proceedings*, vol. 5855, 102–105. Bruges, Belgium: SPIE.

25 Milczewski, M.S., da Silva, J.C.C., Martelli, C. et al. (2012). Force monitoring in a maxilla model and dentition using optical fibre Bragg gratings. *Sensors* 12: 11957–11965.

26 Al-Fakih, E., Osman, N., Eshraghi, A., and Adikan, F. (2013). The capability of fibre Bragg grating sensors to measure amputees' transtibial stump/socket interface pressures. *Sensors* 13 (8): 10348–10357.

27 Al-Fakih, E.A., Arifin, N.B., Pirouzi, G. et al. (2017). Optical fibre Bragg grating-instrumented silicone liner for interface pressure measurement within prosthetic sockets of lower-limb amputees. *Journal of Biomedical Optics* 22 (8): 087001.

28 Butz, R.C. and Dennison, C.R. (2015). In-fibre Bragg grating impact force transducer for studying head-helmet mechanical interaction in head impact. *Journal of Lightwave Technology* 33 (13): 2831–2838.

29 Leal-Junior, A.G., Frizera, A., Marques, C. et al. (2018). Polymer optical fibre strain gauge for human-robot interaction forces assessment on an active knee orthosis. *Optical Fibre Technology* 41: 205–211.

30 Wang, W.-C., Panergo, R.R., Galvanin, C.M. et al. (2003). A flexible micromachined optical sensor for simultaneous measurement of pressure and shear force distribution on foot. In: *Smart Nondestructive Evaluation and Health Monitoring of Structural and Biological Systems II*, vol. 5047, 275–286. International Society for Optics and Photonics.

31 Danisch, L.A., Englehart, K., and Trivett, A. (1999). Spatially continuous six-degrees-of-freedom position and orientation sensor. In: *Fibre Optic and Laser Sensors and Applications; Including Distributed and Multiplexed Fibre Optic Sensors VII*, vol. 3541, 48–57. International Society for Optics and Photonics.

32 Cowie, B.M., Webb, D.J., Tam, B. et al. (2006). Distributive tactile sensing using fibre Bragg grating sensors for biomedical applications. In: *The First IEEE/RAS-EMBS International Conference on Biomedical Robotics and Biomechatronics*, 312–317. IEEE.

33 Hong, L. and Prohaska, O.J. (1988). Fibre optic sensor for static pressure measurements. In: *Microsensors and Catheter-Based Imaging Technology*, vol. 904, 71–75. International Society for Optics and Photonics.

34 Matsunaga, T., Ito, A., Osaki, S. et al. (2017). Local internal pressure measurement system of cerebral aneurysm model using ultra-miniature fibre-optic pressure sensor. In: *International Symposium on Micro-NanoMechatronics and Human Science (MHS)*, 1–3. IEEE.

35 Matsunaga, T., Haga, Y., Osaki, S. et al. (2017). Multipoint pressure measurement in blood vessel model for evaluation of intravascular treatment of cerebral aneurysm using fibre-optic pressure sensors. In: *2017 IEEE International Conference on Cyborg and Bionic Systems (CBS)*, 136–139. IEEE.

36 Shi, C., Li, T., and Ren, H. (2017). A millinewton resolution fibre Bragg grating-based catheter two-dimensional distal force sensor for cardiac catheterization. *IEEE Sensors Journal* 18 (4): 1539–1546.

37 Zhongbao, Q. and Ningbo, G. (2018). Mechanical properties of pressure vessel based on fibre Bragg grating. In: *International Conference on Electronics Technology (ICET)*, 1–8. IEEE.

38 Gurkan, D., Starodubov, D., and Yuan, X. (2005). Monitoring of the heartbeat sounds using an optical fibre Bragg grating sensor. In: *Sensors*, 306–309. IEEE.

39 Allsop, T., Earthrowl-Gould, T., Webb, D., and Bennion, I. (2003). Embedded progressive-three- layered fibre long-period gratings for respiratory monitoring. *Journal of Biomedical Optics* 8 (3): 552–558.

40 J. De Jonckheere, Narbonneau, F., D'angelo, L., et al. (2010). FBG-based smart textiles for continuous monitoring of respiratory movements for healthcare applications. *The 12th IEEE International Conference on e-Health Networking, Applications and Services.* IEEE, pp. 277–282.

41 Moore, S.W. (1994). A fibre optic system for measuring dynamic mechanical properties of embryonic tissues. *IEEE Transactions on Biomedical Engineering* 41 (1): 45–50.

42 Erdemir, A., Piazza, S.J., and Sharkey, N.A. (2002). Influence of loading rate and cable migration on fibreoptic measurement of tendon force. *Journal of Biomechanics* 35: 857–862.

43 Mishra, V., Singh, N., Rai, D. et al. (2010). Fibre Bragg grating sensor for monitoring bone decalcification. *Orthopaedics & Traumatology: Surgery & Research* 96 (6): 646–651.

44 Mohanty, L., Tjin, S.C., Lie, D.T. et al. (2007). Fibre grating sensor for pressure mapping during total knee arthroplasty. *Sensors and Actuators A: Physical* 135 (2): 323–328.

45 Dennison, C.R., Wild, P.M., Wilson, D.R., and Cripton, P.A. (2008). A minimally invasive in-fibre Bragg grating sensor for intervertebral disc pressure measurements. *Measurement Science and Technology* 19 (8): 085201.

46 Dennison, C.R., Wild, P.M., Dvorak, M. et al. (2008). Validation of a novel minimally invasive intervertebral disc pressure sensor utilizing in-fibre Bragg gratings in a porcine model: an *ex vivo* study. *Spine* 33: E589–E594.

47 Roriz, P., Abe, I., Schiller, M.W. et al. (2011). Ex vivo intervertebral disc bulging measurement using a fibre Bragg grating sensor. *Experimental Mechanics* 51 (9): 1573–1577.

48 Dennison, C.R., Wild, P.M., Byrnes, P.W. et al. (2008). *Ex vivo* measurement of lumbar intervertebral disc pressure using fibre-Bragg gratings. *Journal of Biomechanics* 41 (1): 221–225.

49 Dennison, C.R., Wild, P.M., Wilson, D.R., and Gilbart, M.K. (2010). An in-fibre Bragg grating sensor for contact force and stress measurements in articular joints. *Measurement Science and Technology* 21 (11): 115803.

50 Esmonde-White, K.A., Esmonde-White, F.W., Morris, M.D., and Roessler, B.J. (2011). Fibre-optic Raman spectroscopy of joint tissues. *Analyst* 136 (8): 1675–1685.

51 Baier, V., Marchi, G., Foehr, P. et al. (2017). Characterization of bovine cartilage by fibre Bragg grating-based stress relaxation measurements. In: *2017 25th Optical Fibre Sensors Conference (OFS)*, 1–4. IEEE.

52 Reikeras, O., Aarnes, G.T., Steen, H. et al. (2011). Differences in external and internal cortical strain with prosthesis in the femur. *The Open Orthopaedics Journal* 5: 379–384.

53 Komi, P., Belli, A., Huttunen, V. et al. (1996). Optic fibre as a transducer of tendomuscular forces. *European Journal of Applied Physiology and Occupational Physiology* 72 (3): 278–280.

54 Erdemir, A., Hamel, A.J., Piazza, S.J., and Sharkey, N.A. (2003). Fibreoptic measurement of tendon forces is influenced by skin movement artifact. *Journal of Biomechanics* 36 (3): 449–455.

55 Ren, L., Song, G., Conditt, M. et al. (2007). Fibre Bragg grating displacement sensor for movement measurement of tendons and ligaments. *Applied Optics* 46 (28): 6867–6871.

56 Behrmann, G.P., Hidler, J., and Mirotznik, M.S. (2012). Fibre optic micro sensor for the measurement of tendon forces. *Biomedical Engineering Online* 11 (1): 77.

57 Silva, J.C., Carvalho, L., Nogueira, R.N. et al. (2004). FBG applied in dynamic analysis of an implanted cadaveric mandible. In: *Second European Workshop on Optical Fibre Sensors*, vol. 5502, 226–230. International Society for Optics and Photonics.

58 Karam, L.Z., Milczewski, M.S., and Kalinowski, H.J. (2012). Strain monitoring of the periodontal ligament in pig's mandibles. In: *Proceedings of SPIE. OFS2012 22nd International Conference on Optical Fibre Sensors, 84215W*, vol. 8421. Beijing, China: SPIE.

59 Romanyk, D.L., Guan, R., Major, P.W., and Dennison, C.R. (2017). Repeatability of strain magnitude and strain rate measurements in the periodontal ligament using fibre Bragg gratings: an *ex vivo* study in a swine model. *Journal of Biomechanics* 54: 117–122.

60 Tjin, S.C., Tan, Y.K., Yow, M. et al. (2001). Recording compliance of dental splint use in obstructive sleep apnoea patients by force and temperature modelling. *Medical and Biological Engineering and Computing* 39 (2): 182–184.

61 Wosniak, C., Silva, W.J., Cardoso, R. et al. (2012). Determination of chewing patterns in goats using fibre Bragg gratings. In: *OFS2012 22nd International Conference on Optical Fibre Sensors, SPIE Proceedings*, vol. 8421, 84214F. Beijing, China: SPIE.

62 Karam, L.Z., Pegorini, V., Pitta, C.S.R. et al. (2014). *Ex vivo* determination of chewing patterns using FBG and artificial neural networks. In: *23rd International Conference on Optical Fibre Sensors, SPIE Proceedings*, vol. 9157, 91573Z. Santander, Spain: SPIE.

63 Agah, M.R., Laksari, K., and Darvish, K. (2012). Investigating the hyperelasticity of porcine aorta under sub-failure loading. In: *2012 38th Annual Northeast Bioengineering Conference (NEBEC)*, 432–433. IEEE.

64 Mo, Z., Xu, W., and Broderick, N. (2017). Epidural space identification exploration by a fibre optic tip-force sensing needle. In: *2017 24th International Conference on Mechatronics and Machine Vision in Practice (M2VIP)*, 1–5. IEEE.

65 Mo, Z., Xu, W., and Broderick, N.G. (2017). Capability characterization via *ex vivo* experiments of a fibre optical tip force sensing needle for tissue identification. *IEEE Sensors Journal* 18 (3): 1195–1202.

66 Hoejer, S., Krantz, M., Ekstroem, L. et al. (1999). Microstructure-based fibre optic pressure sensor for measurements in lumbar intervertebral discs. In: *Biomedical Sensors, Fibres, and Optical Delivery Systems*, vol. 3570, 115–123. International Society for Optics and Photonics.

67 Ferreira, L.A., Araújo, F.M., Mascarenhas, T. et al. (2006). Dynamic assessment of women pelvic floor function by using a fibre Bragg grating sensor system. In: *Optical Fibres and Sensors for Medical Diagnostics and Treatment Applications VI*, vol. 6083, 60830H. International Society for Optics and Photonics.

68 Arkwright, J.W., Blenman, N.G., Underhill, I. et al. (2009). *In vivo* demonstration of a high resolution optical fibre manometry catheter for diagnosis of gastrointestinal motility disorders. *Optics Express* 17 (6): 4500–4508.

69 Arkwright, J.W., Underhill, I.D., Maunder, S.A. et al. (2009). Design of a high-sensor count fibre optic manometry catheter for *in vivo* colonic diagnostics. *Optics Express* 17 (25): 22 423–22 431.

70 Finni, T., Komi, P., and Lukkariniemi, J. (1998). Achilles tendon loading during walking: application of a novel optic fibre technique. *European Journal of Applied Physiology and Occupational Physiology* 77 (3): 289–291.

71 Elvin, N., Elvin, A., Scheffer, C. et al. (2009). A preliminary study of patellar tendon torques during jumping. *Journal of Applied Biomechanics* 25 (4): 360–368.

72 Nascimento, P.F., Franco, A.P., Fiorin, R. et al. (2017). Characterization of the occlusal splints using optical fibre sensors. In: *2017 SBMO/IEEE MTT-S International Microwave and Optoelectronics Conference (IMOC)*, 1–4. IEEE.

73 Umesh, S., Padma, S., Asokan, S., and Srinivas, T. (2016). Fibre Bragg grating based bite force measurement. *Journal of Biomechanics* 49 (13): 2877–2881.

74 Karam, L.Z., Kalinowski, A., Pegorini, V. et al. (2015). *In vivo* analysis of bone strain using fibre Bragg grating sensor and decision tree algorithm in bovine during masticatory movements. *International Microwave and Optoelectronics Conference* (3–6 November 2015). Porto de Galinhas, Pernambuco, Brazil: IMOC.

75 Pegorini, V., Karam, L.Z., Pitta, C.S.R. et al. (2015). *In vivo* pattern classification of ingestive behavior in ruminants using FBG sensors and machine learning. *Sensors* 15: 28456–28471.

76 Kalinowski, A., Karam, L.Z., Pegorini, V. et al. (2017). Optical fibre Bragg grating strain sensor for bone stress analysis in bovine during masticatory movements. *IEEE Sensors Journal* 17 (8): 2385–2392.

77 Hansen, T.-E. (1983). A fibreoptic micro-tip pressure transducer for medical applications. *Sensors and Actuators* 4: 545–554.

78 Totsu, K., Haga, Y., and Esashi, M. (2004). Ultra-miniature fibre-optic pressure sensor using white light interferometry. *Journal of Micromechanics and Microengineering* 15 (1): 71.

79 Tian, Y., Wu, N., Zou, X. et al. (2013). A study on packaging of miniature fibre optic sensors for *in vivo* blood pressure measurements in a swine model. *IEEE Sensors Journal* 14 (3): 629–635.

80 Cavaiola, C., Saccomandi, P., Massaroni, C. et al. (2016). Error of a temperature probe for cancer ablation monitoring caused by respiratory movements: *ex vivo* and *in vivo* analysis. *IEEE Sensors Journal* 16 (15): 5934–5941.

81 Wise, S., Gardner, W., Sabelman, E. et al. (1990). Evaluation of a fibre optic glove for se-automated goniometric measurements. *Journal of Rehabilitation Research and Development* 27 (4).

82 Nishiyama, M. and Watanabe, K. (2009). Wearable sensing glove with embedded hetero-core fibre-optic nerves for unconstrained hand motion capture. *IEEE Transactions on Instrumentation and Measurement* 58 (12): 3995–4000.

83 da Silva, A.F., Goncalves, F., Mendes, P., and Correia, J. (2011). FBG sensing glove for monitoring hand posture. *IEEE Sensors Journal* 11 (10): 2442–2448.

84 Fujiwara, E., dos Santos, M.F.M., and Suzuki, C.K. (2014). Flexible optical fibre bending transducer for application in glove-based sensors. *IEEE Sensors Journal* 14 (10): 3631–3636.

85 Bell, J. and Stigant, M. (2007). Development of a fibre optic goniometer system to measure lumbar and hip movement to detect activities and their lumbar postures. *Journal of Medical Engineering & Technology* 31 (5): 361–366.

86 Bell, J.A. and Stigant, M. (2008). Validation of a fibre-optic goniometer system to investigate the relationship between sedentary work and low back pain. *International Journal of Industrial Ergonomics* 38 (11–12): 934–941.

87 Cloud, B.A., Zhao, K.D., Ellingson, A.M. et al. (2017). Increased seat dump angle in a manual wheelchair is associated with changes in thoracolumbar lordosis and scapular kinematics during propulsion. *Archives of Physical Medicine and Rehabilitation* 98 (10): 2021–2027.

88 Cloud, B.A., Zhao, K.D., Breighner, R. et al. (2014). Agreement between fibre optic and optoelectronic systems for quantifying sagittal plane spinal curvature in sitting. *Gait & Posture* 40 (3): 369–374.

89 Koyama, Y., Nishiyama, M., and Watanabe, K. (2011). Wearable motion capturing with the flexing and turning based on a hetero-core fibre optic stretching sensor. In: *21st International Conference on Optical Fibre Sensors*, vol. 7753, 77534B. International Society for Optics and Photonics.

90 Koyama, Y., Nishiyama, M., and Watanabe, K. (2013). A motion monitor using hetero-core optical fibre sensors sewed in sportswear to trace trunk motion. *IEEE Transactions on Instrumentation and Measurement* 62 (4): 828–836.

91 Nishiyama, M., Sasaki, H., and Watanabe, K. (2007). Restraint-free wearable sensing clothes using a hetero-core optic fibre for measurements of arm motion and walking action. In: *Sensors and Smart Structures Technologies for Civil, Mechanical,* *and Aerospace Systems 2007*, vol. 6529, 65291Y. International Society for Optics and Photonics.

92 Donno, M., Palange, E., Di Nicola, F. et al. (2008). A new flexible optical fibre goniometer for dynamic angular measurements: application to human joint movement monitoring. *IEEE Transactions on Instrumentation and Measurement* 57 (8): 1614–1620.

93 Stupar, D.Z., Bajic, J.S., Manojlovic, L.M. et al. (2012). Wearable low-cost system for human joint movements monitoring based on fibre-optic curvature sensor. *IEEE Sensors Journal* 12 (12): 3424–3431.

94 Kim, S.G., Jang, K.W., Yoo, W.J. et al. (2014). Feasibility study on fibre-optic goniometer for measuring knee joint angle. *Optical Review* 21 (5): 694–697.

95 Leal-Junior, A.G., Vargas-Valencia, L., dos Santos, W.M. et al. (2018). POF-IMU sensor system: a fusion between inertial measurement units and POF sensors for low-cost and highly reliable systems. *Optical Fibre Technology* 43: 82–89.

96 Rocha, R., Silva, A., Carmo, J., and Correia, J. (2011). FBG in PVC foils for monitoring the knee joint movement during the rehabilitation process. In: *Engineering in Medicine and Biology Society, EMBC, 2011 Annual International Conference of the IEEE*, 458–461. Boston, MA, USA: IEEE.

97 Fujiwara, E., Wu, Y.T., Santos, M.F. et al. (2015). Identification of hand postures by force myography using an optical fibre specklegram sensor. In: *24th International Conference on Optical Fibre Sensors*, vol. 9634, 96343Z. International Society for Optics and Photonics.

98 Fujiwara, E., Wu, Y.T., Santos, M.F.M. et al. (2017). Optical fibre specklegram sensor for measurement of force myography signals. *IEEE Sensors Journal* 17 (4): 951–958.

99 Fujiwara, E., Wu, Y.T., Suzuki, C.K. et al. (2018). Optical fibre force myography sensor for applications in prosthetic hand control. In: *2018 IEEE 15th International Workshop on Advanced Motion Control (AMC)*, 342–347. IEEE.

100 Guo, J., Niu, M., and Yang, C. (2017). Highly flexible and stretchable optical strain sensing for human motion detection. *Optica* 4 (10): 1285–1288.

101 Karam, L., Patyk, R., Saccon, F., and Kalinowski, H. (2011). Walk frequency measurements using a fibre optic Bragg grating sensor. In: *V Latin American Congress on Biomedical Engineering CLAIB 2011*, 128–131. Habana: Springer.

102 Prasad, A.S.G., Omkar, S.J.N., Anand, K. et al. (2013). Evaluation of airline exercises prescribed to avoid deep vein thrombosis using fibre Bragg grating sensors. *Journal of Biomedical Optics* 18 (9): 097007.

103 Al-Fakih, E.A., Osman, N.A.A., Adikan, F.R.M. et al. (2016). Development and validation of fibre Bragg grating sensing pad for interface pressure measurements within prosthetic sockets. *IEEE Sensors Journal* 16 (4): 965–974.

104 Galvão, J.R., Zamarreño, C.R., Martelli, C. et al. (2017). Strain mapping in carbon-fibre prosthesis using optical fibre sensors. *IEEE Sensors Journal* 17 (1): 3–4.

105 de Bastos, T.P., Galvão, J.R., Martelli, C., and da Silva, J.C.C. (2018). Smart carbon fibre foot prosthesis. In: *Specialty Optical Fibres*, JTu2A–17. Optical Society of America.

106 SpillmanJr, W., Mayer, M., Bennett, J. et al. (2004). A 'smart' bed for non-intrusive monitoring of patient physiological factors. *Measurement Science and Technology* 15 (8): 1614.

107 Hao, J., Jayachandran, M., Kng, P.L. et al. (2010). FBG-based smart bed system for healthcare applications. *Frontiers of Optoelectronics in China* 3 (1): 78–83.

108 Nishyama, M., Miyamoto, M., and Watanabe, K. (2011). Respiration and body movement analysis during sleep in bed using hetero-core fibre optic pressure sensors without constraint to human activity. *Journal of Biomedical Optics* 16 (1): 017002.

109 Sprager, S. and Zazula, D. (2013). Detection of heartbeat and respiration from optical interferometric signal by using wavelet transform. *Computer Methods and Programs in Biomedicine* 111 (1): 41–51.

110 Dziuda, L., Skibniewski, F., Rozanowski, K. et al. (2011). Fibre-optic sensor for monitoring respiration and cardiac activity. In: *Sensors, 2011 IEEE*, 413–416. IEEE.

111 Zyczkowski, M., Uzieblo-Zyczkowska, B., Dziuda, L., and Rozanowski, K. (2011). Using modal-metric fibre optic sensors to monitor the activity of the heart. In: *Optical Fibres, Sensors, and Devices for Biomedical Diagnostics and Treatment XI*, vol. 7894, 789404. International Society for Optics and Photonics.

112 Dziuda, L., Skibniewski, F.W., Krej, M., and Lewandowski, J. (2012). Monitoring respiration and cardiac activity using fibre Bragg grating-based sensor. *IEEE Transactions on Biomedical Engineering* 59 (7): 1934–1942.

113 Chen, Z., Teo, J.T., Ng, S.H., and Yim, H. (2011). Smart pillow for heart-rate monitoring using a fibre optic sensor. In: *Optical Fibres, Sensors, and Devices for Biomedical Diagnostics and Treatment XI*, vol. 7894, 789402. International Society for Optics and Photonics.

114 Dziuda, L., Skibniewski, F.W., Krej, M., and Baran, P.M. (2013). Fibre Bragg grating-based sensor for monitoring respiration and heart activity during magnetic resonance imaging examinations. *Journal of Biomedical Optics* 18 (5): 057006.

115 Chen, Z., Lau, D., Teo, J.T. et al. (2014). Simultaneous measurement of breathing rate and heart rate using a microbend multimode fibre optic sensor. *Journal of Biomedical Optics* 19 (5): 057001.

116 Prasad, A.G., Omkar, S., Vikranth, H. et al. (2014). Design and development of fibre Bragg grating sensing plate for plantar strain measurement and postural stability analysis. *Measurement* 47: 789–793.

117 Otsuka, Y., Koyama, Y., and Watanabe, K. (2014). Monitoring of plantar pressure in gait based on hetero-core optical fibre sensor. *Procedia Engineering* 87: 1465–1468.

118 Domingues, M.F., Tavares, C., Leitão, C. et al. (2017). Insole optical fibre Bragg grating sensors network for dynamic vertical force monitoring. *Journal of Biomedical Optics* 22 (9): 091507.

119 Leal-Junior, A.G., Frizera, A., Avellar, L.M. et al. (2018). Polymer optical fibre for in-shoe monitoring of ground reaction forces during the gait. *IEEE Sensors Journal* 18 (6): 2362–2368.

120 Hao, J., Tan, K., Tjin, S. et al. (2003). Design of a foot-pressure monitoring transducer for diabetic patients based on FBG sensors. In: *The 16th Annual Meeting of the IEEE Lasers and Electro-Optics Society, 2003. LEOS 2003*, vol. 1, 23–24. IEEE.

121 Domingues, M.F., Alberto, N., Leitão, C. et al. (2017). Insole optical fibre sensor architecture for remote gait analysis – an ehealth solution. *IEEE Internet of Things Journal.*

122 Domingues, M.F., Tavares, C., Alberto, N. et al. (2017). Non-invasive insole optical fibre sensor architecture for monitoring foot anomalies. In: *GLOBECOM 2017-2017 IEEE Global Communications Conference*, 1–6. IEEE.

123 Tavares, C., Domingues, M., Frizera-Neto, A. et al. (2018). Gait shear and plantar pressure monitoring: a non-invasive OFS based solution for e-health architectures. *Sensors* 18 (5): 1334.

124 Galvão, J.R., Di Renzo, A.B., Schaphauser, P.E. et al. (2018). Optical fibre Bragg grating instrumentation applied to horse gait detection. *IEEE Sensors Journal* 18 (14): 5778–5785.

125 Li, J.-J., Zhu, Y., and Drzewiecki, G. (1996). Fibre-optic sensor applications to non-invasive cardiovascular diagnosis. In: *Professional Program Proceedings. ELECTRO'96*, 205–207. IEEE.

126 Miyauchi, Y., Koyama, S., and Ishizawa, H. (2013). Basic experiment of blood-pressure measurement which uses FBG sensors. In: *2013 IEEE International Instrumentation and Measurement Technology Conference (I2MTC)*, 1767–1770. IEEE.

127 Katsuragawa, Y. and Ishizawa, H. (2015). Non-invasive blood pressure measurement by pulse wave analysis using FBG sensor. In: *2015 IEEE International Instrumentation and Measurement Technology Conference (I2MTC) Proceedings*, 511–515. IEEE.

128 Pant, S., Umesh, S., Padma, S., and Asokan, S. (2016). Non invasive assessment of brachial arterial stiffness using fibre Bragg grating sensor. In: *2016 IEEE Conference on Recent Advances in Lightwave Technology (CRALT)*, 1–3. IEEE.

129 Leitao, C., Lima, H., Pinto, J.L. et al. (2012). Development and characterization of new sensors for hemodynamic evaluation: fibre Bragg sensor for arterial pulse waveform acquisition. In: *2012 IEEE 2nd Portuguese Meeting in Bioengineering (ENBENG)*, 1–4. IEEE.

130 Jia, D., Chao, J., Li, S. et al. (2018). A fibre Bragg grating sensor for radial artery pulse waveform measurement. *IEEE Transactions on Biomedical Engineering* 65 (4): 839–846.

131 Allsop, T., Lloyd, G., Bhamber, R.S. et al. (2014). Cardiac-induced localized thoracic motion detected by a fibre optic sensing scheme. *Journal of Biomedical Optics* 19 (11): 117006.

132 Chino, S., Ishizawa, H., Hosoya, S. et al. (2017). Research for wearable multiple vital sign sensor using fibre Bragg grating-verification of several pulsate points in human body surface. In: *2017 IEEE International Instrumentation and Measurement Technology Conference (I2MTC)*, 1–6. IEEE.

133 Suaste, E. and Avila, J.L. (1996). Respiration monitoring using a sensor based on mode-single fibre optic in the intensive care unit. In: *Fibre Optic and Laser Sensors XIV*, vol. 2839, 400–409. International Society for Optics and Photonics.

134 Davis, C., Mazzolini, A., Mills, J., and Dargaville, P. (1999). A new sensor for monitoring chest wall motion during high-frequency oscillatory ventilation. *Medical Engineering & Physics* 21 (9): 619–623.

135 Babchenko, A., Khanokh, B., Shomer, Y., and Nitzan, M. (1999). Fibre optic sensor for the measurement of the respiratory chest circumference changes. *Journal of Biomedical Optics* 4 (2): 224–230.

136 Augousti, A., Maletras, F.-X., and Mason, J. (2005). The use of a figure-of-eight coil for fibre optic respiratory plethysmography: geometrical analysis and experimental characterisation. *Optical Fibre Technology* 11 (4): 346–360.

137 Lau, D., Chen, Z., Teo, J.T. et al. (2013). Intensity-modulated microbend fibre optic sensor for respiratory monitoring and gating during MRI. *IEEE Transactions on Biomedical Engineering* 60 (9): 2655–2662.

138 Yoo, W.-J., Jang, K.-W., Seo, J.-K. et al. (2010). Development of respiration sensors using plastic optical fibre for respiratory monitoring inside MRI system. *Journal of the Optical Society of Korea* 14 (3): 235–239.

139 Kam, W., Mohammed, W.S., Leen, G. et al. (2017). Compact and low-cost optical fibre respiratory monitoring sensor based on intensity interrogation. *Journal of Lightwave Technology* 35 (20): 4567–4573.

140 Allsop, T.D., Earthrowl, T., Revees, R. et al. (2004). Application of long-period grating sensors to respiratory function monitoring. In: *Smart Medical and Biomedical Sensor Technology II*, vol. 5588, 148–157. International Society for Optics and Photonics.

141 Petrović, M., Petrovic, J., Simić, G. et al. (2013). A new method for respiratory-volume monitoring based on long-period fibre gratings. In: *2013 35th Annual International Conference of the IEEE Engineering in Medicine and Biology Society (EMBC)*, 2660–2663. IEEE.

142 Wehrle, G., Nohama, P., Kalinowski, H.J. et al. (2001). A fibre optic Bragg grating strain sensor for monitoring ventilatory movements. *Measurement Science and Technology* 12 (7): 805–809.

143 Zhang, C., Miao, C.-Y., Li, H.-Q. et al. (2009). Smart textile sensing system for human respiration monitoring based on fibre Bragg grating. In: *International Symposium on Photoelectronic Detection and Imaging 2009: Material and Device Technology for Sensors*, vol. 7381, 738104. International Society for Optics and Photonics.

144 D'Angelo, L., Weber, S., Honda, Y. et al. (2008). A system for respiratory motion detection using optical fibres embedded into textiles. In: *2008 30th Annual International Conference of the IEEE Engineering in Medicine and Biology Society*, 3694–3697. IEEE.

145 Allsop, T.D., Bhamber, R., Lloyd, G.D. et al. (2012). Respiratory function monitoring using a real-time three-dimensional fibre-optic shaping sensing scheme based upon fibre Bragg gratings. *Journal of Biomedical Optics* 17 (11): 117001.

146 Zyczkowski, M., Szustakowski, M., Ciurapinski, W., and Uzieblo-Zyczkowska, B. (2011). Interferometric fibre optics based sensor for monitoring of the heart activity. *Acta Physica Polonica A* 120 (4): 782–784.

147 Silva, A., Carmo, J., Mendes, P., and Correia, J. (2011). Simultaneous cardiac and respiratory frequency measurement based on a single fibre Bragg grating sensor. *Measurement Science and Technology* 22 (7): 075801.

148 Presti, D.L., Massaroni, C., Formica, D. et al. (2017). Smart textile based on 12 fibre Bragg gratings array for vital signs monitoring. *IEEE Sensors Journal* 17 (18): 6037–6043.

149 Suaste-Gómez, E., Hernández-Rivera, D., Sánchez-Sánchez, A., and Villarreal-Calva, E. (2014). Electrically insulated sensing of respiratory rate and heartbeat using optical fibres. *Sensors* 14 (11): 21 523–21 534.

150 Allsop, T.D., Carroll, K., Lloyd, G. et al. (2007). Application of long-period-grating sensors to respiratory plethysmography. *Journal of Biomedical Optics* 12 (6): 064003.

151 Dario, P., Femi, D., and Vivaldi, F. (1987). Fibre-optic catheter-tip sensor based on the photoelastic effect. *Sensors and Actuators* 12 (1): 35–47.

152 Hong, L., Prohaska, O.J., and Nara, A.R. (1988). Fibre-optic transducer for blood pressure measurements. In: *Proceedings of the Annual International Conference of the IEEE Engineering in Medicine and Biology Society*, 810–811. IEEE.

153 Singh, M., Li, J., Sigel, G., and Amory, D. (1990). Fibre optic pulse sensor for noninvasive cardiovascular applications. In: *Sixteenth Annual Northeast Conference on Bioengineering*, 103–104. IEEE.

154 Lindberg, L.-G., Ugnell, H., and Oberg, P. (1992). Monitoring of respiratory and heart rates using a fibre-optic sensor. *Medical and Biological Engineering and Computing* 30 (5): 533–537.

155 Augousti, A.T., Raza, A., and Graves, M. (1996). Design and characterization of a fibre optic respiratory plethysmograph (FORP). In: *Biomedical Sensing, Imaging, and Tracking Technologies I*, vol. 2676, 250–258. International Society for Optics and Photonics.

156 Munoz, R., Leija, L., Diaz, F., and Alvarez, J. (1995). 3D continuous monitoring system for localization of upper limb based on optical fibre. In: *Proceedings of 17th International Conference of the Engineering in Medicine and Biology Society*, vol. 2, 1595–1596. IEEE.

157 Arndt, A., Komi, P., Brüggemann, G.-P., and Lukkariniemi, J. (1998). Individual muscle contributions to the in vivo achilles tendon force. *Clinical Biomechanics* 13 (7): 532–541.

158 Komi, P.V. (2000). Stretch-shortening cycle: a powerful model to study normal and fatigued muscle. *Journal of Biomechanics* 33 (10): 1197–1206.

159 Finni, T., Komi, P.V., and Lepola, V. (2000). *In vivo* human triceps surae and quadriceps femoris muscle function in a squat jump and counter movement jump. *European Journal of Applied Physiology* 83 (4–5): 416–426.

160 Wehrle, G. (2000). Fibre optic Bragg grating strain sensor used to monitor the respiratory spectrum. In: *Fourteenth International Conference on Optical Fibre Sensors*, vol. 4185, 41854R. International Society for Optics and Photonics.

161 Finni, T., Komi, P.V., and Lepola, V. (2001). *In vivo* muscle mechanics during locomotion depend on movement amplitude and contraction intensity. *European Journal of Applied Physiology* 85 (1–2): 170–176.

162 Wang, W.-C., Ledoux, W.R., Sangeorzan, B.J., and Reinhall, P.G. (2005). A shear and plantar pressure sensor based on fibre-optic bend loss. *Journal of Rehabilitation Research & Development* 42 (3).

163 Augousti, A., Maletras, F., and Mason, J. (2005). Improved fibre optic respiratory monitoring using a figure-of-eight coil. *Physiological Measurement* 26 (5): 585.

164 Nishiyama, M., Sasaki, H., and Watanabe, K. (2006). Wearable sensing clothes embedding a hetero-core optic fibre for recognizing arm segment posture and motion. In: *Sensors*, 1519–1522. IEEE.

165 Carvalho, L.M.R., da Silva, J.C.C., Nogueira, R.N. et al. (2006). Application of Bragg grating sensors in dental biomechanics. *The Journal of Strain Analysis for Engineering Design* 41 (6): 411–416.

166 Jeanne, M., Grillet, A., Weber, S. et al. (2007). Ofseth: optical fibre embedded into technical textile for healthcare, an efficient way to monitor patient under magnetic resonance imaging. In: *2007 29th Annual International Conference of the IEEE Engineering in Medicine and Biology Society*, 3950–3953. IEEE.

167 Nesson, S.C., Yu, M., Zhang, X., and Hsieh, A.H. (2008). Miniature fibre optic pressure sensor with composite polymer-metal diaphragm for intradiscal pressure measurements. *Journal of Biomedical Optics* 13 (4): 044040.

168 Yoo, W., Jang, K., Seo, J. et al. (2010). Development of nasal-cavity-and abdomen-attached fibre-optic respiration sensors. *Journal of the Korean Physical Society* 57: 1550–1554.

169 Sprager, S. and Zazula, D. (2012). Heartbeat and respiration detection from optical interferometric signals by using a multimethod approach. *IEEE Transactions on Biomedical Engineering* 59 (10): 2922–2929.

170 Rao, S.S. (2017). *The Finite Element Method in Engineering*. Butterworth-Heinemann.

171 Hall, S.J. (2015). *Basic Biomechanics*. New York: McGraw-Hill Education.

172 Wolff, J. (1892). *The Law of Bone Remodelling*, vol. 1986. Berlin-Heidelberg, German: Springer-Verlag, Translation of the German edition.

173 Bayliss, L., Mahoney, D.J., and Monk, P. (2011). Normal bone physiology, remodelling and its hormonal regulation. *Surgery – Oxford International Edition* 30 (2): 47–53.

174 Milczewski, M.S., Kalinowski, H.J., da Silva, J.C.C. et al. (2011). Stress monitoring in a maxilla model and dentition. In: *21st International Conference on Optical Fibre Sensors*, vol. 7753, 77534V. International Society for Optics and Photonics.

175 Paterno, L., Ibrahimi, M., Gruppioni, E. et al. (2018). Sockets for limb prostheses: a review of existing technologies and open challenges. *IEEE Transactions on Biomedical Engineering* 65 (9): 1996–2010.

176 Ding, L., Tong, X., and Yu, L. (2017). Quantitative method for gait pattern detection based on fibre Bragg grating sensors. *Journal of Biomedical Optics* 22 (3): 037005.

177 Yang, X., Chen, Z., Elvin, C.S.M. et al. (2014). Textile fibre optic microbend sensor used for heartbeat and respiration monitoring. *IEEE Sensors Journal* 15 (2): 757–761.

178 Toriumi, D.M., Raslan, W.F., Friedman, M., and Tardy, M.E. (1990). Histotoxicity of cyanoacrylate tissue adhesives: a comparative study. *Archives of Otolaryngology–Head & Neck Surgery* 116 (5): 546–550.

179 Zhang, C.-H., Song, X.-M., He, Y.-L. et al. (2012). Use of absorbable hemostatic gauze with medical adhesive is effective for achieving hemostasis in presacral hemorrhage. *The American Journal of Surgery* 203 (4): e5–e8.

180 Bugden, S., Shean, K., Scott, M. et al. (2016). Skin glue reduces the failure rate of emergency department–inserted peripheral intravenous catheters: a randomized controlled trial. *Annals of Emergency Medicine* 68 (2): 196–201.

181 Callister, W.D. and Rethwisch, D.G. (2011). *Materials Science and Engineering*. New York: Wiley.

182 Milczewski, M.S., da Silva, J.C.C., Abe, I. et al. (2006). Determination of setting expansion of dental materials using fibre optical sensing. *Measurement Science and Technology* 17 (5): 1152–1156.

183 Milczewski, M., Silva, J., Paterno, A. et al. (2007). Measurement of composite shrinkage using a fibre optic Bragg grating sensor. *Journal of Biomaterials Science, Polymer Edition* 18 (4): 383–392.

184 Kinet, D., Mégret, P., Goossen, K. et al. (2014). Fibre Bragg grating sensors toward structural health monitoring in composite materials: challenges and solutions. *Sensors* 14 (4): 7394–7419.

185 Galvão, J.R., Zamarreño, C.R., Martelli, C. et al. (2017). Smart carbon fibre transtibial prosthesis based on embedded fibre Bragg gratings. *IEEE Sensors Journal* 18 (4): 1520–1527.

186 Chen, Z., Teo, J.T., Ng, S.H., and Yang, X. (2012). Portable fibre optic ballistocardiogram sensor for home use. In: *Optical Fibres and Sensors for Medical Diagnostics and Treatment Applications XII*, vol. 8218, 82180X. International Society for Optics and Photonics.

187 Miyauchi, Y., Ishizawa, H., Koyama, S., and Sato, S. (2012). Verification of the systolic blood-pressure measurement principle by FBG sensors. In: *2012 Proceedings of SICE Annual Conference (SICE)*, 619–622. IEEE.

188 Fujiwara, E., Wu, Y.T., Santos, M.F. et al. (2015). Development of an optical fibre FMG sensor for the assessment of hand movements and forces. In: *2015 IEEE International Conference on Mechatronics (ICM)*, 176–181. IEEE.

8

Optical Fibre Chemical Sensors

T. Hien Nguyen and Tong Sun

Photonics and Instrumentation Research Centre, City University of London, London, UK

8.1 Introduction

Chemical sensing is a process where chemical information (composition, concentration, chemical activity, etc.) of a species in a sample of interest is transformed into an analytically useful (e.g. electrical) output signal. Generally, two main steps are involved in the functioning of a chemical sensor: recognition and transduction. In the recognition step, the analyte interacts selectively with the recognition element of the sensor, causing a change in one or more physicochemical parameters. This variation is then transduced into the output signal that is amplified and monitored (Figure 8.1). Ideally, the sensing measurement is performed in a continuous and reversible way and in remote conditions if necessary, looking to the ideal chemical sensing system with its ability to provide in real time the spatial and temporal distributions of a molecular or ionic species.

Fibre-optic chemical sensing is a subclass of optical chemical sensing in which the radiation guided by the optical fibre is modulated in one or more of its characteristics by the measurand, with the modulated light carried out to the detection and processing unit by the same or a second fibre. Therefore, this sensing approach eliminates the need of dedicated power and telemetry channels. Generally, fibre-optic chemical sensors are based on either direct or indirect sensing scheme. In the first one, a spectroscopically physical property of the analyte can be measured directly through its interaction with the fibre-guided optical field, not requiring a specific chemical recognition element. In the second one, a recognition element

Optical Fibre Sensors: Fundamentals for Development of Optimized Devices, First Edition.
Edited by Ignacio Del Villar and Ignacio R. Matias.
© 2021 The Institute of Electrical and Electronics Engineers, Inc.
Published 2021 by John Wiley & Sons, Inc.

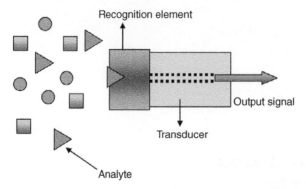

Figure 8.1 Schematic representation of a chemical sensor.

is used to generate an analyte-dependent spectroscopically detectable signal within the sensing region of the fibre. The chemical changes induced by the interactions of the analyte with the immobilized reagents are measured spectroscopically by analysing the radiation that returns from the sensing head.

In this chapter, various approaches to fibre-optic-based chemical sensing are described. The first part summarizes the basic principles and mechanisms of each approach, while the second part describes specific applications in detail, which highlights the critical issues associated with this sensing technology.

8.2 Principles and Mechanisms of Fibre-Optic-Based Chemical Sensing

In this section, the basic principles and mechanisms of main approaches to fibre-optic chemical sensing including absorption, luminescence, and surface plasmon resonance (SPR) are outlined. There exist also chemical sensors based on interferometers and gratings, but they are described in Chapter 4, where the reader can find information about the sensing principles.

8.2.1 Principle of Chemical Sensor Response

The response of an optical chemical sensor depends on the interactions between the analyte and the reagent [1]. When a sensing reagent (R) reacts with an analyte species (A), a product (AR) is formed; thus the process can be represented as

$$A + R \leftrightarrow AR \tag{8.1}$$

where R or AR is usually absorbing or luminescent to enable the target analyte A to be measured in an optical form. The reagent (sensor material) employed is required to be both analyte selective and analyte sensitive so that it can produce a distinctive optical signal change for the given analyte species.

The chemical transduction is normally based on the equilibrium established during the chemical reaction between A and R, and this equilibrium can be described as

$$K = \frac{[AR]}{[A][R]} \qquad (8.2)$$

where K is the equilibrium constant and the square brackets indicate the equilibrium concentration of the species involved. The change in absorption or luminescence signal can be caused either by the decrease of the reagent R consumed in the chemical reaction or by the formation of product AR, which increases its absorption or fluorescence. Under both circumstances, the change in optical property of R or AR can be related to their concentration and, in turn, related to the concentration of the analyte A, causing changes in the measured optical property.

Assuming the reagent is consumed during the reaction, its total initial concentration (c_R) at any time will be given by

$$c_R = [AR] + [R] \qquad (8.3)$$

Thus, from Eqs. (8.2) and (8.3), the analyte A concentration can be expressed as

$$[A] = \frac{1}{K} \times \left(\frac{c_R}{[R]} - 1 \right) \qquad (8.4)$$

If the optical property of AR is being measured, the analyte concentration can thus be related to the concentration of AR as follows:

$$\frac{1}{[A]} = K \times \left(\frac{c_R}{[AR]} - 1 \right) \qquad (8.5)$$

The concentrations of R in Eq. (8.4) and of AR in Eq. (8.5) are related to the measured optical property.

Based on the Eqs. (8.1), (8.4), and (8.5), the underpinning sensing mechanism is based on the use of a direct indicator (reagent R), which requires an appropriate equilibrium constant for the measurement of the desired analyte concentration range. The response of the sensor, however, is also dependent on the total amount of the indicator used. As a result, any uncontrolled variable that affects the equilibrium constant will be a potential source of error. For example, for any reaction involving ions, variations in ionic strength (IS) will affect the K values.

Indicators showing reversible reactions with analytes are generally preferred for use in optical sensors because they can provide continuous and unperturbed measurements. The equilibrium response of indicators does not depend on mass transfer; however, the response time (i.e. the time required to reach equilibrium) is dependent on mass transfer. Therefore, the sensors that do not involve chemical reactions to enable the reversible interactions of analytes with indicators usually provide much shorter response time.

If the measured optical parameter is dependent on the ratio of the concentrations of the two forms of the indicator, i.e. [AR]/[R], the response no longer depends on the total amount of the indicator, although the dependence on the equilibrium constant remains. As a result, measurements of the ratio of optical intensities at two wavelengths can be used to determine the analyte concentration. Such ratio measurements are inherently more stable and thus more favoured for long-term measurements as it is able to minimize the drift caused by the light source fluctuation or the environmental influence. This type of ratio measurements can be described as follows by rearranging Eq. (8.2):

$$[A] = \frac{1}{K} \times \frac{[AR]}{[R]} \tag{8.6}$$

A range of 'pre-calibrated' sensors based on ratiometric intensity measurements have been widely reported for various applications.

8.2.2 Absorption-Based Sensors

Many molecules absorb ultraviolet (UV) or visible light, and different molecules absorb radiation of different wavelengths. Absorption-based optical sensors can be *colorimetric* or *spectroscopic* in nature. Colorimetric sensors are based upon detection of an analyte-induced colour change in the sensor material, while spectroscopic absorption-based sensors rely on detection of the analyte by probing its intrinsic molecular absorption [2, 3].

Absorption of optical energy arises from transitions in the electronic, vibrational, and/or rotational energy states of the atoms and molecules and occurs only if the difference in the energy states involved matches exactly the same amount of energy of the exciting photons. Visible and UV radiation induces electronic excitation, infrared radiation promotes vibrational excitation, and microwave radiation gives rise to rotational transitions.

Thus absorption leads to a loss of the power of the radiation as it passes through the target sample. Therefore, after encountering a number of absorbing species, a light beam of initial intensity, I_0, will be transmitted by the sample with a reduced intensity I. It should be noted that only those frequencies that are absorbed will be attenuated and all other frequencies will pass through with no power loss. The decrease in the light intensity is determined by the number of absorbing species in the light path and is related to the concentration, c, of the absorbing species through the Beer–Lambert equation:

$$A_b = \log\left(\frac{I_0}{I}\right) = \varepsilon lc \tag{8.7}$$

where A_b is the absorbance, l is the length of the light path, and ε is the molar absorptivity, which is characteristic of the analyte substance at a given wavelength.

Optical fibre chemical sensors can be designed to enable the guided light in the fibre to interact with the target analyte, therefore inducing either the colour change on the sensor material (e.g. indicator) or the change of intensity of the light as a result of the spectroscopic absorption expressed in Eq. (8.7).

8.2.3 Luminescence-Based Sensors

When a molecule absorbs light energy, electrons are promoted from the ground state (S_0) to excited states; this phenomenon is called excitation. Subsequently, electrons drop in energy from the electronically excited states to the ground state. This may involve emission of light. This phenomenon is called luminescence (Figure 8.2). Luminescence is formally divided into two categories, fluorescence and phosphorescence, depending on the nature of the excited state. Phosphorescence is emission of light from triplet excited states, in which the unpaired electron in the excited orbital has the same spin orientation as the ground state unpaired electron. Transitions to the ground state are forbidden, and the emission rates are slow (10^3–10^0 s^{-1}) so that phosphorescence lifetimes are typically milliseconds to seconds. Fluorescence is emission of light from excited singlet states, in which the unpaired electron in the excited orbital is paired (of opposite spin) to the unpaired electron in the ground state orbital. Consequently, return to the ground state is

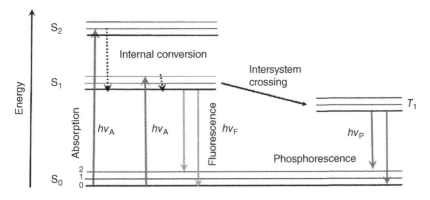

Figure 8.2 Partial energy diagram for a photoluminescence system.

spin allowed and occurs rapidly by emission of a photon. The emission rates of fluorescence are typically $10^8 \, s^{-1}$, so a typical fluorescence lifetime is near 10 ns.

In both cases, the emitted radiation is of a different frequency to that of the exciting radiation, and its intensity, I_L, is dependent on the intensity of the exciting radiation, I_0, and the concentration, c, of the luminescent species.

For weakly absorbing species, i.e. $A_b < 0.05$, the intensity of luminescence can be expressed by the following equation:

$$I_L = k' I_0 \theta \varepsilon l c \tag{8.8}$$

where l is the length of the light path in the sample, ε is the molar absorptivity, θ is the quantum efficiency of the luminescence, and k' is the fraction of the emission that can be measured. When the excitation radiation I_0 is a constant, Eq. (8.8) can be simplified to be

$$I_L = k_L c \tag{8.9}$$

where $k_L = k' I_0 \theta \varepsilon l$.

In the presence of some species (e.g. oxygen), the luminescence decay of an activated species could compete with a collisional quenching decay mode. The mean lifetime of the activated species is decreased, and the luminescence intensity is reduced. In this case, the luminescence intensity, I_L, is related to the concentration of the quenching species, c_q, by the Stern–Volmer equation:

$$I_0/I_L = 1 + K_{SV} c_q \tag{8.10}$$

where I_0 is the luminescence intensity in the absence of the quencher and K_{SV} is the Stern–Volmer constant.

In addition to the luminescence quenching mechanism discussed above, excited species can also be generated through a chemical reaction. The measured light that is emitted as the excited species returns to the ground state is then known as chemiluminescence, which can be quantitatively related to the concentration of the analyte species.

Due to the different ways of measuring absorbance (where light is measured as the difference in intensity between light passing through the reference and the sample) and fluorescence (where the intensity is measured directly), fluorescence is a lot more sensitive than absorption and is a preferred technique for high-sensitivity detection.

Fluorescence-based optical fibre sensors can be realized through the immobilization of analyte-sensitive fluorescence dye onto the fibre platform. One of the possible sensor designs is to replace a portion of a fibre cladding with a solid matrix that contains the analyte-sensitive fluorescent compound. When the evanescent wave ('leaked' from the section where the fibre cladding is removed) interacts with the target analyte, the fluorescence signal will change accordingly. Alternatively, the analyte-sensitive fluorescent compound can be immobilized directly onto the fibre end surface to form a fibre sensor tip. When the sensor probe interacts with the target analyte, the fluorescence signal generated can be directly collected by the fibre, which also sends the 'excitation' light.

8.2.4 Surface Plasmon Resonance (SPR)-Based Sensors

SPR is a resonance phenomenon where the resonant oscillation of conduction electrons at the interface of two media with dielectric constants of opposite signs, for instance, a metal and a dielectric, is stimulated by incident light [4, 5]. The charge density wave is associated with an electromagnetic (EM) wave, the field vectors of which reach their maxima at the interface and decay evanescently into both media. This surface plasma wave (SPW) is a TM-polarized wave (magnetic vector is perpendicular to the direction of propagation of the SPW and parallel to the plane of interface). The propagation constant of the SPW propagating at the interface between a semi-infinite dielectric and metal is given by the following expression:

$$\beta = k\sqrt{\frac{\varepsilon_m n_s^2}{\varepsilon_m + n_s^2}} \tag{8.11}$$

where k is the free-space wavenumber, ε_m is the dielectric constant of the metal, and n_s is the refractive index of the dielectric.

At the resonance angle θ_{sp} (the angle at which coupling of energy occurs between the incident light and the surface plasmon waves), the propagation constant of the incident beam parallel to the prism base is equal to the real part of the SPR propagation constant. This equality condition is expressed by

$$\beta = k n_p \sin \theta_{sp} \tag{8.12}$$

where n_p is the refractive index of the prism.

Equations (8.11) and (8.12) provide for the theoretical transduction mechanism for SPR as a sensor. Specifically, an increase in refractive index of the sensed dielectric will cause a shift in the resonance spectrum towards larger SPR coupling angles. By measuring the SPR resonance parameters, one can determine the complex refractive index of the dielectric using the calculated SPR propagation constant, β.

Traditionally, SPR is measured using the Kretschmann configuration (Figure 8.3a). Currently, optical fibre SPR probes present the highest level of miniaturization of SPR devices, allowing for chemical and biological sensing in inaccessible locations where the mechanical flexibility and the ability to transmit optical signals over a long distance make the use of optical fibres very attractive.

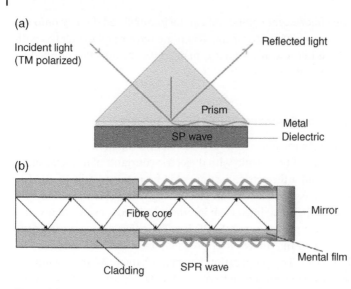

Figure 8.3 (a) The Kretschmann configuration. (b) The fibre-optic configuration.

In the fibre-optic configuration (Figure 8.3b), the light coupling prism is not required. The sensing element is a segment of fibre in which the cladding is removed and a metal film or a layer of metal nanoparticles (localized surface plasmon resonance [LSPR]) is symmetrically deposited on the fibre core. The length of the cladding removed is calculated to optimize the average number of reflections for all the propagating rays of light in the fibre. Unlike traditional SPR measurements, which employ a discrete excitation wavelength while modulating the angle of incidence, the SPR fibre-optic sensor system uses a white light source, resulting in a large range of excitation wavelengths. The range of incident angles is limited to only those angles that propagate in the optical fibre. The sensed parameters can then be determined from the measured resonance spectrum in the transmitted spectral intensity distribution [5].

Differently from the absorption- and luminescence-based sensors, sensors based on SPR do not require the use of immobilized dyes (label-free sensors). However, it is essential that binding of the analyte to the receptor (usually the dielectric) causes a change in the refractive index of the dielectric for the sensing to happen.

Apart from SPRs, there exist other phenomena that can be used in label-free sensing platforms, such as LSPR [6], which is a specific type of SPR confined in a space comparable to or smaller than the wavelength, typically a nanoparticle, leading to enhancement of electric field in the proximity of the particle's surface, and evanescent wave sensors (also known and guided or lossy mode resonance sensors), which are based on the attenuation of the light transmission when a mode is guided in the thin film deposited on the optical fibre [7]. All this phenomena, along with SPRs, can also be used in lab on tip, i.e. the generation of an optical structure on the end face of an optical fibre, as it will be described in Chapter 9.

8.3 Sensor Design and Applications

In this section, examples of chemical sensor design and their respective underpinning sensing technologies projected to fulfil specific application requirements are presented. More specifically, fibre-optic sensing systems for pH, mercury, and cocaine will be described in detail.

8.3.1 Optical Fibre pH Sensors

Optical fibre pH sensors have been actively investigated in recent years because of their importance for both *in situ* and *in vivo* pH measurements in various aspects of scientific research and in a range of practical applications, in particular those where available conventional glass electrodes are not suitable [8–11]. As indicated before, optical fibre-based sensors have shown many advantageous characteristics such as small size, immunity to EM interference, remote sensing capability, resistance to chemicals, and biocompatibility [12, 13]. Several types of pH optrodes (the optical fibre analogue of electrode) have been proposed and demonstrated over the years. A small number of them are based on refractive index changes in certain parts of the fibre caused by pH-induced swelling/shrinking of a deposited nanostructured film, which lead to changes in the fibre transmission or reflection spectra [14–16]. This type of pH sensors can be highly accurate in principle. However, they can be influenced by temperature fluctuations, stress, vibrations, and certain chemical interferences. Therefore, the majority of pH optrodes have been developed to function through monitoring the changes in the absorbance or fluorescence properties of certain pH-sensitive indicators that are immobilized on/in proton-permeable solid substrates [17].

Although there have been a number of reports on the development of fibre-optic pH sensors in the literature, most of them were constructed to operate, with certain limitations, in the physiological or near-neutral pH region. Very limited research has been undertaken to explore the pH measurements at both extremes, i.e. either at the low or high pH region where the pH response of most glass pH electrodes is imperfect [18, 19]. The reasons for limited success with many previous designs are quite varied, but the use of appropriate pH indicators and the effective immobilization of the indicators are probably the key factors in the development of an optimum optical pH sensor, as they govern the lifetime and signal stability of the sensor. Poor immobilization tends to result in dye leaching and consequently a drifting of the calibration of the probe, which leads to the gradual breakdown of its useful sensing ability [17, 20]. Among several widely used immobilization methods are absorption or entrapment [13, 21, 22], layer-by-layer (LbL) electrostatic self-assembly [23, 24], and covalent binding [20, 25–29]. The covalent binding method can produce more reliable and durable sensors, as the indicators are virtually bonded to the substrate. Therefore, they are unlikely to leach out under normal conditions, although the fabrication process is relatively complicated and time consuming [30]. Under extreme conditions, however, the sensor reliability and durability are not just determined by the immobilization method used but also by other factors, such as the stability of the pH indicators themselves, the stability of the substrates, and the linking bonds between the fibre substrate and the sensor material. For example, the commonly used ester linkage and acid amide linkage are not very stable in acidic or alkaline aqueous conditions [9, 17].

In this section, different optical fibre pH sensors, created specifically for different pH ranges, are described. The new approach aims to overcome the limitations of the existing sensor technologies highlighted above, thus creating more stable and therefore more useful devices.

8.3.1.1 Principle of Fluorescence-Based pH Measurements

The development of the present pH optrodes is basically based on the fluorometric determination of pH. It makes use of a fluorescent dye as an indicator, HA, to induce pH-sensitive changes in the measured fluorescence intensity. In aqueous solution, the following equilibrium can be reached:

$$HA \leftrightarrow H^+ + A^-$$

The relationship between the protonation state of the indicator and the pH is governed by the Henderson–Hasselbalch equation:

$$pH = pK_a + \log \frac{[A^-]}{[HA]} \tag{8.13}$$

where $[A^-]$ and $[HA]$ are the concentrations of the dissociated and undissociated forms of the indicator and pK_a is the acid–base constant. $[A^-]$ and $[HA]$ are related to fluorescence intensities by $[A^-] = F - F_{max}$ and $[HA] = F_{min} - F$, where F is a measured fluorescence intensity of the system, F_{max} is the fluorescence intensity of the fully protonated system, and F_{min} is the fluorescence intensity of the deprotonated system. The expressions are then substituted into Eq. (8.13) to provide Eq. (8.14):

$$pH = pK_a + \log \frac{F - F_{max}}{F_{min} - F} \tag{8.14}$$

Equation (8.14) can be rewritten in terms of F to give

$$F = \frac{F_{max} + F_{min} \times 10^{(pH - pK_a)}}{10^{(pH - pK_a)} + 1} \tag{8.15}$$

This results in an 'S-shaped' relation of the fluorescence intensity versus pH graph, centred on the pK_a value. Equation (8.15) is used as a model for a non-linear fitting method to calculate the pK_a value, which is the pH where 50% of the dye population in solution is protonated.

For polymer-bound dyes, due to the changes in the local micro-environment – the Boltzmann model – Eq. (8.16) has proven to give better fittings:

$$F = \frac{F_{max} + F_{min} \times e^{(pH - pK_a)/dpH}}{e^{(pH - pK_a)/dpH} + 1} \tag{8.16}$$

where dpH is the slope of the curve within its linear zone.

8.3.1.2 pH Sensor Design

As discussed above, the sensor reliability and durability are determined by many factors, which include the immobilization method used, the stability of the pH indicators themselves, the stability of the solid support, and the linkage between the fibre substrate and the sensor material. Thus, in the following pH sensor design, all these aspects have been taken into account.

8.3.1.2.1 Choice of Fluorescent Dyes

Although there has been a variety of pH indicators, both commercially available and reported in the literature, that are known to be useful for the spectroscopic determination of pH, the majority of them are absorptive dyes rather than fluorescent dyes, and only a few meet the requirements of being stable for a long period of time in extreme conditions such as highly alkaline or highly acidic media, having reasonable fluorescence quantum yield and good photostability, being fully compatible with LED light sources, and bearing one or more functional chemical groups suitable for covalent immobilization of the indicator. Novel polymerizable coumarin dyes bearing a carboxylic acid group (for low pH measurements) or an imidazole group (for high pH measurements) have been designed and synthesized. Coumarin-based indicators have been chosen for this application as coumarins are widely used as laser dyes for single-molecule fluorescence, and so they are 'tried and tested' in terms of the key property of being photostable [31, 32]. The dissociation of the carboxylic acid group allows for the determination of pH in the acidic region of the pH scale, which makes it suitable for gastric measurements [33, 34] and acidic soil measurements [35] as well as the measurement of pH in certain chemical reactors, whereas the protonation/deprotonation of the nitrogen on the imidazolyl group allows for the determination of pH in the alkaline region, making it suitable for monitoring pH changes in concrete structures. For the neutral pH region, acrylamidofluorescein (AAF) has been used as the pH indicator since the protonation/deprotonation of fluorescein occurs in the neutral pH range. The preparation of the fluorescent dyes is outlined in Figure 8.4, and their data are summarized in Table 8.1.

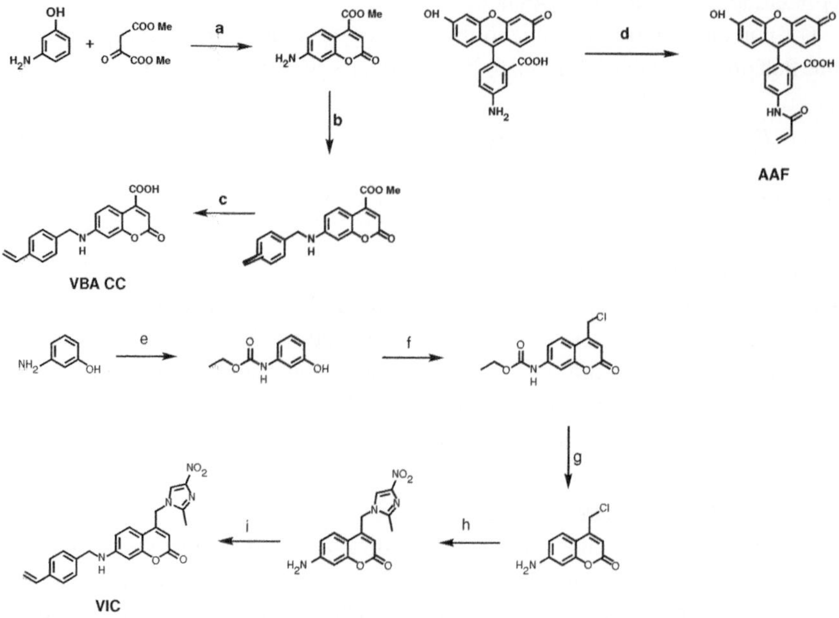

Figure 8.4 Preparation of fluorescent monomers VBACC, AAF, and VIC. (a) 110 °C, 2 hours, 42%; (b) CH$_2$=CHC$_6$H$_4$CH$_2$Cl, K$_2$CO$_3$, KI, MeCN, 85 °C, 50 hours, 44%; (c) 1 M NaOH, EtOH/THF (2 : 1), 50 °C, 12 hours, 94%; (d) acryloyl chloride, dry acetone; (e) ClCOOEt, EtOAc, 100 °C, 93%; (f) ethyl 4-chloroacetoacetate, H$_2$SO$_4$, H$_2$O, r.t, 19 hours, 71%: (g) conc. H$_2$SO$_4$, glacial AcOH, 125 °C, 2 hours, 87%; (h) NaH (60%), 2-methyl-4-nitroimidazole, DMF, 100–60 °C, 18 hours, 87%; (i) vinylbenzyl chloride, K$_2$CO$_3$, KI, MeCN, 80 °C, 2 days, 10%.

Table 8.1 Spectral data and pK_a values of the fluorescent dyes.

Dye	UV max (nm)	Emission max (nm)	pK_a	Φ (%)	Working pH range
VBACC	387	534	2.40 ± 0.05	14.6	0.5–6.0
AAF	450	515	6.45 ± 0.06	—	4.0–8.0
VIC	370	470	12.48 ± 0.06	0.8	10.0–13.2

8.3.1.2.2 *pH Sensor Probe Design and Fabrication*

Building on the successful synthesis of the fluorescent dyes shown in Figure 8.4, the next step in the development of the sensors was the creation of appropriate pH sensing probes incorporating the dye developed.

In the intrinsic probe design (low pH and neutral pH sensors), the dyes were covalently bound to the fibre surface by polymerization, in an approach similar to the method reported by Uttamlal et al. [20, 36], but allyltriethoxysilane (ATES) was used to functionalize the fibre surface with polymerizable groups rather than 3-(trimethoxysilyl)propyl methacrylate to avoid the unstable ester linkage. This requires a multistep process, and the fabrication of the pH sensing probes used in the work is shown schematically in Figure 8.5.

The distal end of a 1000 µm diameter UV multimode fibre was polished in succession with 5, 3, and 1 µm polishing pads and washed with acetone to create a clean, polished surface. The distal end was then immersed in 10% KOH in isopropanol for 30 minutes with subsequent rinsing in copious amounts of distilled water and dried with compressed nitrogen. After that, it was treated in a 30 : 70 (v/v) mixture of H_2O_2 (30%) and H_2SO_4 (conc.) (piranha solution) for 60 minutes, rinsed in distilled water for 15 minutes, and dried in an oven at 100 °C for 30 minutes. This procedure leaves the surface with exposed hydroxyl groups that facilitate bonding of ATES.

The fibre surface was then modified by silanizing for two hours in a 10% solution of ATES in ethanol. The fibre was washed with methanol and distilled water, respectively, in an ultrasonic bath. Subsequently, it was dried in an oven at 60 °C

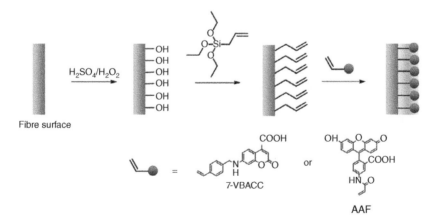

Figure 8.5 Preparation of an intrinsic pH sensor probe: schematic of the processes involved.

for two hours. This procedure functionalizes the fibre surface with polymerizable allyl groups.

Monomer stock solution was prepared by dissolving the fluorescent dye, 1,4-bis (acryloyl)piperazine cross-linker, acrylamide co-monomer, and AIBN initiator in dimethylformamide (DMF). The stock solutions were purged thoroughly with argon for 10 minutes. A small volume of the solution was placed into a capillary tube via syringe, and the distal end of the fibre was inserted. They were sealed quickly with polytetrafluoroethylene (PTFE) tape and polymerized in an oven at 80 °C for 18 hours. This procedure forms a polymer layer of the dye that is covalently bound to both the cylindrical surface and the distal end surface of the fibre. However, only the polymer on the distal end surface is responsible for the fluorescence signal, which is produced by direct excitation from the light source. The polymer on the side plays no role in the sensing process since evanescent wave excitation is eliminated by keeping the cladding of the fibre intact. A typical pH probe prepared by this procedure is shown in Figure 8.6 where it can be seen that the distal end of the probe shows a distinctive coloration due to the presence of the dye. The sensor tip was placed in pH 7 buffer for 24 hours to remove all unreacted materials and the excess amount of polymer formed that was not directly bound to the fibre. The probe was then stored in a cool and dark place until use.

This intrinsic optical fibre sensor design, despite offering superior performance and fast response, is not suitable for certain applications, such as the application in concrete structures (high pH sensor), due to the fragile nature of optical fibres, especially in highly alkaline media. The polymer was, therefore, prepared separately and packed in a tablet form in between a quartz disc and a nylon membrane to provide a robust mechanical design for safe embedment in harsh and corrosive environments as shown in Figure 8.7. The tablet with the quartz disc side facing upward was then put in a holder made from PTFE with an SMA thread on the top

Figure 8.6 Typical pH sensor tip prepared in this work showing the active distal end of the sensor.

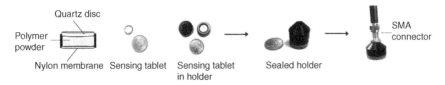

Figure 8.7 Preparation of a high pH sensor probe in a tablet form.

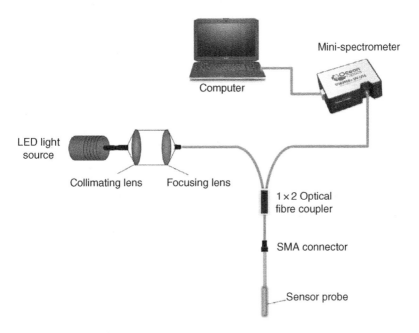

Figure 8.8 pH sensor system set-up.

to enable its connection to the fibre using an SMA connector. The porous bottom of the tablet enables the direct contact between the sensing tablet and the surrounding environment. All the parts were kept tightly together by placing a thin layer of superglue over the joints.

8.3.1.3 Set-Up of a pH Sensor System

Further to the completion of the sensor fabrication, Figure 8.8 shows the set-up of a pH sensor system. As shown in the figure, light from a LED, emitting at a centre wavelength of 375 nm, is coupled through a multimode UV/visible fibre (with hard polymer cladding, 1000 µm silica core, and numerical aperture [NA] of

0.37), using collimation and focusing lenses, into one branch of a 2×1 multimode fibre coupler. The other end of the fibre coupler is connected to the sensor probe with the active sensing region being located at the distal end of the fibre or with the sensing tablet being located at the bottom. Following pH interaction with the active region, a portion of the total light emitted from the sensing layer is collected and guided through the other branch of the fibre coupler to an Ocean Optics USB2000 spectrometer, with the output being displayed on a computer screen.

8.3.1.4 Evaluation of the pH Sensor Systems

8.3.1.4.1 Response Times of the pH Sensors

Figure 8.9a shows the dynamic response obtained from the spectrofluorometer of the VBACC pH sensing probe (low pH sensor, intrinsic sensor design) to a step change from pH 0.5 to 6, while Figure 8.9b shows the response of the VIC pH sensing probe (high pH sensor, non-intrinsic sensor design) obtained from the same spectrometer to a step change from pH 10 to 13. It can be seen from the figure that the intrinsic sensor design offers much faster response rate, 25 seconds for 95% of the total signal change, in comparison with other pH sensors such as the sensor reported by Wallace et al. [29], which showed a response time of around 500 seconds, or the device reported by Netto et al. [33], which showed a response time of few minutes. The faster response of this pH sensor is likely due to its key design features: both the relatively low thickness of the polymer film and its hydrophilicity. All this has obvious advantages where a rapid change of pH is to be monitored and a real-time measurement to be achieved. For the non-intrinsic sensor design, the response time was rather long, around 50 minutes for 95% of the total signal change, and at the initial stage depended on the direction in which the pH of the solution was varied. This response time, which is long in comparison with those of previously developed pH sensor systems, could be due to diffusion in the rather thick polymer layer. However, this is not a problem for concrete monitoring as pH changes in concrete structures occur over much longer periods of time.

8.3.1.4.2 Response of the Sensors to Different pHs

The calibration measurements of the sensor characteristics were performed in either 50 mM citrate buffer (for pH ≤ 6) or 50 mM phosphate buffer (for pH > 6). The titration curves are shown in Figure 8.10. The VBACC sensor probe exhibited an increase in fluorescence intensity with increasing pH in the range from 0.5 to 6.0. The AAF sensor probe showed a similar response in the range from 4.0 to 8.0, while the VIC sensor probe exhibited a decrease in fluorescence intensity with increasing pH in the range from 10 to 13.2. The response ranges of the sensors are conveniently wider than the dynamic response ranges of the free dyes. The pK_a values calculated using Eq. (8.16) for the three probes are 3.72 ± 0.06, 6.48 ± 0.06, and 11.9 ± 0.2, respectively. These values for the immobilized form of the

(a)

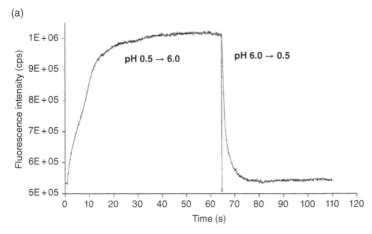

(b)

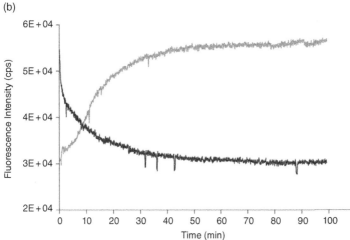

Figure 8.9 Dynamic response of the VBACC pH sensor (a) and VIC pH sensor (b).

dyes are slightly different from those for their free form in solution, and this arises probably because of the change in the polarity of the micro-environment [10].

8.3.1.4.3 *Effect of Ionic Strength (IS)*

Sensitivity to IS can be a serious problem in the cases of optical fibre sensors as it affects pK_a values, thus resulting in errors in pH determination. Most optrodes reported in the literature so far suffer from cross-sensitivity to IS to a certain degree, especially those that are based on indicators with charged group(s) [37] or those that make use of sol–gels [38] or cellulose membrane [39] as the solid support. The effect of IS was investigated with the VBACC pH sensor probe using the

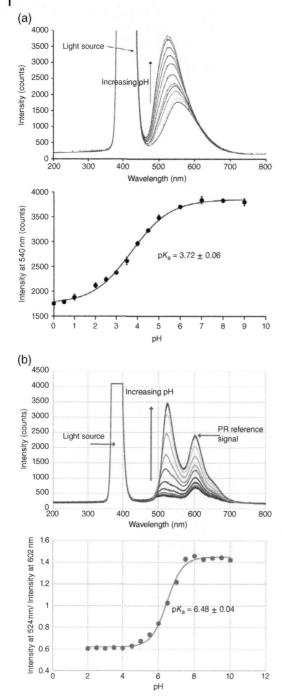

Figure 8.10 Response of the VBACC pH sensor (a), AAF pH sensor (b), and VIC pH sensor (c) to pH. The titration plots are shown on the right.

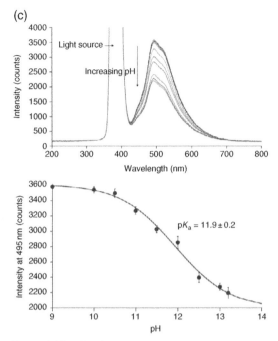

Figure 8.10 (Continued)

prepared pH 4 and 7 buffer solutions adjusted with NaCl to different IS ranging from 10 to 2000 mM. The fluorescence intensity obtained for each solution was converted to a pH value using the calibration curve, and this is presented in Table 8.2. As can be seen from the table, there appears to be no sensitivity to IS for the sensor, even at very high concentrations of NaCl. The insignificant errors caused are probably due to the system error rather than the change in IS.

8.3.1.4.4 Reproducibility and Photostability of the pH Sensors

The stability of the probes in terms of storage, their susceptibility to error due to intense irradiation of the sample, and their reproducibility in use are all very critical to the successful application of the systems. An evaluation of these parameters was made in order to understand better the performance of the sensors and establish their suitability for industrial applications. The stability of the sensors was tested by calibrating them with buffer solutions at different pH values and recalibrating them after a certain period of time. After each calibration, the probes were washed thoroughly with a pH 7.0 buffer or immersed in a pH 7.0 buffer for a few hours in case of the non-intrinsic sensor design, followed by the same procedure with distilled water, and then they were stored in the dark until next use. No

Table 8.2 pH measurements of pH 4 and 7 buffer solutions with different IS using the VBACC pH sensor probe.

IS (mM)	pH 4.00 buffer solution	Difference	pH 7.00 buffer solution	Difference
0	4.05	0.05	7.02	0.02
10	4.06	0.06	7.11	0.11
50	4.01	0.01	7.09	0.09
100	4.11	0.11	7.04	0.04
200	4.02	0.02	7.04	0.04
500	4.05	0.05	7.07	0.07
1000	4.07	0.07	7.10	0.10
2000	4.10	0.10	7.07	0.07

significant difference was observed between the measurements and the pK_a values calculated: it is very pleasing to note that these are almost the same even after months, illustrating the high stability of the sensor scheme produced (Figure 8.11).

Photostability is one of the critical properties of fluorescent indicators and thus of the dyes used in this sensor application. In order to test the photostability of the dye, the probe was coupled into the fluorimeter through a dichroic mirror using a fibre bundle. The excitation light (at a wavelength of 375 nm) was launched to the distal end of the probe illuminating the sensing material with light from the intense, high-power Xe lamp of the fluorimeter continuously for one hour. The fluorescence intensity data from the probe were collected over that period and displayed. As can be seen from Figure 8.12, no photobleaching was observed for the VBACC probe over the time investigated and with the high flux of photons onto the probe. For the VIC probe, the intensity of fluorescence was reduced by 7–8% for the dry state. However, under the same conditions, no photobleaching was seen for the similar probe, prepared in exactly the same way using the same polymer, which was immersed in a pH 10 buffer solution. The reason for the difference in photostability between the dry probe and wet probe is unclear. It could be because the excited single state of the dye is stabilized by solvation. It should be noted that in actual measurements, the material is not excited continuously, but only for 30 seconds each time when data need to be collected, after the sensor reaches its equilibrium, by a much weaker 3 mW LED light source. Therefore, a little photobleaching of the material in dry state when being illuminated by a strong light should not cause any problem for the monitoring. When compared with the results of other materials, these sensing materials offer excellent

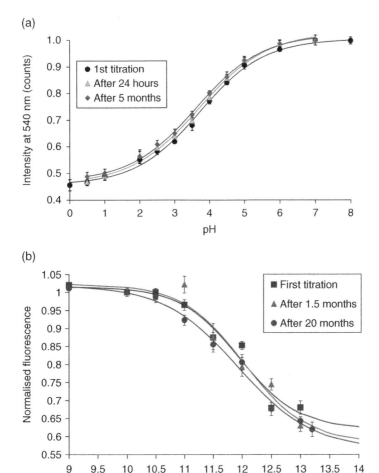

Figure 8.11 Repeated titration curves for the VBACC sensor probe (a) and the VIC sensor probe (b) obtained after a certain period of time.

performance: the decrease observed in the fluorescence intensity was 65% for carboxyfluorescein and 10–13% for iminocoumarin derivatives, again after 60 minutes of continuous illumination using a mercury lamp [10]. Thus an important conclusion is that the materials prepared using the coumarin fluorophores and synthesized specifically for this application in this work possess superior photostability, a feature that is critically important with excitation of sensor probes by high-intensity solid-state sources.

(a)

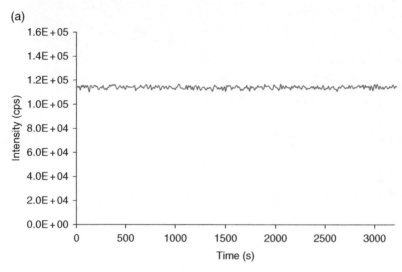

(b)

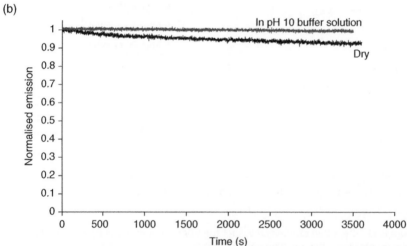

Figure 8.12 Fluorescence intensity of the VBACC probe (a) and VIC probe (b) at the emission wavelength as function of time during 60 minutes of continuous illumination by light from a high-power Xe lamp.

8.3.1.5 Comments

The examples of fibre-optic pH sensors presented above indicate the potential of this technology. A further important feature of these types of sensors is that they are potentially inexpensive to produce in quantity and the large Stokes shifts shown allow for more accurate measurements due to the minimum level of interference between light source and fluorescence signals generated. For industrial applications, it is necessary to develop a suitable 'packaging' to withstand use by inexperienced operators. However prior work with optical fibre relative humidity sensors has shown an effective design that could be employed to protect the sensitive fibre tip. In addition, the intrinsic sensor design discussed here has enabled direct light coupling between the fibre and the sensor material; therefore, there is a minimum loss caused by the excitation or the fluorescence signal collection. Also, sensors of this design can readily be used together: multiplexed along a single optical fibre or along a parallel optical network using various techniques to identify each individual sensor probe. Thus, there is considerable flexibility in the approach, and, as various applications are considered, the sensor scheme can be tailored for different uses, thus emphasizing the versatility of the design discussed here.

8.3.2 Optical Fibre Mercury Sensor

Mercury pollution in soil due to mining and industrial activities poses a serious problem across the world both from an economic and health perspective [40–42]. Common methods for the remediation of mercury-contaminated soil include excavation and disposal, but these methods are often costly and crude. Mercury, either in the inorganic form as Hg^{2+} or in the organic form as methyl mercury, once introduced into the body, can accumulate and cause serious irreversible damage to the immune system, the central nervous system, and kidneys [41, 43]. It is also believed that mercury causes various neurodegenerative diseases such as Alzheimer's diseases and Parkinson's disease [42]. Therefore, the detection of mercury is very important for the protection of human health and the minimization of its exposure in the environment. Such sensors would also provide a warning of exposure and thus act as a trigger for treatment.

In recent years, a number of chemical sensors for the detection of Hg^{2+} have been reported, based either on electrochemical methods [44–46] or fluorescence [47–60]. However, there are weaknesses with them and their applications, where most are not suitable for use in the field. Biosensors based on whole bacterial cells or bacterial heavy metal binding proteins [61, 62], which can be considered as alternative devices, also suffer from certain limitations due to the fragile and unstable nature of the biological recognition elements. Commercial heavy metal sensors for use in soil are very limited and are typically either very expensive or require the extraction of soil prior to its manipulation and analysis, which could allow sample degradation to occur prior to measurement. Consequently, there is a strong industrial need for the development of a low-cost and portable alternative for mercury detection, thus providing a fast screening solution to yield new information on what is an important aspect of improving the environment. The sensor system described in this section is an example of a fibre-optic chemical sensor developed to fulfil this objective.

8.3.2.1 Sensor Design and Mechanism

In this sensor design, a novel fluorescent polymeric material for Hg^{2+} detection based on a derivative of coumarin (acting as the fluorophore) and an azathia crown ether moiety (acting as the mercury ion receptor) has been synthesized using the ion imprinting technique and covalently attached to the distal end surface of an optical fibre. Azathia crown ethers have been reported to have strong ability to coordinate with heavy and transition metal ions and have previously been used as receptors in the design of fluorescent sensors for Hg^{2+} [48, 63]. However, the signalling mechanism employed in those sensors was based on intramolecular charge transfer (ICT), leading to the quenching of fluorescence upon analyte binding. It is strongly desirable that the fluorescence modulation is in the 'off–on' direction, since this will lead to the best signal-to-noise characteristics and avoid potential false positive responses due to degradation of the system. In this sensor design, the presence of the amine significantly reduces the fluorescence of the fluorophore due to the quenching of its fluorescence by the nitrogen lone pair electrons through photoinduced electron transfer (PET). Upon complex formation with Hg^{2+}, the nitrogen lone pair electrons are donated to Hg^{2+}, which therefore abolishes or tremendously reduces the fluorescence quenching. Consequently, binding of Hg^{2+} switches on fluorescence, as illustrated in Figure 8.13.

8.3.2.1.1 Synthesis of the Mercury-Sensitive Fluorophore

The fluorophore was designed to contain suitable functional group(s) for covalent immobilization of the molecule to a substrate/optical fibre to avoid dye leaching, a common problem resulting from poor immobilization that consequently causes a drifting of the calibration of the probe, which leads to the gradual breakdown of its useful sensing ability [17]. To meet the desired requirements, a novel polymerizable coumarin, styryl (1,4,7,10-tetrathia-13-azacyclopentadecanyl)methyl coumarin (STAMC, **4**), was prepared in multiple steps starting from a commercially available phenolic compound as outlined in Figure 8.14. The Pechmann reaction of

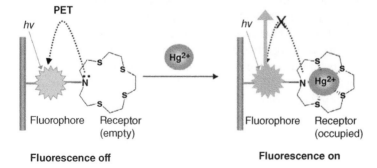

Figure 8.13 Illustration of fluorescence switching on by Hg^{2+} binding via a PET mechanism.

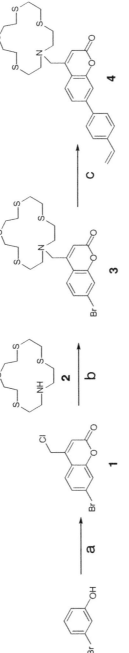

Figure 8.14 Preparation of mercury-sensitive fluorophore STAMC. (a) Ethyl 4-chloroacetoacetate, H_2SO_4, H_2O, r.t, 22 hours, 68%; (b) vinylbenzylchloride, K_2CO_3, KI, MeCN, 80 °C, 18 hours, 69%; (c) $CH_2=CHC_6H_4B(OH)_2$, $Pd(PPh_3)_4$, K_2CO_3, dioxane, 85 °C, 18 hours, 52%.

a phenol with a β-carbonyl ester is a versatile approach for the synthesis of 4-substituted coumarins. The substitution of chlorine for an azathia crown ether moiety was achieved by performing a reaction with the azathia crown ether **2**. A polymerizable group was introduced into the coumarin structure via a Suzuki coupling of the Br-substituted coumarin **3** with vinylphenylboronic acid using K_2CO_3 in dioxane as a base/solvent mixture.

8.3.2.1.2 *Sensor Probe Fabrication*

STAMC was covalently immobilized onto the fibre surface by polymerization, in an approach similar to the method previously reported [64]. Several factors need to be taken into consideration when preparing this type of sensing material, designed to work in aqueous/moisture environments: (i) the material has to be sufficiently hydrophilic to allow for water and ions transport, (ii) it should not be too rigid (preventing accessibility of ions to the receptors) nor insufficiently robust, (iii) and it has to show long-term stability to allow for multiple measurements. A number of co-monomers (methyl methacrylate, vinylalinine, vinylpiridine, acrylamide), cross-linkers (1,4-bis(acryloyl)piperazine, bisacrylamide, ethylene glycol dimethacrylate, poly(ethylene glycol) diacrylate), solvents (acetonitrile, ethanol, dimethylformamide, *N,N*-dimethylacetamide [DMA], pH 7 phosphate buffer), initiators (ammonium persulphate, azobisisobutyronitrile, diphenyl(2,4,6-trimethylbenzoyl)phosphine oxide), and initiation methods (thermal or photochemical) have been investigated in an attempt to find the most sensitive polymer that satisfies the above requirements. However, all the resulting polymers, despite being very good in quality, showed a very poor sensitivity to Hg^{2+} in comparison with the free fluorophore in solution. This could be because the specific orientation of the metal ion and receptor required for the interactions to happen, particularly when the binding is on a 1 : 2 basis, is difficult to achieve in polymer when conformational flexibility of the receptor is partly compromised. In addition, self-association and self-quenching of the fluorophore in the polymer also reduce its availability to interact with Hg^{2+} ions. The ion imprinting technique was then employed to create binding sites since the polymer mimics the composition of the pre-polymerization mixture and maintains the specific orientation of the functional group(s) by the imprinting process. Molecular imprinting has been demonstrated widely as a versatile technique for the preparation of molecular receptors capable of the selective recognition of given target molecules [65–67]. However, in this application, ion imprinting was solely used as a method for improving the binding properties and sensitivity of the polymer as the fluorophore is inherently selective. This arrangement enabled the binding sites to match the charge, size, and coordination number of the ion. Moreover, the complex geometry could be preserved through the cross-linking and leaching steps, generating a favourable environment for the template ion rebinding [67].

The polymer was prepared directly on the surface of the optical fibre in H_2O/DMA (1 : 1, v/v) using acrylamide as the co-monomer, poly(ethylene glycol) diacrylate (Mn 575) as the cross-linker, and ammonium persulphate as the initiator. The molar amount of fluorescent monomer (STAMC) used was fixed at 1 : 16 of the cross-linker. More fluorescent monomer would be expected to give more binding sites, but too high a concentration of fluorophore could also result in fluorescence quenching by the inner filter effect. A template–fluorescent monomer ratio of 1 : 1 or higher was employed to ensure that the fluorescent monomer was fully complexed. Hg^{2+} ions were then removed from the polymer using a Tris-EDTA buffer solution. A typical probe prepared by this procedure is shown in Figure 8.15, where it can be seen that the distal end of the probe shows a coloration due to the presence of the fluorophore.

Normal   ex 375 nm

Figure 8.15 Typical mercury sensor probe prepared in this work showing the active distal end of the sensor.

8.3.2.2 Evaluation of the Mercury Sensor System
8.3.2.2.1 Properties and Fluorescence Response of the Fluorophore in Solution
Absorption and fluorescence studies of free STAMC (**4**) were performed in $H_2O/$ MeCN (7 : 3, v/v) as the compound is not soluble in H_2O alone. The absorption spectrum of STAMC shows only one main absorption band in the UV region at 342 nm. Emission spectrum for the compound recorded in the same solvent using excitation at the absorbance maximum includes one band at 472 nm. It is noted that STAMC exhibits a very large Stokes shift (the difference in wavelength between the absorption and the fluorescence spectral peaks) of 130 nm, which is very important for the sensor system design to minimize the interference of the excitation light with the fluorescence emission. The quantum yield of STAMC was calculated to be 1.5%, which is relatively low as the fluorescence of the compound is quenched by the nitrogen lone pair through PET.

Adding Hg^{2+} to STAMC in solution resulted in a significant increase in fluorescence intensity of the fluorophore. Figure 8.16 shows the changes in fluorescence of STAMC with different concentrations of Hg^{2+}. It can be seen that the emission intensity at 472 nm increases sharply, nearly sixfold after the addition of only 25 µM (0.5 equiv.) of Hg^{2+}. No significant change was observed after further addition of Hg^{2+}, indicating that an approximate saturation was reached. This suggests that the interactions of the fluorophore with Hg^{2+} are very strong but not on a 1 : 1 basis, rather on a 2 : 1 basis.

The binding stoichiometry of STAMC and Hg^{2+} was determined using Job's method of continuous variation [68]. The total concentration of STAMC and

(a)

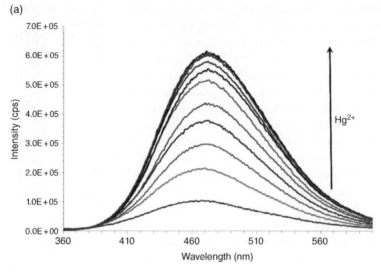

(b)

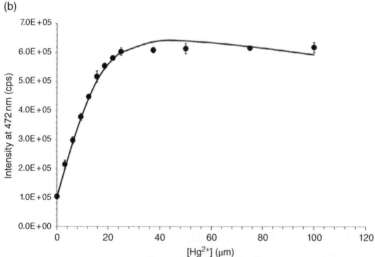

Figure 8.16 (a) Emission spectra of STAMC (50 μM, λ_{ex} = 345 nm) in H$_2$O/MeCN (7 : 3, v/v) with the addition of Hg^{2+} (0–100 μM). (b) Titration plot at 472 nm. The solid line shows the computer fit.

Hg^{2+} was maintained constant at 50 μM, with a continuous variation of the molar fraction of STAMC. Figure 8.17 shows the fluorescence intensity variation at 472 nm as a function of $f4$ ($f4$ being the fraction of STAMC, **4** in a fixed total concentration ([STAMC] + [Hg^{2+}] = 50 μM). The sharp peak in the plot at $f4$ = 0.67 confirms that the dominant species is the 2 : 1 complex, which is also in agreement with the titration data.

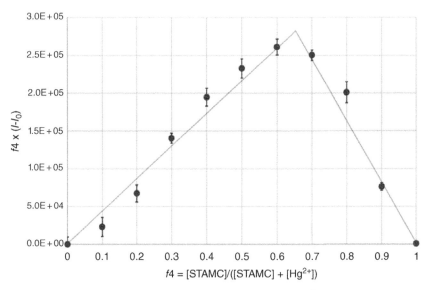

Figure 8.17 Job plot for the association between STAMC and Hg^{2+} in $H_2O/MeCN$ (7 : 3, v/v).

The titration data were then fitted to the 2 : 1 association model using the Bindfit program [69], giving an extremely high binding constant $K_{2:1}$ of $1.29 \pm 0.13 \times 10^9$. There are a number of complicating factors, including self-association and self-quenching of the fluorophore, which means the true association constant is likely to be less than this. Nonetheless, this indicates that the association between STAMC and Hg^{2+} is very strong and STAMC is an excellent receptor for Hg^{2+}.

The selectivity of STAMC for Hg^{2+} was studied in $H_2O/MeCN$ (7 : 3, v/v). Different heavy metal, transition metal, alkali earth, and alkali ions including Cd^{2+}, Pb^{2+}, Cu^{2+}, Ni^{2+}, Zn^{2+}, Ca^{2+}, Mg^{2+}, Co^{2+}, Cr^{3+}, Mn^{2+}, Fe^{3+}, Al^{3+}, Ag^+, and Na^+ were used for the investigation. The concentration of all the metal ions was fixed at $50 \mu M$ (1 equiv.), where only Hg^{2+} caused a dramatic change in the fluorescence intensity, indicating that STAMC is highly selective for the detection of Hg^{2+} (Figure 8.18), which is a matter of necessity for an excellent chemosensor.

8.3.2.2.2 Response of the Sensor to Hg²⁺

The calibration of the sensor was performed using a series of solutions of mercury chloride in deionized water. The probe was immersed in the Hg^{2+} solutions, and the signals were allowed to reach constant values before being recorded. The sensor was rinsed with Tris-EDTA buffer solution (10 mM Tris-HCl; 1 mM EDTA), followed by deionized water between measurements. In a way that is similar to that seen for the free fluorophore, the sensor exhibited an increase in fluorescence intensity with increasing Hg^{2+} concentration in the range of $0-28 \mu M$

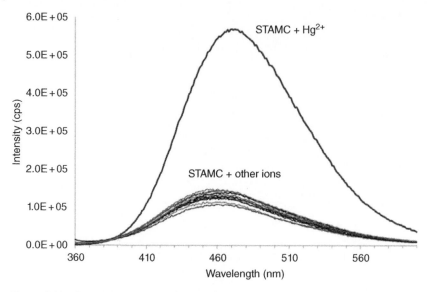

Figure 8.18 Emission spectra of STAMC (50 μM, λ_{ex} = 345 nm) in H$_2$O/MeCN (7 : 3, v/v) in the presence of 1 equiv. of different metal ions.

(Figure 8.19). At higher concentrations of Hg^{2+}, no further change of intensity was observed due to the saturation of all available binding sites. It has also been noted that the emission peak of the immobilized form of the fluorophore is slightly 'red-shifted' compared with that of its free form in solution. It seems that this probably can be attributed to the change in the polarity of the micro-environment. The lower limit of detection of the system may vary since it depends on the type and sensitivity of detector used. With the Ocean Optics mini-spectrometer used in this work, the lowest concentration of Hg^{2+} that can cause a distinguishable change in fluorescence intensity is around 0.15 μM. The sensitivity of the sensor is suitable for use in soil where the total concentration of Hg normally ranges from 0.5 to 3000 ppm (2.5 μM–15 mM) with acceptable limits of up to 12 ppm [70–72]. In certain applications where a lower detection range is required, a thinner layer of the sensing material can be fabricated together with the use of a more sensitive detector, which to some extent could lower the detection limit of the sensor.

8.3.2.2.3 Reusability and Photostability

Sensor reusability is important for the development of a tool that can allow multiple, rapid, and real-time measurements in the field. Consequently, a range of known mercury binders and washing agents were screened to identify a method for probe regeneration. *N,N,N′,N′*-Tetrakis(2-pyridylmethyl)ethylenediamine (TPEN) has been reported as a good ion chelator to remove metal ions and restore

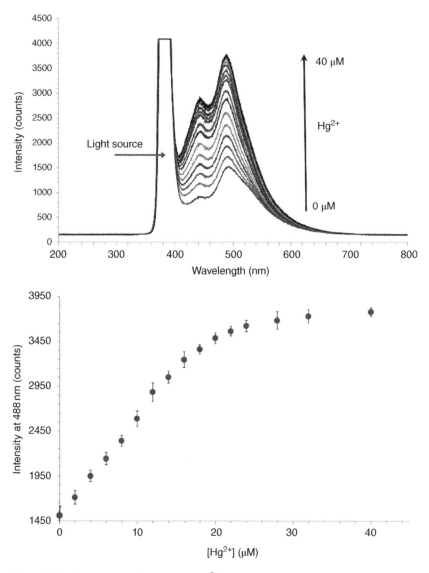

Figure 8.19 Response of the sensor to Hg^{2+} in the concentration range from 0 to 40 µM in H_2O (left). Titration plot at 488 nm (right).

the systems back to baseline levels [47, 73, 74]. However, in this study, a Tris-EDTA buffer solution proved superior to TPEN in terms of regeneration time and efficiency. Figure 8.20 shows the dynamic response of the sensor obtained from the spectrofluorometer to a step change from no Hg^{2+} present (0 µM) to 10 µM Hg^{2+} in H_2O and back again by removal of the mercury using a Tris-EDTA

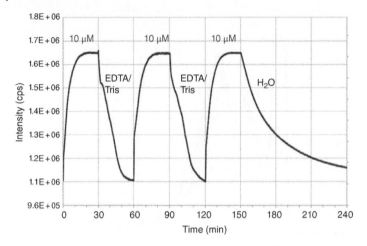

Figure 8.20 Forward and reverse dynamic response of the sensor probe with the addition of 10 μM Hg^{2+} and removal of Hg^{2+} using a Tris-EDTA buffer solution or pure water for three cycles (λ_{ex} = 375 nm, λ_{em} = 440 nm).

buffer solution (10 mM Tris-HCl; 1 mM EDTA) for three cycles. It can be seen from the figure that the metal binding process is completely reversible and the Tris-EDTA solution restores fluorescence of the sensor back to the baseline level. The sensor can also be recovered in H_2O, but the process requires a greater amount of time, almost two hours to attain a 90% recovery compared with just 15 minutes with the use of EDTA.

The response time of the sensor upon adding Hg^{2+} can also be seen from Figure 8.20. Although around 75% of the total signal change occurred within 5 minutes, it took around 11 minutes for the sensor to attain 95% of the total change and 18 minutes to reach equilibrium. The recovery of the sensor after Hg^{2+} binding required a greater amount of time, almost 30 minutes to restore the sensor back to the baseline level using the said Tris-EDTA solution. A stronger EDTA buffer solution can be used to make the process faster, but too strong solutions may damage the system.

As mentioned in Section 8.3.1.4, photostability is one of the critical properties of fluorescent sensors and thus requires careful investigation. The method used to test the photostability of the mercury sensing material was similar to that of the pH sensor, and the data are shown in Figure 8.21.

It was interesting to observe that the intensity of fluorescence was reduced by 3–4% for the dry state and in H_2O over the time investigated, whereas under the same conditions a small increase in fluorescence intensity was seen for the similar probe, prepared in exactly the same way using the same polymer, which

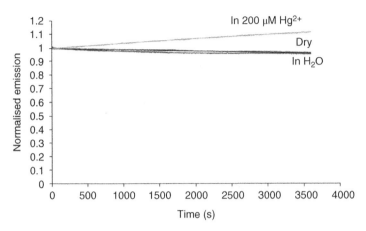

Figure 8.21 Fluorescence intensity of the sensor probe (in dry state, in H_2O, and in a 200 μM Hg^{2+} solution) as function of time during 60 minutes of continuous illumination by light from a high-power Xe lamp (λ_{ex} = 375 nm; λ_{em} = 440 nm).

was immersed in a 200 μM Hg^{2+} solution (after equilibrium being reached). The reason for this is unclear. It could be because the photodecomposed product of Hg^{2+}-fluorophore complex is more fluorescent than the original fluorophore. It should be noted that in actual measurements, the material is not excited continuously, but only for 30 seconds each time when data need to be collected, after the sensor reaches its equilibrium, by a much weaker 3 mW LED light source. Therefore, a little photobleaching of the material when being illuminated by a strong light source should not cause any significant problem for the monitoring, and the material prepared for this application still offers excellent performance.

8.3.2.3 Comments

The high-quality performance of the fibre-optic sensor system presented above based on fluorescence turn-on of a coumarin bearing an azathia crown ether moiety in the presence of Hg^{2+} via PET indicates the potential of this technology in a range of practical applications. The sensor is potentially well suited for *in situ* long-term monitoring of mercury in the environment. The highly effective approach developed in this work can also be used for the preparation of sensors for other heavy metals by substituting a suitable receptor for the mercury receptor – azathia crown ether.

8.3.3 Optical Fibre Cocaine Sensor

The total value of trade from illicit drugs is at an estimated $321 billion/yr, which is higher than the cumulative GDP of 88 countries across the world (UN report 2005).

Illicit drug use in the United Kingdom remains very high, and costs run to £15.4 billion each year [75], in areas ranging across policing, detection, related crime, medical care, and family support. Cocaine is one of the most commonly abused drugs, and this has led to extensive investigative research efforts for its detection, due to the adverse health effects and related dangers associated with its illicit use [76, 77].

There are several major analytical methods available for the analysis of cocaine and its metabolites, including gas chromatography–mass spectrometry (GC–MS) [78, 79], high-performance liquid chromatography (HPLC) [80, 81], thin-layer chromatography [82], voltammetry [83], radioimmunoassay [84], and enzyme-linked immunosorbent assay (ELISA) [85]. These traditional methods, despite having achieved very good results, are generally expensive, time consuming, and cumbersome for real-time measurements outside the laboratory, some of which also require sample clean-up and derivatization of cocaine prior to analysis.

Biosensors, which rely on the specificities of the binding sites of receptors, enzymes, antibodies, or DNA as biological sensing elements, have been considered as alternative analytical devices due to their specificity, portability, speed, and low cost [86]. Biosensors for cocaine based on monoclonal antibodies [87, 88] and especially aptamers [89–93] have been developed in recent years. However, these sensors suffer from certain limitations in light of their potential practical applications in the field due to the fragile and unstable nature of the biological recognition elements. Therefore, the development of stable, compact, and portable sensing systems that are capable of real-time detection of the target drug remains a compelling goal. The system described in this section illustrates how the fibre-optic sensing technology can contribute to this objective.

8.3.3.1 Sensing Methodology

Molecular imprinting has been extensively demonstrated over the last three decades as a versatile technique for the preparation of synthetic molecular receptors capable of the selective recognition of given target molecules. The approach is based on the self-assembly of a template molecule with polymerizable monomers possessing functional group(s) interacting with the template [65, 66]. After polymerization, the template is removed, leaving vacant recognition sites that are complementary in shape and functional groups to the original template. Molecularly imprinted polymers (MIPs) provide an exciting alternative to biological receptors as recognition elements in chemical sensors [94].

Here it is presented an example of a robust fibre-optic chemical sensor for cocaine detection based on the combination of molecular imprinting (as a method for generating chemically selective binding sites) and fluorescence modulation (as a means of signalling the presence and concentration of the analyte). The attraction of this approach lies in the advantages offered both by the optical fibre in

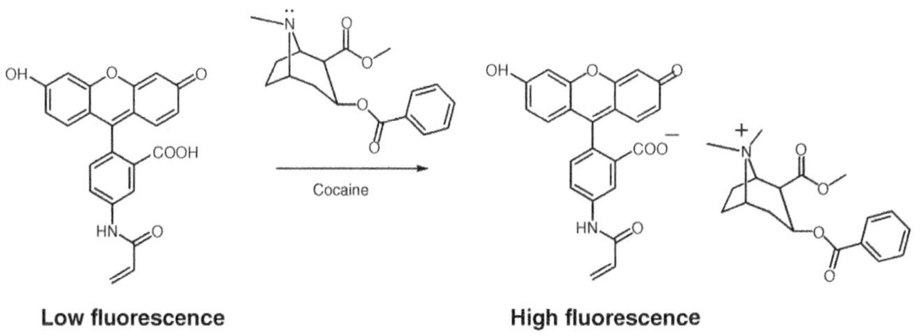

Low fluorescence   **High fluorescence**

Figure 8.22 Interaction between AAF and cocaine.

terms of small size, immunity to EM interference, remote sensing capability, resistance to chemicals, and biocompatibility [12, 13] and by the synthetic polymer receptor in terms of robustness, thermal and chemical stabilities, low cost, and long shelf life [66].

The MIP receptor that is selective for cocaine was covalently bonded to the distal end of the optical fibre, which facilitated rapid and highly sensitive detection. AAF was used as fluorescent functional monomer interacting with the template cocaine. The sensing mechanism depends on changes in the frontier orbitals of fluorescein, which occur when it is deprotonated by a base. The deprotonated form is fluorescent, and the protonated form is much less so. In the presence of cocaine, the carboxylate group of AAF is deprotonated. Cocaine acts as a base in the ion pair complex, accepting a proton from AAF and leading to an increase in the observed fluorescence intensity (Figure 8.22).

The imprinting and sensing strategy is illustrated in Figure 8.23. A complex is formed between the functional group –COOH on the fluorophore and the amine group on the template/analyte. The complex is copolymerized with a cross-linking monomer and co-monomer on the surface of the fibre, which has been functionalized with polymerizable groups. Then the template/analyte is extracted from the polymer. The resulting MIP formed on the fibre contains recognition sites incorporating the fluorophore and exhibits an increase in fluorescence intensity selectively in the presence of the template/analyte. As a result, the selectivity of the sensor has been designed to arise from the functional group of the fluorophore and from the shape of the cavity.

8.3.3.2 Design and Fabrication of a Cocaine Sensor System

AAF was prepared from fluoresceinamine according to the literature procedure [95] as shown in Figure 8.24.

The fabrication of the cocaine sensing probe requires a multistep process where the fibre preparation step was similar to the method used for pH and mercury

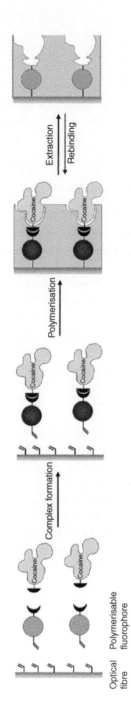

Figure 8.23 The preparation of a cocaine sensing MIP on the surface of the optical fibre that exhibits fluorescence changes upon template binding.

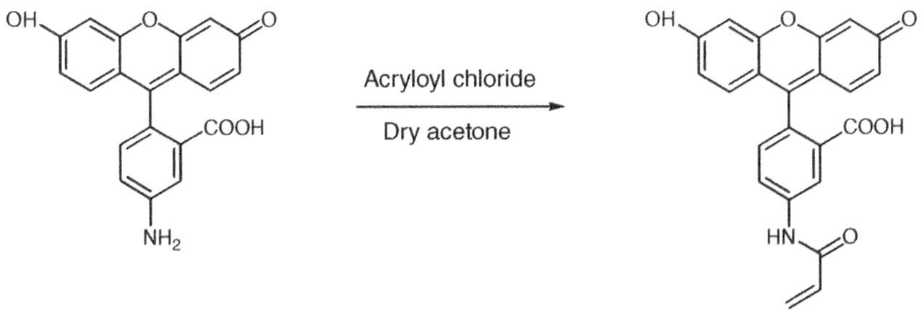

Figure 8.24 Preparation of acrylamidofluorescein (AAF).

sensors. The pre-polymerization mixture was prepared by dissolving cocaine (6.1 mg, 0.02 mmol), AAF (4.0 mg, 0.01 mmol), ethylene glycol dimethacrylate cross-linker (150.9 µL, 0.8 mmol), acrylamide co-monomer (10.0 mg, 0.14 mmol), and 2,2′-azobisisobutyronitrile initiator (1.1 mg) in 222 µL dry MeCN. The solution was purged thoroughly with argon for 10 minutes. A small volume of the solution was placed into a capillary tube via syringe, and the distal end of the fibre was inserted. They were sealed quickly with PTFE tape and polymerized in an oven at 70 °C for 16 hours. This procedure forms an MIP layer on both the cylindrical surface and the distal end surface of the fibre. However, only the MIP on the distal end surface is responsible for the fluorescence signal that is produced by direct excitation from the light source. The MIP on the side plays no role in the sensing process since evanescent wave excitation is eliminated by keeping the cladding of the fibre intact. The probe prepared by this procedure is shown in Figure 8.25 where it can be seen that the distal end of the probe shows a distinctive coloration due to the presence of the fluorophore. The sensor tip was washed repeatedly with MeOH–AcOH (8 : 2, v/v) in an ultrasonic bath, followed by the same procedure with MeOH alone to remove the template and all unreacted materials and the excess amount of polymer formed that was not directly bound to the fibre. The probe was then stored in a cool and dark place until use. A control probe (non-imprinted polymer [NIP]) was prepared at the same time under identical conditions using the same recipe but without the addition of the template cocaine.

The sensor system set-up used for the measurements undertaken to calibrate the cocaine probe was the same as that used for the pH and mercury probes, as presented in Figure 8.8. .

8.3.3.3 Evaluation of the Cocaine Sensor System
8.3.3.3.1 *Response Time of the Cocaine Sensor*
Before performing measurements to calibrate the sensor, its response time was investigated. Figure 8.26 shows the dynamic response of the sensor obtained from the spectrofluorometer to a step change from no cocaine present (0 µM) to 25 µM

(a)

(b)

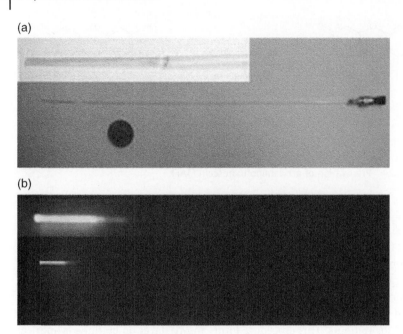

Figure 8.25 Cocaine probe prepared in this work showing the active distal end of the sensor (a) under normal conditions and (b) when 375 nm ultraviolet (UV) light was launched to the end of the fibre.

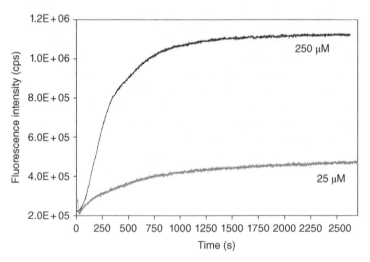

Figure 8.26 Dynamic response of the sensor probe at 515 nm (excitation at 375 nm) showing the 15 minutes response time (to 95%).

and to 250 µM cocaine in H_2O/MeCN 9 : 1. Although around 70% of the total signal change occurred within 5 minutes, it took around 15 minutes for the sensor to attain equilibrium (to 95%) in 250 µM cocaine and 20 minutes in 25 µM cocaine. The higher concentration of cocaine appeared to give a slightly quicker response time. However, the difference was not significant. This response time is considered to be rapid compared with other MIP sensor systems where a few hours incubation is required for the interaction between the template/analyte and the binding sites in the MIP to reach equilibrium [96, 97]. This important result is most probably due to both the intrinsic sensor design and the thickness of the polymer film, since the thicker the polymer layer, the longer it takes for the target compound to penetrate into the polymer network to interact with the binding sites.

8.3.3.3.2 *Cocaine Sensor Calibration*
The calibration measurements were performed by immersing the probe in different cocaine solutions at various concentrations. The signals were allowed to reach constant values and then recorded. After each measurement, the probe was washed with MeOH–AcOH (8 : 2, v/v) in an ultrasonic bath, followed by the same procedure with MeOH alone to remove bound cocaine. Initially, experiments were carried out in MeCN/H_2O 9 : 1. MeCN was used because the MIP was prepared in MeCN, so its recognition properties would be expected to be best in MeCN (since this should result in no loss of selectivity due to MIP swelling) [98]. H_2O was added at 10% (v/v) in order to reduce non-specific binding. The sensor exhibited an increase in fluorescence intensity with increasing cocaine concentration in the range from 0 to 250 µM (Figure 8.27a). At higher concentrations of cocaine, no further change of intensity was observed due to the saturation of all available binding sites. It was also interesting to see if the sensor could work in aqueous media where biological recognition mainly occurs. Measurements were carried out in a manner similar to those of Figure 8.27a, but the solvent system was replaced by H_2O/MeCN 9 : 1 (MeCN was added to solubilize the analyte). The sensor showed a greater increase in fluorescence in the aqueous than in the organic solution (Figure 8.27b), which is attributed to the difference between the photophysical properties of the fluorophore in aqueous and in organic media. The dynamic response range of the sensor in aqueous solution is also wider, from 0 up to 500 µM. This arises because non-covalent interactions between cocaine and the functional groups in the MIP were weaker in H_2O and thus the available binding sites were not fully occupied until higher concentrations of cocaine were used.

The lower limit of detection of the system may vary since it depends on the type and sensitivity of detector used. With the Ocean Optics mini-spectrometer used in this work, the lowest concentration of cocaine that can cause a distinguishable change in fluorescence intensity is 2 µM. The response of the control probe (NIP) to cocaine was also studied, and it was observed that the NIP probe showed

(a)

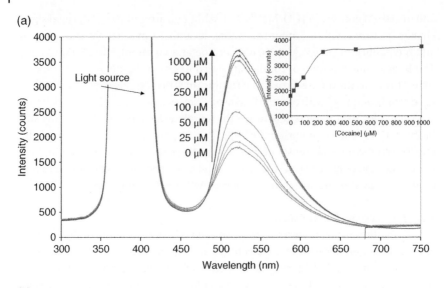

(b)

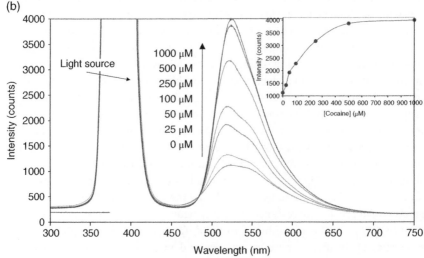

Figure 8.27 Response of the sensor to cocaine in the concentration range from 0 to 1000 μM in (a) MeCN/H$_2$O 9 : 1 and (b) in H$_2$O/MeCN 9 : 1. Insets show the dependence of emission maximum on cocaine concentration.

a lesser increase in fluorescence upon cocaine addition of 0.1 mM H$_2$O/MeCN 9 : 1 than the MIP probe (139% compared with 52%; Figure 8.28), suggesting that the analyte bound to the MIP more strongly than to the NIP and confirming the existence of recognition sites in the MIP.

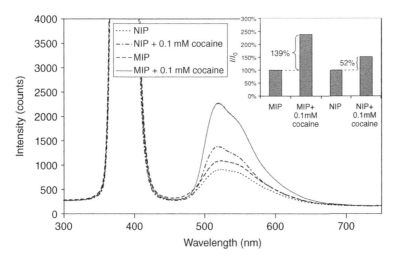

Figure 8.28 Response of the sensor probe and control probe to 0.1 mM cocaine in H_2O/ MeCN 9 : 1.

8.3.3.3.3 Selectivity of the Cocaine Sensor Towards Different Drugs

Different drugs including cocaine, ketamine, amphetamine sulphate, ecgonine methyl ester, and buprenorphineHCl were used for an investigation into the selectivity of the probe developed to cocaine, as it is often seen in the presence of other agents. The concentration of all the drugs considered was fixed at 500 µM in H_2O/ MeCN 9 : 1, where the most significant increase in the fluorescence signal intensity was seen for cocaine. It can thus be observed from Figure 8.29 that the sensor responds less to any of these drugs than to the template cocaine. This once again indicates successful imprinting and selective recognition sites in the MIP. The difference in fluorescence response of the sensor to different competitors can be explained in terms of the difference in their basicities and the similarity in shape and functional groups of their structures to that of cocaine. Significantly higher reactivity of the sensor for codeine compared with that for other competitors may also be due to the availability of more functional groups on the codeine molecule that are able to interact non-covalently with the binding sites in the MIP. It should also be noted that some of the drugs tested were in the salt forms, not free bases, and the presence of acids might affect the test results obtained.

8.3.3.3.4 Reproducibility and Photostability

The stability of the probe in terms of storage, its susceptibility to error due to intense irradiation of the sample, and its reproducibility in use is very critical to the successful application of the system. A preliminary evaluation of these

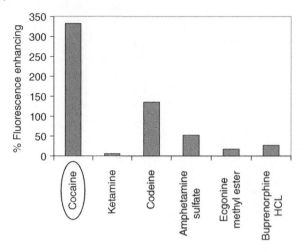

Figure 8.29 Response of the sensor probe to different drugs with concentrations of 500 μM in H$_2$O/MeCN 9 : 1.

parameters was made in order to understand better the performance of the sensor. The stability of the sensor was tested by calibrating it with different cocaine concentrations ranging from 0 to 500 μM and recalibrating it after 24 hours and then after one month. After each calibration, the probe was washed thoroughly with MeOH–AcOH (8 : 2, v/v) in an ultrasonic bath, followed by the same procedure with MeOH alone to remove bound cocaine, and then it was stored in the dark until next use. No significant difference was observed between the measurements, and the results obtained were found to be fairly reproducible even after one month. In order to test the photostability of the sensor, it was coupled into the fluorimeter through a dichroic mirror using a fibre bundle. The excitation light at 375 nm was launched to the distal end of the probe consisting of the sensing material by the high-power Xe lamp of the fluorimeter continuously for one hour. The fluorescence intensity of the probe was dynamically collected. As can be seen from Figure 8.30, very little photobleaching (less than 1%) was observed over the time investigated, suggesting that the MIP prepared in this work possesses superior photostability, a feature that is critically important with excitation by high-intensity solid-state sources.

8.3.3.4 Comments

The fibre-optic sensing system described above is an example of an effective approach using the molecular imprinting technique for the development of a specific drug sensor for cocaine detection that shows both superior performance and fast response. Indeed, the cocaine sensor developed has demonstrated an increase

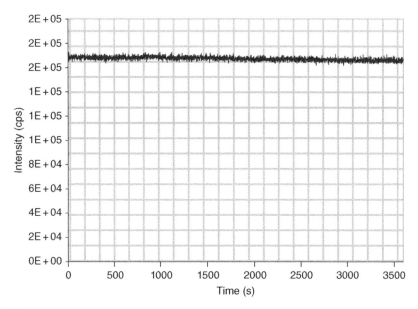

Figure 8.30 Fluorescence intensity of the probe at the emission wavelength as function of time during 60 minutes of continuous illumination by a high-power Xe lamp.

in fluorescence intensity in response to cocaine in the concentration range of 0–500 μM in aqueous acetonitrile mixtures with good selectivity and reproducibility over one month. Once its performance is further refined, this type of sensors will potentially make a significant impact on the homeland security enhancement as they can provide technical evidence on the spot with minimum invasion. Based on the above successful evaluation of this novel, robust, compact, and portable cocaine sensor, the molecular imprinting technique has shown to be promising both for the expansion of drug detection and for next-stage commercial exploitation.

8.4 Conclusions and Future Outlook

This chapter has addressed generic principles of fibre-optic chemical sensing, as well as some chemical sensor core designs using an optical fibre as a generic sensor substrate (based upon a generic spectroscopic sensor design protocol), illustrated with examples derived from City University London R&D developments in this field. The underpinning sensing mechanism has varied from the induced physical change to the chemical interaction between the target molecules and the sensor materials, thus inducing the fluorescence signal change. The corresponding spectral signal change, as a result, has been demonstrated either through the spectral shift or the fluorescence intensity variation.

These optical fibre chemical sensors developed have shown a huge potential to address various industrial challenges. Compared with electrochemical sensors, optical fibre chemical sensors have shown stronger resistance to chemical attacks due to the use of glass sensor substrate. Therefore, they are suitable to be used in harsh environment where contamination may pose a threat to electrochemical sensors. In addition, optical fibre chemical sensors are immune to electromagnetic (EM) interference and thus are suitable to be used in conditions where EM interference is an issue. If the sensor signals are wavelength encoded, it is possible to create a multiplexed sensor array within a single length of a fibre using the wavelength-division-multiplexing technique. All these merits offered by optical fibre chemical sensors have given them a unique selling point and a bright future to be expanded to meet various industrial needs.

Acknowledgements

The authors would like to thank the UK Engineering and Physical Sciences Research Council and the EU for the funding support via a number of research projects.

References

1 Narayanaswamy, R. (1993). Optical chemical sensors – transduction and signal-processing. *Analyst* 118 (4): 317–322.

2 McDonagh, C., Burke, C.S., and MacCraith, B.D. (2008). Optical chemical sensors. *Chem. Rev.* 108 (2): 400–422.

3 Wolfbeis, O.S. (2008). Fibre-optic chemical sensors and biosensors. *Anal. Chem.* 80 (12): 4269–4283.

4 Homola, J., Yee, S.S., and Gauglitz, G. (1999). Surface plasmon resonance sensors: review. *Sensors Actuators B Chem.* 54 (1–2): 3–15.

5 Jorgenson, R.C. and Yee, S.S. (1993). A fibre-optic chemical sensor based on surface plasmon resonance. *Sensors Actuators B Chem.* 12: 213–220.

6 Chau, L.K., Lin, Y.F., Cheng, S.F., and Lin, T.J. (2006). Fibre-optic chemical and biochemical probes based on localized surface plasmon resonance. *Sensors Actuators B Chem.* 113 (1): 100–105.

7 Del Villar, I., Arregui, F.J., Zamarreño, C.R. et al. (2017). Optical sensors based on lossy-mode resonances. *Sensors Actuators B Chem.* 240: 174–185.

8 Grant, S.A., Bettencourt, K., Krulevitch, P. et al. (2001). In vitro and in vivo measurements of fibre optic and electrochemical sensors to monitor brain tissue pH. *Sensors Actuators B Chem.* 72 (2): 174–179.

9 Peterson, J.I., Goldstein, S.R., Fitzgerald, R.V., and Buckhold, D.K. (1980). Fibre optic pH probe for physiological use. *Anal. Chem.* 52 (6): 864–869.

10 Vasylevska, A.S., Karasyov, A.A., Borisov, S.M., and Krause, C. (2007). Novel coumarin-based fluorescent pH indicators, probes and membranes covering a broad pH range. *Anal. Bioanal. Chem.* 387 (6): 2131–2141.

11 Wolfbeis, O.S. (2002). Fibre-optic chemical sensors and biosensors. *Anal. Chem.* 74 (12): 2663–2677.

12 Grattan, K.T.V. and Meggitt, B.T. (1999). *Chemical and Environmental Sensing*, vol. 4. Kluwer Academic Publishers.

13 Lee, S.T., Gin, J., Nampoori, V.P.N. et al. (2001). A sensitive fibre optic pH sensor using multiple sol-gel coatings. *J. Opt. A Pure Appl. Opt.* 3 (5): 355–359.

14 Gu, B., Yin, M.-J., Zhang, A.P. et al. (2009). Low-cost high-performance fibre-optic pH sensor based on thin-core fibre modal interferometer. *Opt. Express* 17: 22296–22302.

15 Hu, P., Dong, X., Wong, W.C. et al. (2015). Photonic crystal fibre interferometric pH sensor based on polyvinyl alcohol/polyacrylic acid hydrogel coating. *Appl. Opt.* 54: 2647–2652.

16 Goicoechea, J., Zamarreño, C.R., Matias, I.R., and Arregui, F.J. (2009). Utilization of white light interferometry in pH sensing applications by mean of the fabrication of nanostructured cavities. *Sensors Actuators B Chem.* 138: 613–618.

17 Liu, Z.H., Liu, J.F., and Chen, T.L. (2005). Phenol red immobilized PVA membrane for an optical pH sensor with two determination ranges and long-term stability. *Sensors Actuators B Chem.* 107 (1): 311–316.

18 Safavi, A. and Bagheri, M. (2003). Novel optical pH sensor for high and low pH values. *Sensors Actuators B Chem.* 90 (1-3): 143–150.

19 Carey, W.P., Degrandpre, M.D., and Jorgensen, B.S. (1989). Polymer-coated cylindrical wave-guide absorption sensor for high acidities. *Anal. Chem.* 61 (15): 1674–1678.

20 Uttamlal, M., Sloan, W.D., and Millar, D. (2002). Covalent immobilization of fluorescent indicators in photo- and electropolymers for the preparation of fibreoptic chemical sensors. *Polym. Int.* 51 (11): 1198–1206.

21 Fujii, T., Ishii, A., Kurihara, Y., and Anpo, M. (1993). Multiple fluorescence-spectra of fluorescein molecules encapsulated in the silica xerogel prepared by the sol-gel reaction. *Res. Chem. Intermed.* 19 (4): 333–342.

22 Arregui, F.J., Otano, M., Fernandez-Valdivielso, C., and Matias, I.R. (2002). An experimental study about the utilization of Liquicoat solutions for the fabrication of pH optical fibre sensors. *Sensors Actuators B Chem.* 87 (2): 289–295.

23 Goicoechea, J., Zamarreno, C.R., Matias, I.R., and Arregui, F.J. (2008). Optical fibre pH sensors based on layer-by-layer electrostatic self-assembled Neutral Red. *Sensors Actuators B Chem.* 132 (1): 305–311.

24 Egawa, Y., Hayashida, R., and Anzai, J.I. (2006). Multilayered assemblies composed of brilliant yellow and poly(allylamine) for an optical pH sensor. *Anal. Sci.* 22 (8): 1117–1119.

25 Ensafi, A.A. and Kazemzadeh, A. (1999). Optical pH sensor based on chemical modification of polymer film. *Microchem. J.* 63 (3): 381–388.

26 Kostov, Y., Tzonkov, S., Yotova, L., and Krysteva, M. (1993). Membranes for optical pH sensors. *Anal. Chim. Acta* 280 (1): 15–19.

27 Saari, L.A. and Seitz, W.R. (1982). pH sensor based on immobilized fluoresceinamine. *Anal. Chem.* 54 (4): 821–823.

28 Baldini, F., Giannetti, A., and Mencaglia, A.A. (2007). Optical sensor for interstitial pH measurements. *J. Biomed. Opt.* 12 (2): 024024.

29 Wallace, P.A., Elliott, N., Uttamlal, M. et al. (2001). Development of a quasi-distributed optical fibre pH sensor using a covalently bound indicator. *Meas. Sci. Technol.* 12 (7): 882–886.

30 Lin, J. (2000). Recent development and applications of optical and fibre-optic pH sensors. *TrAC Trends Anal. Chem.* 19 (9): 541–552.

31 Eggeling, C., Widengren, J., Rigler, R., and Seidel, C.A.M. (1998). Photobleaching of fluorescent dyes under conditions used for single-molecule detection: evidence of two-step photolysis. *Anal. Chem.* 70 (13): 2651–2659.

32 Drexhage, K.H. (1976). Fluorescence efficiency of laser-dyes. *J. Res. Nat. Bur. Stand. Sec. A* 80 (3): 421–428.

33 Netto, E.J., Peterson, J.I., McShane, M., and Hampshire, V. (1995). A fibreoptic broad-range pH sensor system for gastric measurements. *Sensors Actuators B Chem.* 29 (1–3): 157–163.

34 Wiczling, P., Markuszewski, M.J., Kaliszan, M. et al. (2005). Combined pH/organic solvent gradient HPLC in analysis of forensic material. *J. Pharm. Biomed. Anal.* 37 (5): 871–875.

35 Simek, M., Jisova, L., and Hopkins, D.W. (2002). What is the so-called optimum pH for denitrification in soil? *Soil Biol. Biochem.* 34 (9): 1227–1234.

36 Sloan, W.D. and Uttamlal, M. (2001). A fibre-optic calcium ion sensor using a calcein derivative. *Luminescence* 16 (2): 179–186.

37 Opitz, N. and Lübbers, D.W. (1983). New fluorescence photometrical techniques for simultaneous and continuous measurements of ionic strength and hydrogen ion activities. *Sensors Actuators B Chem.* 4: 473–479.

38 Duong, H.D., Sohn, O.J., Lam, H.T., and Rhee, J.I. (2006). An optical pH sensor with extended detection range based on fluoresceinamine covalently bound to sol-gel support. *Microchem. J.* 84 (1–2): 50–55.

39 Werner, T. and Wolfbeis, O.S. (1993). Optical sensor for the pH 10–13 range using a new support material. *Fresenius J. Anal. Chem.* 346 (6–9): 564–568.

40 Wang, J.X., Feng, X.B., Anderson, C.W.N. et al. (2012). Remediation of mercury contaminated sites – a review. *J. Hazard. Mater.* 221: 1–18.

41 Holmes, P., James, K.A.F., and Levy, L.S. (2009). Is low-level environmental mercury exposure of concern to human health? *Sci. Total Environ.* 408 (2): 171–182.

42 Mutter, J., Naumann, J., Sadaghiani, C. et al. (2004). Amalgam studies: disregarding basic principles of mercury toxicity. *Int. J. Hyg. Environ. Health* 207 (4): 391–397.

43 Houston, M.C. (2011). Role of mercury toxicity in hypertension, cardiovascular disease, and stroke. *J. Clin. Hypertens.* 13 (8): 621–627.

44 Liu, Z.H., Huan, S.Y., Jiang, J.H. et al. (2006). Molecularly imprinted TiO_2 thin film using stable ground-state complex as template as applied to selective electrochemical determination of mercury. *Talanta* 68 (4): 1120–1125.

45 Bui, M.P.N., Brockgreitens, J., Ahmed, S., and Abbas, A. (2016). Dual detection of nitrate and mercury in water using disposable electrochemical sensors. *Biosens. Bioelectron.* 85: 280–286.

46 Noh, M.F.M. and Tothill, I.E. (2011). Determination of lead(II), cadmium(II) and copper(II) in waste-water and soil extracts on mercury film screen-printed carbon electrodes sensor. *Sains Malays.* 40 (10): 1153–1163.

47 Nolan, E.M. and Lippard, S.J. (2007). Turn-on and ratiometric mercury sensing in water with a red-emitting probe. *J. Am. Chem. Soc.* 129 (18): 5910–5918.

48 Isaad, J. and El Achari, A. (2013). Azathia crown ether possessing a dansyl fluorophore moiety functionalized silica nanoparticles as hybrid material for mercury detection in aqueous medium. *Tetrahedron* 69 (24): 4866–4874.

49 Aydin, Z.Y., Wei, Y.B., and Guo, M.L. (2014). An "off-on" optical sensor for mercury ion detection in aqueous solution and living cells. *Inorg. Chem. Commun.* 50: 84–87.

50 Bera, K., Das, A.K., Nag, M., and Basak, S. (2014). Development of a rhodamine-rhodanine-based fluorescent mercury sensor and its use to monitor real-time uptake and distribution of inorganic mercury in live zebrafish larvae. *Anal. Chem.* 86 (5): 2740–2746.

51 Dai, H.L. and Xu, H. (2011). A water-soluble 1,8-naphthalimide-based 'turn on' fluorescent chemosensor for selective and sensitive recognition of mercury ion in water. *Bioorg. Med. Chem. Lett.* 21 (18): 5141–5144.

52 Hou, C., Urbanec, A.M., and Cao, H.S. (2011). A rapid Hg^{2+} sensor based on aza-15-crown-5 ether functionalized 1,8-naphthalimide. *Tetrahedron Lett.* 52 (38): 4903–4905.

53 Kaewtong, C., Niamsa, N., Wanno, B. et al. (2014). Optical chemosensors for Hg^{2+} from terthiophene appended rhodamine derivatives: FRET based molecular and in situ hybrid gold nanoparticle sensors. *New J. Chem.* 38 (8): 3831–3839.

54 Kaewtong, C., Wanno, B., Uppa, Y. et al. (2011). Facile synthesis of rhodamine-based highly sensitive and fast responsive colorimetric and off-on fluorescent reversible chemosensors for Hg^{2+}: preparation of a fluorescent thin film sensor. *Dalton Trans.* 40 (46): 12578–12583.

55 Tharmaraj, V. and Pitchumani, K. (2012). An acyclic, dansyl based colorimetric and fluorescent chemosensor for Hg(II) via twisted intramolecular charge transfer (TICT). *Anal. Chim. Acta* 751: 171–175.

56 Tian, M.Z., Liu, L.B., Li, Y.J. et al. (2014). An unusual OFF-ON fluorescence sensor for detecting mercury ions in aqueous media and living cells. *Chem. Commun.* 50 (16): 2055–2057.

57 Wang, X.Y., Zhao, J.J., Guo, C.X. et al. (2014). Simple hydrazide-based fluorescent sensors for highly sensitive and selective optical signaling of Cu^{2+} and Hg^{2+} in aqueous solution. *Sensors Actuators B Chem.* 193: 157–165.

58 Yang, R., Guo, X.F., Wang, W. et al. (2012). Highly selective and sensitive chemosensor for Hg^{2+} based on the naphthalimide fluorophore. *J. Fluoresc.* 22 (4): 1065–1071.

59 Zhang, X.B., Guo, C.C., Li, Z.Z. et al. (2002). An optical fibre chemical sensor for mercury ions based on a porphyrin dimer. *Anal. Chem.* 74 (4): 821–825.

60 Ruan, S., Ebendorff-Heidepriem, H., and Ruan, Y. (2018). Optical fibre turn-on sensor for the detection of mercury based on immobilized fluorophore. *Measurement* 121: 122–126.

61 Bontidean, L., Mortari, A., Leth, S. et al. (2004). Biosensors for detection of mercury in contaminated soils. *Environ. Pollut.* 131 (2): 255–262.

62 Ivask, A., Virta, M., and Kahru, A. (2002). Construction and use of specific luminescent recombinant bacterial sensors for the assessment of bioavailable fraction of cadmium, zinc, mercury and chromium in the soil. *Soil Biol. Biochem.* 34 (10): 1439–1447.

63 Dai, H.J., Liu, F., Gao, Q.Q. et al. (2011). A highly selective fluorescent sensor for mercury ion (II) based on azathia-crown ether possessing a dansyl moiety. *Luminescence* 26 (6): 523–530.

64 Nguyen, T.H., Venugopalan, T., Sun, T., and Grattan, K.T.V. (2016). Intrinsic fibre optic pH sensor for measurement of pH values in the range of 0.5–6. *IEEE Sensors J.* 16 (4): 881–887.

65 Wulff, G. (1995). Molecular imprinting in cross-linked materials with the aid of molecular templates – a way towards artificial antibodies. *Angew. Chem. Int. Ed. Engl.* 34 (17): 1812–1832.

66 Haupt, K. and Mosbach, K. (2000). Molecularly imprinted polymers and their use in biomimetic sensors. *Chem. Rev.* 100 (7): 2495–2504.

67 Branger, C., Meouche, W., and Margaillan, A. (2013). Recent advances on ion-imprinted polymers. *React. Funct. Polym.* 73 (6): 859–875.

68 Huang, C.Y. (1982). Determination of binding stoichiometry by the continuous variation method – the job plot. *Methods Enzymol.* 87: 509–525.

69 Thordarson, P. (2011). Determining association constants from titration experiments in supramolecular chemistry. *Chem. Soc. Rev.* 40 (3): 1305–1323.

70 Revis, N.W., Osborne, T.R., Holdsworth, G., and Hadden, C. (1990). Mercury in soil – a method for assessing acceptable limits. *Arch. Environ. Contam. Toxicol.* 19 (2): 221–226.

71 Bashor, B.S. and Turri, P.A. (1986). A method for determining an allowable concentration of mercury in soil. *Arch. Environ. Contam. Toxicol.* 15 (4): 435–438.

72 Gray, J.E., Theodorakos, P.M., Fey, D.L., and Krabbenhoft, D.P. (2015). Mercury concentrations and distribution in soil, water, mine waste leachates, and air in and around mercury mines in the Big Bend region, Texas, USA. *Environ. Geochem. Health* 37 (1): 35–48.

73 Dodani, S.C., He, Q.W., and Chang, C.J. (2009). A turn-on fluorescent sensor for detecting nickel in living cells. *J. Am. Chem. Soc.* 131 (50): 18020–18021.

74 Xia, S., Shen, J.J., Wang, J.B. et al. (2018). Ratiometric fluorescent and colorimetric BODIPY-based sensor for zinc ions in solution and living cells. *Sensors Actuators B Chem.* 258: 1279–1286.

75 Murray, R. and Tinsley, L. (2006). *Measuring Different Aspects of Problem Drug Use: Methodological Developments*. London: Home Office Online Report.

76 Degenhardt, L., Chiu, W.T., Sampson, N. et al. (2008). Toward a global view of alcohol, tobacco, cannabis, and cocaine use: findings from the WHO World Mental Health Surveys. *PLoS Med.* 5 (7): 1053–1067.

77 Gilloteaux, J. and Ekwedike, N.N. (2010). Cocaine causes atrial purkinje fibre damage. *Ultrastruct. Pathol.* 34 (2): 90–98.

78 Strano-Rossi, S., Molaioni, F., Rossi, F., and Botre, F. (2005). Rapid screening of drugs of abuse and their metabolites by gas chromatography/mass spectrometry: application to urinalysis. *Rapid Commun. Mass Spectrom.* 19 (11): 1529–1535.

79 Popa, D.S., Vlase, L., Leucuta, S.E., and Loghin, F. (2009). Determination of cocaine and benzoylecgonine in human plasma by Lc-Ms/Ms. *Farmacia* 57 (3): 301–308.

80 Tagliaro, F., Antonioli, C., Debattisti, Z. et al. (1994). Reversed-phase high-performance liquid-chromatographic determination of cocaine in plasma and human hair with direct fluorometric detection. *J. Chromatogr. A* 674 (1-2): 207–215.

81 Trachta, G., Schwarze, B., Sagmuller, B. et al. (2004). Combination of high-performance liquid chromatography and SERS detection applied to the analysis of drugs in human blood and urine. *J. Mol. Struct.* 693 (1–3): 175–185.

82 Antonilli, L., Suriano, C., Grassi, M.C., and Nencini, P. (2001). Analysis of cocaethylene, benzoylecgonine and cocaine in human urine by high-performance thin-layer chromatography with ultraviolet detection: a comparison with high-performance liquid chromatography. *J. Chromatogr. B* 751 (1): 19–27.

83 Oiye, E.N., de Figueiredo, N.B., de Andrade, J.F. et al. (2009). Voltammetric determination of cocaine in confiscated samples using a cobalt hexacyanoferrate film-modified electrode. *Forensic Sci. Int.* 192 (1–3): 94–97.

84 Baumgartner, W.A., Black, C.T., Jones, P.F., and Blahd, W.H. (1982). Radioimmunoassay of cocaine in hair – concise communication. *J. Nucl. Med.* 23 (9): 790–792.

85 Verebey, K. and Depace, A. (1989). Rapid confirmation of enzyme multiplied immunoassay technique (Emit) cocaine positive urine samples by capillary gas-liquid-chromatography nitrogen phosphorus detection (Glc Npd). *J. Forensic Sci.* 34 (1): 46–52.

86 Wang, J. (2006). Electrochemical biosensors: towards point-of-care cancer diagnostics. *Biosens. Bioelectron.* 21 (10): 1887–1892.

87 Devine, P.J., Anis, N.A., Wright, J. et al. (1995). A fibreoptic cocaine biosensor. *Anal. Biochem.* 227 (1): 216–224.

88 Meijler, M.M., Kaufmann, G.F., Qi, L.W. et al. (2005). Fluorescent cocaine probes: a tool for the selection and engineering of therapeutic antibodies. *J. Am. Chem. Soc.* 127 (8): 2477–2484.

89 Liu, J.W. and Lu, Y. (2006). Fast colorimetric sensing of adenosine and cocaine based on a general sensor design involving aptamers and nanoparticles. *Angew. Chem. Int. Ed.* 45 (1): 90–94.

90 Zhang, C.Y. and Johnson, L.W. (2009). Single quantum-dot-based aptameric nanosensor for cocaine. *Anal. Chem.* 81 (8): 3051–3055.

91 Li, X.X., Qi, H.L., Shen, L.H. et al. (2008). Electrochemical aptasensor for the determination of cocaine incorporating gold nanoparticles modification. *Electroanalysis* 20 (13): 1475–1482.

92 Baker, B.R., Lai, R.Y., Wood, E.H. et al. (2006). An electronic, aptamer-based small-molecule sensor for the rapid, label-free detection of cocaine in adulterated samples and biological fluids. *J. Am. Chem. Soc.* 128 (10): 3138–3139.

93 Li, Y., Qi, H.L., Peng, Y. et al. (2007). Electrogenerated chemiluminescence aptamer-based biosensor for the determination of cocaine. *Electrochem. Commun.* 9: 2571–2575.

94 Alexander, C., Andersson, H.S., Andersson, L.I. et al. (2006). Molecular imprinting science and technology: a survey of the literature for the years up to and including 2003. *J. Mol. Recognit.* 19 (2): 106–180.

95 Munkholm, C., Parkinson, D.R., and Walt, D.R. (1990). Intramolecular fluorescence self-quenching of fluoresceinamine. *J. Am. Chem. Soc.* 112 (7): 2608–2612.

96 Kriz, D., Ramstrom, O., Svensson, A., and Mosbach, K. (1995). Introducing biomimetic sensors based on molecularly imprinted polymers as recognition elements. *Anal. Chem.* 67 (13): 2142–2144.

97 Turkewitsch, P., Wandelt, B., Darling, G.D., and Powell, W.S. (1998). Fluorescent functional recognition sites through molecular imprinting. A polymer-based fluorescent chemosensor for aqueous cAMP. *Anal. Chem.* 70 (10): 2025–2030.

98 Kempe, M. and Mosbach, K. (1991). Binding-studies on substrate-and enantio-selective molecularly imprinted polymers. *Anal. Lett.* 24 (7): 1137–1145.

9

Application of Nanotechnology to Optical Fibre Sensors

Recent Advancements and New Trends

Armando Ricciardi, Marco Consales, Marco Pisco, and Andrea Cusano

Optoelectronics Group, Department of Engineering, University of Sannio, Benevento, Italy

This chapter presents an overall picture pertaining to the 'Lab-on-FIBRE technology' vision illustrating the main milestones set along the technological roadmap aimed to achieve as ultimate objective the development of flexible, multifunctional plug-and-play fibre-optic sensors for specific applications. Main achievements in the identification of nanofabrication strategies properly working onto not conventional substrates as the case of optical fibres are here collected and discussed. Perspectives and challenges that lie ahead are highlighted with a special focus towards full spatial control at nanoscale and high-throughput production scenarios. The rapid progress at the fabrication stage has now opened new avenues towards the development of multifunctional plug-and-play platforms, here discussed with particular emphasis on their new functionalities and performances, demonstrating the potentials of this powerful technology in many strategic application scenarios.

9.1 Introduction

Lab-on-fibre (LOF) technology is an emerging research field that essentially envisions the integration of functional materials at micro- and nanoscales (i.e. the 'labs') onto optical fibres and is aimed at developing a future generation of advanced all-in-fibre miniaturized devices and components exploitable in many

All authors contributed equally to this work.

Optical Fibre Sensors: Fundamentals for Development of Optimized Devices, First Edition.
Edited by Ignacio Del Villar and Ignacio R. Matias.
© 2021 The Institute of Electrical and Electronics Engineers, Inc.
Published 2021 by John Wiley & Sons, Inc.

strategic sectors ranging from optical processing to environmental monitoring, life science, and safety and security applications [1]. The key concept is to transform an inert optical fibre into a multifunctional system where ultra-compact labs are developed and miniaturized into a single optical fibre, thus disruptively enlarging the conventional optical fibre functionalities and performances [2–4].

The integration of advanced functional materials at micro- and nanoscale, exhibiting the more disparate properties, combined with suitable light–matter mechanisms, is the key for the development of highly integrated and multifunctional technological platforms completely realized in a single optical fibre. This achievement would be the cornerstone of a new photonics technological revolution that would lead to the definition of a novel generation of micro- and nanophotonic devices 'all-in-fibre' [5, 6]. Optical fibres are well suited to support this revolution also in virtue of the dynamicity and versatility of the related fabrication processes. The recent development of fibre processing technologies has indeed enabled the fabrication of fibrous structures with increasingly complex functionalities [7–10].

The first significant breakthroughs in optical fibre processing, which contrasted with conventional step-index solid core silica fibres, were the fabrication of 1D and two-dimensional (2D) photonic crystal (PC) fibres [11–13]. This marked the beginning of a deeper interest, not only in the functionality of the fabricated fibres and, in particular, the engineering of their optical properties but also in the materials and the physical processes at play behind the fabrication technique.

One of the major impacts was the realization of miniaturized optofluidic platforms offering potential for achieving more functional and more compact devices. Such integrated systems bring fluid and light together and exploit their microscale interaction for a large variety of applications [14]. Moreover, the understanding of the viscous flow and surface science at play in this approach has also led to the design and fabrication of fibres with new materials [15]. This has been exploited in optics for the fabrication of solid core polycrystalline semiconductor fibres. It has also led to the design of optoelectronic fibres that can not only guide light but also exhibit a variety of novel functionalities, such as optical, heat, or chemical sensing, piezoelectric actuation, surface-emitting fibre lasers, advanced optical probes, or field effect and phase change-based devices. Polymer fibres with electrically conducting domains can also be used in optical imaging systems or in purely electronic functions, such as touch sensing or capacitors [10, 16, 17]. At the same time, recent studies of the interactions of light with nanostructures have revealed a weirder and wonderful behaviour, and application of nanophotonics research has already demonstrated promise for new devices aimed to slowing, filtering, trapping, confining, processing, and enhancing light.

Everybody can recognize that nanoplasmonics [18], metasurfaces [19], optomechanics [20], PC [21], ultra-high-resolution near-field optical imaging [22], optofluidics [23], magneto-plasmonics [24], and quantum optics [25], just to name a few, have recently opened new avenues towards the dream of untapped device functionality.

As matter of fact, the combination of the impressive performances of on-chip nanophotonics with the unique advantages of optical fibres led to the birth of a new fascinating field of research called 'lab-on-fibre' technology.

Novel 'intelligent fibres' with unique diagnostics features for life science applications as well as new capabilities to elaborate, process, exchange, and manipulate the carried information triggered by internal or external stimuli may become a close reality.

Nanotechnology and nanophotonics have been considered the most promising key enabling technologies to meet this dream; however, relevant barriers and limitations have to be considered in order to translate the vision to reality [26–28].

The most relevant challenges include:

- Integration of functional materials with the more disparate electromagnetic, mechanical, and biochemical properties onto not conventional substrates as the case of and its control at nanoscale.
- Pushing light–matter interaction within optical fibres at its ultimate limit opening new avenues for easy to use plug-and-play multifunctional devices designed for specific applications.
- Although stringent and severe, LOF needs and requirements did not find the optical fibre technology totally unprepared. The integration of materials and structural modification onto this not conventional substrate has indeed a quite long tradition [29]. Depending on the specific location where functional materials are integrated to, LOF technology umbrella can be subdivided in three main subclasses [5] as schematically reported in Figure 9.1:
  - Lab-around-fibre devices where functional materials are integrated onto the (outer) cylindrical surface of optical fibres.
  - Lab-on-tip (LOT) devices where the intrinsic light-coupled termination of optical fibres plays the role of substrate for materials integration.
  - LOF devices where functional materials are integrated within the holey structure pertaining to specialty fibres.

Among the three subcategories, the LOT platforms are very attractive platforms for sensing, especially for their inherently light-coupled micro-sized active area. The all-optical working principle and interrogation system make fibre tip probes very easy to use, opening new avenues for plug-and-play device for access to remote, confined, and hostile environments. Although the potential to translate the paradigm of nanophotonic 'lab-on-chip' in new ultra-compact and ready to use fibre-optic nano-devices was immediately recognized by the scientific community, a great amount of work was required to set the first milestones of the LOT roadmap dealing with the identification of suitable fabrication strategies able to correctly engineer the fibre facet.

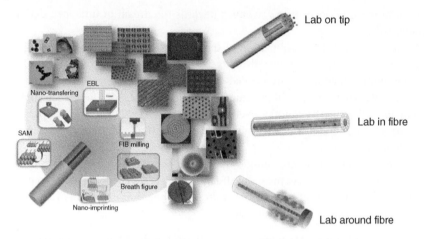

Figure 9.1 The lab-on-fibre paradigm. *Source:* Reproduced with permission from [5]. Copyright Wiley-VCH Verlag GmbH & Co. KGaA. Reproduced with permission of IEEE.

9.2 A View Back

In LOT devices, the interaction between light and the parameters to be measured takes place onto the optical fibre end facet. The intrinsic light-coupled termination of optical fibres plays the role of substrate for materials integration. The flat tip of an optical fibre thus becomes an inherently light-coupled micro-sized active area that can be effectively exploited for remote sensing applications. The unique properties offered by the LOT technology pushed researchers to focus their efforts in finding effective strategies to engineer the fibre tip, by integrating functional materials and nanostructures, suitably chosen for the specific application.

Since the early 1990s, fibre tips were abrasively modified or textured and subsequently coated with a metallic film supporting plasmonic resonances [30]. In addition to 'surface' decorations, preliminary attempts of modifying the fibre tip were also induced through high-temperature processes such as arc discharge, used to create silica microspheres supporting whispering gallery mode resonators with high-quality factors [31]. However, in most of the cases, the poor control and the modest technological readiness of the integration methodologies posed severe limitations in their applicability in real scenarios. Repeatability and disorder-induced frustration of the performances resulted in severe barriers limiting the effective development of valuable technological platforms able to compete with more mature technologies. The demand of improving optical fibre probe functionalities was continuously growing. Moreover, the light control at nanoscale can be achieved only by controlling the geometrical features of the structures in which the light flows. The fibre tip thus was integrated with functional materials and structures with sub-wavelength dimensions (such as photonic and plasmonic structures, just to name a few) supporting resonant effects at very specific wavelengths.

To this aim, most of the modern micro- and nano-fabrication approaches arising from the microelectronic industry have been suitably adapted to work onto the unconventional substrate such as the fibre tip [6]. With the passing years, fabrication approaches followed different directions that can be mainly classified in four different categories: the first one includes the direct approaches in which nanostructuring processes (deposition and patterning of the fibre tip) are directly realized onto the optical fibre facet. In the direct approaches fall all the 'top-down' approaches such as electron beam lithography (EBL) and focused ion beam (FIB) milling, interference lithography (IL), and femtosecond (fs) laser micromachining. The second category embraces indirect approaches that deal with the preventive fabrication of nanostructures on planar substrates via standard nanolithographic processes and their subsequent transfer onto the fibre tip. Indirect approaches include nanoimprint lithography (NIL) and nanotransferring. The third group, self-assembly (SA) methodologies, essentially uses the chemical properties of single atomic or molecular species in order to create self-assembled clusters into valuable conformations created on the fibre tip. These methodologies involve the synthesis of materials, through chemical reactions, permitting to the forerunner elements to grow in dimension. Finally, the last group includes the major advances in three-dimensional (3D) microprinting of arbitrary 3D micro/ nanostructures with a surface roughness acceptable for optical applications.

Although the aforementioned micro- and nanofabrication tools were judiciously used to customize and pattern the optical fibre tip to ensure the ultimate light–matter interaction level for specific applications, further functionalities and additional degrees of freedom can be achieved through smart materials integration processes, in which the presence of active compounds and advanced materials on the fibre tip is exploited for achieving responsivity and sensitivity enhancement.

In the following, we review the main fabrication routes reported so far for LOT devices. Within the above-mentioned classification, we discuss both strength and weakness of each approach by presenting a broad overview on the realized device in both biochemical sensing and optomechanics applications.

9.3 Nanofabrication Techniques on the Fibre Tip for Biochemical Applications

In the particular context of (bio)chemical sensing applications, lab design typically involves the exploitation of advanced photonic systems providing light control at the nanoscale combined with sophisticated chemistry, with the ultimate goal of enhancing the light–matter interaction in specific spatial locations. Starting from mid-2000s, different fabrication approaches based on both bottom-up and bottom-down technologies have been successfully demonstrated. In the following, in the wake of our previous review paper [5], we discuss on the main fabrication routes reported so far with a particular emphasis on the last achievements. Within this classification, we also present an overview on the realized device and their general performances for biochemical sensing applications.

9.3.1 Direct Approaches

FIB milling has become one of the most popular and precise maskless direct patterning tools at the present time, being able to directly pattern or 'mill' the specimen surface with nanometre precision. However, FIB suffers from some undesired effects, mainly the angled sidewalls of realized holes and the ion implantation inside the specimen during the milling procedure [32]. Both aspects need to be carefully taken into consideration in the design stage to guarantee a correct development of the device [33]. In addition, FIB milling processes are generally time consuming, and only a small area (typically of the order of a few hundreds of μm^2) can be processed with one fabrication step.

The first investigations on the FIB milling capability to micromachine the tip of an optical fibre date back to 2006, when small aperture patterns were directly written for the first time on the fibre facet for photonics applications [34]. In the same year, a high-sensitivity microcantilever-based sensor was demonstrated [35], especially useful for atomic force microscopy. Since then, several attempts aimed at direct patterning the optical fibre tip by FIB have been proposed for the development of novel optical fibre components [36–39] and surface plasmon resonance (SPR)-based sensors [33, 40–47], the latter mainly relying on metallic nanostructures, supporting plasmonic resonances, obtained by milling a thin gold layer previously deposited on the fibre tip.

Dhawan et al. fabricated plasmon resonance-based nanosensors by creating periodic arrays of nano-apertures on the gold (Au)-coated cleaved (or tapered) terminations of step-index multimode optical fibres (MMF) [40, 41]. Realized probes exhibited a bulk refractive index (RI) sensitivity of 533 nm/RIU as well as the ability to detect biomolecular interactions in a biotin–streptavidin binding experiment. The same plasmonic optical fibre probes have been also exploited for surface-enhanced Raman spectroscopy (SERS) measurements [42].

With a similar approach, fibre-optic nanosensors based on Au nanohole arrays milled on the tip of single-mode optical fibres (SMF) were fabricated by Andrade et al. and exploited for SERS measurements [43]. The effect of the hole shape on the SERS performance was also investigated by realizing both circular and bow-tie nano-apertures, the last ones being able to promote an improvement in the collected SERS intensity.

More recently, Micco et al. proposed a technique for templating the optical fibre tip [33]. The process is based on a direct writing process followed by a high RI (SiO_x) overlay deposition, giving rise to a 'double-layer' PC slab supporting guided resonances. Fibre tip templating is particularly useful since it allows to increase the process flexibility; after a suitable optimization of the recipe for patterning the fibre, different materials (metals, dielectric, and semiconductors) may be deposited to create complex nanostructures.

In 2017, a pioneering work on the integration of optical metasurfaces on the tip of SMFs was presented, giving the birth to the first 'optical fibre metatips (OFMTs)' [44]. OFMTs mainly rely on plasmonic phase gradient metasurfaces realized by FIB milling inhomogeneous arrays of rectangular aperture nano-antennas (each one rotated by $\pm45°$ in the plane of the fibre tip) in thin Au layer previously deposited on the fibre end face. As a proof of concept, the designed plasmonic OFMTs were able to impress, to the normally impinging beam, a linear phase (and constant amplitude) profile, so as to steer the output beam in desired directions by an arbitrary deflection angle. It was also shown that OFMTs deflecting the output beam by 90° are able to efficiently couple normally incident light to surface waves and exhibit a sensitivity to local RI variations higher than that of the corresponding gradient-free uniform array [44, 45]. Specifically, the authors investigated (both numerically and experimentally) the surface sensitivity to the presence of 40 nm thick SiO_x overlay (having RI of about 1.7) and compared it with that obtained by the corresponding gradient-free structure (consisting of a periodic pattern of the same nano-antennas but equally rotated). Notably, the phase gradient OFMTs exhibited a greater enhancement of the electric field at the metatip surface and, consequently, an improved surface sensitivity (the resonant dip in the reflectance spectrum of the devices redshifted ~224 nm) with respect to that obtained by the benchmark (for which a resonant wavelength shift of ~132 nm was obtained).

In the same year, a high-quality all-dielectric metamaterial dispersive grating was realized on the facet of an SMF [46]. The nanostructure was realized by FIB milling alternating deep and shallow grooves into the silica constituting the fibre, followed by the deposition of a thin amorphous silicon layer. By illuminating the nanostructure with normally incident linearly polarized light (throughout the fibre), a sharp dip in the transmission spectrum was obtained at $\lambda = 1385.5$ nm, having a Q-factor as high as 310, corresponding to the excitation of a Fano resonance in the metamaterial structure. Results reported by the authors showed the ability of the all-dielectric structure to detect changes in the surrounding medium RI, with a sensitivity of about 400 nm/RIU accompanied by a very narrow line width of ~5 nm.

Finally, an SPR-based LOT multi-parameter nanoprobe for the simultaneous RI and temperature measurements was recently proposed by Kim and Yu [47]. The nanoprobe employs a plasmonic crystal cavity consisting of a FIB-milled 2D rectangular Au grating in the fibre core, surrounded by four distributed Bragg reflector gratings in the cladding area (designed to enhance the Q-factor of the plasmonic resonances of the SMF device). The crystal cavity supports two spatially separated Rayleigh anomaly (RA) surface plasmon polariton (SPP) resonances (see Figure 9.2): one more located in the solution medium and the other located in the fibre silica. The simultaneous measurement of temperature and RI is made

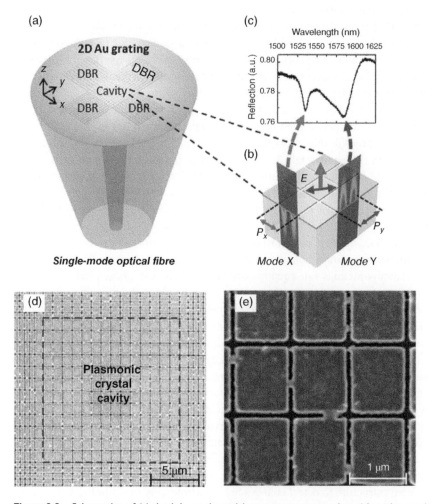

Figure 9.2 Schematics of (a) the lab-on-tip multi-parameter nanoprobe with a plasmonic crystal cavity on the fibre end face; (b) two spatially separated RA-SPP resonance modes excited in the plasmonic crystal cavity; (c) reflection spectrum from the nanoprobe with two spatially separated resonances; and (d, e) SEM images of the plasmonic crystal cavity of the fabricated multi-parameter nanoprobe. *Source:* Adapted with permission [47]. Reproduced with permission of Springer Nature.

possible by the fact that while both resonances are sensitive to thermal variations (even if with different sensitivities depending on the thermo-optic coefficient of the two media, i.e. solution and silica), only the resonance located in the solution medium is sensitive to RI variation (the sensitivity is about 1150 nm/RIU).

IL is another facile, low-cost direct writing nanolithography technique, widely used in the recent years. It is based on the maskless exposure of a photoresist (PR) layer with two or more coherent light beams, which allows to obtain regular arrays of fine features over large areas, with periodicity in one, two, or three dimensions. The main drawback of this technique consists in the limited resolution, that is, diffraction limited.

The first application of the IL technique to the fibre tip was demonstrated in 1999 [48], when waveguide gratings with sub-wavelength period were integrated onto the end face of a standard optical fibre for the realization of diffractive fibre-optic devices. By using the same approach, more recently, Feng et al. fabricated a PR grating on a ZnO waveguide layer previously deposited on the polished end face of a 600 μm core MMF [49].

A couple of years later, Yang et al. realized an array of nanopillars (with diameter of 160 nm) on the fibre tip by using PR as sacrificial layer for etching the fibre tip, which was then coated with silver (Ag) for achieving a SERS-active platform [50]. The probe, working in the optrode configuration, was successfully used for the detection of toluene vapours.

In addition, more recently, Kim and Jeong introduced a maskless direct fabrication method for the integration of microstructures on the optical fibre end face [51]. It is based on the combination of PR spin coating and programmable digital micromirror device (DMD)-based ultraviolet (UV) photolithography (details in Figure 9.3). It allows for rapid prototyping and low-cost fabrication ensuring a spatial resolution of 2.2 μm on the fibre surface. Different kinds of microstructures (lines, rings, holes, checker arrays, etc.) were fabricated on the tip of either SMF or MMF by using this process.

Finally, Hemmati et al. recently proposed a fibre facet-mounted device consisting of a silicon nitride (Si_3N_4)-based resonator directly fabricated on the tip of an MMF [52]. The device was realized by combining different microfabrication processes, such as holographic IL and reactive ion etching (RIE). The interaction of light coming from the MMF with the silicon nitride resonator results in a sharp dip in the transmission spectrum at a wavelength around 1550 nm, which redshifts with the external RI, with a sensitivity of about 200 nm/RIU.

Among the direct patterning approaches, EBL has also been intensively used in the last decade for the integration of nanostructures onto the facet of an optical fibre. EBL basically consists of the chemical modification of electron beam (EB) resist (previously deposited on the fibre tip) caused by EB irradiation. Charge dissipation issue, imposed by the electrically insulating nature of silica glass, is

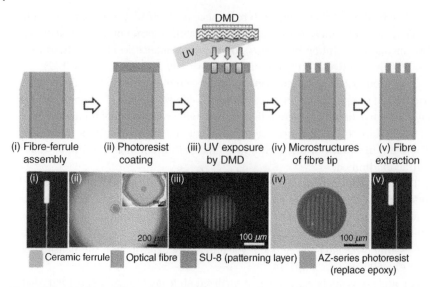

(i) Fibre-ferrule assembly (ii) Photoresist coating (iii) UV exposure by DMD (iv) Microstructures of fibre tip (v) Fibre extraction

Ceramic ferrule Optical fibre SU-8 (patterning layer) AZ-series photoresist (replace epoxy)

Figure 9.3 Maskless fabrication procedure for functional optical elements on the optical fibre top surface: (i) insertion of the optical fibre in a ceramic ferrule assembly, (ii) spin coating of PR over the entire fibre (inset: multilayer PR coating result), (iii) UV exposure by DMD-based maskless lithography, (iv) photolithographic definition of SU-8 resist on the fibre ferrule assembly, and (v) extraction of the optical fibre from the ferrule. *Source:* Adapted with permission [51]. Reproduced with permission of OSA Publishing.

overcome by depositing onto the resist sacrificial layers such as conducting polymers [53] or metals [54, 55]. Typical patterned areas are of the order of $100 \times 100\,\mu m^2$; the resist can be used as an active part of optical components [55–57] or as sacrificial layer for etching or lift-off procedures [58, 59]. In the first case, it is possible to give rise to hybrid metallo-dielectric nanostructures. For example, Consales et al. integrated on the fibre tip a double plasmonic structure constituted by a gold pillar array on the bottom and a patterned metallic slab on the top separated by a ZEP layer [55]. EBL was used to pattern the resist ZEP 520A layer (Zeon Chemicals), which was successively covered by a thin Au layer (deposited on both the ridges and the grooves of the patterned ZEP layer). On the other hand, the patterned EB resist can be employed as mask for etching metallic layers previously deposited over the fibre facet [53]. By following this approach, an array of Au pillars supporting localized SPR (LSPR) has been effectively integrated on the fibre tip [60]. This plasmonic device shows a high RI sensitivity of 195.72 nm/RIU and a label-free affinity sensing capability, preliminarily demonstrated in the case of biotin–streptavidin interaction. Alternatively, a lift-off approach can be employed, consisting of the deposition of materials over the patterned resist (acting as a mask) followed by its removal. Sanders et al. realized a biosensor constituted by an Au pillar array directly fabricated via EBL and Au lift-off process, capable of detecting femtomolar concentrations of free prostate-specific antigen (f-PSA) [61]. With a similar approach, Ricciardi et al. demonstrated an optical fibre biosensor for detecting human thyroglobulin (Tg) at nanomolar concentrations [59].

In all the fabrication approaches discussed so far, the process was applied to a single optical fibre. Wang et al. recently proposed a novel fabrication approach for implementing nanostructures with deep sub-wavelength dimensions on the end faces of multiple fibres at once [62]. The workflow of the fabrication procedure is shown in Figure 9.4.

To demonstrate their approach, a square array (pitch 1.1 μm) of gold trimers (radius and thickness of 190 and 40 nm, respectively) with precisely engineered deep sub-wavelength gaps (75 nm) has been successfully patterned on the end faces of step-index fibres. Dense arrays of nano-trimers optical resonances are located within the single-mode regime of typical telecommunication fibres, showing asymmetrical transmission spectra with strong dips at around 1.65 μm wavelength that results from a coupling of the individual trimer resonance with those of its neighbouring unit. RI sensing experiments yielded a sensitivity of about 390 nm/RIU, representing the state of the art for such a device type.

A few attempts of direct fabrication of LOT devices by means of fs laser micromachining have also been reported so far [63–69]. While this process is generally faster and cheaper than the ones described above, the surface of the patterned areas is generally rougher [63]. Fs laser ablation was demonstrated to create fibre end surface grating [64], Fresnel zone plate lenses [65], and SERS probes tested for Rhodamine 6G (R6G) detection (Figure 9.5) [66, 67].

Direct fs laser ablated nanostructures were also fabricated on the tip of sapphire optical fibres [68]. Specifically, Yuan et al. fabricated a reflection-based SERS probe by creating nanometre-scale structures on the end face of a single-crystal sapphire fibre, followed by the chemical plating of a thin Ag layer. The performances of the realized probe were analysed using a R6G solution (with a concentration of 10^{-7} M) and were compared with those obtained by using the same SERS-active substrate but realized on the tip of fused silica SMF and MMF. Results showed that the sapphire fibre SERS probe produced a much weaker background Raman signal compared with the fused silica counterparts. In addition, it was able to retrieve the Raman fingerprint of the R6G solution, performing much better than standard glass fibres, that on the contrary were not able to pick up any recognizable Raman signature [68].

Finally, SERS probes based on polymer fibres produced by fs laser ablation and subsequent photoreduced deposition of Ag nanoparticles (NPs) were also recently proposed by Youfu et al. [69]. Specifically, in Ref. [69], surface grating structures realized on the end face of polymethyl methacrylate (PMMA) optical fibres revealed to be able to improve the SERS performances (in case of R6G detection)

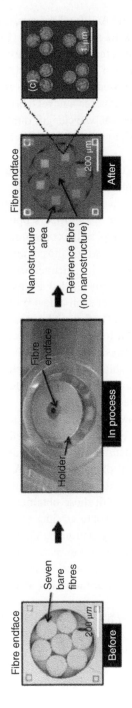

Figure 9.4 Workflow of the nanostructure fabrication process described in [62]: a hexagonal bundle of seven SMF-28 fibres is glued into a silica capillary that is then fixed inside a metal holder. This is mechanically polished before spin coating and any lithographic steps, thus providing a base for spin coating films of resists of precise thickness. This planarized configuration is inserted into a customized holder (diameter: 75 mm, height 12 mm) that is compatible with the EBL machine and contains the seven optical fibres with their free ends being winded inside the central holey section of the container. (c) SEM image of four trimer units on the fibre facet. *Source:* Adapted with permission [62]. Reproduced with permission of OSA Publishing.

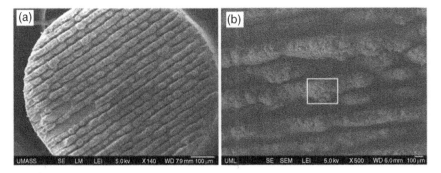

Figure 9.5 SEM images of (a) the facet of the optical fibre SERS probe and (b) micrograting-like structure of the SERS substrate (enlarged image of the boxed area of (a)). *Source:* Adapted with permission [67]. Reproduced with permission of IEEE.

by several times with respect to ordinarily roughened counterparts. In particular, Youfu et al. demonstrated the possibility to optimize the sensing properties of the polymer-based SERS probe by judiciously tailoring the fs laser pulse energy and scan surface period during the fabrication process. Optimized probes revealed their potentiality to detect R6G solutions with concentrations down to 10^{-8} M.

9.3.2 Indirect Approaches

The NIL is a reliable fabrication technique for high-throughput, low-cost, and high-resolution indirect manufacturing of micro/nanometre-scale patterns on the fibre termination [70, 71]. This technique is able to produce surface relief geometrical features by physically deforming a material in a temperature- and pressure-controlled printing process. To this aim, a stamp (or mould/master/template) is used. The imprint could be implemented directly onto optical fibres or onto polymers previously deposited upon their facet.

The direct imprinting of fibre ends has been demonstrated with optical fibres made of all materials characterized by transition temperatures lower than those of common silica, such as polymer [72, 73], polycrystalline, silver halide [74], and chalcogenide glass [75]. Unfortunately, imprinting the fibre material produces a sort of swelling of the fibre tip due to its initial thermal softening [73, 75, 76]. In order to avoid this drawback, the NIL method has been applied onto polymer pre-deposited on the fibre tip. Periodic patterns with dimensions of hundreds of nanometres have been successfully demonstrated [77–79]. The imprinting step requires a careful alignment of the fibre, especially if the pattern on the mould has small dimensions [79].

To increase the throughput, Kostovski et al. developed a technique for parallel, self-aligned, and portable nanoimprint [6, 80]. They applied NIL to replicate biological nanostructures (anti-reflection nanostructure on a cicada wing) onto the fibre tip, along with a curved, large area mould based on self-assembled nanostructure of anodized aluminium oxide pattern as a master template [6]. The probes were metalized with Ag and tested for SERS measurements.

Calafiore et al. recently reported the fabrication of complex 3D structures for light wavefront manipulation on the facet of an optical fibre by UV-NIL [81]. The process was designed to perform fibre imprinting with a transparent mould, which allows simplifying the alignment of the mould and the fibre thanks to the inspection of the imprinted site. It was tested for the fabrication of wavelength selective 3D beam splitter, but it can be extended to the fabrication of a multitude of other optical components [81, 82].

By using a similar approach (details in Figure 9.6), the same authors demonstrated the UV-NIL fabrication of functional 'campanile' near-field probes with sub-70 nm scale resolution and sub-100 nm positioning precision [83]. The campanile probe was composed of a 3D pyramidal metal–insulator–metal (MIM) geometry and was used in an actual near-field scanning optical microscope (NSOM) measurement performing hyperspectral photoluminescence mapping of standard fluorescent beads.

The nanotransferring is a further indirect approach to realize micro-nanostructures on the fibre cleaved end. As the word suggests, it is based on transferring, on the fibre facet, nanostructures previously realized onto planar substrates. The proposed approaches can be essentially classified in two main categories, namely, 'release and attach' or 'contact and separate' techniques. In the first case, the structure fabricated on the planar substrate is first released and then transferred over the fibre facet where it is bound or melted. In this framework, Smythe et al. have used a thin, sacrificial thiolene film to strip the metallic nanostructures (realized by means of EBL) from the patterned substrate [84]. Successively, the film is transferred onto the fibre facet, and the sacrificial layer is removed. By using this process, a bidirectional optical fibre probe was employed for *in situ* SERS detection [69]. The same group also conceived a different approach that conjugates the nano-skiving technique (the thin sectioning of epoxy nanostructures covered by thin metallic layers) with manual relocation of the sectioned slabs onto the fibre tips through dipping in water environment [85].

Recently, Wang and co-workers demonstrated a flexible transfer technique for the fabrication of plasmonic structures on the end facets of fibres [86]. The structures were first realized on planar substrates before they were 'flexibilized' into thin films and transferred directly onto fibre tips. Figure 9.7 shows the fabrication procedure. The success of this fabrication was verified by the microscopic characterization of the metallic photonic structures on the fibre end facets.

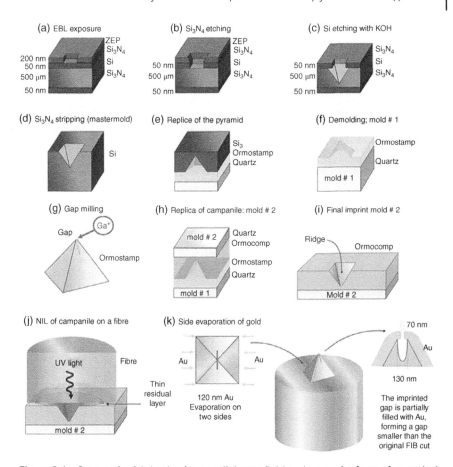

Figure 9.6 Process for fabricating 'campanile' near-field probes on the facet of an optical fibre by UV-NIL. (a) EBL exposure and development of ZEP. (b) Pattern transfer into Si_3N_4 by RIE. (c) Undercut of silicon in KOH to form inverted pyramids. (d) Si_3N_4 stripping and completion of the mastermould. (e) Replication of the mastermould into OrmoStamp to form pyramids. (f) Demoulding and completion of mould #1. (g) Ga+ FIB milling of the gap at the apex of the pyramid. (h) Replication of milled pyramids into Ormocomp. (i) Demoulding and completion of mould #2 on a quartz substrate. (j) Imprint on a fibre. (k) Evaporation of 120 nm Au on two of the four sides of the pyramid. The imprinted slit at the apex prevents gold from shortcutting the two sides and creates a plasmonic gap at the tip of the probe. *Source:* Adapted with permission [83]. Reproduced with permission of Springer Nature.

In the 'contact and separate' approach, the fibre tip, coated with an epoxy layer, is put in contact with the planar substrate containing the master structure and then released. By following this approach, Shambat et al. integrated a PC cavity onto the facet of a standard optical fibre [87, 88]. Similarly, Lerma Arce et al.

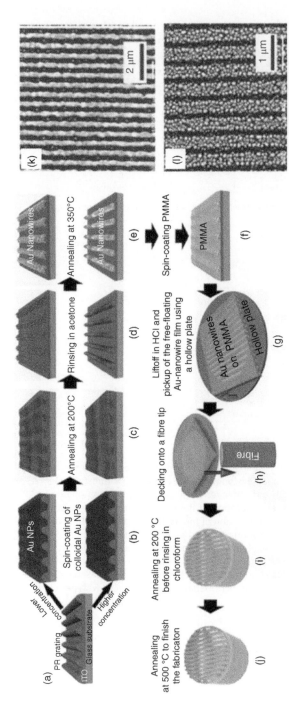

Figure 9.7 Steps of the fabrication procedure described in [86]. (a) Template PR grating fabricated by IL [49] on a glass substrate coated with 200 nm indium tin oxide. (b) Spin coating of colloidal gold nanoparticles (AuNPs) (5–10 nm in diameter). (c) Annealing at 200 °C. (d) Rinsing the sample in acetone to remove the PR template grating and the ligands on the AuNPs. (e) Annealing at 350 °C to produce gold nanowires. (f) Spin coating of PMMA to supply a supporting film for the AuNPs. (g) Lift-off of the AuNP nanowires supported by the PMMA film in HCl solution and pickup of the free-floating Au-nanowire film using a hollow plate. (h) Decking of the gold nanowire film onto the end facet of an optical fibre. (i) Annealing at 200 °C and rinsing the fibre tip in chloroform to remove PMMA film. (j) Fixing the gold nanowires onto the end facet of the fibre after annealing at 500 °C to finish the fabrication. (k, l) SEM images of the grating structures with each grating line consisting of continuous gold nanowires and randomly distributed gold nanoparticles. *Source:* Reproduced with permission of RSC publishing.

integrated a silicon-on-insulator (SOI) ring resonator circuit onto the cleaved end of an optical fibre [89]. Moreover, Jia et al. realized a plasmonic sensor by transferring periodic Au hexagonal nanohole array onto MMF tips by means of an epoxy glue deposited along the whole fibre tip [90, 91].

Recently, He et al. proposed a glue-and-strip fabrication approach for transferring SPR devices upon SMF end facets [92]. First, an amount of optical epoxy adhesive is applied on the tip of the fibre. Then, the fibre is mounted on a multi-axis translation stage and moved towards the substrate that carries the gold film with the nanopatterns (previously fabricated by means of standard EBL process). The fibre is observed under a microscope. The alignment marks on the gold film enable a coarse alignment while the precise alignment is achieved by sending a broadband light into the fibre, monitoring its reflection spectrum. After alignment, the epoxy is either cured by a UV lamp or heating the substrate. Finally, the fibre tip is moved off the substrate quickly to strip the gold film. With this technique an in-plane second-order distributed feedback (DFB) SPR cavity was integrated on an SMF end facet. Experimental RI sensitivity of 628 nm RIU-1, with an LOD of 7×10^{-6} RIU, was demonstrated [93]. By chemically functionalizing the sensor surface, the measurement of the real-time interaction processes between hIgG and anti-hIgG molecules was preliminarily demonstrated.

9.3.3 Self-Assembly

Self-assembly (SA) is a process by which disordered pre-existing components organize themselves in an ordered structure through only local interactions [94]. Taking inspiration from nature, materials scientists have attempted to exploit SA principles to create artificial materials, with hierarchical structures and tailored properties, for the fabrication of functional devices [94, 95]. In fact, SA in manufacturing could lead to cheaper and more efficient fabrication by using inexpensive spontaneous processes. This is undoubtedly the main advantage of bottom-up engineering compared with more established top-down fabrication approaches. Moreover, under specific conditions, SA in principle may be adapted to work at arbitrary physical scales, whereas such scale selectivity is difficult to achieve by conventional manufacturing tools [94], enabling the parallel realization and thus a cost-effective mass production of LOF devices.

The earliest attempts of using SA techniques to decorate optical fibre tips were driven by SERS applications [96]. From then on, different approaches and applications have been proposed by using the controlled arrangement of metal NPs on the fibre tip. Jeong et al. realized an optical fibre plasmonic biosensor using spherical Au NPs integrated on an MMF tip [97]. Such a biosensor was exploited for detecting antibody–antigen reaction of interferon-gamma. Sciacca and Monro proposed a solution for achieving multiplex detection on a probe by exploiting different metallic NPs (Au and Ag) [98]. Since Au and Ag nanospheres (NSs) present distinct LSPR signatures with limited overlap, a multiplexed dip biosensor has been fabricated by functionalizing NPs with different antibodies. Metallic layer plays an important role in achieving high sensitivity through SERS enhancement. Consequently, great efforts have been made to achieve high enhancement factors by adopting various noble metal nanoplatforms, including alumina, silver, and gold. Star-shape gold nanostructures called 'gold nanostars' (GNS) have emerged as an ideal platform for SERS due to their outstanding optical properties [99]. Unlike gold NSs, for the GNS, aggregation is no longer necessary to generate strong brighter SERS signals, therefore increasing reproducibility and stability of the sensing platform. Fales and co-workers demonstrated that coating with silver the GNS (thus achieving the silver-coated GNP) endows the GNS with over an order of magnitude of signal enhancement compared with uncoated ones [100]. Silver-coated GNP were successfully integrated on the fibre facet of a 400/440 MMF with numerical aperture (NA) of 0.25 [101]. Raman spectra were recorded using different optrodes dipped into 1 mM p-MBA solution around 1600 cm^{-1}. It was demonstrated that the silver-coated GNP-based optrodes presented more significant signal compared with single metallic NPs. Moreover, the proposed optrode was capable to detect different types of analytes with an LOD for rhodamine B between 10^{-7} and 10^{-8} M, indicating that it could be a promising candidate in SERS chemical and biological sensing applications.

Many different SA processes have been exploited so far also for making nano- and microstructures in a regular ensemble. Some of them are well-consolidated fabrication techniques like block-copolymer SA and NS lithography based on colloidal assembly. Others are based on less known processes, such as breath figure formation. A very recent review collected the main advanced and future trends related to the use of SA phenomena for the development of 'LOF' optrodes [94]. In example, by using the breath figure formation, Pisco et al. proposed a lithographic process to create metallo-dielectric patterns on the optical fibre tip [102–105]. With this procedure, water droplets are auto-organized into a close hexagonal arrangement at liquid polymer/air interface. The droplet arrangement acts as a template and imprints a hexagonal pattern of holes after solvent evaporation of the polymeric film. Several configurations based on metallo-dielectric crystals with different geometrical features have been successfully demonstrated [106]. Similarly, Pisco et al. adapted the NS lithography to operate on the optical fibre tip in order to obtain regular and repeatable patterns [106]. Starting from the promising results achieved in terms of reproducible metallo-dielectric patterns on the optical fibre tip, Quero et al. engineered an LOF optrode for SERS application. Recently, Ni et al. performed a systematic study aimed at the integration of opal and inverse opal films onto bundles of optical fibres [107]. PS opal films and silica inverse opal films were fabricated on single-mode optical fibre bundle by isothermal heating evaporation-induced SA and sol–gel co-assembly methods, respectively. Such analysis demonstrates the potential extension of SA approaches to multiple fibres and thus their potential impact for the mass production of devices.

9.3.4 Smart Materials Integration

As discussed in the previous sections, LOF technology has led to the development of advanced optical biosensors capable of detecting and discriminating among large classes of molecules. The fibre tips provide a platform where it is possible to realize nanostructures, ranging from semiconductor PC to plasmonic antennas, which precisely confine highly concentrated optical fields and facilitate the interaction of these fields with local chemical and biological variations. In this way, fibre-coupled spectroscopic measurements can be performed within the fibre optrode itself, providing information about superficial environment changes even at sub-wavelength scale. Responsivity and sensitivity enhancement of the LOF optrodes is achieved not only by using advanced chemistry built on the nanostructure but also by exploiting the integration of active compounds and advanced materials. In this respect, in the last years, scientific community has focused the attention on 'smart' or stimuli-responsive polymers, i.e. materials that undergo reversible large physical or chemical changes in response to small external changes in the environmental conditions, such as temperature, pH, light, magnetic or electric field, ionic strength, and biological molecules. In this context, Tierney and co-workers have successfully deposited on the fibre tip a Fabry–Pérot cavity made of a ~50 μm thick layer of hydrogels (HGs), i.e. a cross-linked 3D polymeric network structure, which can absorb and retain considerable amounts of water [108]. Variations in the optical length of the HG cavity cause a change in the phase of the interference wave arising from light reflected at the fibre–gel and gel–solution interfaces. Glucose- [109, 110] and DNA-sensitive HG- [111] based optical fibre probes have been reported. Following this approach, Muri et al. recently reported on an LOT device fabricated by immobilizing gold nanoparticles (GNPs) in an HG droplet polymerized on the fibre end face [112]. This system allows increasing the number of NPs available for sensing, it offers precise control over the NP density, and the NPs are positioned in a true 3D aqueous environment. The proposed platform can measure volumetric changes in a stimuli-responsive HG or measure binding to receptors on the NP surface. It can also be used as a two-parameter sensor by utilizing both effects. Successively, the same authors proposed a similar device integrated on the facet of a cleaved double-clad optical fibre [113]. The swelling degree of the HG is measured interferometrically using the single-mode inner core, while the LSPR signal is measured using the multimode inner cladding.

In the last years, also microgels (MGs), which are colloidal HG particles with diameters ranging from 100 nm to µm, have been successfully integrated on the fibre tip. Aliberti et al. recently proposed a device characterized by an MG layer baked by a metal nanostructure directly realized on the optical fibre tip (see Figure 9.8) [114]. The nanostructure consists of a thin gold layer patterned (by means of FIB milling process) with a square lattice of holes supporting plasmonic resonance. The excitation of a plasmonic resonance gives rise to a reflection spectrum dip highly sensitive to both RI and thickness of the MG layer deposited on top. Interaction of light with MG particles during their swelling/shrinking dynamics is not trivial because of the inverse relationships between their size and RI. The relationships between these two quantities have been derived from experimental observations based on conventional morphological analysis [115]. In response to binding events of glucose molecule, the MG network concentrates the target molecule and amplifies the optical response, leading to remarkable sensitivity enhancement. Moreover, by acting on the MG degrees of freedom, such as concentration and operating temperature, it is possible to control the limit of detection and tune the working range as well as the response time of the probe. These unique characteristics pave the way for advanced label-free biosensing platforms, suitably reconfigurable depending on the specific application. The MGs were integrated on the fibre tip by means of a customized fabrication path, based on a dip coating procedure, that provides enhanced control on density distribution and coverage factor of MG particles. In addition to MG concentration in solution, also temperature and pH of the solution are key parameters, which strongly influence the particle

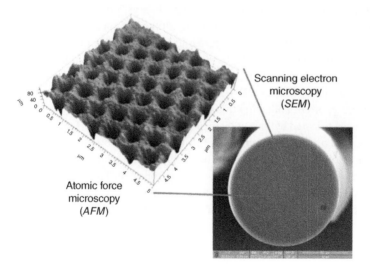

Figure 9.8 SEM and AFM images of the MG-assisted lab on-fibre probe.

density on the fibre tip. This is not a trivial point since different degrees of MG layer compactness give rise to different responsivities to both physical and chemical parameters [116]. The effectiveness of the deposition protocol was recently improved and optimized in such a way to allow for the creation of MG layers with coverage factors larger than 90% in only 30 minutes [117]. The procedure is general, and it is valid independently from the typology of both substrate and MG used. In fact, the fabrication procedure was successfully validated on both gold-coated and uncoated optical fibre tips. Overall, the assessment of MG deposition technique on the fibre tip represents a solid foundation for developing advanced LOF probes with interesting fabrication throughput (the probe production can be also parallelized) and thus ready to be exploited at the industrial application level. Combination of MGs with LOF technology also could open a plethora of applications in the biomedical field as drug delivery systems, cell culture supports, bioseparation devices, sensors, or actuator systems [118].

9.4 Nanofabrication Techniques on the Fibre Tip for Optomechanical Applications

Optomechanics investigates the interaction between mechanical structures and light in order to engineer advanced optomechanical transducers [20], allowing the manipulation of the light flow through optomechanical actuators or the detection of weak mechanical effects through advanced sensing configurations. Merging the extremely intriguing features of optomechanics with LOF technology could enable the development of optomechanical LOF devices operating on a very small scale with new functionalities and unprecedented performances. Indeed, optomechanical devices can offer alternative opportunities by exploiting the optical and mechanical physical interaction between photons and mechanical structures starting from displacement-based optical interactions down to the quantum level interactions mediated by radiation-pressure coupling or gradient forces.

The implementation of optomechanics-assisted LOF devices requires the employment and the development of specific fabrication techniques enabling the printing of arbitrary 3D micro/nanostructures and the inclusion of materials with different elastic properties on optical fibres. Most of the platforms demonstrated so far under the LOF umbrella mainly rely on plasmonic interactions occurring when metallic nanostructures are integrated onto optical fibres. Consequently, the fabrication techniques developed so far have been specialized in defining 2D micro- and nanostructures on the optical fibre.

In the following we report on the main developments to date, concerning the merging of LOF technology with the Opto-mechanics by highlighting the fabrication approaches already implemented and outlining the main results in terms of achieved performances and in-field applications.

The first demonstrations of miniaturized optomechanical LOF devices date back to about one decade ago when Kilic et al. proposed a micromachined optical fibre microphone consisting of a tiny Fabry–Pérot microcavity formed by a PC diaphragm placed at short distance from a mirror deposited on the tip of an SMF [119]. This miniaturized fibre microphone was used to measure air pressures with a limit of detection as low as $18\,\mu Pa/\sqrt{Hz}$ at 30 kHz. Successively, the same authors fabricated an optical hydrophone based on a compliant membrane suitably integrated on the optical fibre tip to be exploited in underwater applications with the impressing capability to measure acoustic pressures with a resolution close to the background noise in the ocean over a bandwidth of 10 kHz [120].

In the same years (2006), Iannuzzi et al. first launched the so-called fibre-top cantilever envisioning multifunctional optomechanic platforms arising from the integration of a cantilever carved on the tip of single-mode optical fibre [121–123]. The proposed approach relied on the inherent Fabry–Perot cavity formed between the fibre and the cantilever as main transducing mechanism. Applications of fibre-top cantilever probes were demonstrated for sensing and atomic force microscopy applications [35]. First fibre-top probes were initially fabricated by means of FIB milling. Successively, Iannuzzi et al. proposed a more versatile and sustainable fabrication process involving laser micromachining acting onto larger optical substrates as the case of glass ferrules [124, 125]. This successful translation transformed the initial 'fibre-top cantilever' technology in the 'ferrule-top technology'. Also, a top-down approach via align-and-shine photolithography [126] was proposed to fabricate low mass gold fibre-top cantilevers.

Following these technological improvements, a wide variety of novel fibre-optic microsensors were demonstrated, including pressure sensors [127], acoustic emission detectors [128], and humidity and force transducers [129, 130]. Sensing systems based on micromechanical transducers directly created on the top of a glass ferrule were also demonstrated for surface topography reconstruction and photoacoustic spectrometry [131].

By observing the results of these pioneering studies, everybody may recognize the enormous potential arising from the judicious connection of LOF and optomechanics technologies for the development of a novel class of fibre devices with new functionalities and unparalleled performances when compared with traditional mechanical transducers.

As an example, a ferrule-top approach was recently used to integrate a complex micromechanical structure, featuring a dual-beam cantilever with a properly designed proof mass, on the optical fibre tip. The innovative optomechanical structure was designed to act as seismic accelerometer, exhibiting competitive performance for seismic applications when compared with commercial platforms [132].

Prototypes were fabricated, and their responsivity was preliminarily measured in the laboratory, demonstrating a resolution down to 0.44 μg/$\sqrt{\text{Hz}}$ over a 3 dB frequency band of 60 Hz. Successively, the LOF seismic accelerometers were integrated in a conventional seismic station and used for seismic surveillance applications. First prototypes were able to sense and register the ground acceleration associated with the earthquakes that occurred near the end of October and early November 2016 in Central Italy (see Figure 9.9). The wave traces were compared with the recordings of a traditional sensor. The time correlation and spatial coherence have been used to show the high fidelity of the LOF sensors in seismic wave detection. The lightness and small size of the LOF sensors and the immunity of the optical fibre towards electromagnetic interferences enable a simpler installation when compared with traditional sensors. The overall performances and the effectiveness in detecting ground acceleration during earthquakes, combined with the benefits of technology, make the proposed seismic accelerometer as a valuable and promising alternative to standard seismic accelerometers.

A significant milestone achieved along the roadmap pertaining to optomechanics-assisted LOF platforms was set by the first demonstration of a ferrule-top 'nano-indenter' [133], now commercially distributed. The indenter uses a force microprobe obtained by carving a cantilever on the top of a glass ferrule that hosts an optical fibre for read-out purposes. The ferrule-top probe uses wavelength modulation in a Fabry–Pérot cavity configuration to detect cantilever deflection and to drive a feedback-controlled piezoelectric actuator. By superposing a small oscillatory load to the probe and recording the indentation depth at the frequency of oscillation, the optomechanical instrument allows to control the static load applied onto the sample and consequently to extract the local elastic and viscous moduli of the samples. The resulting indenter instrument was proposed to probe the viscoelastic properties of soft hydrated tissues and biomaterials as well [134–137]. A similar prototype was also recently realized to operate *in situ* by means of a needle [138]. Basically, the cantilever was mounted on an 8 cm long borosilicate capillary (diameter: 1 mm, wall thickness: 0.21 mm). As shown in Figure 9.10, the optical fibre is supported by a second borosilicate capillary (diameter: 0.55 mm, wall thickness: 0.075 mm), rigidly mounted inside the first capillary. A Fabry–Pérot cavity was created between the cleaved facet of the fibre and the cantilever by coupling the distal end of the fibre to an interferometric interrogation system. The probe can be used to perform minimally invasive measurements retrieving the mechanical properties of a biological tissue in its local environment by enabling a deeper understanding of the role of mechanics in physiology and tissue engineering [138].

These first successful demonstrations combined with a fruitful industrial exploitation of the developed platforms posed the basis for further developments aimed to provide more complex and powerful optomechanical LOF platforms for *in vivo* biomedical applications.

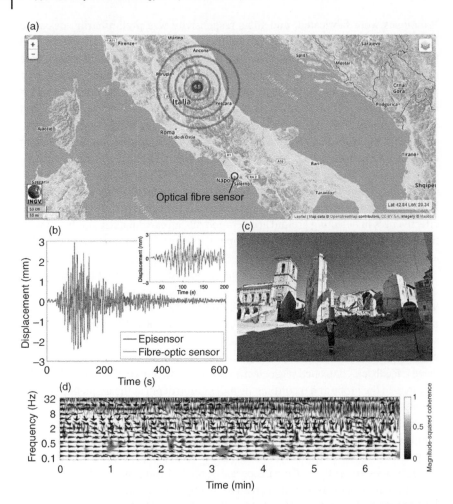

Figure 9.9 (a) Map of the Norcia earthquake epicentre retrieved from the INGV official website (http://cnt.rm.ingv.it; the map is released under a CC-BY-SA licence [https:// creativecommons.org/licenses/by-sa/3.0]); (b) comparison between the ground displacement traces of the LOF sensors and the reference sensor. (c) photograph of St. Benedict Cathedral in Norcia after the earthquake; in the inset, a picture of the cathedral before the earthquake is displayed (the picture is released under a CC-BY-SA licence https:// creativecommons.org/licenses/by-sa/3.0 by Wikimedia). (d) Wavelet coherence between the optical and Episensor displacement signal over a 400 seconds window. *Source:* Reproduced with permission from [132]. The article is licensed under a Creative Commons Attribution 4.0 International License: http://creativecommons.org/licenses/by/4.0. Reproduced with permission of Springer Nature.

(a) (b) (c)

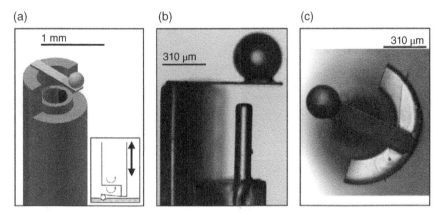

Figure 9.10 Schematic model of the probe, which consists of a cantilever indentation spring, an optical fibre for the interferometric read-out of the displacement of the cantilever (dark color), and a borosilicate sphere to create a spherical contact with the indented surface. The inset shows a schematic of the indentation procedure, where emphasis is put on the interferometric read-out and the movement of the piezoelectric transducer (not to scale). (b) Microscope image of the probe, showing the interferometric cavity. (c) Top view of the sensor. *Source:* Reproduced with permission from [138]. The article is licensed under a Creative Commons Attribution 4.0 International License: http://creativecommons.org/licenses/by/4.0. Reproduced with permission of Springer Nature.

On this line, Guggenheim et al. developed a novel plano-concave polymer microresonators [139] integrated on the optical fibre tip. The sensors were based upon a solid plano-concave polymer microcavity formed between two highly reflective mirrors, which is embedded in a layer of matching polymer so as to create an acoustically homogeneous planar structure as schematically illustrated in Figure 9.11a. The cavity is constructed by depositing a droplet of optically clear UV-curable liquid polymer onto a dielectric mirror-coated polymer substrate. The droplet stabilizes to form a smooth spherical cap under surface tension and is subsequently cured under UV light (see Figure 9.11b). The second dielectric mirror coating is then applied, followed by the addition and curing of further polymer to create the encapsulating layer (see Figure 9.11c).

The sensor is operated by illuminating it with a laser tuned to the edge of the cavity resonance. Under these conditions, the stress due to an incident acoustic wave modulates the cavity optical thickness producing a corresponding modulation in the reflected optical power that is detected by a photodiode.

This technique enables the realization of advanced LOF ultrasound transducers with unprecedented performances (i.e. omnidirectionality, bandwidth up to 40 MHz, and a noise equivalent pressure as low as 2.1 mPa/$\sqrt{\text{Hz}}$) when compared with traditional piezoelectric transducers [139].

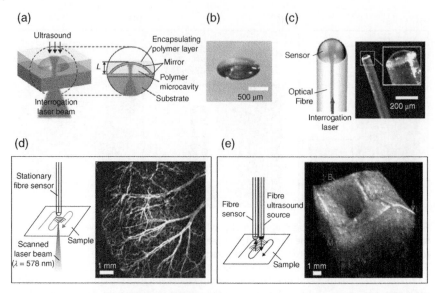

Figure 9.11 (a) Schematic of a plano-concave optical microresonator (*L* = cavity thickness). The sensor comprises a plano-concave polymer microcavity encapsulated in a planar polymer layer. (b) Photograph of polymer microcavity prior to application of the encapsulating polymer layer. (c) Schematic (left) and photograph (right) of the optical fibre microresonator sensor. (d) Schematic of fibre microresonator sensor-based optical resolution photoacoustic microscopy experiment and image of mouse ear vasculature *in vivo*. (e) Schematic of all-fibre pulse-echo ultrasound experiment and 3D pulse-echo ultrasound image of *ex vivo* porcine aorta. *Source:* Reproduced with permission from Macmillan Publishers Ltd. [139]. Copyright 2017. Reproduced with permission of Springer Nature.

In light of the extraordinary performances, these ultrasound probes have been successfully demonstrated for high-resolution photoacoustic and ultrasound imaging in clinically relevant scenarios.

In Figure 9.11d an optical resolution photoacoustic microscopy image of the mouse ear acquired *in vivo* showing the microvasculature at the level of individual capillaries is shown. This image was obtained by scanning a pulsed focussed photoacoustic excitation laser beam over an 8 mm × 8 mm area and recording the photoacoustic signals at each scan point with the sensor in a fixed position at the centre of the scan area. The fibre sensor was located at a distance of 1.2 mm from the skin surface, and an ultrasound gel was used as acoustic coupling medium. In Figure 9.11e, a 3D high-resolution pulse-echo ultrasound image of an *ex vivo* porcine aorta sample is reported. The image was obtained by raster scanning a fibre microresonator sensor and a fibre-based laser ultrasound source emitting broadband ultrasound pulses with a frequency content extending to 30 MHz.

The returning echoes from subsurface tissue structures are recorded by the fibre sensor and used to reconstruct the 3D image.

The mentioned achievements are solid enough to realistically envision, in the near future, the development of smart needle with unprecedented functionalities and integration and miniaturization levels, able to monitor clinically relevant parameters in real time and *in vivo*, opening new ways to monitor the disease appearing and its progression/regression, by imaging and differentiating cancer cells and quantifying cancer biomarker concentrations in close proximity to the tumour itself.

Another relevant milestone was achieved few years ago, in 2017, when Yao et al. proposed a novel optical microfabrication technology to directly print polymeric optomechanical structures in 3D arbitrary formats on the end face of fibre-optic ferrules. This established a sustainable and cost-effective fabrication root for a novel generation of optomechanics-assisted LOF platforms designed for specific applications.

The optical 3D µ-printing system consists of a high-power UV source (365 nm), a UV-grade digital mirror device (DMD) for the generation of optical patterns, projection optics for scaling down the optical images, and a digital camera for machine vision metrology.

The authors fabricated several optomechanical structures, namely, suspended-mirror devices (SMDs), on the fibre tip. A dynamic optical exposure process was used to reproduce the 3D SMD image by successive slices, and the predefined SMDs are fabricated through photopolymerization of the SU-8 PR. The application of such fabricated ferrule-top SMD as displacement microsensors was experimentally demonstrated [140]. Successively, the same authors used the proposed fabrication approach also to obtain a chemical sensor for simultaneous detection of CO_2 concentration and temperature [141].

In 2018, Yao et al. presented an improved version of the optical µ-printing technology able to directly fabricate suspended microbeams on the end face of a standard SMF [142]. Here an ultrasonic nozzle was used for precise deposition of thin layers of SU-8 on the fibre tip, and a digital microheater was used to rapidly perform the post-baking stage, as shown in Figure 9.12a. In Figure 9.12b the main fabrication steps are illustrated. Figure 9.12c shows the SEM images of three SU-8 suspended microbeams fabricated on the end faces of optical fibres. Such optical fibre-tip sensors have been demonstrated to detect the RI change of liquid and the gas pressure of ambient environment with sensitivities of 917.3 nm/RIU to RI change and 4.29 nm/MPa to gas-pressure change.

As envisaged by the same authors, the developed fabrication process can be extended for integrating different functional polymers in order to detect several parameters, such as humidity or other gas components [141]. Furthermore, since the Young modulus of SU-8 and similar PR is relatively low (with respect to glass,

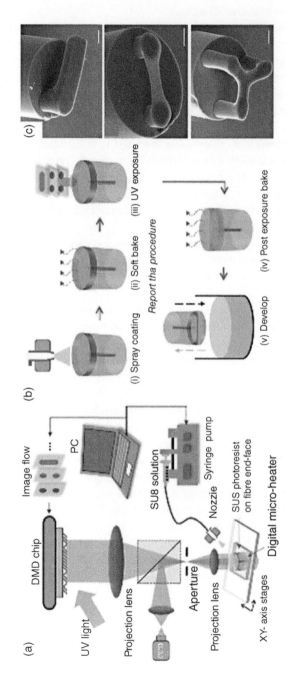

Figure 9.12 (a) Schematic diagram of the optical 3D μ-printing technology. (b) Flow chart for printing the optical fibre tip sensors: (i) the optical fibre was moved below the ultrasonic nozzle for spray coating of a thin layer of SU-8, (ii) the film was *in situ* soft-baked to remove solvent, (iii) selective optical exposure is carried out with the assistance of the digital camera, (iv) the sample was post-baked *in situ* by using the integrated digital microheater, and (v) the sample was developed by using propylene glycol methyl ether acetate (PGMEA). (c) Scanning electron microscope (SEM) images of SU-8 suspended microbeams printed on the end face of optical fibres. All scale bars are 20 μm. *Source:* Reproduced with permission from [142]. The article is licensed under a Creative Commons Attribution 4.0 International License: http://creativecommons.org/licenses/by/4.0. Reproduced with permission of MDPI.

silicon, etc.), the optical 3D printing is a valuable option to fabricate optomechanical cavities with low resonant frequencies and/or critical damping. The combined control of the resonant frequency and damping can be greatly useful in tailoring the response of SMD for a specific application.

Additionally, the proposed optical microfabrication technology results to be able to rapidly print complex 3D mechanical structures on the end faces of optical fibres. The reported 3D structures indeed were modelled by using 200 slices, and each structure was fabricated by dynamically generating optical patterns, frame by frame, in order to irradiate the SU-8 film on the optical fibre end face. Since the typical reported exposure time [142] for each frame is about 10 seconds, the proposed 3D printing process is a promising candidate for high-throughput fabrication of micro-optomechanical LOF devices.

Definitively, by taking advantages from the availability of a wider class of constituent materials and by the process versatility, the 3D microprinting technology can be considered a valuable cost-effective strategy to fabricate multifunctional plug-and-play devices based on optomechanical cavities with precise and controlled geometries onto optical fibre substrates.

9.5 Conclusions

In this chapter, we provided an overall picture of the 'LOF technology' vision illustrating the main technological advances and new trends along the technological roadmap aimed to achieve as ultimate objective the development of flexible, multifunctional plug-and-play fibre-optic platforms designed for the specific application.

Although our group pioneered the LOF concept, its foundation was essentially the result of a natural and common viewpoint pertaining to many scientists involved in fields tightly linked to nanotechnology and optical fibres. The main driving force was the common and profound conviction that the connection between optical fibres and nanotechnology had the real potential to open the door for a novel generation of fibre-optic platforms with new functionalities (not possible before) and unprecedented figures of merit in terms of performances, miniaturization levels, and power consumption, just to name a few.

The idea of disruptively enlarging the functionalities of optical fibres and achieving unprecedented performances strongly depends on the capability to manage reliable and cost-effective nanofabrication strategies operating onto not conventional substrates as the case of optical fibres. Several nanotechnology-assisted fabrication techniques demonstrated to be able to integrate specific materials and ingenious nanostructures with precise and controlled geometries onto optical fibre substrates, but such approaches often require expensive equipment, featuring serial production. These issues pose severe limitations towards the achievements of high-throughput processes needed to translate 'Lab' prototypes from lab curiosities in real devices and platforms, properly working in real application scenarios.

The SA processes, being based on an 'autonomous organization of materials and structures', are inherently low-cost procedures. Additionally, they can be potentially extended to multiple fibres for parallel production, resulting suitable for the mass production of fibre devices. Therefore, they look as a promising tool to find concrete market opportunities for LOF devices. On the same line, novel techniques (nano-apertures without lithography and 3D light printing) have been recently proposed to address mass and rapid production issues.

Thanks to these efforts, many new functionalities have been already demonstrated, opening new application fields and market options, while the first achievements of unparalleled performances will soon provide concrete opportunities for the realistic replacement of existing technologies with engineered LOF platforms in many strategic sectors.

Despite the not trivial nature of the optical fibre substrate, nowadays, it is possible to conceive any arbitrary 3D object integrated onto its facet as demonstrated through the construction of the smallest micro-house in the world as shown in Figure 9.13 [143].

This disruptive breakthrough definitively confirms the potential to significantly impact the technological roadmap pertaining to LOF fabrication thanks to the development of new robotics fabs where robotic arms and micro/nano technologies are judiciously combined to provide unique capabilities for cutting, etching, folding, assembling, and welding thin membranes on the fibre facets with unprecedented spatial control.

Conceiving any arbitrary object on an intrinsically light-coupled substrate as the optical fibre tip is equivalent to manage a large set of degrees of freedom in order to realize any kind of nanophotonic architectures on fibre substrates, pushing light–matter interaction at its ultimate limit.

Considering the technological maturity achieved so far, it is now expected a new step-ahead along the technological roadmap, consisting in the fundamental

Figure 9.13 Micro-house assembled by origami and welded on top of the facet of an optical fibre in a SEM and FIB dual beam. *Source:* Reproduced with permission from [143]. Reproduced with permission of IEEE.

transition from 'component and device' level towards the 'system' level where advanced LOF platforms are conceived and properly integrated in complex systems designed for specific applications.

Clear evidences of this technological transition are provided by the first demonstrations of LOF assisted OCT nano-endoscopes and *in vivo* ultrasonic imaging tools specially designed for *in vivo* imaging [144]. Similar considerations can be easily extended to the first optical systems incorporating LOF meta-devices designed for optical processing and computing [145].

In the next future, it is thus not surprising to expect the development of 'Intelligent Needles' where LOF platforms are judiciously integrated and specialized for bio-manipulation, tissue and liquid biopsies, loco-regional echography, and drug delivery with a disruptive impact in precision medicine scenarios and clinically relevant investigations.

While most of the target applications have their focus on life science applications, taking advantage for the extraordinary added value linked to the easy integration of LOF platforms in catheters and hypodermic needles, there is no real limit for the real exploitation of the technology in a large number of other industrial scenarios.

Optomechanics-assisted LOF technologies seems to be the most promising route to open new applications fields as well as identify alternative market options.

The idea to merge mechanical micro- and nanocavities onto optical fibre substrates envisions the development of plug-and-play LOF platforms able to disruptively extend their functionalities.

Indeed, most of the optomechanical devices demonstrated so far are based on the displacement of a compliant mechanical element strongly or intrinsically coupled with a photonic cavity. Since the most natural and immediate application of this configuration is the detection of the force exerted on the mechanical element, first demonstrations of optomechanical LOF devices have been mainly focused on force and pressure sensing or other related physical magnitudes (acoustic wave, acceleration, etc.). At this stage, this technological connection has not yet expressed its full potential.

Target milestones envision now a new technological step towards the development of LOF devices operating on a very small scale and incorporating integrated mechanical and optical systems devoted to actively changing and controlling optical signals by optomechanical actuation. Indeed, optomechanical devices can offer alternative opportunities by exploiting the optical and mechanical physical interaction between photons and mechanical structures down to the quantum level mediated by radiation-pressure coupling or on gradient forces. Therefore, by a proper engineering of the optomechanical system and judiciously managing the integration with optical fibre platforms, a wide variety of unexplored devices could be realized including advanced nanosensors but also optical signal processing devices such as optical switches, optical delay lines, wavelength converters, tunable optical filters, and tunable sources.

These achievements would bring 'LOF' technology towards a central role not only in scenarios related to diagnostics and monitoring but also in the ICT field, where optical fibres have already found their main remarkable accomplishments.

References

1 Cusano, A., Consales, M., Crescitelli, A., and Ricciardi, A. (2015). *Lab-on-Fibre Technology*. Springer.

2 Consales, M., Pisco, M., and Cusano, A. (2012). Lab-on-fibre technology: a new avenue for optical nanosensors. *Photonic Sensors* 2 (4): 289–314.

3 Ricciardi, A., Crescitelli, A., Vaiano, P. et al. (2015). Lab-on-fibre technology: a new vision for chemical and biological sensing. *Analyst* 140: 8068–8079.

4 Albert, J. (2014). A lab on fibre. *IEEE Spectrum* 51: 48–53.

5 Vaiano, P., Carotenuto, B., Pisco, M. et al. (2016). Lab on fibre technology for biological sensing applications. *Laser & Photonics Reviews* 10: 922–961.

6 Kostovski, G., Stoddart, P.R., and Mitchell, A. (2014). The optical fibre tip: an inherently light-coupled microscopic platform for micro- and nanotechnologies. *Advanced Materials* 26: 3798–3820.

7 Schmidt, M.A., Argyros, A., and Sorin, F. (2016). Hybrid optical fibres – an innovative platform for in-fibre photonic devices. *Advanced Optical Materials* 4: 13–36.

8 Gumennik, A., Stolyarov, A.M., Schell, B.R. et al. (2012). All-in-fibre chemical sensing. *Advanced Materials* 24: 6005–6009.

9 Tao, G., Stolyarov, A.M., and Abouraddy, A.F. (2012). Multimaterial fibres. *International Journal of Applied Glass Science* 3: 349–368.

10 Nguyen-Dang, T., Page, A.G., Qu, Y. et al. (2017). Multi-material micro-electromechanical fibres with bendable functional domains. *Journal of Physics D: Applied Physics* 50: 144001.

11 Temelkuran, B., Hart, S.D., Benoit, G. et al. (2002). Wavelength-scalable hollow optical fibres with large photonic bandgaps for CO_2 laser transmission. *Nature* 420: 650.

12 Russell, P. (2003). Photonic crystal fibres. *Science* 299: 358–362.

13 Argyros, A. (2009). Microstructured polymer optical fibres. *Journal of Lightwave Technology* 27: 1571–1579.

14 Deng, D., Orf, N., Abouraddy, A. et al. (2008). In-fibre semiconductor filament arrays. *Nano Letters* 8: 4265–4269.

15 Abouraddy, A.F., Bayindir, M., Benoit, G. et al. (2007). Towards multimaterial multifunctional fibres that see, hear, sense and communicate. *Nature Materials* 6: 336–347.

16 Bayindir, M., Sorin, F., Abouraddy, A.F. et al. (2004). Metal–insulator–semiconductor optoelectronic fibres. *Nature* 431: 826–829.

17 Monat, C., Domachuk, P., and Eggleton, B.J. (2007). Integrated optofluidics: a new river of light. *Nature Photonics* 1: 106–114.

18 Mayer, K.M. and Hafner, J.H. (2011). Localized surface plasmon resonance sensors. *Chemical Reviews* 111: 3828–3857.

19 Yu, N., Genevet, P., Kats, M.A. et al. (2011). Light propagation with phase discontinuities: generalized laws of reflection and refraction. *Science* 334: 333–337.

20 Aspelmeyer, M., Kippenberg, T.J., and Marquardt, F. (2014). Cavity optomechanics. *Reviews of Modern Physics* 86: 1391.

21 Yablonovitch, E. (1994). Photonic crystals. *Journal of Modern Optics* 41: 173–194.

22 R. J. Hermann and M. J. Gordon, "Nanoscale optical microscopy and spectroscopy using near-field probes," *Annual Review of Chemical and Biomolecular Engineering*, vo. 9, pp. 365–387, 2018.

23 Fan, X. and White, I.M. (2011). Optofluidic microsystems for chemical and biological analysis. *Nature Photonics* 5: 591–597.

24 Maksymov, I.S. (2015). Magneto-plasmonics and resonant interaction of light with dynamic magnetisation in metallic and all-magneto-dielectric nanostructures. *Nanomaterials* 5: 577–613.

25 O'brien, J.L., Furusawa, A., and Vučković, J. (2009). Photonic quantum technologies. *Nature Photonics* 3: 687–695.

26 Prasad, P.N. (2004). *Nanophotonics*. Wiley.

27 Kik, P.G. and Brongersma, M.L. (2007). *Surface Plasmon Nanophotonics*, 1–9. Springer.

28 Koenderink, A.F., Alù, A., and Polman, A. (2015). Nanophotonics: shrinking light-based technology. *Science* 348: 516–521.

29 Cusano, A., Consales, M., Pisco, M. et al. (2011). Lab on fibre technology and related devices, part I: a new technological scenario; lab on fibre technology and related devices, part II: the impact of the nanotechnologies. *Proceedings of SPIE* 8001: 800122.

30 Maier, S.A. (2007). *Plasmonics: Fundamentals and Applications*. New York: Springer Science &Business Media LLC.

31 Laine, J.P., Little, B.E., and Haus, H.A. (1999). Etch-eroded fibre coupler for whispering-gallery-mode excitation in high-Q silica microspheres. *IEEE Photonics Technology Letters* 11: 1429–1430.

32 Rajput, N.S. and Luo, X. (2015). *Micro and Nano Technologies, Micromanufacturing Engineering and Technology*, 2e, 61–80. William Andrew Publishing.

33 Micco, A., Ricciardi, A., Pisco, M. et al. (2015). Optical fibre tip templating using direct focused ion beam milling. *Scientific Reports* 5: 15935.

34 Nellen, P.M. and Brönnimann, R. (2006). Milling micro-structures using focused ion beams and its application to photonic components. *Measurement Science and Technology* 17: 943–948.

35 Iannuzzi, D., Deladi, S., Berenschot, J.W. et al. (2006). Fibre-top atomic force microscope. *Review of Scientific Instruments* 77 (10): 106105.

36 Liberale, C., Minzioni, P., Bragheri, F. et al. (2007). Miniaturized all-fibre probe for three-dimensional optical trapping and manipulation. *Nature Photonics* 1 (12): 723–727.

37 Mivelle, M., Ibrahim, I.A., Baida, F. et al. (2010). Bowtie nano-aperture as interface between near-fields and a single-mode fibre. *Optics Express* 18: 15964–15974.

38 Wang, F., Yuan, W., Hansen, O., and Bang, O. (2011). Selective filling of photonic crystal fibres using focused ion beam milled microchannels. *Optics Express* 19: 17585–17590.

39 Kang, S., Joe, H.E., Kim, J. et al. (2011). Subwavelength plasmonic lens patterned on a composite optical fibre facet for quasi-one-dimensional Bessel beam generation. *Applied Physics Letters* 98 (24): 241103.

40 Dhawan, A., Gerhold, M.D., and Muth, J.F. (2008). Plasmonic structures based on subwavelength apertures for chemical and biological sensing applications. *IEEE Sensors Journal* 8 (6): 942–950.

41 Dhawan, A., Muth, J.F., Leonard, D.N. et al. (2008). Focused ion beam fabrication of metallic nanostructures on end faces of optical fibres for chemical sensing applications. *Journal of Vacuum Science and Technology* 26: 2168–2173.

42 Dhawan, A., Du, Y., Yan, F. et al. (2010). Methodologies for developing surface-enhanced Raman scattering (SERS) substrates for detection of chemical and biological molecules. *IEEE Sensors Journal* 10 (3): 608–616.

43 Andrade, G., Hayashi, J.G., Rahman, M.M. et al. (2013). Surface-enhanced resonance Raman scattering (SERRS) using Au nanohole arrays on optical fibre tips. *Plasmonics* 8 (2): 1113–1121.

44 Principe, M., Consales, M., Micco, A. et al. (2017). Optical fibre meta-tips. *Light: Science and Applications* 6 (3): e16226.

45 Principe, M., Consales, M., Castaldi, G. et al. (2019). Evaluation of fibre-optic phase-gradient meta-tips for sensing applications. *Nanomaterials and Nanotechnology* 9: 1–8.

46 Savinov, V. and Zheludev, N.I. (2017). High-quality metamaterial dispersive grating on the facet of an optical fibre. *Applied Physics Letters* 111 (9): 091106.

47 Kim, H.T. and Yu, M. (2019). Lab-on-fibre nanoprobe with dual high-Q rayleigh anomaly-surface plasmon polariton resonances for multiparameter sensing. *Scientific Reports* 9: 1922.

48 Tibuleac, S., Wawro, D., and Magnusson, R. (1999). Resonant diffractive structures integrating waveguide-gratings on optical fibre end faces. *Proceedings of the Lasers and Electro-Optics Society Annual Meeting-LEOS* 2: 874–875.

49 Feng, S., Zhang, X., Wang, H. et al. (2010). Fibre coupled waveguide grating structures. *Applied Physics Letters* 96 (13): 133101.

50 Yang, X., Ileri, N., Larson, C.C. et al. (2012). Nanopillar array on a fibre facet for highly sensitive surface-enhanced Raman scattering. *Optics Express* 20 (22): 24819–24826.

51 Kim, J.-B. and Jeong, K.-H. (2017). Batch fabrication of functional optical elements on a fibre facet using DMD based maskless lithography. *Optics Express* 25: 16854–16859.

52 Hemmati, H., Ko, Y.H., and Magnusson, R. (2018). Fibre-facet-integrated guided-mode resonance filters and sensors: experimental realization. *Optics Letters* 43 (3): 358–361.

53 Lin, Y., Guo, J., and Lindquist, R.G. (2009). Demonstration of an ultra-wideband optical fibre inline polarizer with metal nano-grid on the fibre tip. *Optics Express* 17 (20): 17849–17854.

54 Prasciolu, M., Cojoc, D., Cabrini, S. et al. (2003). Design and fabrication of on-fibre diffractive elements for fibre-waveguide coupling by means of e-beam lithography. *Microelectronic Engineering* 67: 169–174.

55 Consales, M., Ricciardi, A., Crescitelli, A. et al. (2012). Lab-on-fibre technology: toward multifunctional optical nanoprobes. *ACS Nano* 6 (4): 3163–3170.

56 Ricciardi, A., Consales, M., Quero, G. et al. (2013). Lab-on-fibre devices as an all around platform for sensing. *Optical Fibre Technology* 19 (6): 772–784.

57 Ricciardi, A., Consales, M., Quero, G. et al. (2014). Lab-on-fibre technology: toward multifunctional optical nanoprobes. *ACS Photonics* 1 (1): 69–78.

58 Lin, Y., Zou, Y., Mo, Y. et al. (2010). E-beam patterned gold nanodot arrays on optical fibre tips for localized surface plasmon resonance biochemical sensing. *Sensors* 10 (10): 9397–9406.

59 Ricciardi, A., Severino, R., Quero, G. et al. (2015). Lab-on-fibre biosensing for cancer biomarker detection. *Proceedings of the 24th International Conference on Optical Fibre Sensors*, Brazil (28 September–2 October 2015), pp. 963423–963423.

60 Lin, Y., Zou, Y., and Lindquist, R.G. (2011). A reflection-based localized surface plasmon resonance fibre-optic probe for biochemical sensing. *Biomedical Optics Express* 2 (3): 478–484.

61 Sanders, M., Lin, Y., Wei, J. et al. (2014). An enhanced LSPR fibre-optic nanoprobe for ultrasensitive detection of protein biomarkers. *Biosensors and Bioelectronics* 61: 95–101.

62 Wang, N., Zeisberger, M., Hübner, U., and Schmidt, M.A. (2018). Nanotrimer enhanced optical fibre tips implemented by electron beam lithography. *Optical Materials Express* 8: 2246–2255.

63 Said, A.A., Dugan, M., Man, S.D., and Iannuzzi, D. (2008). Carving fibre-top cantilevers with femtosecond laser micromachining. *Journal of Micromechanics and Microengineering* 18 (3): 035005.

64 Shin, W., Sohn, I.B., Yu, B.A. et al. (2007). Microstructured fibre end surface grating for coarse WDM signal monitoring. *IEEE Photonics Technology Letters* 19 (8): 550–552.

65 Kim, J.K., Kim, J., Oh, K. et al. (2009). Fabrication of micro Fresnel zone plate lens on a mode-expanded hybrid optical fibre using a femtosecond laser ablation system. *IEEE Photonics Technology Letters* 21 (1): 21–23.

66 Lan, X., Han, Y., Wei, T. et al. (2009). Surface-enhanced Raman-scattering fibre probe fabricated by femtosecond laser. *Optics Letters* 34 (15): 2285–2287.

67 Ma, X., Huo, H., Wang, W. et al. (2010). Surface-enhanced Raman scattering sensor on an optical fibre probe fabricated with a femtosecond laser. *Sensors* 10 (12): 11064–11071.

68 Yuan, L., Lan, X., Huang, J. et al. (2014). Comparison of silica and sapphire fibre SERS probes fabricated by a femtosecond laser. *IEEE Photonics Technology Letters* 26 (13): 1299–1302.

69 Youfu, G., Zhen, Y., Xiaoling, T. et al. (2016). Femtosecond laser ablated polymer SERS fibre probe with photoreduced deposition of silver nanoparticles. *IEEE Photonics Journal* 8 (5): 4502006.

70 Chou, S.Y., Krauss, P.R., and Renstrom, P.J. (1995). Imprint of sub 25 nm vias and trenches in polymers. *Applied Physics Letters* 67 (21): 3114–3116.

71 Chou, S.Y., Krauss, P.R., and Renstrom, P.J. (1996). Imprint lithography with 25-nanometer resolution. *Science* 272 (5258): 85–87.

72 Chandrappan, J., Jing, Z., Mohan, R.V. et al. (2009). Optical coupling methods for cost-effective polymer optical fibre communication. *IEEE Transactions on Components and Packaging Technologies* 32 (3): 539–599.

73 Sakata, H. and Imada, A. (2002). Lensed plastic optical fibre employing concave end filled with high-index resin. *Journal of Lightwave Technology* 20 (4): 638–642.

74 Volkov, A.V., Golovashkin, D.L., Eropolov, V.A. et al. (2007). Studying fabrication errors of the diffraction grating on the end face of a silver-halide fibre. *Optical Memory and Neural Networks* 16 (4): 263–268.

75 Sanghera, J., Florea, C., Busse, L. et al. (2010). Reduced Fresnel losses in chalcogenide fibres by using anti-reflective surface structures on fibre end faces. *Optics Express* 18 (25): 26760–26768.

76 Florea, C., Sanghera, J., Busse, L. et al. (2011). Reduced Fresnel losses in chalcogenide fibres obtained through fibre-end microstructuring. *Applied Optics* 50 (1): 17–21.

77 Viheriala, J., Niemi, T., Kontio, J. et al. (2007). Fabrication of surface reliefs on facets of singlemode optical fibres using nanoimprint lithography. *Electronics Letters* 43 (3): 1.

78 Scheerlinck, S., Taillaert, D., Thourhout, D.V., and Baets, R. (2008). Flexible metal grating based optical fibre probe for photonic integrated circuits. *Applied Physics Letters* 92 (3): 031104.

79 Scheerlinck, S., Dubruel, P., Bienstman, P. et al. (2009). Metal grating patterning on fibre facets by UV-based nano imprint and transfer lithography using optical alignment. *Journal of Lightwave Technology* 27 (10): 1415–1420.

80 Kostovski, G., Chinnasamy, U., Jayawardhana, S. et al. (2011). Sub-15 nm optical fibre nanoimprint lithography: a parallel, self-aligned and portable approach. *Advanced Materials* 23 (4): 531–535.

81 Calafiore, G., Koshelev, A., Allen, F.I. et al. (2016). Nanoimprint of a 3D structure on an optical fibre for light wavefront manipulation. *Nanotechnology* 27: 375301.

82 Koshelev, A., Calafiore, G., Piña-Hernandez, C. et al. (2016). High refractive index Fresnel lens on a fibre fabricated by nanoimprint lithography for immersion applications. *Optics Letters* 41: 3423–3426.

83 Calafiore, G., Koshelev, A., Darlington, T.P. et al. (2017). Campanile near-field probes fabricated by nanoimprint lithography on the facet of an optical fibre. *Scientific Reports* 7 (1): 1651.

84 Smythe, E.J., Dickey, M.D., Whitesides, G.M., and Capasso, F. (2009). A technique to transfer metallic nanoscale patterns to small and non-planar surfaces. *ACS Nano* 3 (1): 59–65.

85 Lipomi, D.J., Martinez, R.V., Kats, M.A. et al. (2010). Patterning the tips of optical fibres with metallic nanostructures using nanoskiving. *Nano Letters* 11 (2): 632–636.

86 Wang, Y. et al. (2018). Flexible transfer of plasmonic photonic structures onto fibre tips for sensor applications in liquids. *Nanoscale* 10: 16193–16200.

87 Shambat, G., Provine, J., Rivoire, K. et al. (2011). Optical fibre tips functionalized with semiconductor photonic crystal cavities. *Applied Physics Letters* 99 (19): 191102.

88 Shambat, G., Kothapalli, S.R., Khurana, A. et al. (2012). A photonic crystal cavity-optical fibre tip nanoparticle sensor for biomedical applications. *Applied Physics Letters* 100 (21): 213702.

89 Arce, C.L., De Vos, K., Claes, T. et al. (2011). Silicon-on-insulator microring resonator sensor integrated on an optical fibre facet. *IEEE Photonics Technology Letters* 23 (13): 890–892.

90 Jia, P. and Yang, J. (2013). Integration of large-area metallic nanohole arrays with multimode optical fibres for surface plasmon resonance sensing. *Applied Physics Letters* 102 (24): 243107.

91 Jia, P. and Yang, J. (2014). A plasmonic optical fibre patterned by template transfer as a high-performance flexible nanoprobe for real-time biosensing. *Nanoscale* 6 (15): 8836–8843.

92 He, X., Yi, H., Long, J. et al. (2016). Plasmonic crystal cavity on singlemode optical fibre end facet for label-free biosensing. *Applied Physics Letters* 108 (23): 231105.

93 Yang, T., He, X., Zhou, X. et al. (2018). Surface plasmon cavities on optical fibre end-facets for biomolecule and ultrasound detection. *Optics and Laser Technology* 101: 468–478.

94 Galeotti, F., Pisco, M., and Cusano, A. (2018). Self-assembly on optical fibres: a powerful nanofabrication tool for next generation "lab-on-fibre" optrodes. *Nanoscale* 10: 22673–22700.

95 Whitesides, G.M. and Grzybowski, B. (2002). Self-assembly at all scales. *Science* 295 (5564): 2418–2421.

96 Viets, C. and Hill, W. (1998). Comparison of fibre-optic SERS sensors with differently prepared tips. *Sensors and Actuators B: Chemical* 51 (1): 92–99.

97 Jeong, H.H., Erdene, N., Park, J.H. et al. (2013). Real-time label-free immunoassay of interferon-gamma and prostate-specific antigen using a fibre-optic localized surface plasmon resonance sensor. *Biosensors and Bioelectronics* 39 (1): 346–351.

98 Sciacca, B. and Monro, T.M. (2014). Dip biosensor based on localized surface plasmon resonance at the tip of an optical fibre. *Langmuir* 30 (3): 946–954.

99 Register, J.K., Fales, A.M., Wang, H.-N. et al. (2015). In vivo detection of SERS-encoded plasmonic nanostars in human skin grafts and live animal models. *Analytical and Bioanalytical Chemistry* 407: 8215–8224.

100 Fales, A.M., Yuan, H., and Vo-Dinh, T. (2014). Development of hybrid silver-coated gold nanostars for nonaggregated surface-enhanced Raman scattering. *The Journal of Physical Chemistry C* 118: 3708–3715.

101 Ran, Y., Strobbia, P., Cupil-Garcia, V., and Vo-Dinh, T. (2019). Fibre-optrode SERS probes using plasmonic silver-coated gold nanostars. *Sensors and Actuators B: Chemical* 287: 95–101.

102 Pisco, M., Galeotti, F., Quero, G. et al. (2014). Miniaturized sensing probes based on metallic dielectric crystals self-assembled on optical fibre tips. *ACS Photonics* 1: 917–927.

103 Pisco, M., Galeotti, F., Grisci, G., Quero, G., and Cusano, A. (2015). Self-assembled periodic patterns on the optical fibre tip by microsphere arrays. *Proceedings of the 24th International Conference on Optical Fibre Sensors*, Curitiba, Brazil (28 September–2 October 2015).

104 Pisco, M., Galeotti, F., Quero, G. et al. (2016). Nanosphere lithography for advanced all fibre Sers probes. *Proceedings of the Sixth European Workshop on Optical Fibre Sensors*, Limerick, Ireland (31 May–3 June 2016).

105 Quero, G., Zito, G., Managò, S. et al. (2018). Nanosphere lithography on fibre: towards engineered lab-on-fibre SERS optrodes. *Sensors* 18 (3): 680.

106 Pisco, M., Galeotti, F., Quero, G. et al. (2017). Nanosphere lithography for optical fibre tip nanoprobes. *Light: Science & Applications* 6: e16229.

107 Ni, H.B., Wang, M., Li, L. et al. (2013). Photonic-crystal-based optical fibre bundles and their applications. *IEEE Photonics Journal* 5: 2400213.

108 Tierney, S., Hjelme, D.R., and Stokke, B.T. (2008). Determination of swelling of responsive gels with nanometer resolution. Fibre-optic based platform for hydrogels as signal transducers. *Analytical Chemistry* 80: 5086–5093.

109 Tierney, S., Falch, B.M.H., Hjelme, D.R., and Stokke, B.T. (2009). Determination of glucose levels using a functionalized hydrogel-optical fibre biosensor: toward continuous monitoring of blood glucose in vivo. *Analytical Chemistry* 81: 3630–3636.

110 Tierney, S., Volden, S., and Stokke, B.T. (2008). Glucose sensors based on a responsive gel incorporated as a Fabry-Perot cavity on a fibre-optic readout platform. *Biosensors and Bioelectronics* 24: 2034–2039.

111 Gawel, K. and Stokke, B.T. (2011). Logic swelling response of DNA–polymer hybrid hydrogel. *Soft Matter* 7: 4615–4618.

112 Muri, H.I. and Hjelme, D.R. (2017). LSPR coupling and distribution of interparticle distances between nanoparticles in hydrogel on optical fibre end face. *Sensors* 17 (12): 2723.

113 Muri, H.I., Bano, A., and Hjelme, D.R. (2018). LSPR and interferometric sensor modalities combined using a double-clad optical fibre. *Sensors* 18 (1): 187.

114 Aliberti, A., Ricciardi, A., Giaquinto, M. et al. (2017). Microgel assisted lab-on-fibre optrode. *Scientific Reports* 7: 14459.

115 Giaquinto, M., Ricciardi, A., Aliberti, A. et al. (2018). Light-microgel interaction in resonant nanostructures. *Scientific Reports* 8: 9331.

116 Giaquinto, M., Micco, A., Aliberti, A. et al. (2018). Optimization strategies for responsivity control of microgel assisted lab-on-fibre optrodes. *Sensors* 18 (4): 1119.

117 Scherino, L., Giaquinto, M., Micco, A. et al. (2018). A time-efficient dip coating technique for the deposition of microgels onto the optical fibre tip. *Fibres* 6 (4): 72.

118 Plamper, F.A. and Richtering, W. (2017). Functional microgels and microgel systems. *Accounts of Chemical Research* 50 (2): 131–140.

119 Kilic, O., Digonnet, M.J.F., Kino, G.S., and Solgaard, O. (2007). External fibre Fabry–Perot acoustic sensor based on a photonic-crystal mirror. *Measurement Science and Technology* 18 (12): 3049–3054.

120 Kilic, O., Digonnet, M.J.F., Kino, G.S., and Solgaard, O. (2011). Miniature photonic-crystal hydrophone optimized for ocean acoustics. *The Journal of the Acoustical Society of America* 129 (4): 1837–1850.

121 Iannuzzi, D., Deladi, S., Schreuders, H. et al. (2006). Fibre-top cantilevers: a new generation of micromachined sensors for multipurpose applications. *18th International Optical Fibre Sensor Conference, Cancún, Mexico* (23–27 October 2006).

122 Iannuzzi, D., Deladi, S., Gadgil, V.J. et al. (2006). Monolithic fibre-top sensor for critical environments and standard applications. *Applied Physics Letters* 88 (5): 053501.

123 Deladi, S., Iannuzzi, D., Gadgil, V.J. et al. (2006). Carving fibre-top optomechanical transducers from an optical fibre. *Journal of Micromechanics and Microengineering* 16: 886–889.

124 Gruca, G., de Man, S., Slaman, M. et al. (2010). Ferrule-top micromachined devices: design, fabrication, performance. *Measurement Science and Technology* 21 (9): 094033.

125 Chavan, D., Gruca, G., de Man, S. et al. (2010). Ferrule-top atomic force microscope. *Review of Scientific Instruments* 81 (12): 123702.

126 Gavan, K.B., Rector, J.H., Heeck, K. et al. (2011). Top-down approach to fibre-top cantilevers. *Optics Letters* 36 (15): 2898–2900.

127 Gruca, G., Chavan, D., Rector, J. et al. (2013). Demonstration of an optically actuated ferrule-top device for pressure and humidity sensing. *Sensors and Actuators A: Physical* 190: 77–83.

128 Schenato, L., Palmieri, L., Gruca, G. et al. (2012). Fibre optic sensors for precursory acoustic signals detection in rockfall events. *Journal of the European Optical Society – Rapid Publications* 7: 12048.

129 Gruca, G., Rector, J.H., Heeck, K., and Iannuzzi, D. (2011). Optical fibre ferrule-top sensor for humidity measurements. *21st International Conference on Optical Fibre Sensors,*Ottawa, Canada (15–19 May 2011).

130 Zuurbier, P., de Man, S., Gruca, G. et al. (2011). Measurement of the Casimir force with a ferrule-top sensor. *New Journal of Physics* 13 (2): 023027.

131 Gruca, G., Heeck, K., Rector, J., and Iannuzzi, D. (2013). Measurement of the Casimir force with a ferrule-top sensor. *Optics Letters* 38 (10): 1672–1674.

132 Pisco, M., Bruno, F.A., Galluzzo, D. et al. (2018). Opto-mechanical lab-on-fibre seismic sensors detected the Norcia earthquake. *Scientific Reports* 8: 6680.

133 Chavan, D., van de Watering, T.C., Gruca, G. et al. (2012). Ferrule-top nanoindenter: an optomechanical fibre sensor for nanoindentation. *Review of Scientific Instruments* 83: 115110.

134 van Hoorn, H., Kurniawan, N.A., Koenderink, G.H., and Iannuzzi, D. (2016). Local dynamic mechanical analysis for heterogeneous soft matter using ferrule-top indentation. *Soft Matter* 12: 3066–3073.

135 Bos, E.J., van der Laan, K., Helder, M.N. et al. (2017). Noninvasive measurement of ear cartilage elasticity on the cellular level: a new method to provide biomechanical information for tissue engineering. *Plastic and Reconstructive Surgery Global Open* 5 (2): e1147.

136 Kamperman, T., Henke, S., Zoetebier, B. et al. (2017). Nanoemulsion-induced enzymatic crosslinking of tyramine-functionalized polymer droplets. *Journal of Materials Chemistry B* 5: 4835–4844.

137 Martorina, F., Casale, C., Urciuolo, F. et al. (2017). In vitro activation of the neuro-transduction mechanism in sensitive organotypic human skin model. *Biomaterials* 113: 217–229.

138 Beekmans, S.V., Emanuel, K.S., Smit, T.H., and Iannuzzi, D. (2017). Minimally invasive micro-indentation: mapping tissue mechanics at the tip of an 18G needle. *Scientific Reports* 7: 11364.

139 Guggenheim, J.A., Li, J., Allen, T.J. et al. (2017). Ultrasensitive plano-concave optical microresonators for ultrasound sensing. *Nature Photonics* 11: 714–719.

140 Yao, M., Wu, J., Zhang, A.P. et al. (2017). Optically 3-D μ-printed ferrule-top polymer suspended-mirror devices. *IEEE Sensors Journal* 17 (22): 7257–7261.

141 Wu, J., Yin, M., Seefeldt, K. et al. (2018). In situ M-printed optical fibre-tip CO_2 sensor using a photocrosslinkable poly(ionic liquid). *Sensors and Actuators B: Chemical* 259: 833–839.

142 Yao, M., Ouyang, X., Wu, J. et al. (2018). Optical fibre-tip sensors based on in-situ μ-printed polymer suspended-microbeams. *Sensors* 18: 1825.

143 Rauch, J.Y., Lehmann, O., Rougeot, P. et al. (2018). Smallest microhouse in the world, assembled on the facet of an optical fibre by origami and welded in the μRobotex nanofactory. *Journal of Vacuum Science & Technology A* 36: 041601.

144 Pahlevaninezhad, H., Khorasaninejad, M., Huang, Y.W. et al. (2018). Nano-optic endoscope for high-resolution optical coherence tomography in vivo. *Nature Photonics* 12 (9): 540–547.

145 Xomalis, A., Demirtzioglou, I., Plum, E. et al. (2018). Fibre-optic metadevice for all-optical signal modulation based on coherent absorption. *Nature Communications* 9: 182.

10

From Refractometry to Biosensing with Optical Fibres

Francesco Chiavaioli, Ambra Giannetti, and Francesco Baldini

Institute of Applied Physics "Nello Carrara" (IFAC), National Research Council (CNR), Florence, Italy

During the last two decades, many optical fibre sensors (OFSs) have been developed. In this chapter, attention will be focused on OFSs that measure refractive index (RI) changes as main sensing mechanism [1]. The basic sensing concepts and parameters that can be used to assess the performance of OFSs will be deeply analysed. Afterwards, optical fibre refractometers will be firstly taken into account, where the RI changes encompass all the volume around the fibre. Then we will move towards optical fibre biosensors (OFBs), where the RI changes occur just on the functionalized surface of the fibre. In this sense, the distinction between volume or bulk RI changes and surface RI changes plays the crucial role [2]. Therefore, this feature that is strongly related to the penetration depth of the evanescent field excited in the OFS [3] will be highlighted and clarified. As far as the OFBs are concerned, three domains are attracting more and more interest in the scientific community, such as immuno-based, oligonucleotide-based, and whole cell/microorganism-based biosensors, most of them related to healthcare or critical diseases. Indeed, advanced diagnostics is one of the key research fields nowadays. However, very few OFBs have reached the market, while most of them, even with impressive performance, have not overtaken the laboratory stage. Finally, future tracks and perspectives will be highlighted in order to provide some opportunities for OFSs to be used in the field of advanced diagnostics.

Optical Fibre Sensors: Fundamentals for Development of Optimized Devices, First Edition.
Edited by Ignacio Del Villar and Ignacio R. Matias.
© 2021 The Institute of Electrical and Electronics Engineers, Inc.
Published 2021 by John Wiley & Sons, Inc.

10.1 Basic Sensing Concepts and Parameters for OFSs

Given the increasing attention of the scientific community to OFSs, the need for a worldwide acceptable standardization of the sensing performance could be of general interest for the scientific community. The idea is providing a reference to facilitate any comparison, not only within the same class of OFSs but also among different kinds of sensors based on spectral resonance, such as Bloch surface wave sensors [4], surface plasmon resonance (SPR) and localized SPR (LSPR) devices [5, 6], resonating structures [7, 8], and any resonance-based photonic configuration [9].

OFSs that are based on the measurement of a RI change rely on the evanescent field interactions at the boundary between the fibre and the surrounding environment. The evanescent field is generated at the interference between the incident and reflected rays, producing an evanescent wave that travels in the direction of the fibre axis and decays exponentially, perpendicular to the fibre–surrounding environment interface, as shown in Figure 10.1 [10]. Therefore, the response of such platforms depends strongly on the capability of the evanescent field to penetrate the surrounding environment. Hence, the key parameter is the penetration depth δ_p (i.e. the distance from the interface at which the amplitude of the evanescent wave of the electromagnetic field decreases by a factor of $1/e$). Broadly speaking, the larger the δ_p, the greater the portion of radiation in contact with the surrounding environment is, thus enabling a higher sensitivity [11]. Since δ_p depends on the mode travelling in the denser medium (fibre cladding) as well

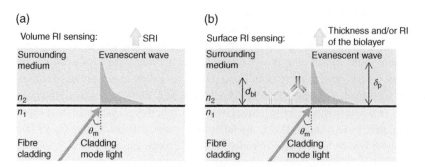

Figure 10.1 (a) Sketch of the RI sensing based on a bulk or volume change. (b) Sketch of the RI sensing based on a surface change in which only a portion of the evanescent wave of extent δ_p interacts with the biological layer of thickness d_{bl}. SRI, surrounding refractive index; n_1, fibre cladding RI (or more generally the RI of the denser medium); n_2, surrounding environment RI (or more generally the RI of the less dense medium); d_{bl}, thickness of the biological layer; δ_p, penetration depth of the evanescent field. *Source:* Reprinted with permissions from MDPI [2].

as on both the surrounding RI (SRI) and the operating wavelength λ, the longer the λ, the greater the δ_p of the evanescent wave in the surrounding environment is [11]. In fact, this phenomenon explains why higher-order modes – which are less bound to the waveguide structure but extend far more into the surrounding environment (larger δ_p) – are more sensitive to the RI changes.

As schematically represented in Figure 10.1, the difference between a bulk or volume RI change (a) and a surface RI change (b) deserves to be clarified since it gives the critical distinction between an optical refractometer, which measures volume RI changes, and an optical biosensor, which measures surface RI changes that occur on the functionalized surface of the device and are related to the inter-actions between the analyte under investigation and a biological recognition ele-ment (BRE) grafted on the surface, the so-called biolayer. Recently, Sinibaldi et al. [4, 12] and Chiavaioli et al. [2] pointed out this distinction, providing also some calculations that relate the volume RI sensitivity to the surface one. In volume RI sensing (Figure 10.1a), the evanescent wave related to the considered mode (or in general the incident light) interacts with all the surrounding environment for all the extent of its δ_p. Conversely, in surface RI sensing (Figure 10.1b), the interaction of interest is between the layer in which the BRE is grafted and a por-tion of the evanescent wave, with the thickness of the biolayer usually much less than the δ_p (typically, tens of nanometres compared in the case of visible window with 200–500 nm). In this second case, since RI changes induced on the sensor sur-face due to the interactions between the BRE and the analyte under investigation are the measurand of interest, a washing step in a buffer solution should be per-formed after the injection of the subsequent concentration of the analyte in order to measure only the signal change caused by the binding interaction [11]. If this procedure is not adopted, it must be clear that, in any case, a signal change occurs and can be measured, even if the variation related to a surface RI change is usually less than the one related to a bulk RI change due to the biomolecule dimension interacting with the BRE [13].

Different strategies have been developed over the years to enhance the light–matter (SRI or biomolecule) interaction in OFSs:

1) The first approach regards the deposition of a thin layer of sub-wavelength thickness with a higher RI than the fibre cladding. This means that most of the optical power carried by the selected mode travels within the deposited coating. Depending on the optical properties (i.e. thickness and RI) of the coat-ing and on the SRI, this leads to a great change in the propagation conditions of the optical signal [14]. Therefore, the evanescent wave of the mode extends more in the surroundings, and, hence, this leads to an increased RI sensitivity. By carefully adjusting the coating thickness, it is possible to tune the maximum sensitivity to a desirable range of SRIs. This approach can be followed in

unmodified fibres or in fibres embedding long-period gratings (LPGs) [15] and tilted fibre Bragg gratings (TFBGs) [16]. In the case of LPG, an increase sensitivity is also achieved, coupling the fundamental mode to a high-order cladding mode at the dispersion turning point (DTP) in its phase matching curve [17]. In this case, the mode details two resonances that merge into a single broader one at the DTP. These dual-peak LPGs showed a higher sensitivity than other LPGs fabricated on a bare fibre. Very recently, by combining the two previous approaches, it was proved a remarkable sensitivity of 10^5 nm/RIU with a DTP LPG working in modal transition, theoretically [18], and a sensitivity of 1.2×10^4 nm/RIU in a narrow RI around 1.33, experimentally [19].

A special case of this first approach involves the deposition of metal oxides, such as zinc oxide, gallium zinc oxide, indium tin oxide, and tin oxide, which allows the exploitation of a novel physical phenomenon named lossy mode resonance (LMR) [20]. LMR was firstly observed in waveguides [21] and, recently, applied to fibre optics [22]. In combination with specially modified geometries of the optical fibres, like cladding removed multimode fibres (CRMMFs) or D-shaped single-mode fibres (SMF), it is possible to greatly enhance this phenomenon and, hence, to attain remarkable performance in biosensing [23, 24].

2) A second approach is based on the combination of OFSs with SPR or LSPR [25]. As for SPR, the fibre is coated with a nm-thick uniform metallic (mostly gold but also silver sometimes) layer able to excite SPR due to the core–cladding efficient coupling mechanism. Usually, this approach is combined with a grating structure within the fibre, i.e. TFBGs [26] or LPG [27], to further enhance sensor performance. LSPR differs from SPR for the fact that a nanostructured pattern or specially designed nanostructures are deposited around the fibre or on the fibre tip [28]. In this case, a more confined but stronger light–matter interaction can be attained, allowing the detection of very small biomolecules and also the comprehension of the different phenomena that occur during the binding interaction [6].

3) The third approach consists of using microstructured optical fibres (MOFs) based on internally modified fibre geometries or photonic crystals [29–31]. These configurations present unique properties in light guiding and are very attractive for biochemical sensing due to their capability to use the fibre cladding as microfluidic channel [32]. In such a case, the evanescent wave of the travelling mode interacts directly with the liquid passing though the holes, thus enabling potentially high sensitivity by using a small volume of solution while preserving the fibre integrity and structure.

4) The last approach relies on removing partially or totally the fibre cladding by means of etching [33], polishing [34], or using directly microfibres or nanofibres [35, 36], i.e. fibres with a diameter of a few micrometres or down to micrometres, respectively. In all these cases, the evanescent wave of the travelling

mode extends outside the fibre, and, consequently, the measured signal depends on the SRI. Therefore, the sensitivity of these configurations is strongly dependent on the diameter of the fibre. In general, the smaller the fibre diameter, the higher the sensitivity, although this mechanism introduces greater fragility and higher difficulties in fibre handling.

Next subsections summarize the most common parameters used in the literature of fibre-optic sensing, starting from the ones of general interest, then passing through the parameters related to volume RI sensing, and, finally, concluding with the parameters related to surface RI sensing.

10.1.1 Parameters of General Interest

There are a number of parameters of generic interest that can be associated with any sensor. It is worth mentioning the uncertainty, accuracy, precision, stability, drift, repeatability, reproducibility, and response time. In fact, with the final aim of realizing a commercial device, most of those should be claimed in the sensor data sheet.

10.1.1.1 Uncertainty

In metrology, any measurement should be presented along with its uncertainty (or margin of error). Uncertainty gives a range of values that are most likely close to the 'true value' of the measurement and is charted by means of the error bar or reported by the classical notation: *measured value $\pm$ uncertainty.* Usually, the uncertainty of a measurement is obtained by repeating the measurement enough times to attain a good estimation of the standard deviation σ of the measurement values. Here, any single value has an uncertainty equal to the standard deviation. Conversely, if the values are averaged, the mean value of the measurement will then have a much smaller uncertainty, equal to the standard error of the mean value, which is σ divided by the square root of the number of measurements n, $\sigma/\sqrt{n}$. When the uncertainty is given with the standard error of the measurement, the measured value will fall within a specific uncertainty range. By considering a normal distribution of the errors, the standard errors can easily be converted to 68.3% (σ), 95.4% (2σ), or 99.7% (3σ) confidence intervals. In general, uncertainty depends on both accuracy and precision of the sensor: the lower the accuracy and precision of the sensor, the larger the measurement uncertainty.

10.1.1.2 Accuracy and Precision

Accuracy and precision are two parameters that are often confused. Accuracy is the deviation between the mean value obtained from a series of measurements (i.e. the data set) and the true value of the measurand assumed as the reference value. Therefore, accuracy represents how close the measurement is to the true value. Conversely, precision is the level of dispersion of the data set (i.e. the variance, σ^2) with respect to its mean value. Therefore, precision represents how every single measurement agrees with the other. In general, a measurement can be accurate but not precise, precise but not accurate, accurate and precise, or neither accurate nor precise.

10.1.1.3 Sensor Drift and Fluctuations
The drift and fluctuations of a sensor represent other two general parameters, which are often overlooked but deserve attention since they determine the degree of sensor stability, after taking into account the best stabilization of the environmental parameters, such as temperature, humidity, mechanical deformation, etc. A stability test can be performed by measuring the time evolution of the signal under a usual experimental condition. Afterwards, the drift can be evaluated if the long-term measured signal details a clear trend during the measurement time. Conversely, fluctuations can directly be observed during the stabilization test in correspondence of specific occurrences happened during the measurement.

10.1.1.4 Repeatability
Repeatability is the level of agreement between a series of measurements of the same measurand, in which every single measurement has been performed with the same sensor under the same experimental conditions. Hence, the repeatability usually expressed in percentage reflects both the long-term stability of the experimental set-up and the sensor lifetime. The error of repeatability influences the uncertainty and represents the lower bound limit of the best precision obtained from the measurement.

10.1.1.5 Reproducibility
Reproducibility is the level of agreement between a series of measurements of the same measurand, in which every single measurement has been performed with the use of different sensors under the same experimental conditions. Hence, the reproducibility usually expressed in percentage reflects the capability of manufacturing and implementing sensors with the same specifications. The error of reproducibility influences the uncertainty and represents the maximum deviation of the measurements.

10.1.1.6 Response Time
The response time of a sensor plays an essential role in real-time applications. For a generic sensor, it is defined as the time required going from the initial value to a certain percentage of the final value. If an exponential behaviour of the sensor response is assumed, the response time is usually referred to as the time constant

and equals to $1/e$. In electronics and chemistry, it is defined as the time required going from 10 up to 90% of the total variation caused by the change of the measurand of interest, also named rise time. In biosensing, this parameter can be defined as the time necessary to reach the equilibrium condition (plateau or saturation) during a binding interaction between the immobilized receptor and the analyte under investigation. Broadly speaking, the response time depends on a first faster phenomenon that is the kinetics and on a second slower one that is the diffusion towards the biolayer. The last one is slower for lower concentrations of analyte [37]. Moreover, when a coating material (i.e. porous materials, polymers, thin films in general) is deposited on an optical fibre, the response time usually increases due to both the electrostatic interactions between the coating and the analyte and the granularity/homogeneity of the coating [38]. In addition, the response time of a biosensor can be also influenced by the matrix in which the analyte is spiked: the higher the matrix complexity (for example, real matrix such as serum, plasma, and blood), the longer the response time given an expected lower mobility of the analyte [15]. A possible approach to overcome this issue involves the measurement of the initial binding rate of the interaction, proposed in the past [39] and recently confirmed [15] as an effective methodology for quantitatively determining the analyte concentration.

10.1.2 Parameters Related to Volume RI Sensing

In this section, the parameters specifically related to devices based on volume RI sensing are discussed. The three most widespread and crucial parameters are RI sensitivity, resolution, and figure of merit (FOM). For each, a definition and a description are provided together with some illustrative figures when necessary to better clarify the concept.

10.1.2.1 Refractive Index Sensitivity

In order to evaluate the RI sensitivity of an OFS, the response curve of the device should be obtained. It is assessed by monitoring the change of the signal (i.e. the resonance peak wavelength λ or the peak intensity I) as a function of the change in SRI. Usually, for each SRI evaluated, the value of the signal is averaged within a certain time slot or within a certain number of measurements, thus resulting in an experimental point (mean value) and the corresponding error (standard deviation). Hence, the response curve is drawn by the set of experimental points with the corresponding error bar. The sensitivity (S) of the sensor to the SRI can be attained by differentiating the response curve in a specific RI range ($\partial\lambda/\partial n_{\mathrm{sur}}$ or $\partial I/\partial n_{\mathrm{sur}}$) and is expressed in nanometer per refractive index unit or decibel per refractive index unit. It is worth pointing out that some devices exhibit a non-linear response with the increment of SRI, and this is the reason of specifying where the sensitivity has been evaluated considering a linear approximation. In the case of fibre gratings (LPGs and FBGs), the sensitivity increases exponentially as it becomes closer and closer to the RI of fibre cladding [2]. As an illustrative example, a typical response curve of an LPG in terms of wavelength shift is shown in Figure 10.2a, while its corresponding sensitivity is shown in Figure 10.2b, as a function of the SRI. The light grey circles highlight the region related to the maximum wavelength shift and thus to the highest RI sensitivity.

10.1.2.2 Resolution

While the RI sensitivity is quite clear-cut in literature, the evaluation of the resolution (R) is more complicated and is often incorrect. Resolution is defined as the smallest change in SRI that a sensing device is able to measure; in other words, R represents the lowest change in SRI that produces a measurable change in wavelength or in intensity. Hence, R is proportional to the RI sensitivity but depends on all the noise sources that influence the signal stability, which can experimentally be obtained through the standard deviation σ of the measurements. In fact, σ collects all the noise sources from the equipment and the environment, including the algorithm used to calculate the peak wavelength or intensity, the repeatability, and also the influence of the interrogation technique. Therefore, the resolution can be calculated with the following equation and is expressed in refractive index units:

$$R = \frac{p\,\sigma}{S} \tag{10.1}$$

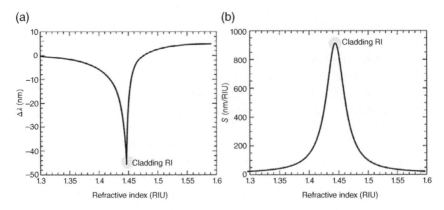

Figure 10.2 (a) A typical response curve of a non-coated LPG to the SRI in the 1.30–1.60 RI range. (b) The corresponding RI sensitivity achieved from the response curve, exhibiting a non-linear trend with increasing sensitivity. *Source:* Reprinted with permissions from MDPI [2].

where σ is calculated by considering a set of measurements in repeatable and reproducible conditions (thus including all the individual sources of noise) and p represents the confidence interval (see Section 10.1.1.1). Usually, R is simply evaluated by dividing the instrument spectral resolution by the sensitivity S. Indeed, this is not the sensor resolution, but represents the *maximum theoretical resolution* (MTR). In addition, another critical concept to be highlighted, which is often disregarded, is the fact that R could not be the same throughout the sensor working range because of the non-linear behaviour of S (see Figure 10.2). This means that it should be considered a minimum and maximum resolution, as occurs for accuracy.

10.1.2.3 Figure of Merit (FOM)

The FOM normalizes the volume RI sensitivity S to the width of resonance peak characterized by the full width at half maximum (FWHM), which describes how precisely it is possible to measure the resonance minimum in terms of wavelength or intensity. FOM can be expressed as [1]

$$\text{FOM} = \frac{S}{\text{FWHM}} \tag{10.2}$$

Generally, the greater the sensitivity, the larger the FOM. However, sensitivity and FWHM are usually related in the sense that it is very hard having large sensitivity with a narrow width of the resonance. Conversely, it is common that devices with very narrow FWHM exhibit a not so high sensitivity and vice versa; this means that a trade-off between sensitivity and FWHM always happens. For example, LMR-based devices usually possess very high S of the order of 10^3–10^4 nm/RIU with a large FWHM of the order of tens of nanometres. On the other hand, FBG-based devices traditionally have a very narrow FWHM of the order of 10^{-2} nm with a not so high sensitivity of the order of 10^2 nm/RIU. Therefore, the second type of sensor is preferable in terms of FOM, while the first one is better as far as the volume sensitivity is concerned. A deep analysis was carried out in past literature [40].

10.1.3 Parameters Related to Surface RI Sensing

In this section, differently from the previous section, the parameters specifically related to devices based on surface RI sensing are discussed. The most important parameters are limit of detection (LOD), specificity or selectivity and regenerability or reusability, which can be obtained starting from the sensorgram and related calibration curve. For each, a definition and a description are provided together with some illustrative figures when necessary to better clarify the concept.

10.1.3.1 Sensorgram and Calibration Curve

A sensorgram is a chart reporting the evolution of the signal as a function of time. More specifically, the sensorgram details the time evolution of the peak wavelength or of the peak intensity. An example is given in Figure 10.3. For the sake of completeness, the value of the resonance peak wavelength was previously obtained from raw data (i.e. the transmission spectrum) by means of a suitable data processing, which involves a fitting procedure or other mathematical approaches used to calculate the minimum of the signal (e.g. centroid analysis). It is worth pointing out that from the sensorgram it is possible to extrapolate a series of other critical parameters of a biosensor, such as response time and calibration curve. Figure 10.3 accounts for all the typical steps performed during the implementation of a bioassay, which are indicated with arrows, starting from the grafting of the receptor (BRE) on the sensor surface, followed by the surface passivation to block free remaining functional groups, so as to discard any non-specific adsorption, and the specificity test (this fundamental part will be deeply discussed later), up to the assay completion performed by injecting increasing concentrations of the analyte under investigation. After each injection of the analyte, a washing step in a running buffer is carried out in order to measure the signal change only due to a surface modification. Figure 10.3 also details some illustrative

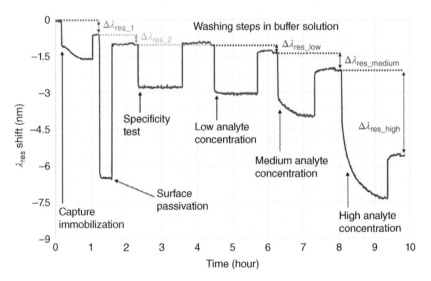

Figure 10.3 An example of a sensorgram accounting for the most significant steps performed during the implementation of a bioassay, highlighted with arrows, and the critical wavelength shifts taken after the washing steps carried out with a runner buffer. *Source:* Modified with permissions from MDPI [2].

shifts of the resonance peak wavelength taken in buffer after the respective washing when the flow is stopped. The higher the analyte concentration, the larger the surface shift is.

For a biosensor, the calibration curve is a chart reporting the signal change as a function of the concentration of the analyte under investigation. More specifically, the calibration curve details the change in the peak wavelength or in the peak intensity with respect to the analyte concentration, which can be reported in linear or logarithmic scale. The signal can be expressed as the absolute value of resonance peak wavelength or the corresponding shift scaled down to the reference concentration that is the concentration at zero analyte, named blank measurement. An example is given in Figure 10.4. The black circles are the experimental points, whereas the grey line represents the sigmoidal curve of the experimental data using the logistic fitting function, a well-accepted mathematical model for describing the sigmoidal behaviour of a biosensor [11, 15].

10.1.3.2 Limit of Detection (LOD) and Limit of Quantification (LOQ)

The LOD represents the most important and critical parameter for assessing the performance of any biosensor. LOD is reported in terms of a concentration value [C] of the analyte under investigation, such as gram per litre (usually in microgram per litre or nanogram per litre considering the molecular weight of the analyte) or molarity M (usually in nanomolars or picomolars considering the molecular

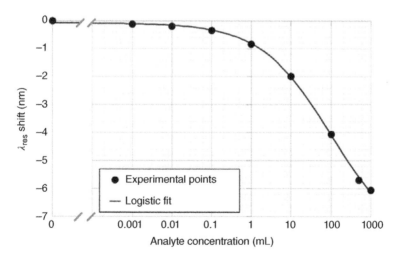

Figure 10.4 An example of a calibration curve obtained from an LPG-based biosensor. The black circles are the experimental points at increasing concentrations (expressed in log scale), whereas the grey line represents the sigmoidal curve of the experimental data using the logistic fitting function. *Source:* Modified with permissions from MDPI [2].

weight of the analyte). Given its importance, it is not so difficult to find different manners for evaluating the LOD. In addition, it is also quite common to find published results where the device has been tested in a certain concentration range, while the claimed LOD was below (even orders of magnitude) the lowest measured concentration. Technical discussions and analytical considerations regarding LOD have already been done in the literature [2, 41], and, hence, this subsection accounts for the different manners to calculate LOD.

In the first approach, LOD is calculated following the recommendation given by the *American Chemical Society* (ACS) [42], which is based on the use of the standard deviations at low concentration. The steps can be summarized as follows:

1) Repeated measurements ($k = 7$–10) of the blank sample (i.e. the one not containing the analyte under investigation).
2) Calculation of the mean value $\bar{y}_{blank}$.
3) Samples containing the target analyte are prepared at a concentration of 1–5 times higher than the suspected LOD.
4) k measurements are performed, and mean value $\bar{y}$ and its standard deviation σ_y are achieved.
5) Signal at LOD can be calculated as

$$y_{LOD} = \bar{y}_{blank} + t_{(\alpha,k-1)} \cdot \sigma_y \tag{10.3}$$

where $t_{(\alpha,k-1)}$ is the α quantile of t-Student function with $k - 1$ degrees of freedom where $(1 - \alpha)$ defines the confidence interval. For the sake of simplicity, $t = 3$ is usually assumed in Eq. (10.3), meaning a minimum of 16 samples (8 blanks and 8 low concentration samples) to be analysed within a confidence interval of 99%.

In the second approach, LOD is determined by using the calibration curve of the biosensor and the *IUPAC* recommendation [43]. This last states that LOD can be obtained from the biosensor calibration curve f by considering the averaged blank signal plus three times the standard deviation of the blank sample ($3\sigma_{blank}$), according to the following equation:

$$x_{LOD} = f^{-1}(\bar{y}_{blank} + 3\sigma_{blank}) \tag{10.4}$$

It is worth pointing out that in the case of a low number of concentrations close to the suspected LOD with a linear response of the biosensor, this approach is very similar to the first one. However, a second modified approach has been proposed as a mixture of the two previous ones [2]: in Eq. (10.4), the maximum standard deviation obtained among all the experimental points ($\sigma_{y_{max}}$) is considered. Nevertheless, the modified method is different from the recommendations of the standardization organizations, and it could represent a valuable option to evaluate the LOD within a certain degree of reliability and repeatability, especially when $k < 3$, as often occurs in the literature.

In the third approach, LOD is calculated by means of both the sensor resolution and the surface sensitivity $S_{surface}$ [40], which is the RI change $\Delta n_{surface}$ divided by the maximum surface density ρ_{max} of the investigated analyte, as expressed by

$$x_{LOD} = \frac{R}{S_{surface}} = \frac{R}{\left(\dfrac{\Delta n_{surface}}{\rho_{max}}\right)} = \frac{R}{\Delta n_{surface}}\rho_{max} \qquad (10.5)$$

By using this last approach, some authors consider the instrumental resolution as R and the slope of the calibration curve as $S_{surface}$. However, this means that all the measurements mostly fall on the calibration curve and the related standard deviation is lower than the instrumental resolution, which appears a low realistic case in practical applications. Broadly speaking, it seems that the best manner for determining LOD would be the first two approaches, even if the last one is quite common for SPR-based devices when further considerations and parameters are taken into account [6].

Figure 10.5 illustrates a typical calibration curve at low analyte concentrations reported in linear scale (lin-lin), where the data are fitted by means of the well-known Beer–Lambert law (Figure 10.5a) [44]. Figure 10.5b details the same experimental points of Figure 10.5a, but reported in logarithmic scale (log-lin); thus the fitting curve becomes linear. The error bars result from repeated measurements at a single concentration. As an example and by using the second approach with $\bar{y}_{blank} = 0.025$ nm, the red lines in Figure 10.5b allow evaluating the LOD equal to 0.23 µg/l ($\sigma_{blank} = 0.0168$ nm). On the other hand, by using the second modified approach, the resulting LOD is 0.38 µg/l given a larger $\sigma_{y_{max}}$ (0.0201 nm).

As a further example, by looking at Figure 10.4 and considering the second and second modified approaches, the LOD for repeated experiments results in 5 µg/l ($\sigma_{blank} = 0.016$ nm) and 7 µg/l ($\sigma_{y_{max}} = 0.0189$ nm), respectively. It is clear that, given the more conservative nature of the second modified approach, LODs are slightly higher than those related to the second approach. Obviously, the corresponding concentration C at the LOD can be also expressed in molarity M by using the following equation with the knowledge of the molecular weight MW of the analyte:

$$M\left[\frac{mol}{l}\right] = \frac{C\left[\frac{mg}{l}\right]}{MW\left[\frac{mg}{mol}\right]} \qquad (10.6)$$

In addition, as described with the third approach and in Refs. [12, 25], LOD can be also expressed in gram per square millimeter (usually in picogram per square millimeter considering the involved MWs), that is to say, in terms of ρ_{max} at the saturation.

(a)

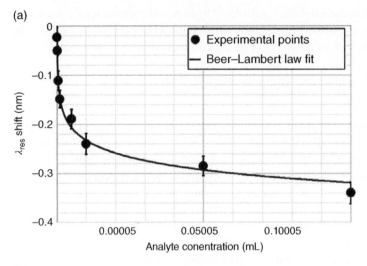

(b)

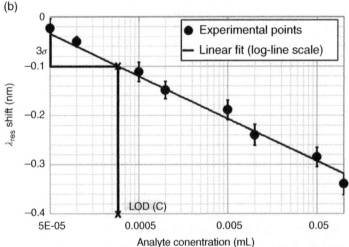

Figure 10.5 (a) An example of calibration curve at low analyte concentrations reported in linear scale (lin-lin), where the data are fitted by means of the Beer–Lambert law. The error bars result from repeated measurements at a single concentration. (b) The same calibration curve reported in logarithmic scale (log-lin); thus the fitting curve becomes linear. The vertical and horizontal lines allow evaluating the LOD according to the second approach.

It is worth pointing out that LOD is different and must not be confused with the minimum detectable concentration (MDC), described in the *ISO recommendation* [45], which provides the minimum concentration of the analyte that the biosensor is able to detect reliably. Therefore, the calculation of this parameter does not imply the use of the complete calibration curve, but just few experimental points plus the blank point at 0 analyte concentration.

The limit of quantification (LOQ) represents a more effective, thus real parameter for assessing the performance of any biosensor in practical conditions since LOQ takes into account 10 times σ and, hence, a greater level of repeatability and reproducibility. Therefore, to calculate LOQ, one can use Eq. (10.4) with 10σ instead of 3σ.

10.1.3.3   Specificity (or Selectivity)

Specificity or selectivity is undoubtedly an essential feature for any sensor but plays a key role in chemical and/or biochemical sensors. Specificity represents the device ability to sense a specific or a selective biomolecule or chemical compound, instead of all the other interfering biomolecules or chemical compounds, also named cross-sensitivity. For the sake of completeness, the term 'selectivity' is more used in chemical sensing, such as gas sensors and electronic noses and tongues, while the term 'specificity' is very common in biosensing.

Here, the focus will be paid on the use of this parameter as applied to refractometry and biosensing. In the former case, fibre-optic sensors are usually sensitive to a series of measurands, such as strain (axial deformation), temperature, and SRI. For instance, if the device should be sensitive to SRI changes only, the effect of cross-sensitivities (strain and temperature) must be taken into account and eliminated by means of the realization of modified or hybrid sensing configurations. It should be pointed out that, in the case of physical sensors, the scientific community usually speaks about cross-sensitivity, rather than selectivity or specificity. In the latter case, specificity is exactly the device ability to sense a specific target analyte. If the measurement is performed by spiking the analyte in a simple running buffer solution (it contains just the analyte under investigation), the biosensor specificity can be tested by injecting a solution containing analytes different from the one under investigation, also known as negative control. Conversely, if the measurement is carried out by spiking the analyte into a more complex matrix that also contains other biomolecules (i.e. serum, plasma, blood), the biosensor specificity can easily be tested by injecting this solution free of the target analyte and then measuring the baseline recovering after the washing. In particular, the specificity depends on the affinity of the receptor/analyte pair, the surface functionalization, the assay protocol, and the chemistry used in general [2, 11, 23].

10.1.3.4   Regeneration (or Reusability)

The last critical parameter in biosensing is represented by regeneration or sensor reusability, from which derives the possibility of the device usage in real applications. The use of integrated systems would allow the implementation of a model assay that encompasses a regeneration step in order to recreate a situation in which the receptors are free to interact again with the specific target analytes in consecutive cycles, instead of a disposable or one-time device. The first type of regeneration involves the restoration of the entire sensing layer, thus unsticking not only the target analyte but also the grafted receptor and the functional layer. However, the second type of regeneration is based on the detaching of the target analyte only in order to evaluate the repeatability and reproducibility too. Given the time-consuming procedure and the requirement of tight treatments often performed outside the microfluidic system, such as electrochemical cleaning, UV/ozone exposure, and ammonia–hydrogen peroxide mixture solution stripping, the latter type of regeneration is the most common, practical, and used one in biosensing technology platforms.

The first step of regeneration entails the selection of the best regeneration solution as a function of the receptor/analyte pair used [23, 38]. It is proved that the same regeneration solution could just work for a specific receptor/analyte pair, but not for any other. To assess the effect of regeneration, a standard procedure encompasses the repetition (at least 3 times) of the measurement of the same analyte concentration after the respective cycle of regeneration. Figure 10.6 accounts for an

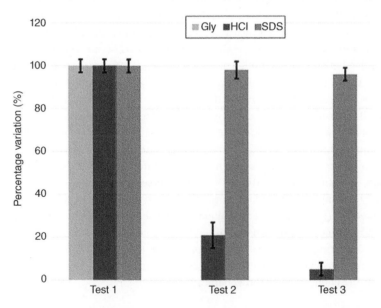

Figure 10.6 An illustrative example of the regeneration test repeated 3 times with the same device that is used to detect IgG/anti-IgG interactions. A concentration of the analyte of 1 mg/l under the same experimental conditions is injected, and three different regeneration solutions are tested: Gly, HCl, and SDS. The signal recovery in each test is reported together with the error bars resulting in multiple (*n* = 3) measurements. *Source:* Reprinted with permissions from MDPI [2].

illustrative example, in which the same device is used to detect IgG/anti-IgG interactions. A defined concentration of the analyte of 1 mg/l (roughly 7 nM) under the same experimental conditions is injected, and three different regeneration solutions are tested: glycine (Gly), hydrochloric acid (HCl), and sodium dodecyl sulphate (SDS). The signal recovery in each test is reported together with the error bars resulting in multiple ($n = 3$) measurements. It should be noted that, for each sensor, the first value is set to 100% in order to clearly evaluate the amount of signal loss in each cycle due to the regeneration process. Usually, a loss below 5% after the third cycle of regeneration is considered a very good result. In addition, it can easily be observed that the best regeneration solution for the selected receptor/analyte pair is SDS, while Gly seems ineffective (this is why the graph reports 0% in the second and third tests) and HCl removes almost all the functionalities.

10.2 Optical Fibre Refractometers

This section reports on the results presented in the literature using OFSs as optical refractometers. This means that the fibre optics is used to measure the shift of the resonance peak wavelength or the change in intensity as a function of volume SRI changes (see Figure 10.1a). In this case, the device consists of a bare fibre or of a modified or coated fibre, which is used only to enhance the RI sensitivity and does not serve as a selective sensing layer able to capture a BRE or to change its optical characteristics (e.g. the overlay RI and thickness) as a function of a chemical/biochemical interaction, as in the case of surface RI changes (see Figure 10.1b).

The interfering cross-sensitivities (temperature, strain, and curvature) represent a crucial issue related to OFS-based refractometry, especially for OFS based on gratings, because they influence the detection capabilities and performances of these devices. Cross-sensitive issues are often solved by utilizing more elaborated schemes, for instance, by using more than one sensing head to measure and compensate the changes of the interfering physical parameters or by using a differential arrangement. In any case, the use of a mechano- and thermo-stabilized microfluidic systems helps to dramatically reduce this issue to the minimum [46]. Though this aspect has been discussed in detail in Ref. [2], it deserves to be mentioned in any case.

The literature accounts for three fibre-optic sensing structures that are employed in refractometry: **interferometers**, **grating-based structures**, and **other resonance-based structures**. In the latter case, the resonance is generated by nm-thick coating deposited around the fibre as discussed in Section 10.1. The RI working range provides an idea of the application field of the device. Three main regions can typically be distinguished:

- Gaseous ($n_{sur} = 1$).
- Aqueous ($n_{sur} = 1.32$–1.34).
- Dense liquids with RIs approaching or exceeding the silica RI ($n_{sur} = 1.44$–1.47).

Moreover, as pointed out in Section 10.1.2, the sensitivity of fibre-optic refractometers is strongly dependent on both the RI working range and sensing configuration and scales down with the decrease of the working wavelength [3]. The performance of these different configurations is assessed by means of the most common parameters, such as sensitivity, FWHM, FOM, etc.

10.2.1 Optical Interferometers

Interferometers can be classified into four main kinds explained in Chapter 4 and in [47]:

- Fabry–Pérot interferometers (FPIs)
- Mach–Zehnder interferometers (MZIs)
- Michelson interferometers (MIs)
- Sagnac interferometers (SIs)

Some of them can be analysed by tracking intensity changes at a specific wavelength, while in others the interferometric fringe pattern is analysed in the spectral and/or phase domains. It is worth keeping in mind that any accurate measurement can just be attained for gradual SRI changes, where one fringe maximum or minimum can be followed during the entire RI test. On the other hand, any sudden change in SRI cannot be measured if it results in a wavelength shift greater than the fringe spacing. FPIs typically exhibit a RI sensitivity of the order of 10–10^3 nm/RIU with a minimum FWHM of few nanometres and a maximum FOM of the order of 10^3/RIU. As far as the MZIs are concerned, RI sensitivity of the order of 10^2–10^4 nm/RIU with minimum FWHM of very few nanometres and maximum FOM of 10^3/RIU can be attained. On the other hand, MIs can achieve a RI sensitivity of the order of 10–10^3 nm/RIU with a minimum FWHM of few nanometres and a maximum FOM of the order of 10^2/RIU. Finally, SIs can reach very remarkable performances with RI sensitivity exceeding 10^4 nm/RIU, a minimum FWHM of 10^{-2} nm, and a maximum FOM of the order of 10^4/RIU.

10.2.2 Grating-Based Structures

Fibre gratings were also presented in Chapter 4, the most important groups being FBGs, TFBGs, and LPGs, with a significantly higher grating period compared with FBGs and TFBGs.

Regarding refractometry, given the inherent coupling mechanism of light in FBGs, they cannot be sensitive to SRI changes. However, differently from the inherent SRI sensitivity of TFBGs and LPGs, FBGs can be made sensitive to the SRI by reducing the diameter of the fibre cladding by etching so that the effective index of the core mode is affected by the SRI: the evanescent field related to the core mode starts penetrating the surrounding medium when the cladding diameter is decreased. The smaller the fibre cladding, the higher the SRI sensitivity although the device handling and fragility dramatically drop down. Etched FBGs in combination with interferometric structures are also reported in the literature for specific applications [47]. Modified FBGs typically exhibit a RI sensitivity that ranges from few nanometer per refractive index unit up to 10^3 nm/RIU when an MZI is developed. A common FWHM of FBGs is 10^{-1}–10^{-2} nm, thus attaining an FOM that lies of the order of 10–10^2/RIU up to 10^3/RIU in the best case. In general, there is a trade-off between RI sensitivity and FWHM: when an higher RI sensitivity is obtained, a less FWHM is expected and vice versa.

As far as the TFBGs are concerned, counter-propagating cladding modes are used to detect SRI changes, while the Bragg resonance related to the core mode is used to compensate any change in temperature. One detection scheme is based on tracking the wavelength shift or intensity changes related to one selected cladding mode. Conversely, a second detection scheme is based on tracking the optical changes of the entire envelope of the spectral comb generated by all the cladding modes. TFBGs can achieve a RI sensitivity of the order of 10–10^2 nm/RIU with a minimum FWHM of 10^{-1} nm and a maximum FOM of the order of 10^3/RIU. Gold-coated TFBGs that also exploit the effect of the excitation of SPR phenomenon usually exhibit better performances with respect to standard TFBGs [48]. In addition, gold-coated highly tilted (tilt angle $>30°$) FBGs allow the detection of gas and volatile molecules having a RI similar to that of air ($n_{sur} = 1$) [49].

Theoretically speaking, LPGs can reach the highest RI sensitivity and, hence, could be the best candidate in optical refractometry. However, given all the cross-sensitivity issues of them [2], LPGs are sometimes disregarded with respect to other fibre-based configurations when accurate and high-resolution SRI measurements have to be carried out. In LPGs, the excitation of co-propagating cladding modes allow directly sensing SRI changes both in shift of the resonance wavelength and in transmission intensity. According to the literature [2, 3, 11], the penetration depth of the evanescent wave related to each cladding mode that ranges within 50–500 nm increases with the order of the cladding mode. Therefore, cladding modes of higher order permit to attain higher RI sensitivity. Overall, three phenomena have been explored in order to enhance the RI sensitivity of this kind of device during the last years:

• The DTP or also named turn-around point (TAP).
• The mode transition.
• The cladding diameter reduction.

The use of specialty fibres (photonic crystal fibres, microstructured fibres, etc.) together with the deposition of a nanometer-thick coating can improve the intrinsic device sensitivity beyond the theoretical limit. Moreover, the combination of different approaches based on an LPG can theoretically attain a RI sensitivity of 1.43×10^5 nm/RIU [50]. In any case, LPG-based configurations can achieve a RI sensitivity of the order of 10^2–10^4 nm/RIU with a typical FWHM of tens of nm and FOM of the order of 10–10^2/RIU.

10.2.3 Other Resonance-Based Structures

In last years, refractometers based on the generation of optical resonances due to the deposition of thin films have become research hot topics given their broad operating wavelength range, real-time response, simplicity of implementation, and, what is more important, very high RI sensitivity to surrounding environment. Prism-based configurations, which derive from the well-known Kretschmann configuration, remain as the gold standard sensing platform for commercial devices in biochemical applications, although many of their limitations, such as bulky size or complexity, prevent them from the creation of simple, compact, portable, low-cost, and remote sensing instrumentation. Some of the previously mentioned constraints can be overcome by the use of a waveguide or, even better, of an optical fibre acting as the high RI prism that couples the incident light for the excitation of the optical resonances. Both Kretschmann and fibre-optic configurations, which are depicted in Figure 10.7, permit to develop a high-performance label-free sensing platform that can be used for biochemical applications on the basis of monitoring surface RI changes through resonance wavelength shifts.

The conditions for the excitation of optical resonances are primarily established by the dielectric properties of both the materials that support the resonances, or resonance-supporting coatings given that a thin-film is usually used to this purpose, and that of the medium surrounding the coating [47]. One of these resonances is given by the well-known SPR generated by coupling the evanescent light to surface plasmon polaritons (SPPs), explained in Chapter 8, whereas LMR were presented in Chapters 2 and 4.

In order to improve the light–matter interaction related to the resonance phenomenon, most fibre-optic resonance-based sensors need to modify the fibre geometry by means of chemical etching, burning, UV lithography, laser writing, mechanical abrasion, arc discharge, and so on. Different fibre-optic configurations can be implemented by the previously mentioned processes, which include cladding removal, fibre bending or tapering, side polishing, or fibre microstructuring. All these structures have been widely studied using plastic, polymeric or silica fibres, and single-mode or multimode fibres. In addition, many authors have also explored the application of novel strategies to enhance the sensitivity and

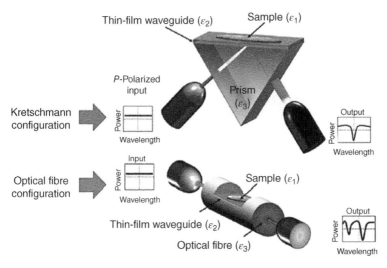

Figure 10.7 Sketch of Kretschmann (top) and fibre-optic (bottom) configurations used to develop an optical resonance interrogation system. *Source:* Reprinted with permissions from Elsevier [51].

resolution of these devices [52], such as the combination of different fibre architectures and materials, the use of combined fibre interrogation technologies, and other more sophisticated designs using multilayers, nanoparticles, and microstructured thin films of conductive and dielectric materials [1].

The RI working range and resonance wavelength are established by the dielectric properties of the fibre and the coating material, which also determine the applications of the device (e.g. biological applications require RI working range around 1.33) and the interrogation equipment to be used (e.g. instrumentation in the visible region is cheaper than that in the infrared region). Obviously, the sensitivity depends also on the dielectric properties of the fibre and the coating material. Therefore, the suitable combination of the fibre (plastic, silica, chalcogenide glasses, sapphire) and the resonance-supporting coating material (metal or dielectric) permits to enhance the device sensitivity, more than the other parameters related to refractometry, such as the resolution, the FWHM, etc. [47]. The reason lies in the fact that sensitivity increases as well as the resonance width, that is to say, the FWHM. Consequently, it is crucial to find the best way of enhancing either the FWHM or the sensitivity without the other parameters being affected by this improvement. In this sense, a decrease in the number of modes propagating through the fibre by means of the use of reduced core diameter, polarization-maintaining, or tapered fibres can attain the goal of both a less FWHM and, hence, of a narrower resonance. The same conclusions can be reached considering the FOM [47].

Resonance-based configurations can typically attain a RI sensitivity of the order of 10^3 nm/RIU with some remarkable examples that reported on RI sensitivity of the order of 10^4 nm/RIU around $n_{sur} = 1.33$. The values of FWHM range from 10^{-1} nm (SPR-based TFBG) up to 10^2 nm (SPR or LMR in multimode fibres), whereas the values of FOM lie in the order of $10-10^2$/RIU with some remarkable examples that reported on FOM of the order of 10^3/RIU (LMR in D-shaped fibres, SPR in photonic crystals fibres, and SPR-based TFBG and LPG).

10.3 Optical Fibre Biosensors

An OFB is an analytical device applied in medical, food, or environmental contest, which uses the principles mainly of optical spectroscopy in conjunction with a biological system or BRE [53].

Since in this chapter the attention is focused on OFBs, where the RI changes occur just on the functionalized surface of the fibre, in this contest, according to the BRE used as sensing element, OFBs could be classified in three main different groups:

- Immuno-based biosensors
- Oligonucleotide-based biosensors
- Whole cell/microorganism-based biosensors

However, all of those are also classified as affinity sensors. The affinity sensors are usually compared with the catalytic sensors, which are based on enzymatic reactions. The latter kind of sensors foresees the immobilization onto the fibre of an enzyme, which causes the conversion of the specific substrate of interest to the formation of related products. This reaction props up change in the SRI where the reaction occurs, not directly on the surface of the optical fibre. Therefore, they will not be taken into account in this chapter.

The immuno-based OFBs are based on the specificity of the interaction occurring between the antibody and its antigen (or target analyte). Variations of the concentration of the antigen are monitored by observing a RI change at the surface of the fibre.

The oligonucleotide-based OFBs are again classified also as affinity sensors. In fact, for example, they use the affinity between a single-stranded DNA (ssDNA) to form a double-stranded DNA (dsDNA) with complementary sequences or the affinity that an aptamer has for its specific target molecule.

Finally, whole cell/microorganism-based OFBs are those biosensors where the intact cells (predominantly microbial) are used as BREs and the presence of the analyte is measured by its effect on the living cell.

10.3.1 Immuno-Based Biosensors

The immuno-based biosensors are those devices where the BRE is an antibody. The antibodies are proteins produced by the immune system with the characteristic to be highly selective for the specific antigen. Being proteins, some care needs to be taken in order to avoid denaturation; on the other hand, they require few shrewdness, such as pH, definitive ion concentration, temperature, etc., to work properly. The most common approach is to use them as BRE covalently immobilized onto the optical fibre. Depending on the functional groups present onto the surface of the fibre, they are attached using amine or carboxylic groups; otherwise, if the fibre is covered with gold, for example, the free thiol groups of the antibodies are used. For instance, graphene oxide (GO) that has recently attracted much attention because of its unique structural, mechanical, and electronic properties provides a direct and convenient conjugation for the antibody via a peptide bond between the surface carboxyl groups and the amino groups of the antibody [54].

In a recent paper of 2019, four different TFBG-based configurations sketched in Figure 10.8 have been developed for immunosensing purposes [55]: (i) bare TFBGs (where the antibody is directly absorbed onto a silanized surface), (ii) coated TFBGs with a sputtered gold layer (where the antibody is covalently attached after the immobilization on the gold layer of a self-assembled monolayer of dithiols), (iii) gold-coated TFBGs with electroless deposition (where the silanized fibres were immersed in a solution of 10 nm diameter Au nanoparticles and then placed in a plating solution; afterwards the protocol described in (ii) is followed for the antibody immobilization), and (iv) a hybrid coating combining sputtered and electroless plated (ELP) gold layers. The antibody-immobilization strategies are of course adapted and optimized according to the used substrate, but the same analyte, anti-CK17, which is a relevant biomarker selected for cancer diagnosis, is used for the performance assessment. The first case, the bare TFBG set-up, has showed a higher (worse) LOD compared with the other set-ups, because of the absence of SPR enhancement, but presents some advantages in any case, i.e. it can be used straightforwardly without taking care of polarization effects. The case (ii), the gold-spattered TFBG, has showed the lowest (better) LOD close to 1 ng/l.

Another example of immuno-based biosensor is given by the use of a single-mode-tapered-multimode-single-mode (SMS) structure for bacteria detection by immunoassay [56]. In particular, the SMS biosensor has been developed for the quantification of *Salmonella typhimurium* by the use of a multimode fibre from which a cladding region (thickness $\sim$25 µm) has been removed through chemical etching. This action allows increasing the interaction between the propagating modes of guided light and the fibre-surrounding medium. Greater interaction between them results in higher sensitivity of the SMS biosensor (Figure 10.9). The RI characterization of such device reported a sensitivity of 275.86 nm/RIU.

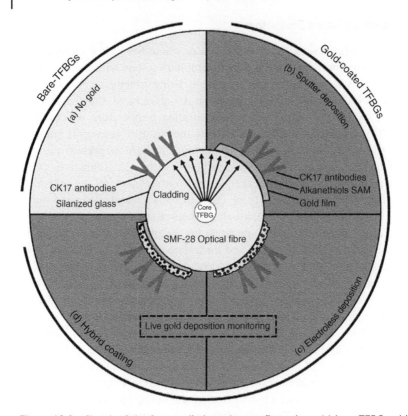

Figure 10.8 Sketch of the four studied sensing configurations: (a) bare TFBGs with adsorbed antibodies on the silica fibre surface, (b) sputter deposition under vacuum and covalent bonding of antibodies on the gold film, (c) electroless deposition of gold and covalent bonding of antibodies on the agglomerated Au particles, and (d) hybrid coating coupling a deposition of 4 nm Au by sputter coater followed by an ELP deposition, which is functionalized using covalent bonding of antibodies. *Source:* Reprinted with permissions from MDPI [55].

10.3.2 Oligonucleotide-Based Biosensors

The use of oligonucleotide sequences as BRE has received special attention in analytical chemistry applications since many years already. What makes this kind of biosensors so interesting is the possibility of a big variety of DNA detection, such as detection of viral DNA, DNA from cancer cells, especially detection of mutated or altered DNA sequences, with even single mismatch mutations. The introduction of highly performing optical devices has pushed their development, in particular for the possibility of detecting specific DNA sequences without the need for nucleic acid amplification [57].

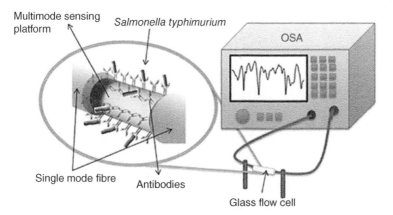

Multimode sensing platform

Salmonella typhimurium

OSA

Single mode fibre

Antibodies

Glass flow cell

Figure 10.9 Schematic demonstration of SMS biosensor with expanded view of anti-*Salmonella typhimurium* antibodies immobilized onto the sensing platform employed for detection of *S. typhimurium*. *Source:* Reprinted with permissions from Elsevier [56].

A quite novel application of this kind of devices has recently been proposed [58], which uses a poly-lysine (PLL) solution covered on an LPG for DNA sequence immobilization (the probe) with the purpose of wine authenticity assessment. The probes and targets used in this work are designed based on the single nucleotide polymorphisms (SNPs) detected within the F3H gene of *Vitis vinifera* L.; in particular, the SNPs are located at three different positions (1039, 1040, and 1065 bp) of the selected probe.

When an oligonucleotide-based biosensor is used to detect targets, which are not nucleic acids, the BRE is called aptamer. Aptamers are synthetic single-strand oligonucleotides (produced by the molecular biology approach called SELEX) [59], which can fold in a stable three-dimensional configuration that can bind different targets, such as small molecules, proteins, hormones, etc. [60–62]. Recently, an aptamer used as BRE has been proposed for specific detection of dopamine molecules [63]. The aptamer is immobilized onto a hybrid graphene–gold coating deposited on a TFBG, photowritten in a commercial single-mode optical fibre as sketched in Figure 10.10. The aptamer is randomly adsorbed onto the graphene, but in the presence of dopamine molecules, the ssDNA aptamer conformation switches from the random flexible structure into a rigid stem-loop folded structure (Figure 10.11). This configuration change of the aptamer affects the SPR spectrum, and the presence of the TFBG increases even more the sensitivity of the device. The sensor accounted for an extremely low LOD of 0.16 pM and a linear response between 10^{-13} and 10^{-8} M.

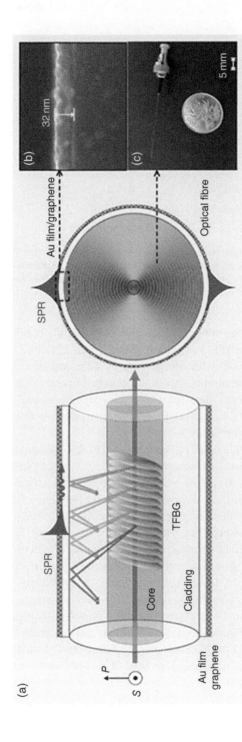

Figure 10.10 (a) Schematic diagram of gold-coated TFBG sensor with graphene surface functionalization (polarimetric TFBG with linearly polarized light emitting for SPR excitations and its energy distribution along fibre cross section). (b) SEM image of the cross section of fibre with hybrid graphene–gold coating. (c) Photograph of the fibre device. *Source:* Reprinted with permissions from Elsevier [63].

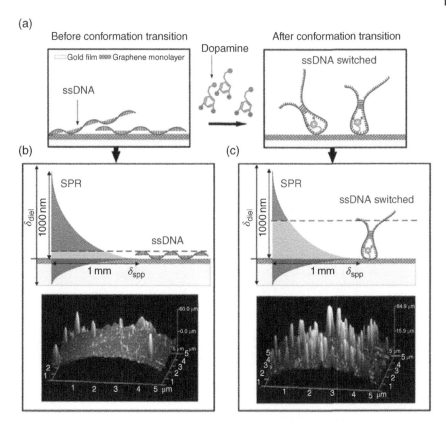

Figure 10.11 (a) Sketch of the aptamer-based dopamine sensing mechanism: the conformation of the ssDNA aptamer is transformed by small dopamine molecules. (b, c) Illustration of the key point of such sensing mechanism: the effective overlap area between the plasmon energy field and ssDNA aptamer molecules has been strongly modified in the presence of dopamine molecules and, hence, induces a much larger plasmon phase velocity change. The bottom pictures are the experimentally achieved AFM images of the aptasensor surfaces before and after capturing dopamine molecules. *Source:* Reprinted with permissions from Elsevier [63].

10.3.3 Whole Cell/Microorganism-Based Biosensors

In contrast to the molecular-based biosensors, BREs constituted by whole cells or microorganisms can detect a wider range of substances and, hence, are more sensitive to the changes in the environment of the cells. The consumption of oxygen or carbon dioxide by some microorganisms or their capability of genetic modification makes them interesting sensing elements [64, 65].

Halkare et al. [66] have reported on the importance of having a robust and reliable sensor for heavy metal detection in food and drinking water. They have used *Escherichia coli* B40 bacterial cells as BRE for the detection of heavy metal ions (Hg^{2+} and Cd^{2+}), since they can bind on the surface of bacterial cells [67]. The cells are immobilized on top of gold nanoparticles (AuNPs) deposited onto an optical fibre, thus forming an LSPR-based biosensor (Figure 10.12a). A light source (a LED with spectral emission in 490–600 nm), an objective lens, the sensor probe, a bare fibre terminator unit, and a portable spectrometer constitute the experimental set-up as schematized in Figure 10.12b.

The binding of the metal ions onto the surface of the bacterial cells produces a change in the localized RI based on the absorbance changes occurring at the peak plasmon resonance wavelength of AuNPs. The sensor measures a combined effect

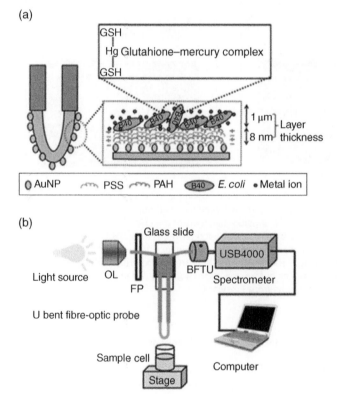

Figure 10.12 (a) Schematic of the proposed bacteria based heavy metal detection scheme. (b) Sketch of the experimental set-up. OL, objective lens; FP, fibre positioner; BFTU, bare fibre terminator unit (not to scale). *Source:* Reprinted with permissions from Elsevier [66].

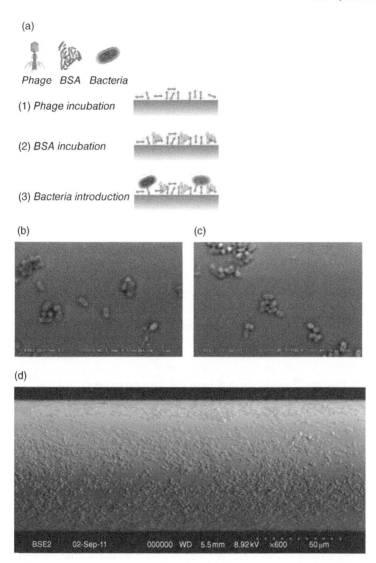

Figure 10.13 (a) Schematic of the main steps of the experiment: phage immobilization, BSA incubation, and bacteria detection (not to scale). Between these steps, the LPG-based fibre device is rinsed with PBS. SEM images of different concentrations of *E. coli* immobilized on the surface of the LPG sensitive region: (b) 10^3 cfu/ml, (c) 10^5 cfu/ml and (d) 10^9 cfu/ml. *Source:* Modified with permissions from OSA [70] and Elsevier [71].

given by the refraction loss due to the changes in SRI and the evanescent wave absorption [68].

Another example of whole cell-based biosensors is given by the use of T4 bacteriophage as BRE. Bacteriophages are viruses that infect bacteria [69]. The T4 virus initiates an *E. coli* infection by binding outer membrane porin C (OmpC) proteins and lipopolysaccharide (LPS) on the surface of *E. coli* cells with its long tail. Starting from Smietana et al. in 2011 [70] and then following other research groups [71, 72], LPG-based devices have been proposed to monitor bacteriophage–bacteria interactions. The T4 phage is immobilized onto the surface of the LPG, and then its specificity for *E. coli* K12 cells makes it useful as BRE for bacterial detection (Figure 10.13a). From SEM images, it is possible to observe the increase in *E. coli* concentrations from 10^3 up to 10^9 cfu/ml (Figure 10.13b).

Brzozowska et al. [73, 74] proposed an LPG-based sensor, where the phage adhesion is applied for bacteria and their endotoxin sensing application. In particular, the LPG surface is coated with nickel ions, which can bind histidine-tag, present in the adhesin (gp37) of T4 bacteriophage in a reversible way. The results are very promising by using bacteriophage-adhesin-coated LPGs for bacterial recognition. Subsequently, in 2016, the adhesin binding mechanism and the same LPG-based device have been used to detect the whole *E. coli* K-12.

10.4 Fibre Optics Towards Advanced Diagnostics and Future Perspectives

It is clear that OFBs represent a valid and widespread option for label-free biosensing, with the ability of obtaining outstanding performance in terms of sensitivity and LOD. Obviously, the advent of nanotechnologies with the possibility of depositing nanomaterials and nanostructured coatings on the surface of the fibres is broadening these chances, leading to the development of biosensing platforms characterized by features unexpected up to a few years ago.

What is not still completely exploited with these kinds of sensors is related to the inherent advantages lying in the use of fibre optics, such as the miniaturization that can pave the way to their use in harsh environments – first of all in clinical settings for *in vivo* applications – and the chance of multiplexing, for example, with a series of sensing heads located along the same fibre segment with different optical characteristics in terms of resonance peak wavelengths and, obviously, with different BREs deposited in correspondence of each sensing head towards a multi-target detection. Considering these aspects, OFSs can really become a unique and irreplaceable optical platform in biosensing.

Advanced diagnostics is one of the key research fields nowadays. However, a very few number of fibre-based devices have reached the market, while most of them, even with impressive performance, have not overtaken the laboratory stage. In fact, while high-performance metrics are starting to be attained experimentally, there are still open questions pertaining to an effective and reliable detection of small biomolecules, possibly up to single molecule, *in vivo* sensing and multi-target detection using OFS-based technology platforms. However, we can confidently foresee a reduction in the gap between fibre-based biosensing platform and market, thanks to the huge improvement in nanotechnologies and in manufacturing reliability and reproducibility.

References

1 Xu, Y., Bai, P., Zhou, X. et al. (2019). Optical refractive index sensors with plasmonic and photonic structures: promising and inconvenient truth. *Adv. Opt Mater.* 7: 1801433.

2 Chiavaioli, F., Gouveia, C., Jorge, P., and Baldini, F. (2017). Towards a uniform metrological assessment of grating-based optical fibre sensors: from refractometers to biosensors. *Biosensors (Switzerland)* 7 (2): 23.

3 Chiavaioli, F., Baldini, F., Tombelli, S. et al. (2017). Biosensing with optical fibre gratings. *Nanophotonics* 6 (4): 663–679.

4 Sinibaldi, A., Montaño-Machado, V., Danz, N. et al. (2018). Real-time study of the adsorption and grafting process of biomolecules by means of Bloch surface wave biosensors. *ACS Appl. Mater. Interfaces* 10 (39): 33611–33618.

5 Pathak, A. and Gupta, B.D. (2019). Ultra-selective fibre optic SPR platform for the sensing of dopamine in synthetic cerebrospinal fluid incorporating permselective nafion membrane and surface imprinted MWCNTs-PPy matrix. *Biosens. Bioelectron.* 133: 205–241.

6 Špačková, B., Scott Lynn, N. Jr., Slabý, J. et al. (2018). A route to superior performance of a nanoplasmonic biosensor: consideration of both photonic and mass transport aspects. *ACS Photon.* 5: 1019–1025.

7 Farnesi, D., Chiavaioli, F., Baldini, F. et al. (2015). Quasi-distributed and wavelength selective addressing of optical micro-resonators based on long period fibre gratings. *Opt. Express* 23 (16): 21175–21180.

8 Farnesi, D., Chiavaioli, F., Righini, G.C. et al. (2014). Long period grating-based fibre coupler to whispering gallery mode resonators. *Opt. Lett.* 39 (22): 6525–6528.

9 Markiewicz, K. and Wasylczyk, P. (2019). Photonic-chip-on-tip: compound photonic devices fabricated on optical fibres. *Opt. Express* 27 (6): 8440–8445.

10 Stewart, G., Jin, W., and Culshaw, B. (1997). Prospects for fibre-optic evanescent-field gas sensors using absorption in the near-infrared. *Sens. Actuators B Chem.* 38: 42–47.

11 Baldini, F., Brenci, M., Chiavaioli, F. et al. (2011). Optical fibre gratings as tools for chemical and biochemical sensing. *Anal. Bioanal. Chem.* 402: 109–116.

12 Sinibaldi, A., Danz, N., Anopchenko, A. et al. (2015). Label-free detection of tumor angiogenesis biomarker angiopoietin 2 using Bloch surface waves on one dimensional photonic crystals. *J. Lightwave Technol.* 33: 3385–3393.

13 Reth, M. (2013). Matching cellular dimensions with molecular sizes. *Nat. Immunol.* 14: 765–767.

14 Del Villar, I., Matias, I.R., Arregui, F.J., and Lalanne, P. (2005). Optimization of sensitivity in long period fibre gratings with overlay deposition. *Opt. Express* 13 (1): 56–69.

15 Chiavaioli, F., Biswas, P., Trono, C. et al. (2015). Sol–gel-based titania–silica thin film overlay for long period fibre grating-based biosensors. *ACS Anal. Chem.* 87: 12024–12031.

16 Renoirt, J.-M., Zhang, C., Debliquy, M. et al. (2013). High-refractive-index transparent coatings enhance the optical fibre cladding modes refractometric sensitivity. *Opt. Express* 21: 29073.

17 Chiavaioli, F., Biswas, P., Trono, C. et al. (2014). Towards sensitive label-free immunosensing by means of turn-around point long period fibre gratings. *Biosens. Bioelectron.* 60: 305–310.

18 Del Villar, I. (2015). Ultrahigh-sensitivity sensors based on thin-film coated long period gratings with reduced diameter, in transition mode and near the dispersion turning point. *Opt. Express* 23: 8389–8398.

19 Smietana, M., Koba, M., Mikulic, P., and Bock, W.J. (2016). Combined plasma-based fibre etching and diamond-like carbon nano-overlay deposition for enhancing sensitivity of long-period gratings. *J. Lightwave Technol.* 34: 4615–4619.

20 Yang, F. and Sambles, J.R. (1997). Determination of the optical permittivity and thickness of absorbing films using long range modes. *J. Mod. Opt.* 44: 1155–1164.

21 Batchman, T.E. and McWright, G.M. (1982). Mode coupling between dielectric and semiconductor planar waveguides. *IEEE J. Quantum Electron.* 18: 782–788.

22 Andreev, A., Pantchev, B., Danesh, P. et al. (2005). A refractometric sensor using index-sensitive mode resonance between single-mode fibre and thin film amorphous silicon waveguide. *Sens. Actuators B Chem.* 106: 484–488.

23 Chiavaioli, F., Zubiate, P., Del Villar, I. et al. (2018). Femtomolar detection by nanocoated fibre label-free biosensors. *ACS Sens.* 3 (5): 936–943.

24 Zubiate, P., Urrutia, A., Zamarreño, C.R. et al. (2019). Fibre-based early diagnosis of venous thromboembolic disease by label-free D-dimer detection. *Biosens. Bioelectron. X* 2: 100026.

25 Homola, J., Yee, S.S., and Gauglitz, G. (1999). Surface plasmon resonance sensors: review. *Sens. Actuators B Chem.* 54: 3–15.

26 Shevchenko, Y.Y. and Albert, J. (2007). Plasmon resonances in gold-coated tilted fibre bragg gratings. *Opt. Lett.* 32: 211–213.

27 Schuster, T., Herschel, R., Neumann, N., and Schäffer, C.G. (2012). Miniaturized long-period fibre grating assisted surface plasmon resonance sensor. *J. Lightwave Technol.* 30: 1003–1008.

28 Consales, M., Ricciardi, A., Crescitelli, A. et al. (2012). Lab on fibre technology: towards multifunctional optical nanoprobes. *ACS Nano* 6 (4): 3163–3170.

29 Rindorf, L., Jensen, J.B., Dufva, M. et al. (2016). Photonic crystal fibre long-period gratings for biochemical sensing. *Opt. Express* 14 (18): 8224–8231.

30 Huy, M.C.P., Laffont, G., Dewynter, V. et al. (2006). Tilted fibre bragg grating photowritten in microstructured optical fibre for improved refractive index measurement. *Opt. Express* 14 (22): 10359–10370.

31 Huy, M.C.P., Laffont, G., Dewynter, V. et al. (2007). Three-hole microstructured optical fibre for efficient fibre bragg grating refractometer. *Opt. Lett.* 32 (16): 2390–2392.

32 Calcerrada, M., García-Ruiz, C., and González-Herráez, M. (2015). Chemical and biochemical sensing applications of microstructured optical fibre-based systems. *Laser Photon. Rev.* 9: 604–627.

33 Schroeder, K., Ecke, W., Mueller, R. et al. (2001). A fibre Bragg grating refractometer. *Meas. Sci. Technol.* 12 (7): 757–764.

34 Liu, Y., Meng, C., Zhang, A.P. et al. (2011). Compact microfibre Bragg gratings with high-index contrast. *Opt. Lett.* 36 (16): 3115–3117.

35 Sun, D., Guo, T., Ran, Y. et al. (2014). In-situ DNA hybridization detection with a reflective microfibre grating biosensor. *Biosens. Bioelectron.* 61: 541–546.

36 Brambilla, G., Xu, F., Horak, P. et al. (2009). Optical fibre nanowires and microwires: fabrication and applications. *Adv. Opt. Photon.* 1 (1): 107–161.

37 Squires, T.M., Messinger, R.J., and Manalis, S.R. (2008). Making it stick: convection, reaction and diffusion in surface-based biosensors. *Nat. Biotechnol.* 26: 417–426.

38 MacCraith, B., McDonagh, C., O'Keeffe, G. et al. (1995). Sol-gel coatings for optical chemical sensors and biosensors. *Sens. Actuators B Chem.* 29: 51–57.

39 DeLisa, M.P., Zhang, Z., Shiloach, M. et al. (2000). Evanescent wave long-period fibre bragg grating as an immobilized antibody biosensor. *ACS Anal. Chem.* 72: 2895–2900.

40 White, I.M. and Fan, X. (2008). On the performance quantification of resonant refractive index sensors. *Opt. Express* 16 (2): 1020–1028.

41 Loock, H.P. and Wentzell, P.D. (2012). Detection limits of chemical sensors: applications and misapplications. *Sens. Actuators B Chem.* 173: 157–163.

42 McCormick, R.M. and Karger, B.L. (1980). Guidelines for data acquisition and data quality evaluation in environmental chemistry. *ACS Anal. Chem.* 52 (14): 2242–2249.

43 International Union of Pure and Applied Chemistry (2014). Compendium of Chemical Terminology – Gold book, version 2.3.3, 1301–1302. https://www.google. com/url?sa=t&rct=j&q=&esrc=s&source=web&cd=1&cad=rja&uact=8& ved=2ahUKEwiwxbyqscDpAhUzQkEAHf3CDr4QFjAAegQIARAB&url=http% 3A%2F%2Fgoldbook.iupac.org%2Ffiles%2Fpdf%2Fgoldbook. pdf&usg=AOvVaw1gTRUxHNof1-OSlF7rKolv.

44 Trumbo, T.A., Schultz, E., Borland, M.G., and Pugh, M.E. (2013). Applied spectrophotometry: analysis of a biochemical mixture. *Biochem. Mol. Biol. Educ.* 41 (4): 242–250.

45 Janiga, I., Mocak, J., and Garaj, J. (2008). Comparison of minimum detectable concentration with the iupac detection limit. *Meas. Sci. Rev.* 8 (5): 108–110.

46 Trono, C., Baldini, F., Brenci, M. et al. (2011). Flow cell for strain-and temperature-compensated refractive index measurements by means of cascaded optical fibre long period and bragg gratings. *Meas. Sci. Technol.* 22: 075204.

47 Urrutia, A., Del Villar, I., Zubiate, P., and Zamarreño, C.R. (2019). A comprehensive review of optical fibre refractometers: toward a standard comparative criterion. *Laser Photon. Rev.* 13: 1900094.

48 Albert, J., Shao, L.-Y., and Caucheteur, C. (2013). Tilted fibre Bragg grating sensors. *Laser Photon. Rev.* 7: 83–108.

49 Liu, Y., Liang, B., Zhang, X. et al. (2019). Plasmonic fibre-optic photothermal anemometers with carbon nanotube coatings. *J. Lightwave Technol.* 37 (123): 3373–3380.

50 Del Villar, I. (2015). Ultra-high-sensitivity sensors based on thin-film coated long period gratings with reduced diameter, in transition mode and near the dispersion turning point. *Opt. Express* 23 (7): 8389–8398.

51 Del Villar, I., Arregui, F.J., Zamarreño, C.R. et al. (2017). Optical sensors based on lossy-mode resonances. *Sens. Actuators B Chem.* 240: 174–185.

52 Caucheteur, C., Voisin, V., and Albert, J. (2015). Near-infrared grating-assisted SPR optical fibre sensors: design rules for ultimate refractometric sensitivity. *Opt. Express* 23 (3): 2918–2932.

53 Lowe, C.R. (2008). Overview of biosensor and bioarray technologies. In: *Handbook of Biosensors and Biochips*. New York, NY: Wiley.

54 Esposito, F., Sansone, L., Taddei, C. et al. (2018). Ultrasensitive biosensor based on long period grating coated with polycarbonate-graphene oxide multilayer. *Sens. Actuators B Chem.* 274: 517–526.

55 Loyez, M., Lobry, M., Wattiez, R., and Caucheteur, C. (2019). Optical fibre gratings immunoassays. *Sensors (Switzerland)* 19 (11): 2595.

56 Kaushik, S., Pandey, A., Tiwari, U.K., and Sinha, R.K. (2018). A label-free fibre optic biosensor for Salmonella Typhimurium detection. *Opt. Fibre Technol.* 46: 95–103.

57 Chen, X., Zhang, L., Zhou, K. et al. (2007). Real-time detection of DNA interactions with long-period fibre-grating-based biosensor. *Opt. Lett.* 32 (17): 2541–2543.

58 Barrias, S., Fernandes, J.R., Eiras-Dias, J.E. et al. (2019). Label free DNA-based optical biosensor as a potential system for wine authenticity. *Food Chem.* 270: 299–304.

59 Gold, L. (2015). SELEX: how it happened and where it will go. *J. Mol. Evol.* 81: 140–143.

60 Ellington, A.D. and Szostak, J.W. (1990). In vitro selection of RNA molecules that bind specific ligands. *Nature* 346: 818–822.

61 Nord, K., Nilsson, J., Nilsson, B. et al. (1995). A combinatorial library of an alpha-helical bacterial receptor domain. *Protein Eng.* 8: 601–608.

62 Colas, P., Cohen, B., Jessen, T. et al. (1996). Genetic selection of peptide aptamers that recognize and inhibit cyclin-dependent kinase 2. *Nature* 380: 548–550.

63 Hu, W., Huang, Y., Chen, C. et al. (2018). Highly sensitive detection of dopamine using a graphene-functionalized plasmonic fibre-optic sensor with aptamer conformational amplification. *Sens. Actuators B Chem.* 264: 440–447.

64 Liu, Q., Wu, C., Cai, H. et al. (2014). Cell-based biosensors and their application in biomedicine. *Chem. Rev.* 114: 6423–6461.

65 Mowbray, S.E. and Amiri, A.M. (2019). A brief overview of medical fibre optic biosensors and techniques in the modification for enhanced sensing ability. *Diagnostics (Switzerland)* 9 (1): 23.

66 Halkare, P., Punjabi, N., Wangchuk, J. et al. (2019). Bacteria functionalized gold nanoparticle matrix based fibre-optic sensor for monitoring heavy metal pollution in water. *Sens. Actuators B Chem.* 281: 643–651.

67 Hobot, J.A., Carlemalm, E., Villiger, W., and Kellenberger, E. (1984). Periplasmic gel: new concept resulting from the reinvestigation of bacterial cell envelope ultrastructure by new methods. *J. Bacteriol.* 160: 143–152.

68 Satija, J., Punjabi, N.S., Sai, V.V.R., and Mukherji, S. (2014). Optimal design for U-bent fibre-optic LSPR sensor probes. *Plasmonics* 9: 251–260.

69 Rakhuba, D.V., Kolomiets, E.I., Szwajcer-Dey, E., and Novik, G.I. (2010). Bacteriophage receptors, mechanisms of phage adsorption and penetration into the host cell. *Pol. J. Microbiol.* 59: 145–115.

70 Smietana, M., Bock, W.J., Mikulic, P. et al. (2011). Detection of bacteria using bacteriophages as recognition elements immobilized on long-period fibre gratings. *Opt. Express* 19 (9): 7971–7978.

71 Tripathi, S.M., Bock, W.J., Mikulic, P. et al. (2012). Long period grating based biosensor for the detection of *Escherichia coli* bacteria. *Biosens. Bioelectron.* 35: 308–312.

72 Dandapat, K., Tripathi, S.M., Chinifooroshan, Y. et al. (2016). Compact and cost-effective temperature insensitive bio-sensor based on long-period fibre gratings for accurate detection of E. coli bacteria in water. *Opt. Lett.* 41 (18): 4198–4201.

73 Brzozowska, E., Śmietana, M., Koba, M. et al. (2015). Recognition of bacterial lipopolysaccharide using bacteriophage-adhesin-coated long-period gratings. *Biosens. Bioelectron.* 67: 93–99.

74 Brzozowska, E., Koba, M., Śmietana, M. et al. (2016). Label-free Gram-negative bacteria detection using bacteriophage-adhesin-coated long-period gratings. *Biomed. Opt. Express* 7 (3): 829–840.

11

Humidity, Gas, and Volatile Organic Compound Sensors

Diego Lopez-Torres and César Elosua

Department of Electrical, Electronic and Communications Engineering, Public University of Navarre, Pamplona, Spain

11.1 Introduction

Nowadays, the development of applications related to the detection of gases and volatile organic compounds (VOCs) is experiencing a great growth mainly due to two very important reasons. Firstly, several gases and VOCs are toxic to humans, and they pose a permanent risk to the environment [1]. Secondly, the evolution of the technology has also caused changes in certain sectors such as industrial, medical, or chemical, which produce polluting gases and VOCs due to the processes that they develop in order to satisfy the necessities of the society [2]. For these reasons, the early detection of these agents is a very important point in a wide range of applications based on gases and VOCs detection.

Moreover, an interesting and new investigation line based on the possibility to use biomarkers of gases to detect the presence of a specific diseases [3] is causing the interest of many research groups due to the opportunity of using the breath of humans as a means of detecting in a quickly and easy way [4]. The detection and quantification of trace volatile metabolites in exhaled human breath, and to quantify them with adequate accuracy for medical diagnosis, is a technical challenge that has only been realized during the last part of the twentieth century, but the results allow to be optimism about the future of this topic. For example, nowadays it is widely known that nitric oxide (NO), ethane, or ammonia (NH_3) can be used as markers of different diseases with different marker origins such as nitric oxide synthases, lipid peroxidation (LPO), or protein metabolism, respectively [5].

Optical Fibre Sensors: Fundamentals for Development of Optimized Devices, First Edition.
Edited by Ignacio Del Villar and Ignacio R. Matias.
© 2021 The Institute of Electrical and Electronics Engineers, Inc.
Published 2021 by John Wiley & Sons, Inc.

This chapter of the book has been structured into six different sections organized as follows. In Section 11.2, the advantages of using optical fibre as technology for gas or VOC sensing are listed and studied, emphasizing the possibility of multiplexing. The next section of the chapter, Section 11.3, has been dedicated to describe sensing materials used to develop optical fibre sensors (OFSs) with the aim of detecting gases and VOCs. In Section 11.4, the different mechanisms of transduction of OFSs are explained. Section 11.5 is exclusively dedicated to optical fibre humidity sensors and their most recent researches. Section 11.6 encompasses the devices used for the identification and sensing of VOCs. In Section 11.7 the artificial systems used for the detection of complex mixtures of VOCs (optoelectronic noses) are studied and explained. Finally, the chapter ends with the main conclusions of the topic covered in this chapter.

11.2 Optical Fibre Sensor Specific Features for Gas and VOC Detection

As it has been mentioned above, it is necessary in a high number of applications related to industrial, chemical, or environmental sectors to detect, with a high accuracy (ppm, ppb, or even ppt), the exact value or concentration of gases and VOCs. It is at this point where the use of OFSs can take advantage over other electronic counterparts. OFSs present interesting properties such as lightweight, low transmission losses in the communication window, or low cost. Moreover, they are immune to electromagnetic interferences, which make them adequate for harsh conditions, i.e. in environments with a high explosion risk or with high levels of humidity, avoiding in this manner the possibility of short circuits and corrosion resistance, which are some examples [6].

Another important feature of OFSs is that they can be multiplexed [7]. A commonly accepted definition of multiplexing is the simultaneous transmission of multiple information channels along a common path. However, sometimes in sensor multiplexing networks, the path or a part of this path is not shared. Therefore, a more general definition of multiplexing, based on an optical multiplexed sensing system, is a sensor network where the number of sources, channels, or detectors is smaller than those that would be required if an equal number of sensing elements were assembled as individual measurement systems; i.e. the main aspect of a multiplexing system is to share elements of the network, reducing, in this manner, the final cost of the entire system and making it more economical and competitive (see Figure 11.1) [8]. The network requirements determine the design of the multiplexing solution. The number of multiplexed sensors, measurement types, noise, cost, dynamic range, and location of the sensors will determine

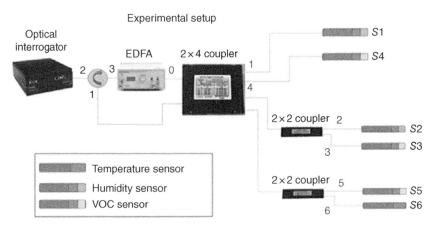

Figure 11.1 One example of a multiplexing experimental set-up based on fibre-optic sensors. *Source:* Reprinted from [8] with permission by IEEE.

the approach used in each multiplexing system. Moreover, there are three main characteristics of multiplexing: the network topology, the type of modulation of the sensor, and the multiplexing technique itself (in other words, the mechanism used for univocally identifying each sensor). Those three concepts are interconnected by choosing the multiplexing technique, while the type of sensors and topologies might be limited. The main multiplexing techniques are listed below: wavelength-division multiplexing (WDM), time-division multiplexing (TDM), polarization-division multiplexing (PDM), and spatial-frequency-division multiplexing (SFDM) [9, 10].

In order to confirm if an OFS can be used to detect VOCs, several parameters related to static and dynamic responses of the sensors have to be studied. The main and most important parameters are:

- *Sensitivity*: It is a parameter to evaluate the detection of the variation of the gas or VOC target concentration values in the surrounding environment. It can be considered as the change in the measurement signal per concentration unit of the gas target; in other words, it is the slope of the calibration graph.
- *Detection limit*: It is the lowest concentration value that can be detected by the sensor at definite and specific conditions.
- *Selectivity*: It is a parameter to evaluate the property of a sensor to respond selectively to a group of gas or VOC targets or even specifically to a single one when it is present with others in the measuring environment.
- *Resolution*: It is the lowest difference of concentration that can be measured by continuously varying the gas or VOC target concentration in the mixture. This

parameter is largely influenced by the electronic devices used in the measurement set-up.

- *Response time*: It is the time required for sensor to respond to a step concentration change from zero to a certain concentration value. Usually, it is specified as the time to reach a specific ratio of the final value in the presence of target gas (usually 90%).
- *Recovery time*: It is the time that the sensor signal takes to return to its initial value after a step concentration change from a certain value to zero. Usually, it is specified as the time to reach a specific ratio of the final value (usually 90%).
- *Stability*: This is a parameter to evaluate the feature of a sensor to maintain its properties when it is continuously employed in hostile environments for a long time.

11.3 Sensing Materials

11.3.1 Organic Chemical Dyes

The term dye refers to chemical compounds that show a specific colour, and in the case of sensing, it is affected by the surrounding medium; specifically organic solvents can induce changes in the intensity and location of their absorption bands. This phenomenon is known as **solvatochromism**, and it is very frequent among transition metal complexes [11]. The absorption band spectra can be blue- or red-shifted: the first case is called *hypsochromism* and the second one *batochromism* (see Figure 11.2). In this manner, it can be said that most of the chemical dyes used

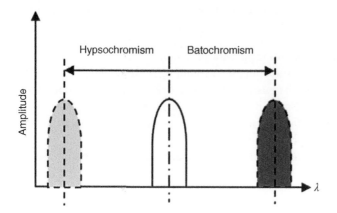

Figure 11.2 Spectral shift produced by solvatochromism.

for sensing are solvatochromic (if the compounds show this behaviour with VOCs, the phenomenon is called **vapochromism)**.

Vapochromism is produced by a shift in an electronic absorption band of a molecule because of a change in the polarity of the surrounding environment. In the case of VOCs, molecules can act as Lewis base or acid (more interactions will be explained later), coordinating electrons with different functional groups of the vapochromic molecules. The place where the coordination takes place is donor or acceptor atoms such as nitrogen, oxygen, or transition metals (silver, gold, platinum, etc.). After this reaction, the electronic absorption bands of the molecule are shifted and, as a consequence, the colour of the product (see Figure 11.3): this kind of change is measurable by an OFS. Moreover, these changes are reversible once the organic vapours are removed. Some specific chemical dyes show a metallic centre that interacts with the target molecules, and in order to generate the transduction, an intense chromophore has to be attached to the structure, producing the colour change. The combination of organic groups with metallic atoms will be described later in this section.

The detection of the VOCs by the dyes is determined by the intermolecular interactions between the sensing material and the target molecule. These reactions show different enthalpy changes: the higher they are, the stronger is the interaction and, in this way, the chemical response. Ordering them from the weakest to the strongest, these forces are van der Waals, dipole–dipole interactions, π–π molecular interaction, hydrogen bonding, charge transfer, hydrogen bonding, acid–base interaction (including Lewis electron donor–acceptor), and coordination bonds. Many electronic sensors use materials whose response is based on van der Waals forces, so their sensitivity is low; the opposite case, with a high sensitivity, could imply a lower robustness as the material would be affected by too many external factors.

A relevant example is Reichardt's dye (RDye), which is commercially available and widely used to detect VOCs [13]. This molecule shows a pyridine group with a positive group, whereas a phenoxide group exhibits an electron-donating capacity (see Figure 11.4). This product shows a negative solvatochromism (blue shift) in

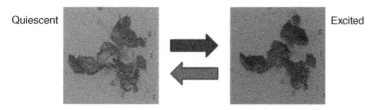

Quiescent ⇄ Excited

Figure 11.3 Colour change of a vapochromic material when exposed to ethanol vapours [12]. *Source:* Reproduced with permission of IEEE.

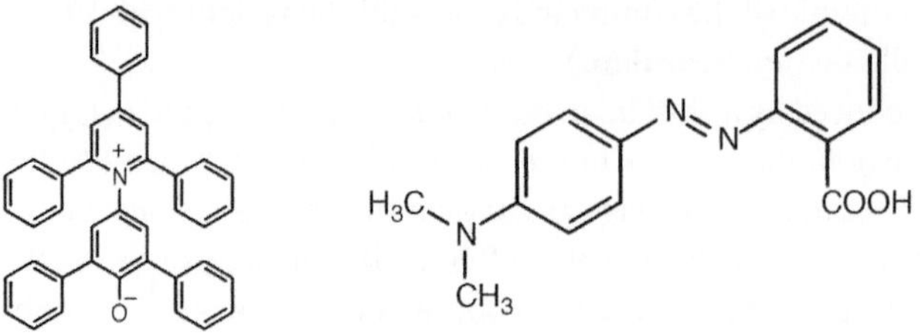

Figure 11.4 Molecular structure of (left) Reichardt's dye and (right) methyl red.

presence of organic solvents. Another example of materials that show a colour change is well known: pH indicators. Considering that Lewis electron donor and acceptor forces are the ones that yield to higher sensitivities, pH indicators are not only strictly affected by acidity; they can also react with VOCs that show specific electron distribution as well as acid or base behaviour (ammonia, acetic acid, pyridine, etc.). One of the most sensitive to organic vapours is methyl red (see Figure 11.4).

The selectivity of the dyes is defined by physic and chemical parameters that yield to compounds that are not typically sensitive to just one VOC. On the contrary, dyes with overlapped selectivities are combined to conform arrays, and their whole response is used to identify the sample. This kind of colorimetric matrices is combined with CCD cameras, and the colour pattern is used as sensing response. As it will be explained later, this idea is applied when working with OFSs by multiplexing the response of the different sensing materials towards the signal processing unit.

11.3.2 Metal–Organic Framework (MOF) Materials

Although metal–organic frameworks (MOFs) can be considered as a specific case of chemical dyes, this family of materials is becoming increasingly important. MOF is a generalization of chemical compounds that show crystalline structures formed by cations or clusters of cations that are linked to organic molecules. There is software available to design the structure of the frameworks depending on the final features, which makes many researching groups to specialize their synthesis for different applications [14]. MOFs were firstly used for separation of catalysis [15] and for gas storage [16], and, during the last years, they have been also prepared for sensing applications, specifically the detection of gases and vapours. Compared with other sensing materials, MOFs can offer specific selectivity, and they can resist temperatures up to 300 °C [17]. Moreover, aromatic groups are typically present in their structure, which produces luminescence under ultraviolet (UV) or visible (VIS) illumination. Following with this phenomenon, the emission can be modulated by absorbing analytes that get easily oxidized or reduced. This transduction can show a lower selectivity and a loss of the emission in time, while a well-known example is ruthenium metalloporphyrin, which is sensitive to oxygen. Other materials show a change in their absorption spectrum when they are exposed to organic vapours: if it takes place in the VIS, it produces a colour change. These principles will be explained in the following paragraphs.

MOFs are highly porous, which is a relevant feature when detecting gases and vapours because it makes their adsorption easier and improves the final sensitivity [18]. MOFs show also the ability of linking with themselves, following a polymer structure, which yields to smaller pores. Once again, the framework can be previously designed to follow a certain template that does not show this polymeric structure.

A key parameter, the final selectivity, is affected by several factors. One of them is based on the molecular size of the analyte to be detected. In this manner, only molecules smaller than the porous can get into the framework [19]. The specific chemical interactions with the gases or VOCs, such as electron coordination or hydrogen bonding, also determine the final selectivity. Functional groups present in the MOF play a key role in this kind or reactions: the polarity of the analyte is commonly used to define the selectivity [20].

Once the general features of the MOFs have been presented, a couple of the most relevant MOFs will be described. As it was pointed out in the previous section, the UV/VIS/MIR spectra of certain chemical compounds are affected by solvatochromism (MOFs showing this behaviour have been reported since 1967 [21]), although they have been used mainly in solid state, not to develop sensors. However, a family of products based on a similar backbone whose core is formed by gold and silver has been used to develop several OFSs [22]. The sensing signal can be either a spectral change or an optical power variation. On the one hand, some authors report changes in optical resonances observed in transmission sensors [23], while vapochromism has been also used to modulate in time optical signals in order to detect organic vapours for low concentrations.

The next group of MOFs to be described shows a metallic atom in the core of the molecule structure, known as metallophthalocyanines (MPc). This group has been used to develop VOC sensors thanks to the intense absorption in the visible spectrum, as well as the chemical stability. Moreover, the chemical changes produced by organic vapours are reversible.

The chemical structure is shown in Figure 11.5, where the metallic atom can be zinc (ZnPc), copper (CuPc), or ruthenium (RuPc), among other metals.

The sensing mechanism is based on a reversible change on the electronic states of the material, specifically between the highest and lowest occupied orbitals ($\pi-\pi$ electronic level transitions). As a consequence, the absorption bands are altered, which can be registered by the OFS. Once again, the polarity of the VOC plays a relevant role [24]. In some cases, the modification of the chemical structure also affects the dimensions of the sensing film, producing swelling effects [25].

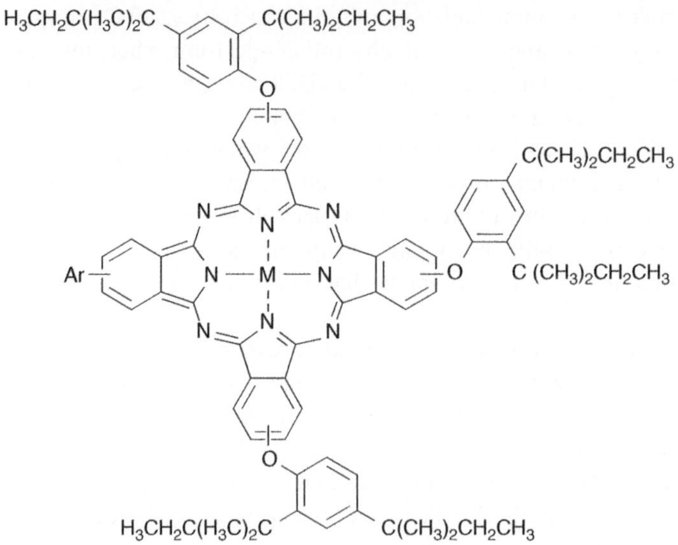

Figure 11.5 Chemical structure of a phthalocyanine. In the core of the molecule, M represents a transition metal ion.

Metalloporphyrins is another type of MOFs whose structure is like MPc: an organic ring with nitrogen atoms in whose core a metallic atom is coordinated with nitrogen atoms. Typical metallic atoms are Zn (see Figure 11.6a), Cu, Fe (see Figure 11.6b), Pd, or Pt. The large family of metalloporphyrins is used in many applications related with VOCs and gases such as O_2 [26] due to their excellent chemical and thermal stability, as well as the spectral shifts observed in presence of organic vapours. Actually, biological molecules that are essential to gas exchange show metalloporphyrins in their structure: haemoglobin and chlorophyll are two examples. Moreover, it has been found out that the mammal odour sense is based on metalloproteins.

The strong coordination link between the organic structure and the metallic core yields into intense colours: in presence of organic vapours, the molecules of the solvents get coordinated with the metallic core, producing significant colour changes or variations in luminescence emissions. Also, in this case, polar solvents (which act as electron donors) are more reactive; in the case of non-polar solvents, they get adsorbed, altering the dimensions of the molecule film.

11.3.3 Metallic Oxides

Seiyama [27] proposed, in the early 1960s, the use of metallic oxides as gas sensors. These architectures have been extensively studied and developed. There are

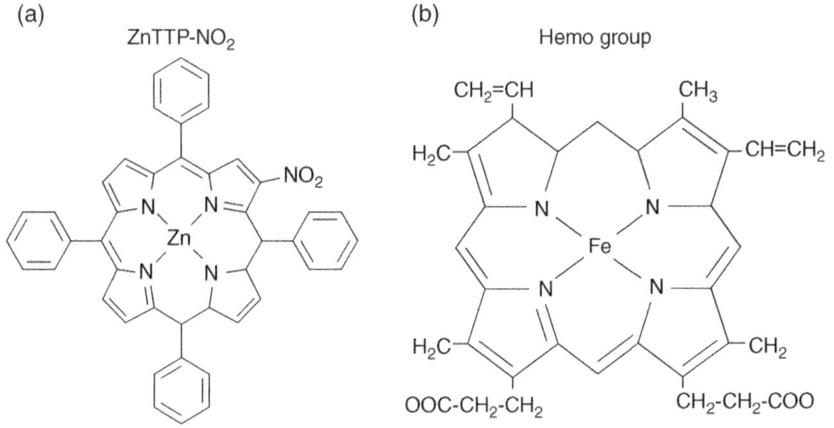

(a)
ZnTTP-NO$_2$

(b)
Hemo group

Figure 11.6 Structure of two metalloporphyrins with different metallic cores: (a) zinc and (b) a haem group (Fe porphyrin).

several reasons that explain why metallic oxides are a promising material for sensor development: they are thermally robust, and they also show an optimal mechanical robustness [27]. In addition, their properties are not affected by ageing, they have a good resistance to chemical degradation simple operation, and they are inexpensive and can work in harsh environments, such as explosive zones or places with high levels of humidity [28].

One of the most interesting features of metallic oxides is that they allow the detection of many gases or VOCs [29]. In other words, they are not selective. But this characteristic does not always have to be taken as a drawback. The low selectivity of the metallic oxides can be very useful in applications such as arrays of sensors or multiplexing. In these cases, with the utilization of post-processing techniques, for example, principal component analysis (PCA) [30], the identification and detection of several gases in complex mixture at the same time can be possible. Moreover, one way to solve the low selectivity of this type of sensors is the surface molecular functionalization [31]. Several works have been demonstrated that, thanks to this method, the sensors can be considered as selective towards a specific gas [32].

In order to explain the mechanism of interaction between gases or VOCs and metallic oxides, many models have been proposed. Currently, it is well established that this mechanism is based on the principle of conductance change between gas/VOC molecules and the sensing layer, due to the chemisorption [33]. Metallic oxide films consist of a large number of grains, contacting at their boundaries. Its electrical behaviour is governed by the formation of double Schottky potential barriers at the interface of adjacent grains caused by charge trapping at the interface. Due to this fact, chemisorption of oxygen from air occurs, and it causes the immobilization of the electrons of the conduction band of the sensing layer, creating a depletion layer in the near surface. When the sensing layer, formed by metallic oxides, is exposed to a chemically reducing gas/VOC, co-adsorption and mutual interaction between the gas/VOC and the oxygen result in oxidation of the gas/VOC at the surface. The removal of oxygen from the grain surface results in a decrease in barrier height. This is a well-established and accepted model for electronic gas/VOC sensors based on metallic oxides, which assumes that the molecules of gases or VOCs penetrate in the sensing layer. But at the same time, instead of taking advantage of the electronic properties of the metallic oxides, their optical transparency is exploited. In this chapter, it is assumed that the chemisorption of gas by the metallic oxide will affect also its optical properties, generally speaking its refractive index.

The previous reaction between metallic oxide surface and gases/VOCs explained above is a crucial factor on the final response of the fibre-optic sensor, and, for this reason, a study of this parameter is necessary in order to optimize the final sensitivity of the sensor [34]. It can be affected by several factors, such as chemical components, surface state, and the gas/VOC concentration inside the sensing layer. This last factor is not constant and decreases as the distance from the external surface increases. In other words, a diffusion of the gas/VOC concentration occurs inside the sensing layer [35]. This means that the effect of the gas on the sensing material is dependent on the depth from the external surface.

The temperature is one of the main parameters to affect the sensitivity of the metallic oxides. In fact, in addition to influencing the type adsorbed oxygen species, the temperature changes the kinetic of adsorption/desorption processes for oxygen and gases/VOCs, making them competitive with each other. This means that, in general, the maximum sensor response position shifts with temperature by changing the activation energy for adsorption of detected species [36]. This strategy can be used to tune the sensitivity of the sensors towards a species rather than to others, improving in this manner the sensor selectivity. At this time, it is important to mention that, based on the explanation above, most of metallic oxide gas sensors (especially electronic sensors) have to be heated up to temperatures over 250 °C to achieve good sensitivities. However, with regard to OFSs, some papers have been proposed to work with metallic oxide thin films at room temperature, avoiding the necessity to heat the sensors and obtaining very promising results and sensitivities [37].

The most popular metallic oxides used to develop fibre-optic sensors are:

- *Tin dioxide (SnO₂)*: It is a widely used metal oxide in electronic sensors for VOCs and gases and moisture measuring. As it has been mentioned, these devices require the material to be heated up to promote the oxidation–reduction on its surface (in the majority of the cases), which is not suitable for certain environments. In addition, this material shows a relatively high refractive index: this property has been employed to induce lossy mode resonances (LMRs) by depositing thin layers of the oxide around optical fibre segments where the cladding has been removed. The technique followed is sputtering so that the final thickness of the layer can be controlled mainly by the deposition time, among other factors. In any case, the final thickness is usually below 1 μm to ensure the transduction.

The resulting coating induces lossy cladding modes that are affected by the refractive index of the layer itself and the environment: these changes produce a registrable spectral shift in the sensor's optical spectrum. OFSs have been reported to detect and measure relative humidity (RH) based on LMRs [38]; interferometers prepared with different optical fibres, for example, microstructure optical fibres (photonic crystal fibre [PCF] and suspended core fibres), were also prepared with SnO_2 with this aim [39]. Moreover, a high number of research groups are working on its potential use for VOC and gases detection with promising results.

- *Indium tin oxide (ITO)*: This metallic oxide was one of the first oxides used to prepare sensing devices. This conducting material shares some properties with others but, due to its high refractive index, also induces LMRs. The first studies followed the dip coating procedure to deposit the oxide using a sol/gel-based mixture: the thickness of the final thin layer was optimized to 170 nm. ITO coatings are affected by refractive index changes. Consequently, coatings sensitive to different parameters can be deposited onto the ITO thin layer to implement sensing devices: in this manner, layer-by-layer nanostructures were used to develop humidity sensors [40]. Due to the same fact, surrounding medium refractive index changes can be also measured directly using the adequate substrate.

  Thanks to the chemical nature of the oxide, it can be also deposited by sputtering, which shortens the construction process. Other compounds can be coated around the sputtered ITO film as it is the case of polyvinylidene (PVdF), a conducting polymer that is sandwiched between two ITO layers to get a tunable optical filter by electric signals: it can be used for electric field sensing too [41].

  Another relevant property of ITO coatings is that the morphology can be adjusted not only by the deposition time but also by a post-thermal treatment under different conditions. This effect was studied by preparing four different ITO-coated surfaces [42]. After the metal oxide sputtering, they were annealed at 500 °C for four hours under different conditions (vacuum, nitrogen, atmosphere), and the fourth one was not treated. As a result, the fourth coating was amorphous and shows the poorest performance; the morphology of the other three ones is crystalline-like, which makes easier the interaction with the environment.

- *Zinc oxide (ZnO)*: It is a transparent semiconductor metal oxide with a strong absorption band in UV/VIS region of the electromagnetic spectrum. Consequently, it can be used for optical applications. ZnO can be found in the form of zincite, but mostly it is prepared synthetically by doping oxygen in zinc metal. Zinc oxide has numerous properties, for example, good transparency, wide and direct bandgap, high thermal conductivity, and electron mobility. Moreover, ZnO holds several additional advantages of biocompatibility, high surface adsorption capabilities, presence of oxygen vacancies, inertness, high chemical stability, surface reactivity, non-swelling in both aqueous and non-aqueous environments, etc. Thanks to these features, ZnO is a suitable candidate for chemical and biosensing applications. Its physisorption and chemisorption properties make it suitable as a recognition candidate for the detection of various gases such as chlorine and hydrogen sulphide. In addition, ZnO also works as a plasmonic material in the IR region, supporting its role as transducer element in various plasmonic sensors. Finally, ZnO also satisfies the conditions required for the LMR to occur [43]. Based on this phenomenon, several fibre-optic sensors have been developed with the aim of detecting gases and VOCs [44].
- *Tungsten trioxide (WO_3)*: It is a transition metallic oxide with wide bandgap from 2.60 to 3.25 eV [45]. It has been reported in several papers that WO_3 film has a colour change when it is exposed to hydrogen, due to the chemical interaction between WO_3 and hydrogen, making its use very interesting to detect it. This gasochromic effect makes it possible that WO_3 detects hydrogen. In addition, there are other ways of detecting changes in the WO_3 film when it is exposed to hydrogen environment by detecting variations in reflectance, transmittance, absorption, and refractive index. However, as it has been explained in this section with other metallic oxides, this chemical interaction between WO_3 and hydrogen is not strong enough because the WO_3 film can also chemically interact with other gases, such as hydrogen sulphide and acetylene. Therefore, bulk WO_3 film does not have a selective sensitivity to hydrogen, which limits its application as a hydrogen sensor. To overcome this drawback, a common method is to modify WO_3 film by surface functionalization or doping metal catalyst: one of the most used is platinum (Pt) as it has been explained in [46].

11.3.4 Graphene

Nowadays, graphene is considered as a promising material for gas and VOC sensing, since its electronic properties are strongly affected by the adsorption of gas and VOC molecules, respectively. Due to its excellent overall performance, this material has caused a great deal of enthusiasm in the scientific community since 2004 [47]. Moreover, its high conductivity and ballistic transport ensures that graphene exhibits very small interferences as a chemical sensor [48] and does not require an auxiliary heating device, in contrast with what happens with several metallic oxides [48]. All these features make graphene an ideal material for gas and VOC detection. Besides, since Novoselov et al. discovered the electrical properties of graphene nanosheets [47], many researchers have begun to study this material, increasing in this manner the number of published papers based on graphene gas/VOC sensors.

However, there is a drawback; the complexity and price of its synthesis make the use of graphene unrealistic in some fields, being very difficult in this way to be competitive with their counterparts. In this regard, graphene oxide (GO) and reduced graphene oxide (rGO) can be good substitutes of graphene in the majority of the applications with a lower cost of production. GO is an analogue of graphene with many functional groups so that the physical and chemical properties of GO are quite different from those of graphene. GO, an intermediate for the synthesis of graphene, can be obtained simply by using permanganate oxidation and ultrasonication of inexpensive graphite materials. This method is well known as the Hummers method or the Hummers and Offeman method. Regarding to rGO, it is produced by the reduction of GO, which contains many functional groups and defects, providing great potential because it is easy and inexpensive for large-scale production. In addition, it can be easily functionalized, improving the selectivity of the optical fibre gas and VOC sensors. For these reasons, it is no wonder that many researchers are interested in exploring GO and rGO as a candidate for gas and VOC sensors [49].

11.4 Detection of Single Gases

The term 'mechanism of transduction of an optical fibre sensor' encompasses the explanation of how the light that travels through the optical fibre is altered by the presence of a certain parameter to measure, in this case VOCs or gases. Due to its presence, the conditions of the surrounding medium in which the sensor is located change and, as a consequence, the optical properties of the OFS. These properties can be measured in different ways by also applying different techniques [50].

One of the most common mechanisms of transduction used in the development of OFSs is the mechanism based on absorption bands. Each chemical element, gas or VOC, has certain absorption bands at certain wavelengths, i.e. a random optical spectrum in presence of a gas or a VOC will have specific optical losses at the absorption wavelengths of the gas or VOC exposed. In this sense, if these losses are measured, using different techniques, the quantification and identification of the gas or VOC is possible (acetylene [51] and methane [52] are two of the most common). Due to this effect, the vast majority of these sensors base their results on quantifying and measuring the optical power difference that occurs in these bands in presence and absence of the gas or VOC at issue. It must be pointed out that, for this optical phenomenon to occur, a specified proportion of the optical power guided through the fibre optic, called evanescent field, has to interact in some way with the gas or VOC molecules. In case this is not possible, the fibre-optic sensor will not be sensitive to the parameter to be measured, and the final measurement of the sensor will not be correct.

The other important group of OFSs that share the same transduction mechanism is that based on the detection and measurement of variations in the effective refractive index of the surrounding media of the sensor. This variation causes changes in the optical spectrum (mainly wavelength or phase shifts) that can be measured by the OFS applying certain techniques, for example, monitoring of a specific wavelength for a certain time or the application of the fast Fourier transform (FFT) with the aim of detecting the fundamental frequency of the spectrum and later monitoring its phase shifts [39] (see Figure 11.7).

As in the previous case, in order for the sensor to detect these changes or variations in the refractive index of the surrounding media of the sensor, a part of the optical power guided by the OFS, again the evanescent field, has to interact with the environment where these variations are occurring due to the presence of a gas or VOC, while another important point to take into consideration is that sensors that base their working mechanism on this transduction method require a very clear and defined spectral footprint. This starting point is very important because it is necessary to select a specific point as a reference in order to carry out the sensor measurements correctly. For this reason, many of the sensors belonging to this group are those based on interferometers [53]. In this type of sensors, it is easier to find this reference point due to the sinusoidal shape of its interferometric pattern (see Figure 11.8).

By comparing both methods, it can be concluded that the mechanism of transduction based on absorption bands allows the sensors to be more selective since, as mentioned above, each gas or VOC has its particular absorption bands, previously known. However, these absorption bands correspond to very specific wavelengths (in the nanometre range), which implies to work with optical devices that have a high spectral resolution (for example, lasers), which can make the final set-up more expensive. On the other hand, OFSs based on refractive index changes do not need to work with these optical devices, but they have the handicap that many factors can alter or modify the effective refractive index of the medium. Therefore, with the aim of increasing the selectivity of these sensors, films with a specific sensitivity to certain parameters are usually deposited on metallic oxide coating.

Numerous research groups are developing fibre-optic sensors based on both transduction mechanisms. Here are some examples of them:

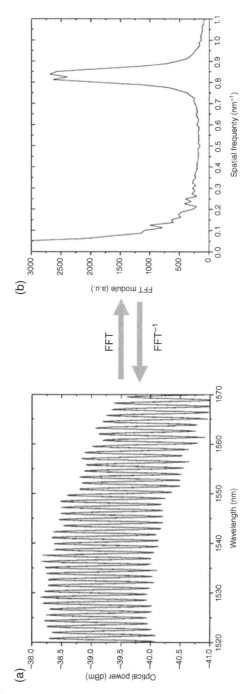

Figure 11.7 Optical spectrum of a low-finesse Fabry–Pérot and its FFT in which the fundamental frequency can be clearly seen. *Source*: Reprinted from [39] with permission by Elsevier. Reproduced with permission of IEEE.

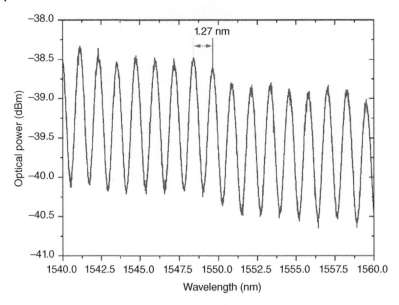

Figure 11.8 Sinusoidal shape of a low-finesse Fabry–Pérot reflected spectrum. *Source:*
Reprinted from [39] with permission by Elsevier. Reproduced with permission of IEEE.

- *OFSs based on absorption lines, examples*: Trying to measure different concentrations of methane, several publications have proposed to detect them by studying the variations of their different absorption lines (mainly in relative optical power). Kornaszewski et al. demonstrated in [54] the detection of methane measuring the variations of the optical power obtained in the transmission spectra when the OFS was exposed to different concentrations of methane at 3.15–3.35 μm, where methane has an absorption region. A. M. Cubillas et al. also demonstrated methane sensing in other different absorption lines located at 1300 and 1670 nm in [52, 55], respectively.
- *OFSs based on wavelength of phase shifts, examples*: Trying to measure different values of RH, A. Lopez Aldaba et al. proposed a multiplexing system for simultaneous interrogation of an OFS based on a low-finesse Fabry–Pérot interferometer, which measures different parameters including VOCs, more concretely methanol [8]. In this papers, both wavelength and phase shifts were measured with very good results. Another example of this working mechanism is explained in [56]. In this paper, an in-reflection PCF interferometer exhibits high sensitivity to different VOCs (acetonitrile and tetrahydrofuran) without the need to deposit any permeable material as sensing layer. When the OFS was exposed to these VOCs, a wavelength shift occurred in the reflection spectrum of the sensor, and the response of the sensor towards VOCs molecules could be characterized and studied.

11.5   Relative Humidity Measurement

RH is an important parameter, and its measurement is commonly used in several aspects of daily life of human beings; continuous RH measurement and control is important for human comfort such as in air-conditioning monitoring and achieving controlled hygienic conditions [57]. RH measurement is also required in a range of areas, including meteorological services, weather forecasting, chemical or medicine industry, food and beverage processing industry, civil engineering, horticulture, and electronic processing. In addition, optical fibre humidity sensors have found new applications in clinical treatment due to the need for humidification of inspired gases in critical respiratory care [58].

RH is defined as the percentage ratio of the current water vapour pressure to the saturated one at a certain temperature [59]. Taking this definition into account, fibre-optic humidity sensors could be considered as gas sensors. Due to the importance of this kind of sensor, a brief summary of the most recently advances in this topic and a classification of them according to their working principle will be presented in the next paragraphs. The first group includes optical fibre humidity sensor based on the optical absorption of materials, because it is one of the most important mechanisms of transduction (as it has been explained in the previous section). These sensors are based on the interaction between the evanescent field of the guided light and the coating deposited used as the sensitive material, providing, in this manner, changes of the transmitted optical power along the whole spectrum. For this kind of sensor, plastic-cladding silica optical fibre or plastic optical fibres (POF) are commonly used, although there are other possibilities such as microstructured optical fibres or the side-polished optical fibre (D-shape). Most recent research has focused on the study of materials (polymers, metallic oxides, hydrogels, etc.) that are becoming common in the development of OFSs. These materials are being studied for their ability to improve certain characteristics of optical fibre humidity sensors, such as response or recovery times or repeatability. Moreover, there are certain materials that provide high sensitivity in extremely low-humidity environments [60], which make them very useful in applications such as lithium-ion battery manufacturing, semiconductor fabrication, and pharmaceutical industry.

The next group includes optical fibre humidity sensors based on fibre Bragg gratings, already explained in Chapter 4. Regarding humidity sensors, several polymeric materials such as polyimide, di-ureasil, or poly(methyl methacrylate), among others, have been used to develop them with very interesting results. Most recent research has focused on studying the performance for this type of sensor when a coating of several kinds of material is deposited along the LPGs. These materials are polymers, hydrogels, gelatin, cobalt-chloride-based materials, and SiO_2 nanospheres [61, 62].

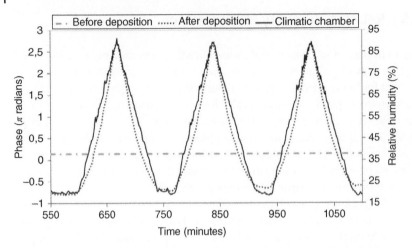

Figure 11.9 Sensor's phase response towards humidity: before deposition and after SnO_2 deposition. *Source:* Reprinted from [39] with permission by Elsevier. Reproduced with permission of IEEE.

Another possibility for the development of optical fibre humidity sensors is based on interferometers, including those also explained in section 4: Fabry–Pérot, Sagnac, Mach–Zehnder, Michelson, and modal interferometers [34, 53] (see Figure 11.9). Currently, these interferometric structures have been studied by numerous research groups providing all of them advantages such as stability, compactness, small size, lightness, etc.

Finally, there is a group of optical fibre humidity sensors that includes microtapers, microring, and micro-knot resonators (MKR), as well as other sensors based on whispering gallery modes (WGM). Due to the versatility of their structures and geometries, this kind of sensors is a good alternative to detect different parameters, such as gases and VOCs, and consequently RH variations. To make the OFS sensitive to RH, it is necessary to deposit a sensing gas layer [63].

11.6 Devices for VOC Sensing and Identification

Materials described in Section 11.3 have been used to develop VOC sensors. As it has been pointed out, overlapped selectivity is not a drawback but a feature to consider when preparing the sensor. The transduction in all the cases is produced by the change in the optical properties of the coating deposited on the fibre, so depending on the set-up, different parameters are used to register their response.

Both the sensing material and the set-up are the criteria followed to classify VOC OFSs. As this chapter is focused on the sensing materials, relevant implementations with the reagents described in Section 11.3 will be described.

The solvatochromism of organic dyes produces a colour change that, as an initial approach, is observed in the VIS spectrum. To achieve that, not only white light sources and spectrometers operating in that range are required: fibres with wider cores than the standard ones are needed to collect the colour from the sensing film. In this manner, plastic-cladding fibres have been used to attach the dye at the end of a cleaved ended pigtail [64]. To make this colorimetry measurement, reflection configuration is used most of the times because colour coupling into the fibre is maximized. The spectral change is determined by the chemical reaction between the dye and the VOC so that it is related to the selectivity of the material and can be used to identify different types of vapours depending on their functional groups. In this manner, short-chain alcohols such as methanol and ethanol will produce a similar spectral change but totally different to ketones (such as acetone) or to organic acids (acetic acid) [65]. Actually, the whole spectral variation can be used as a fingerprint to classify the VOC (see Figure 11.10). The drawback for this approach is that the fibres are not compatible with standard multiplexing, so each sensor requires a source and a spectrometer. In this sense, lab-on-a-fibre solutions overcome this situation by including several indicators on its tip, registering the response using the fibre itself or an external camera [66].

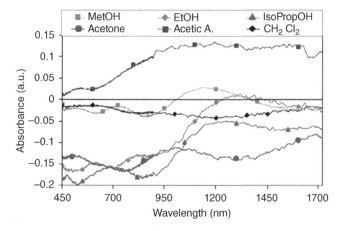

Figure 11.10 Absorption spectra registered from an optical fibre sensor in presence of organic vapours with different organic groups. The shape of the spectra can be used to distinguish between them. *Source:* Reprinted from [65] with permission by Elsevier. Reproduced with permission of IEEE.

Another trend consists of using dyes that also show changes in their NIR spectral range, which makes them compatible with standard fibres and with multiplexing systems [67]. The counterpart of this approach is that the colour change information in the VIS is not used and, although there are materials specifically designed to show their solvatochromism in the NIR region, multiplexing techniques will only use a spectral span of a few nanometres for each sensor [68]. In these cases, transduction mechanisms based on evanescent wave can also be applied, depending on the type of fibre and material, transmitted power attenuation, or the shift of optical resonances [69] (see Figure 11.11).

The configurations of sensors based on MOFs with vapochromism are similar to chemical dye ones. However, this is not the case of luminescent MOFs, where initial approaches were based on extrinsic configurations where the fibre just guided the excitation signal and collected the emitted one [70]. On the other hand, for intrinsic sensors, in which the sensing material is placed directly on the fibre, the interface between fibre and fluorophore has to be optimized even at a higher level compared with colorimetric measurements. Therefore, the configuration that is commonly used is the reflection one but with fibres ended with a shape that maximizes the area of the interface.

The detection of gaseous oxygen is one of the most successful developments so far, using metalloporphyrins as sensing material [71], while the reversible quenching of the luminescence has been successfully used to determine ambient oxygen concentration with a high selectivity [72]. The sensor can be characterized by

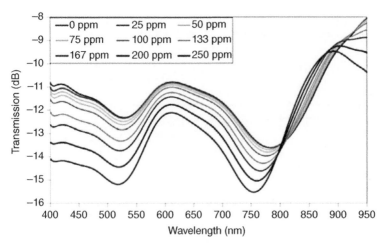

Figure 11.11 Spectral blue shift of a lossy mode resonance of a sensor based on a vapochromic dye when exposed to different methanol concentrations. *Source:* Reprinted from [69] with permission by Elsevier. Reproduced with permission of IEEE.

recording the intensity of the luminescent signal, which might potentially be affected by signal artefacts; a way to overcome these undesired fluctuations consists of measuring the lifetime emission. This last approach is more robust and has yielded to portable prototypes to measure oxygen concentration in speleology applications [73].

11.7 Artificial Systems for Complex Mixtures of VOCs: Optoelectronic Noses

Aroma detection has many applications in different fields: environmental (pollutants control), security (chemical terrorism detection), and food industry are the most significant ones [74]. Using trained workers to detect odours is expensive and potentially harmful for them. Several efforts have been focused along the last decades to develop artificial systems able to identify odours or VOC mixtures. The best prototypes tend to mimic the behaviour of smell sense of animals [75]. The identification of odours is a complex biological process: vertebrate and specifically mammals show the best developed olfactory system. There are more than 1000 receptor genes to codify odour receptors (from 1 to 2% of genome) [76]. In the case of human nose, tens of millions of cells act as chemical receptors that generate information to be processed by thousand neurons.

Biological olfactory sense can distinguish certain volatile compounds with concentrations in the ppb range, as well as identify complex mixtures or odours. In the first case, there are specific receptors for the compounds to be detected, generating a signal depending on its concentration. In the second case, mixtures are formed by lots of compounds. Consequently, they are not detected individually, but its whole response is interpreted as a smell print. This is one of the key factors: one receptor is sensitive to several smells, and a smell is detected by several receptors. In other words, some receptors show a higher sensitivity to certain VOCs (for instance, depending on its functional group), and they are less sensitive to others. The information generated by all the receptors is processed in parallel by the neurons, generating patterns that are learnt, so that the individual will compare future patterns with the ones of its memory.

It is important to keep in mind that there are techniques such as gas chromatography and mass spectrometry, which permit to detect the exact composition of an odour [77], but those solutions are not profitable for applications when the identification is required. Moreover, the equipment required is very expensive, it is not portable, and measurements cannot be performed remotely in real time [78]. In these cases, it is not necessary to know in detail the mixture of the aroma but to identify its composition. One of the most remarkable equipment developed so far is the eNose, an electronic device designed to detect and to distinguish between complex odours employing a sensor array, which is formed by a non-specific sensor [79]. The eNose has been successfully used in applications related with food and beverages [80]. The low-cost sensors and the high integration capacity of electronic technology have positioned this solution as an emerging alternative to other techniques such as GC-MS or human experts.

When optical fibre and electronic sensors are compared, there are some advantages that the first one offers [81]: the information of several sensors can be multiplexed in a single fibre, it is not affected by electromagnetic noise, propagation losses are low, and remote sensing is possible. The main drawback of OFSs is that its maturity is lower than the electronic devices. Therefore, the price of the active devices required is also higher. However, the current tendency of these costs is to decrease because their use in communication and sensing applications is growing every day [82]. In the specific case of the optoelectronic nose, it can be organized is three blocks (see Figure 11.12):

- The optical header, where all the active components (light source, receptors and the data processing element) can be grouped.
- The sensing unit that includes a group or array of sensors (there can be more than one sensing unit). Other magnitudes such as temperature or humidity can be also registered by the array.
- The fibre segment that links the two previous blocks can be considered as an independent part.

This modular configuration makes the system more scalable and reusable: in the case that one of the active components is damaged, or some sensors are unusable, they can be replaced without altering the final performance. As an example, if an improved version of the sensor array is developed, it can be replaced, and the new

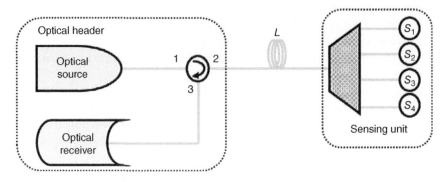

Figure 11.12 Modular configuration for an optoelectronic nose.

device could be interrogated with the same optical header. On the other hand, any active element of the header can be repaired or replaced without affecting the whole system. In addition, in the case of the optical header, the same one could be used to interrogate and analyse the response of some sensor arrays, which would optimize the whole system. Moreover, different sensor arrays could be used in the same system depending on the odours to analyse: the number of arrays could be defined by the characteristics of the sensors and the signal power of the optical source. In Figure 11.13 it is displayed an example of an optoelectronic nose where the sensor array is not at the same place as the optical header. The possibility of using the same header to illuminate several headers is also illustrated.

The information obtained from the sensor array is processed in the optical header by data mining techniques, a multidisciplinary field whose aim is to extract information of specifically patterns from large data sets. The sensor array can be formed by devices with different sensing materials or transduction phenomena or even with distinct parameters or measurements types (amplitude, wavelength, or phase). In any case, the amount of information is significant, and it constitutes the fingerprint of the whole measurement. Therefore, its processing is critical to identify the odour, while the quantification of the compounds present in the aroma is not as relevant in this type of applications as the identification of the whole mixture. The most used methods to process the whole mixture information are PCA and artificial neural networks (ANN).

The processing of the multivariable information is divided into certain stages: in the first one, the data is pre-processed in order to compensate signal drifts, eliminate outliers, and normalize the resulting information. In the second stage, the features are extracted from all the available data, typically, by PCA: it is a linear transformation that redistributes the variance information contained in all the variables so that it is compressed into a few ones [83]. In this manner, the initial multivariable space can be truncated into a three- or two-dimensional one (removing the rest of variables implies a loss of information although it is acceptable

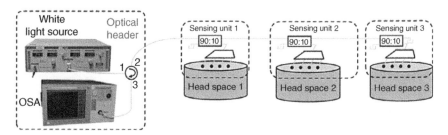

Figure 11.13 Example of an optoelectronic nose configuration.

below 5%). Using an initial set of measurements, a transformation matrix is calculated by applying a linear transformation [84]: this matrix will be used to convert future data into the reduced dimensional space. Therefore, it is important that the initial set contains as much data as possible, including noise, different environmental conditions, and artefacts in order to get a robust transformation. New variables are placed in the transformed dimensional space so that they can be clustered depending on their characteristics [85]. If the clusters are distinguishable, the space can be mathematically divided in different regions so that new measurements can be located in any one of them to be identified. Figure 11.14 shows the PCA transformation of temporal signals into a tri-dimensional space where, initially, the signal is characterized by its temporal samples, which yields into a high-dimensional space. Consequently, after applying PCA, the final space is easier to process.

However, in many situations, PCA is not robust enough to handle with undesired fluctuations or noise: a huge initial data set would be necessary to enhance this behaviour, but it would reduce the performance of the whole system. This problem is treated by using ANN. These non-linear algorithms are based on fuzzy logic, imitating the learning processes of human brain [86]. In a few words, an ANN is a non-linear procedure where variables are processed in parallel to calculate binary outputs, one per the odour or aroma to identify. The ANN has to be also trained also with an initial set of measurements so that it learns to match future data with the patterns learnt. Once the training is over, the algorithm is validated with samples that have not been used during the training; if validation is right, the ANN is ready to process data and classify it. The process is summarized in Figure 11.15.

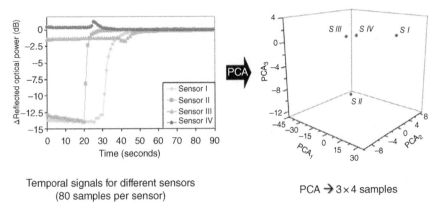

Figure 11.14 PCA transformation of temporal signals into a tri-dimensional space.

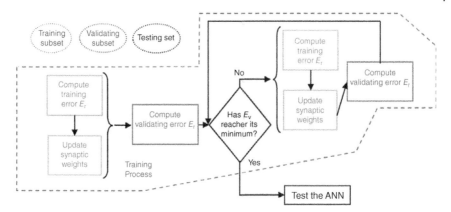

Figure 11.15 Main steps for an ANN training, validation, and final test.

Typically, the information obtained after PCA is used to, first, train an ANN, and later, as input variable for it. Although this kind of algorithms is complex, they are very robust against noise and undesired variations in the sensing signals. Its powerful processing has motivated the improvement of ANN algorithms, so they are easily available in data processing tools such as Matlab®, which simplifies the whole system because the ANN has not to be manually programmed. New types of ANN are being developed, which is the case of non-supervised networks, also known as self-organized map (SOM) [87]: after an initial stabilization step, this kind of algorithms keeps on learning continuously, which makes them very robust. In some cases, ANN is directly applied to the initial data set, although in most applications PCA and ANN are combined to get the synergy of both processes.

11.8 Conclusions

The use of optical fibre to develop sensors is a field that is growing day by day, especially in the area of gas and VOC detection. For this reason, different types of sensors based on different structures and architectures using optical fibre to detect these parameters have been presented in this chapter. The advantages of OFSs over conventional counterparts (mainly, electronic sensors) such as ease of fabrication, low cost, good reproducibility, and excellent optical performance are reported in a high number of publications. Besides, optical fibre also presents some potential advantages: it is versatile, and many different configurations can be achieved to develop different interferometric sensors (MZ, FP, and Michelson) with desirable properties such as good kinetics, small sensor heads, and good stability over time, though other architectures such as gratings and evanescent sensors offer also a good performance.

Thanks to the different mechanisms of transduction of the OFS, the interaction between guided light and gas/VOC molecules or sensing deposited materials is possible, and, consequently, different parameter such as gases or VOCs can be detected. Besides, if these parameters are optimized, the sensitivity and the resolution of the final devices could be improved.

Finally, though it is true that there are features of OFSs, such as selectivity, that must be improved, the majority of the proposed OFSs described here are a good and a real alternative for gas and VOC sensing. For this reason, sensors based on optical fibres are a technology with an outstanding potential in environmental, industrial, chemical, and medical applications.

References

1 Hodgkinson, J. and Tatam, R.P. (2013). Optical gas sensing: a review. *Meas. Sci. Technol.* 24 012004 (59pp).

2 Timmer, B., Olthuis, W., and Van Den Berg, A. (2005). Ammonia sensors and their applications – a review. *Sens. Actuators B Chem.* 107 (2): 666–677.

3 Valenti, D.A. (2010). Alzheimer's disease and glaucoma: imaging the biomarkers of neurodegenerative disease. *Int. J. Alzheimers Dis.* 2010 Article ID 793931, 9pp.

4 Buszewski, B., Kęsy, M., Ligor, T., and Amann, A. (2007). Human exhaled air analytics: biomarkers of diseases. *Biomed. Chromatogr.* 21: 553–556.

5 Pitocco, D., Zaccardi, F., Di Stasio, E. et al. (2010). Oxidative stress, nitric oxide, and diabetes. *Rev. Diabet. Stud.* 7 (1): 15–25.

6 Fidanboylu, K.A. and Efendioğlu, H.S. (2009). Fibre optic displacement sensors and their applications. *5th International Advanced Technologies Symposium,*Karabuk, Turkey (13–15 May 2009).

7 Grattan, K.T.V. and Sun, T. (2000). Fibre optic sensor technology: an overview. *Sens. Actuators A Phys.* 82 (1): 40–61.

8 Lopez-Aldaba, A., Lopez-Torres, D., Aguado, C.E. et al. (2018). Real time measuring system of multiple chemical parameters using microstructured optical fibres based sensors. *IEEE Sens. J.* 18 (13): 5343–5351.

9 Fernandez-Vallejo, M., Bravo, M., and Lopez-Amo, M. (2013). Ultra-long laser systems for remote fibre Bragg gratings arrays interrogation. *IEEE Photonics Technol. Lett.* 25 (14): 1362–1364.

10 Bravo, M. and López-Amo, M. (2012). Remote-time division multiplexing of bending sensors using a broadband light source. *J. Sens.* 2012 Article ID 154586, 6pp.

11 Usón, R., Laguna, A., Laguna, M. et al. (1984). Synthesis and reactivity of bimetallic Au–Ag polyfluorophenyl complexes; crystal and molecular structures of [{AuAg (C6F5)2(SC4H8)}n] and [{AuAg(C6F5)2(C6H6)}n]. *J. Chem. Soc. Dalton Trans.* (2): 285–292.

12 Elosúa, C., Vidondo, I., Arregui, F.J. et al. (2013). Lossy mode resonance optical fibre sensor to detect organic vapors. *Sens. Actuators B Chem.* 187: 65–71.

13 Krech, J.H. and Rose-Pehrsson, S.L. (1997). Detection of volatile organic compounds in the vapor phase using solvatochromic dye-doped polymers. *Anal. Chim. Acta* 341 (1): 53–62.

14 Farha, O.K., Yazaydin, A.Ö., Eryazici, I. et al. (2010). De novo synthesis of a metal-organic framework material featuring ultrahigh surface area and gas storage capacities. *Nat. Chem.* 2 (11): 944–948.

15 Lee, J., Farha, O.K., Roberts, J. et al. (2009). Metal-organic framework materials as catalysts. *Chem. Soc. Rev.* 38 (5): 1450–1459.

16 Li, J.R., Ma, Y., McCarthy, M.C. et al. (2011). Carbon dioxide capture-related gas adsorption and separation in metal-organic frameworks. *Coord. Chem. Rev.* 255 (15–16): 1791–1823.

17 Farha, O.K., Spokoyny, A.M., Mulfort, K.L. et al. (2007). Synthesis and hydrogen sorption properties of carborane based metal-organic framework materials. *J. Am. Chem. Soc.* 129 (42): 12680–12681.

18 Smith, P.A., Koch, D., Hook, G.L. et al. (2004). Detection of gas-phase chemical warfare agents using field-portable gas chromatography-mass spectrometry systems: instrument and sampling strategy considerations. *TrAC – Trends Anal. Chem.* 23 (4): 296–306.

19 Dinc$\overline{?}$, M. and Long, J.R. (2005). Strong H_2 binding and selective gas adsorption within the microporous coordination solid $Mg_3(O_2C-C_{10}H_6-CO_2)_3$. *J. Am. Chem. Soc.* 127 (26): 9376–9377.

20 Lin, X., Blake, A.J., Wilson, C. et al. (2006). A porous framework polymer based on a zinc(II) 4,4′-bipyridine-2,6, 2′,6′-tetracarboxylate: synthesis, structure, and 'zeolite-like' behaviors. *J. Am. Chem. Soc.* 128 (33): 10745–10753.

21 Küsgens, P., Rose, M., Senkovska, I. et al. (2009). Characterization of metal-organic frameworks by water adsorption. *Microporous Mesoporous Mater.* 120 (3): 325–330.

22 Chakraborty, S., Hara, K., and Lai, P.T. (1999). New microhumidity field-effect transistor sensor in ppmv level. *Rev. Sci. Instrum.* 70 (2): 1565–1567.

23 Kitagawa, S. and Matsuda, R. (2007). Chemistry of coordination space of porous coordination polymers. *Coord. Chem. Rev.* 251 (21–24): 2490–2509.

24 Choi, S.Y., Lee, Y.-J., Park, Y.S. et al. (2000). Monolayer assembly of zeolite crystals on glass with fullerene as the covalent linker. *J. Am. Chem. Soc.* 122 (21): 5201–5209.

25 Prestipino, C., Regli, L., Vitillo, J.G. et al. (2006). Local structure of framework Cu(II) in HKUST-1 metallorganic framework: spectroscopic characterization upon activation and interaction with adsorbates. *Chem. Mater.* 18 (5): 1337–1346.

26 Lu, G., Farha, O.K., Kreno, L.E. et al. (2011). Fabrication of metal-organic framework-containing silica-colloidal crystals for vapor sensing. *Adv. Mater.* 23 (38): 4449–4452.

27 Seiyama, T., Fujiishi, K., Nagatani, M., and Kato, A. (1963). A new detector for gaseous components using zinc oxide thin films. *J. Soc. Chem. Ind. Japan* 66 (5): 652–655.

28 Lopez Aldaba, A., Lopez-Torres, D., Campo-Bescós, M. et al. (2018). Comparison between capacitive and microstructured optical fibre soil moisture sensors. *Appl. Sci.* 8 (9): 1499.

29 Elosua, C., Bariain, C., Luquin, A. et al. (2012). Optical fibre sensors array to identify beverages by their odor. *IEEE Sens. J.* 12 (11): 3156–3162.

30 Penza, M., Cassano, G., Aversa, P. et al. (2005). Carbon nanotube acoustic and optical sensors for volatile organic compound detection. *Nanotechnology* 16 (11): 2536–2547.

31 Wang, X., Tabakman, S.M., and Dai, H. (2008). Atomic layer deposition of metal oxides on pristine and functionalized graphene. *J. Am. Chem. Soc.* 130 (26): 8152–8153.

32 Schreiter, M., Gabl, R., Lerchner, J. et al. (2006). Functionalized pyroelectric sensors for gas detection. *Sens. Actuators B Chem.* 119 (1): 255–261.

33 Pijolat, C., Pupier, C., Sauvan, M. et al. (1999). Gas detection for automotive pollution control. *Sens. Actuators B Chem.* 59 (2): 195–202.

34 Lopez-Torres, D., Elosua, C., Villatoro, J. et al. (2017). Enhancing sensitivity of photonic crystal fibre interferometric humidity sensor by the thickness of SnO_2 thin films. *Sens. Actuators B Chem.* 251: 1059–1067.

35 Sakai, G., Matsunaga, N., Shimanoe, K., and Yamazoe, N. (2001). Theory of gas-diffusion controlled sensitivity for thin film semiconductor gas sensor. *Sens. Actuators B Chem.* 80: 125–131.

36 Parks, G.S. and Kelley, K.K. (1926). The heat capacities of some metallic oxides. *J. Phys. Chem.* 30 (1): 47–55.

37 Lopez-Torres, D., Lopez-Aldaba, A., Elosua, C. et al. (2018). Comparison between different structures of suspended-core microstructured optical fibres for volatiles sensing. *Sensors (Switzerland)* 18 (8): 2523.

38 Ascorbe, J., Corres, J.M., Matias, I.R., and Arregui, F. (2016). High sensitivity humidity sensor based on cladding-etched optical fibre and lossy mode resonances. *Sens. Actuators B Chem.* 233: 7–16.

39 Lopez Aldaba, A., Lopez-Torres, D., Elosua, C. et al. (2018). SnO_2-MOF-Fabry-Perot optical sensor for relative humidity measurements. *Sens. Actuators B Chem.* 257: 189–199.

40 Zamarreño, C.R., Hernaez, M., Del Villar, I. et al. (2010). Tunable humidity sensor based on ITO-coated optical fibre. *Sens. Actuators B Chem.* 146 (1): 414–417.

41 Corres, J.M., Ascorbe, J., Arregui, F.J., and Matias, I.R. (2013). Tunable electro-optic wavelength filter based on lossy-guided mode resonances. *Opt. Express* 21 (25): 31668.

42 Del Villar, I., Zamarreño, C.R., Hernaez, M. et al. (2015). Generation of surface plasmon resonance and Lossy mode resonance by thermal treatment of ITO thin-films. *Opt. Laser Technol.* 69: 1–7.

43 Usha, S.P., Shrivastav, A.M., and Gupta, B.D. (2018). Semiconductor metal oxide/polymer based fibre optic lossy mode resonance sensors: a contemporary study. *Opt. Fibre Technol.* 45: 146–166.

44 Tabassum, R. and Gupta, B.D. (2015). Fibre optic hydrogen gas sensor utilizing surface plasmon resonance and native defects of zinc oxide by palladium. *J. Opt. (United Kingdom)* 18 (1): 15004.

45 Zhang, Y., Peng, H., Qian, X. et al. (2017). Recent advancements in optical fibre hydrogen sensors. *Sens. Actuators B Chem.* 244: 393–416.

46 Silva, S.F., Coelho, L., Frazão, O. et al. (2012). A review of palladium-based fibre-optic sensors for molecular hydrogen detection. *IEEE Sens. J.* 12 (1): 93–102.

47 Novoselov, K.S., Geim, A.K., Morozov, S.V. et al. (2004). Electric field in atomically thin carbon films. *Science* 306 (5696): 666–669.

48 Bogue, R. (2014). Nanomaterials for gas sensing: a review of recent research. *Sens. Rev.* 34 (1): 1–8.

49 Hernaez, M., Mayes, A.G., and Melendi-Espina, S. (2018). Graphene oxide in lossy mode resonance-based optical fibre sensors for ethanol detection. *Sensors (Switzerland)* 18 (1): 1–10.

50 Rashleigh, S.C. (1984). Fibre optic sensors. *Electrochem. Soc. Ext. Abstr.* 84–2: 801.

51 Thapa, R., Knabe, K., Faheem, M. et al. (2006). Saturated absorption spectroscopy of acetylene gas inside large-core photonic bandgap fibre. *Opt. Lett.* 31 (16): 2489–2491.

52 Cubillas, A.M., Silva-Lopez, M., Lazaro, J.M. et al. (2008). Methane sensing at 1300 nm band with hollow-core photonic bandgap fibre as gas cell. *Electron. Lett.* 44 (6): 403–405.

53 Lopez-Torres, D., Elosua, C., Villatoro, J. et al. (2016). Photonic crystal fibre interferometer coated with a PAH/PAA nanolayer as humidity sensor. *Sens. Actuators B Chem.* 242: 1065–1072.

54 Kornaszewski, L., Gayraud, N., Stone, J.M. et al. (2007). Mid-infrared methane detection in a photonic bandgap fibre using a broadband optical parametric oscillator. *Opt. Express* 15 (18): 11219–11224.

55 Cubillas, A.M., Silva-Lopez, M., Lazaro, J.M. et al. (2007). Methane detection at 1670-nm band using a hollow-core photonic bandgap fibre and a multiline algorithm. *Opt. Express* 15 (26): 17570–17576.

56 Villatoro, J., Kreuzer, M.P., Jha, R. et al. (2009). Photonic crystal fibre interferometer for chemical vapor detection with high sensitivity. *Opt. Express* 17 (3): 1447.

57 Schirmer, M., Hussein, W.B., Jekle, M. et al. (2011). Impact of air humidity in industrial heating processes on selected quality attributes of bread rolls. *J. Food Eng.* 105 (4): 647–655.

58 Hernandez, F.U., Morgan, S.P., Hayes-Gill, B.R. et al. (2016). Characterization and use of a fibre optic sensor based on PAH/SiO$_2$ film for humidity sensing in ventilator care equipment. *IEEE Trans. Biomed. Eng.* 63 (9): 1985–1992.

59 Hu, D.J.J., Wong, R.Y.-N., and Shum, P.P. (2018). Photonic crystal fibre-based interferometric sensors. In: *Selected Topics on Optical Fibre Technologies and Applications* (eds. F. Xu and C. Mou). IntechOpen.

60 Zhao, Z. and Duan, Y. (2011). A low cost fibre-optic humidity sensor based on silica sol-gel film. *Sens. Actuators B Chem.* 160 (1): 1340–1345.

61 Li, Y., Chen, L., Harris, E., and Bao, X. (2010). Double-pass in-line fibre taper machZehnder interferometer sensor. *IEEE Photonics Technol. Lett.* 22 (23): 1750–1752.

62 Amaral, L.M.N., Frazão, O., Santos, J.L., and Lobo Ribeiro, A.B. (2011). Fibre-optic inclinometer based on taper Michelson interferometer. *IEEE Sens. J.* 11 (9): 1811–1814.

63 Petermann, A.B., Hildebrandt, T., Morgner, U. et al. (2018). Polymer based whispering gallery mode humidity sensor. *Sensors (Switzerland)* 18 (7): 1–9.

64 Bariain, C., Matias, I.R., Fdez-Valdivielso, C. et al. (2005). Optical fibre sensors based on vapochromic gold complexes for environmental applications. *Sens. Actuators B Chem.* 108 (1–2): 535–541. Special issue.

65 Elosua, C., Bariain, C., Matias, I.R. et al. (2009). Optical fibre sensing devices based on organic vapor indicators towards sensor array implementation. *Sens. Actuators B Chem.* 137 (1): 139–146.

66 Ricciardi, A., Crescitelli, A., Vaiano, P. et al. (2015). Lab-on-fibre technology: a new vision for chemical and biological sensing. *Analyst* 140 (24): 8068–8079.

67 Elosua, C., Bariain, C., Luquin, A. et al. (2011). Optimization of single mode fibre sensors to detect organic vapours. *Sens. Actuators B Chem.* 157 (2): 388–394.

68 Lee, J.S., Yoon, N.R., Kang, B.H. et al. (2014). Response characterization of a fibre optic sensor array with dye-coated planar waveguide for detection of volatile organic compounds. *Sensors (Switzerland)* 14 (7): 11659–11671.

69 Elosua, C., Arregui, F.J., Zamarreño, C.R. et al. (2012). Volatile organic compounds optical fibre sensor based on lossy mode resonances. *Sens. Actuators B Chem.* 173: 523–529.

70 Pulido, C. and Esteban, O. (2011). Multiple fluorescence sensing with side-pumped tapered polymer fibre. *Sens. Actuators B Chem.* 157 (2): 560–564.

71 Elosua, C., De Acha, N., Hernaez, M. et al. (2015). Layer-by-Layer assembly of a water-insoluble platinum complex for optical fibre oxygen sensors. *Sens. Actuators B Chem.* 207, no. Part A: 683–689.

72 de Acha, N., Elosúa, C., Matias, I.R., and Arregui, F.J. (2017). Enhancement of luminescence-based optical fibre oxygen sensors by tuning the distance between fluorophore layers. *Sens. Actuators B Chem.* 248: 836–847.

73 Medina-Rodríguez, S., De La Torre-Vega, A., Sainz-Gonzalo, F.J. et al. (2014). Improved multifrequency phase-modulation method that uses rectangular-wave signals to increase accuracy in luminescence spectroscopy. *Anal. Chem.* 86 (11): 5245–5256.

74 Alkasab, T.K., Bozza, T.C., Cleland, T.A. et al. (1999). Characterizing complex chemosensors: information-theoretic analysis of olfactory systems. *Trends Neurosci.* 22 (3): 102–108.

75 White, J., Dickinson, T.A., Walt, D.R., and Kauer, J.S. (1998). An olfactory neuronal network for vapor recognition in an artificial nose. *Biol. Cybern.* 78 (4): 245–251.

76 Baldwin, E.A., Scott, J.W., Shewmaker, C.K., and Schuch, W. (2000). Flavor trivia and tomato aroma: biochemistry and possible mechanisms for control of important aroma components. *HortScience* 35 (6): 1013–1022.

77 Xiaobo, Z. and Jiewen, Z. (2008). Comparative analyses of apple aroma by a tin-oxide gas sensor array device and GC/MS. *Food Chem.* 107: 120–128.

78 Garrigues, S., Talou, T., and Nesa, D. (2004). Comparative study between gas sensors arrays device, sensory evaluation and GC/MS analysis for QC in automotive industry. *Sens. Actuators B Chem.* 103 (1–2): 55–68.

79 Bredie, W.L.P., Lindinger, C., Hall, G. et al. (2006). *Methods for artificial perception: can machine replace man? Dev. Food Sci.* 43: 617–618.

80 Brezmes, J., Llobet, E., Vilanova, X. et al. (2000). Fruit ripeness monitoring using an Electronic Nose. *Sens. Actuators B Chem.* 69 (3): 223–229.

81 Ma, J., Kos, A., Bock, W.J. et al. (2010). Lab-on-a-fibre: building a fibre-optic sensing platform for low-cost and high-performance trace vapor TNT detection. 4th European Workshop on Optical Fibre Sensors, Porto, Portugal (8 September 2010–10 September). Article number 76531E.

82 Giallorenzi, T.G., Bucaro, J.A., Dandridge, A. et al. (1982). Optical fibre sensor technology. *IEEE J. Quantum Electron.* 18 (4): 626–665.

83 O'Farrell, M., Lewis, E., Flanagan, C. et al. (2005). Combining principal component analysis with an artificial neural network to perform online quality assessment of food as it cooks in a large-scale industrial oven. *Sens. Actuators B Chem.* 107 (1): 104–112. Special issue.

84 O'Keeffe, S., Fitzpatrick, C., and Lewis, E. (2007). An optical fibre based ultra violet and visible absorption spectroscopy system for ozone concentration monitoring. *Sens. Actuators B Chem.* 125 (2): 372–378.

85 Capone, S., Epifani, M., Quaranta, F. et al. (2001). Monitoring of rancidity of milk by means of an electronic nose and a dynamic PCA analysis. *Sens. Actuators B Chem.* 78 (1–3): 174–179.

86 Luo, D., Hosseini, H.G., and Stewart, J.R. (2004). Application of ANN with extracted parameters from an electronic nose in cigarette brand identification. *Sens. Actuators B Chem.* 99 (2–3): 253–257.

87 Bauer, J. and Wermter, S. (2013). Self-organized neural learning of statistical inference from high-dimensional data. *IJCAI International Joint Conference on Artificial Intelligence* Beijing, China (3–9 August 2013), pp. 1226–1232.

12

Interaction of Light with Matter in Optical Fibre Sensors

A Biomedical Engineering Perspective

Sillas Hadjiloucas

Department of Biomedical Engineering, University of Reading, Reading, UK

12.1   Introduction

This chapter discusses the main ways light interacts with matter from a biomedical engineering perspective. Basic concepts such as photon energy and physicochemical processes are explained, covering interactions of photons with atoms and molecules (emphasis here is given on photobiology and the multiplicity of the different de-excitation processes). In addition, elastic and inelastic scattering and their application to optical fibre sensing are discussed. Recent developments in integration of optical fibres with other devices are mentioned. These provide new engineering opportunities for innovative interactions of light with matter. Methods to enhance the signal to noise ratio in experiments are proposed, especially localised field enhancement for improved sensitivity. Finally, the intriguing prospect of molecular control using femtosecond pulse shaping to control rather than just observe interactions of light with matter is discussed.

12.2   Energy Content in Light and Its Effect in Chemical Processes

The wavelength range of most interest to optical fibre sensing schemes is between the ultraviolet (UV) and the infrared (IR). The UV is the range of the spectrum with wavelength below 400 nm, and typically around 254 nm, which corresponds

Optical Fibre Sensors: Fundamentals for Development of Optimized Devices, First Edition.
Edited by Ignacio Del Villar and Ignacio R. Matias.
© 2021 The Institute of Electrical and Electronics Engineers, Inc.
Published 2021 by John Wiley & Sons, Inc.

to a frequency of 11.8×10^{14} Hz and energy of 471 kJ/mol. IR on the other hand, has a wavelength above 740 nm, with a representative wavelength of 1.4 μm, frequency of 2.14×10^{14} Hz and energy of 85 kJ/mol. In between the UV and IR, in the visible part of the spectrum, the violet colour has a representative wavelength of 410 nm, frequency of 7.41×10^{14} Hz and energy of 292 kJ/mol. Blue has a representative wavelength of 460 nm, frequency of 6.52×10^{14} Hz and an energy of 260 kJ/mol, the green has a representative wavelength of 520 nm, frequency of 5.77×10^{14} Hz and energy of 230 kJ/mol, and the yellow colour has a representative wavelength of 570 nm, frequency of 5.26×10^{14} Hz and energy of 210 kJ/mol. Near the end of the visible spectrum, are orange and red. Orange has a representative wavelengths of 620 nm with a frequency of 4.84×10^{14} Hz and 4.41×10^{14} Hz and energies of 193 kJ/mol, and the red colour has a representative wavelength of 680 nm with a frequency of 4.41×10^{14} Hz and an energy of 176 kJ/mol. In nature, due to atmospheric absorption, there is very limited radiation at wavelengths below 150 nm. Longer wavelengths can still be supported by specialty fibres such as chalcogenide glass fibres that can support propagation in the 3–5 and the 8–12 μm transmission windows.

Following Planck's proposal in 1901 that light energy carried by a photon is quantized, the light energy carried by a single photon is given from $E = h\omega = hc/\lambda_{vacuum}$, where h is Planck's constant, ω is the frequency, c is the speed of light, and λ_{vacuum} is the wavelength of light in vacuum. From the above expression, a photon of light has an energy directly proportional to its frequency and inversely proportional to its wavelength in vacuum. Since frequency has units of Hz, Planck's constant has units of energy × time, e.g. 6.626×10^{-34} J s, and hc is 1240 eV nm, so the energy per photon of red light is 1240 eV nm/680 nm or 1.82 eV. In chemical and biological/biomedical sensing, instead of energy per photon, a more interesting quantity is the energy per Avogadro's number N (6.022×10^{23}) also known as the energy on a mole basis $E = Nh\omega = Nhc/\lambda_{vacuum}$. For the case of red light, which has a frequency of 4.41×10^{14} Hz, the energy per mole of 680 nm photons is $(6.022 \times 10^{23}\ \text{mol}^{-1})(6.626 \times 10^{-34}\ \text{J s})(4.41 \times 10^{14}\ \text{s}^{-1}) = 176$ kJ/mol. Since 1 eV equals 1.602×10^{-19} J, it is possible to perform the following interconversion: 1 eV/molecule = 96.55 kJ/mol = 23.06 kcal/mol.

One has to place these quantities in some perspective. For example, ATP hydrolysis is around 40–50 kJ/mol, so red light can hydrolyse it. Furthermore, the carbon–carbon bond energy is 348 kJ/mol, and the oxygen–carbon bond is 463 kJ/mol, so UV light can easily break these bonds. The photoelectric effect that describes the ejection of an electron from the surface of a metal is another example where it demonstrates that light energy is not as important as the energy of its photons in order for the phenomenon to occur. Therefore, light provided at the tip of a fibre during a measurement process have some minimum energy facilitating a certain biochemical reaction to occur thus interfering with the process being measured. From the above definition of molar energy it also follows that simple measurement of the total light energy does not indicate how many photons are involved or what their individual energies are unless the wavelength distribution is determined (e.g. spectroscopically).

At this point, however, it must also be clarified that absorption of radiation by an atom or molecule can also be produced by collisions resulting from the random thermal motion of molecules. The higher the temperature, the higher the average kinetic energy, so there is a greater probability that the molecule will achieve a greater energetic state by collisions. The Boltzmann energy distribution describes the frequency with which specific kinetic energies are possessed by molecules at equilibrium at a given temperature $n(E) = n_{total}e^{-E/kT}$ (on a molecule basis) where $n(E)$ is the number of molecules possessing an energy of at least E or more out of the total number n_{total} of molecules, T is Kelvin temperature and k is Boltzmann's constant. One has to clarify, however, that very often in many experiments using optical fibres, we are not interested in the energy per molecule but the energy per mole. To change the Boltzmann distribution from a molecule to a mole basis, one has to multiply Boltzmann's constant [energy/(molecule K)] by Avogadro's number (molecules per mole), which gives us the gas constant, R, [energy/(mole K)] so that $R = kN$. It follows that $n(E) = n_{total}e^{-E/RT}$ (on a mole basis).

Now, if the light energy from the tip of a fibre is sufficient so that there is an accompanying increase of the temperature in the medium, this can also result in a facilitated diffusion process across membranes. There is a minimum kinetic energy U_{min} for molecules to diffuse through a membrane barrier and the velocities in a particular direction are proportional to $\sqrt{T}e^{-U_{min}/RT}$. Hence, it follows that one can define a temperature coefficient Q_{10} for a process to occur, once the surrounding environment has increased by 10 °C, and this is given from

$$Q_{10} = \sqrt{\frac{T + 10}{T}}e^{10U_{min}/[RT(T + 10)]} \tag{12.1}$$

For example, if charged solutes were to passively cross a membrane (U_{min} = 50 kJ/mol or 0.52 eV/mol), at 20 °C, the following temperature coefficient could be calculated:

$$Q_{10} = \sqrt{\frac{303\ K}{293\ K}}e^{(10\ K)(50 \times 10^3\ J/mol)(8.3143\ J/mol/K)(293\ K)(303\ K)} = 2.01 \tag{12.2}$$

which implies a doubling of the passive uptake process with only a 10 °C increase in temperature. Interpreting the above equation, for passive processes where there is no energy barrier to overcome, where U_{min} = 0, the Q_{10} is near unity. In enzymatic reactions observed by optical fibre sensors, where active transport of a solute is involved, it is usually the case that $Q_{10} \geq 2$, denoting that these processes are sensitive to temperature. The above simple calculation clarifies that energy unintentionally transferred from the tip of a fibre in a biomedical sensing experiment can lead to some wavelength non-specific alterations of the environment under observation. This is by increasing the temperature and rates of reactions therefore affecting diffusion processes.

The above discussion leads to the concept of activation energy, which refers to the minimum amount of energy necessary for some reaction to take place. By representing the activation energy per mole by some parameter A, the rate constant κ_R varies with temperature as follows: $\kappa_R = Be^{-A/RT}$ where B is a constant. This is the well-known Arrhenius equation and has the distinct advantage that it enables a plot of the logarithm of the rate constant associated with a reaction against $1/T$ to be equal to $\ln(B) - A/RT$. A particularly useful feature of the Arrhenius equation described above is that the term $(-A/R)$, which is the slope, can be used to determine the activation energy. Enzymes encountered in biochemical reactions reduce the activation energy and permit many more molecules to have enough energy to get over the energy barrier separating the reactants from the products.

For a reaction to take place, the fraction of atoms or molecules that have a kinetic energy above a certain value is important. Based on the kinetic theory of gases, the average kinetic energy of translational motion in a gas phase is $3RT/2 = 3.72\,\text{kJ/mol}$ at $25\,°\text{C}$. As a result the Boltzmann factor becomes $e^{-(3RT/2)/(RT)} = e^{-1.5} = 0.22$ or 22% in other words; 22% of the molecules have kinetic energies that are higher than the average. Conversely, in order to find at what temperature 44% of the molecules would have such kinetic energies, one can solve the following equation:

$$T = \frac{1}{-\ln(0.44)}\frac{E}{R} = \frac{1}{0.82}\frac{(3720\,\text{J/mol})}{(8.3143\,\text{J/mol/K})} = 546\,\text{K} = 273\,°\text{C} \qquad (12.3)$$

From the above discussion, it becomes clear that raising the temperature increases the fraction of molecules that have higher energies.

Following the discussion of the energy content in blue light at 460 nm, which corresponds to a kinetic energy of 260 kJ/mol, the fraction of molecules possessing at least this energy is $e^{-E/RT} = e^{-(260\,\text{kJ/mol})(2.48\,\text{kJ/mol})} = 3 \times 10^{-46}$ at 25 °C, which is a very small number, making a spontaneous activation energy transition a very unlikely event. So it is almost impossible for a particular molecule to gain this amount of energy by means of thermal collisions. As a result, absorption of blue light can lead to the attainment of very improbable energetic states that would not be encountered naturally.

12.3 Relevance of Wien's Law to Physicochemical Processes

When radiometry is performed, the wavelength distribution of a hot object is given by Planck's radiation distribution formula, which states that the relative photon flux density per unit wavelength interval is proportional to $\lambda^{-4}/(e^{hc/\lambda kT} - 1)$, where

T is the temperature of the radiation source observed. The above expression is applicable under the assumption that the object under observation is a perfectly efficient emitter, in other words a black-body emitter. As the sun's temperature is assumed to be 5800 K and tungsten lamps are at 2900 K (they have very similar wavelength distribution, only shifted to longer IR wavelengths), radiometry performed with optical fibres coupled to a tungsten lamp can be referenced to that temperature.

Knowing the temperature of a black-body radiator, it is possible to predict at what temperature the radiation emitted from it will be maximal. A differentiation of Planck's distribution formula with respect to wavelength when setting the derivative to zero produces the desired result. This is known as Wien's displacement law: $\lambda_{max}T = 3.6 \times 10^6$ nm K (photon basis), where λ_{max} is the wavelength position for maximum photon flux density and T is the surface temperature of the source. Using the above values for the sun temperature and tungsten lamp temperature, λ_{max} is 620 nm for the sun and 1240 nm for the tungsten lamp, respectively. Using Wien's displacement law, it follows that objects at a standard temperature of 25 °C (298 K) observed radiometrically have a maximum photon flux density at 12 µm, which is far into the longer IR wavelengths. At this point an important clarification needs to be made. In an optical fibre-based experiment, where fibres are used to excite some pigment molecules, as is the case in photobiology investigations, our interest would be in the spectral distribution of photons per unit wavelength interval, whereas in applications where we are interested in energy gain per interval, e.g. when artificially illuminating plant leaves with optical fibres coupled to LEDs to photosynthesize in a controlled environment for intensive agriculture, we are usually more interested in the spectral distribution of energy per unit spectral interval. So Planck's curve should be expressed as relative energy flux density per unit wavelength interval, which is given from $\lambda^{-5}/(e^{hc/\lambda kT} - 1)$ and Wien's displacement law for maximum energy output is $\lambda_{max}T = 2.9 \times 10^6$ nm K (energy basis).

12.4 Absorption of Light by Molecules

When discussing the absorption of light by molecules, there are two well-established principles: firstly, only light that is actually absorbed can produce a chemical change (this is known as the Grotthus–Draper law of photochemistry), and secondly each absorbed photon activates only one molecule (this is known as the Stark–Einstein law). As light is absorbed by electrons in a molecule, these electrons can either jump to a higher energy orbit, or they can move more rapidly than they did before the photons arrived. Light of the appropriate wavelength can cause a (ground) electron to jump from one possible energetic state to another (excited) state. For this to occur, the energy of the photon $E = Nh\omega = Nhc/\lambda_{\text{vacuum}}$ must equal the difference in energy between some allowed excited state of the atom or molecule and the initial state. When light from an optical fibre interacts with an electron, there is a periodic driving force from the electromagnetic wave acting on the electron, which induces it to move in a particular direction in space. At a particular frequency, the electron is in resonance and absorbs the provided radiation, so the resonating electron behaves as an electric dipole as it is forced to move in an oscillatory motion. Since the electric dipoles that can be induced in a particular molecule are characteristic of that molecule, each molecular species has a unique absorption spectrum. For gases, one generally observes clear resonances, so spectra appear as well-defined narrow absorption lines in the spectrum, whereas because of additional frictional forces and other higher-order interactions between electrons in liquids and solids, the absorption lines tend to broaden, and we instead observe absorption bands.

Since the probability that light will be absorbed depends also on the relative orientation of the electromagnetic field with respect to the possible induced oscillations of the electrons in the molecule, a requirement for absorption is that the electric field vector associated with the light must have a component parallel to the direction of some potential electric dipole in the molecule so that an electron can be induced to oscillate. It turns out actually that the probability for absorption is proportional to the square of the cosine of the angle between the electric field vector of light and the direction of the induced electric dipole in the molecule.

12.5 The Role of Electron Spin and State Multiplicity in Spectroscopy

Light absorption is also affected by the spin of electrons in an atom or molecule. Electrons spin around their axis, and such rotation has an angular momentum. All electrons have the same magnitude of spin, but spin is a vector quantity and thus can have different directions in space. Electron spin can have two orientations: it can either be aligned parallel or antiparallel to the local magnetic field. This local magnetic field aligns the spins and is always present because of moving charges in the nucleus as well as because of the motion of the electrons.

The unit of angular momentum and spin has dimensions of energy × time. As a matter of convenience, the spin of electrons is expressed in units of $h/(2\pi)$ (J s). In these units, the projection of the spin for a single electron along the magnetic field is either $+1/2$ when the spin is parallel to the local magnetic field or $-1/2$ when the spin in antiparallel to it. Furthermore, for an atom or molecule, there is a net spin that is associated with the vector sum of the spins of all the electrons; the magnitude of this net spin, in most textbooks, has the symbol S.

When discussing the spectroscopic properties of molecules, the concept of spin multiplicity is often utilized. An electronic state has a spin multiplicity defined as $2S + 1$. If the net spin S of an atom or molecule is equal to 0 (indicating that all spin projections of all the electrons along the magnetic field cancel each other), $2S + 1 = 1$, and the state is called a singlet. If $S = 1$, then $2S + 1 = 3$, and the state is called a triplet. Singlets and triplets are the most important spin multiplicities encountered in chemical physics. These states are symbolized in most textbooks by S and T, respectively (Figure 12.1). When there is an odd number of electrons in a molecule, $S = 1/2$ so $2S + 1 = 2$. This state is called a doublet.

Furthermore, electrons are confined to certain allowed regions in space referred to as orbitals (these are named in order of proximity to the nucleus as s, p, d, and f orbitals). When an electron with a particular spin direction is in a given atomic orbital, it is not possible for a second electron to have the same spin direction in the same orbital (Pauli's exclusion principle). When a molecule has all of its electrons paired in orbitals and their spins are in opposite directions, the net spin of the molecule is 0, and the molecule is in a singlet state (ground or unexcited state).

From the above discussion, in optical fibre based experiments, when an electron is excited by a photon to an unoccupied orbital, the spins of the two electrons in the different orbitals may be in opposite direction (giving rise to an excited singlet state). Alternatively, the two electrons can have spins in the same direction (giving rise to a triplet state in the same direction but in different orbitals). Friedrich Hund originally postulated that the level with the greatest spin multiplicity has the lowest energy. In other words, an excited triplet state has a lower energy than the correspondent excited singlet state. Such postulation was later on confirmed using quantum mechanical calculations.

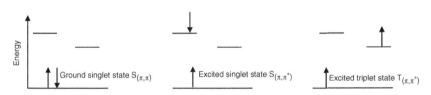

Figure 12.1 The influence of light on a pair of electrons in a molecular orbital. Arrows indicate the directions of electron spins with respect to the local magnetic field.

12.6 Molecular Orbitals, Bond Conjugation, and Photoisomerization

Electrons moving in different orbits around an atom have different probability distributions of being present in various regions of space. In other words, the probability of finding an electron is greatest along a particular axis through the nucleus and is exceedingly small at the nucleus itself. Since molecules may contain several atoms and electrons can be shared between atoms, the spatial localization of electrons in molecules is the result of a superposition (linear combination) of the individual spatial distributions of atomic orbitals centred around the nuclei involved. As a result of the above description, the probability of finding electrons in various regions of space in molecules can be accurately predicted using quantum mechanical calculations on the basis of individual probability distributions of electrons in their atomic orbitals.

In the case where electrons are shared between nuclei, the probability distributions are delocalized. This sharing of electrons is responsible for the chemical bonds that prevent molecules from separating into their constituent atoms. These electrons are more sensitive to perturbations and as a result tend to be more involved in light absorption processes. In an analogy of an s atomic orbital, the lowest energy molecular orbital is called an σ orbital. Furthermore, non-bonding electron pairs such as the ones found in oxygen or nitrogen occur in non-bonding orbitals, also known as n orbitals. The latter do not take part in the bonding between nuclei. The molecular equivalent of p electrons in atoms is the π molecular orbitals. These are associated with delocalized electrons in bonds joining two or more atoms.

When light from the tip of an optical fibre is used to initiate a photochemical reaction, it is the absorption of photons by the π electrons that is responsible for the spectroscopic properties observed. Excitation of the π electrons results in the electron moving into a π^* orbital (the * refers to the excited high energy molecular orbital). Because π orbitals are delocalized between two nuclei, the sharing of electrons in the π orbital assists in the joining of atoms. As a result, this type of molecular orbital is referred to as the bonding orbital. On the other hand, electrons in the π^* orbital tend to decrease bonding between atoms in a molecule (they have an anti-bonding function). The higher energy electronic state of π^* orbitals results in a more unstable molecular state. In many molecules of biological importance, π^* orbitals have an energy that is only a few hundred kJ/mol higher than the corresponding π orbitals. As a result, visible light has sufficient energy to excite π electrons into π^* orbitals. The decrease in energy separation between the π and the π^* orbitals as the number of double bonds in conjugation increases is responsible for the commonly observed shift of the peaks of the absorption bands towards longer wavelengths. The maximum absorption coefficient is approximately proportional to the number of double bonds in the conjugated system being observed.

Another physicochemical process that can result of the absorption from light energy is photoisomerization. It is possible to have (i) cis–trans isomerization of a double bond, (ii) a double bond shift, or (iii) a molecular rearrangement, such as ring cleavage or formation. All these phenomena are often encountered in carbon–carbon bonds. Cis–trans photoisomerization is a result of different spatial distributions of electrons in π and π^* orbitals and enables the rotation of molecular chains around a double bond. Normally, a double bond has two π electrons, so a light absorption event that leads to the excitation of one π electron to a π^* orbital causes an anti-bonding process. As a result of the photoisomerization process, the new electronic configuration permits rotation around the carbon–carbon axis. This explains how the two isomers can have very different physicochemical properties.

The n electrons mentioned earlier tend to interact strongly with water by participating in hydrogen bonding. Since hydrogen bonds are weaker than other types of bonds, the energy level of n electrons is considerably lower in an aqueous environment than in an organic solvent. In addition, the energy of π^* electrons is also lowered by water but to a lesser extent than for n electrons. Contrary to this, π orbitals are the least affected by a solvent. From the above description, it follows that a transition from an n to a π^* orbital takes more energy in water than in an organic solvent, whereas π-to-π^* transitions take less energy in an aqueous environment than in an organic one (Figure 12.2).

12.7 De-excitation Processes Through Competing Pathways: Their Effect on Lifetimes and Quantum Yield

An excitation process from the absorption of light is normally followed by a de-excitation process. The time taken for the absorption of a photon is the time required for one cycle of the light wave to propagate through an electron. This

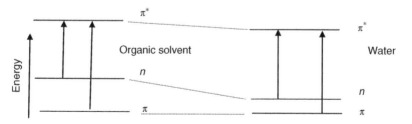

Figure 12.2 The effect of a solvent on the energies of n, π, and π^* orbitals. An n-to-π^* transition requires more energy in water than in an organic solvent, whereas a π-to-π^* transition takes less energy. *Source:* Redrawn from [1]. Reproduced with permission of IEEE.

time is approximately equal to the reciprocal of the frequency of the light (the time required for one complete oscillation of the electromagnetic field). For blue light, at 460 nm, the time for one cycle is $1/(6.52 \times 10^{14}\,\text{s}^{-1})$ or $1.5 \times 10^{-15}\,\text{s}$. The time-scales associated with the de-excitation process are expressed in lifetimes. A lifetime is defined as the time required for the number of molecules in a given excited state to decrease to $1/e$ or 37% of the initial number. Lifetimes are particularly useful in describing first-order processes because the initial species in such processes decay exponentially with time.

In fluorescence light imaging microscopy (FLIM) studies, a particularly useful quantity is a half-time, which is the time necessary for the number of species in a given state to decrease by 50% (for a single exponential decay, $\ln(2) = 0.693 \times$ lifetime). Fluorescence describes the process where electromagnetic radiation is emitted from a molecule when it transitions from an excited singlet state to a singlet ground state. For most organic molecules, fluorescence lifetimes range from 10^{-9} to $10^{-6}\,\text{s}$. De-excitation of an excited state often occurs without the emission of any radiation (radiationless de-excitation). In this case, the energy of the absorbed photon is converted to heat. The radiationless transition is a rapid process (of the order of the $10^{-12}\,\text{s}$). The de-excitation of a triplet state $T_{(\pi,\pi*)}$ to a singlet state $S_{(\pi,\pi)}$ can be radiationless or can be by radiation through phosphorescence (typical lifetimes from 10^{-3} to $10\,\text{s}$). The reason for the long de-excitation lifetimes observed in phosphorescence is that this process requires both the transition from one electronic state to another and the reversal of the electron spin. These two events, however, are not time coincident.

Finally, another molecular de-excitation process is through delayed fluorescence. A molecule, in the relatively long-lived excited triplet state, $T_{(\pi,\pi*)}$, can sometimes be supplied with sufficient thermal energy by collisions to put it into a higher energy excited singlet state $S_{(\pi,\pi*)}$. Subsequent radiation, as the molecule goes from this excited singlet state to the ground state $S_{(\pi,\pi)}$, has the typical characteristics of fluorescence but is considerably delayed following the light absorption event because the excitation has spent some time in the $T_{(\pi,\pi*)}$ state.

From the above discussion, it becomes clear that excited states have a well-defined energy and specific lifetimes. But there are several other processes that also compete for the de-excitation of that state in addition to fluorescence, phosphorescence, and radiationless transitions. For example, the excitation energy can also be transferred to another molecule putting it into an excited state, while the originally excited molecule returns to its ground state. Alternatively, an excited electron may leave the molecule that absorbed the photon. As a result, an absorbed photon can also be reradiated as electromagnetic energy $h\omega$ as follows:

$$S_{(\pi,\pi*)} \xrightarrow{\kappa_1} S_{(\pi,\pi)} + h\omega \tag{12.4}$$

This fluorescence is a first-order process with a rate constant κ_1 and is typically of the order of 10^{-8} s. It must be noted, however, that the stated lifetime associated with this process corresponds to an upper time limit because of competing de-excitation pathways. For example, when there are radiationless transitions, two different states can be reached:

$$S_{(\pi,\pi^*)} \xrightarrow{\kappa_2} S_{(\pi,\pi)} + \text{heat} \tag{12.5}$$

$$S_{(\pi,\pi^*)} \xrightarrow{\kappa_3} T_{(\pi,\pi^*)} + \text{heat} \tag{12.6}$$

Radiationless transitions cause excitation of certain vibrational modes of other pairs of atoms within the molecule (Franck–Condon processes). In this case, excited molecules can return to their ground state, and all the radiant energy of the absorbed light is converted into kinetic energy of the surrounding molecules. Radiationless transitions also obey first-order kinetics. The lowest excited triplet state usually lasts 10^4–10^8 times longer than $S_{(\pi,\pi^*)}$. This increases the probability for more intermolecular collisions. Because each collision increases the opportunity for a given reaction to occur, the $T_{(\pi,\pi^*)}$ state is of significant importance in photobiology.

In addition, it is possible that the absorption of light can lead to a photochemical reaction, which is initiated by a different molecule than the one that actually absorbed the photon. In other words, an electronic excitation can also be transferred between molecules. For example, the excitation energy of $S_{1(\pi,\pi^*)}$ may be transferred to second molecule, which can be in a ground state $S_{2(\pi,\pi)}$:

$$S_{1(\pi,\pi^*)} + S_{2(\pi,\pi)} \xrightarrow{\kappa_4} S_{1(\pi,\pi)} + S_{2(\pi,\pi^*)} \tag{12.7}$$

In the above process, the second molecule becomes excited, whereas the first molecule returns to its ground state. The transfer of electronic excitation from one molecule to another is commonly encountered in pigments associated with biological processes.

Finally, there is a further de-excitation process where the $S_{(\pi,\pi^*)}$ state in a photochemical reaction donates a π^* electron to a suitable acceptor. The electron removed from the $S_{(\pi,\pi^*)}$ state is replaced by another one from a donor compound $D_{(\pi)}$ (which is a doublet, because one of the π orbitals contains an unpaired electron before ending up in its original ground state $S_{(\pi,\pi)}$):

$$S_{(\pi,\pi^*)} \xrightarrow{\kappa_5} D_{(\pi)} + e^* \tag{12.8}$$

From the above description, there are multiple competing pathways for the de-excitation of an excited single state $S_{(\pi,\pi^*)}$. The dynamics of the de-excitation process can be described by the following expression:

$$-\frac{dS_{(\pi,\pi^*)}}{dt} = (\kappa_1 + \kappa_2 + \kappa_3 + \kappa_4 + \kappa_5)S_{(\pi,\pi^*)} \tag{12.9}$$

Solving the above differential equation can provide the number of molecules in the excited singlet state at any subsequent time t on the basis of the initial number of molecules in that singlet state at $t = 0$:

$$S_{(\pi,\pi^*)}\big|_t = S_{(\pi,\pi^*)}\big|_0 e^{-(\kappa_1 + \kappa_2 + \kappa_3 + \kappa_4 + \kappa_5)t} \tag{12.10}$$

Similarly to the above expression, the time required for the number of excited molecules to decrease to $1/e$ of the initial value can also be calculated:

$$S_{(\pi,\pi^*)}\big|_\tau = e^{-1}S_{(\pi,\pi^*)}\big|_0 = S_{(\pi,\pi^*)}\big|_0 e^{-(\kappa_1 + \kappa_2 + \kappa_3 + \kappa_4 + \kappa_5)t} \tag{12.11}$$

An alternative way to describe the collective effect of all these rate constants is through the following expression:

$$(\kappa_1 + \kappa_2 + \kappa_3 + \kappa_4 + \kappa_5)\tau = 1 \tag{12.12}$$

which shows that the greater the rate constant for any particular de-excitation process, the shorter will be the lifetime τ of the excited state. When there are j de-excitation processes with first-order kinetics, one can write

$$\frac{1}{\tau} = \sum_j \kappa_j = \sum_j \frac{1}{\tau_j} \tag{12.13}$$

which leads to

$$S_{(\pi,\pi^*)}\big|_t = S_{(\pi,\pi^*)}\big|_0 e^{-t/\tau} \tag{12.14}$$

When multiple de-excitation processes are possible, τ is smaller than the lifetime of any individual competing reaction. A parameter of interest when observing fluorescent phenomena is that of quantum efficiency Φ_i, which represents the fraction of molecules in some excited state that will decay through all possible competing pathways:

$$\Phi_i = \frac{\text{number of molecules using } i\text{th de-excitation reaction}}{\text{number of excited molecules}} = \frac{\tau}{\tau_i} \tag{12.15}$$

The shorter the lifetime for a particular de-excitation pathway, the larger the fraction of molecules using that pathway will be, and the higher its quantum yield (the fraction of molecules that will use that pathway per photon absorbed by the system). In the engineering discipline, it is convenient to use Laplace transforms to solve the differential equations associated with de-excitation processes. System identification algorithms can be used to infer individual de-excitation lifetimes.

Finally, another relevant phenomenon that needs to be mentioned is that of quenching. This refers to a decreased observed fluorescence intensity, which is encountered in excited state reactions, during energy transfer in a reaction, when there is a complex formation or due to collisions. From a biomedical engineering perspective, Förster resonance energy transfer, Dexter electron transfer and excimer (short-lived excited dimeric state) or exciplex (short-lived excited state of a heterodimeric complex) formation are the most important mechanisms responsible for dynamic quenching that may be observed using optical fibre sensors. Förster resonance energy transfer (FRET or FET) is observed while the donor is in the excited state. It is a dipole-dipole interaction between the transition dipoles of the donor and acceptor and is extremely dependent on the donor-acceptor distance, R, falling off at a rate of $1/R^6$. FRET also depends on the donor-acceptor spectral overlap as well as the relative orientation of the donor and acceptor transition dipole moments. It typically occurs over very short distances up to 100 Å. Dexter electron transfer is a collisional energy transfer at short range that falls off exponentially with distance, and depends on the spatial overlap of donor and quencher molecular orbitals. Dexter electron transfer is encountered in interactions between a dye and a solvent when hydrogen bonds are formed between them. In both Förster and Dexter energy transfer, the shapes of the absorption and fluorescence spectra of the dyes remain unchanged. Excimers are associated with quenched behaviour of mainly diatomic molecules that would not naturally bond if they were both in the ground state. Energetically, the state of an excimer is lower than its excited monomer counterpart. Excimer lifetime is of the order of nanoseconds. The energy gap between the highest occupied molecular orbital (HOMO) and the lowest unoccupied molecular orbital (LUMO) is known as the HOMO-LUMO gap. If the molecule absorbs light whose energy is equal to this gap, an electron in the HOMO may be excited to the LUMO state. In the case of studying excimers with optical fibre sensors, at least one of the two molecules has completely filled valence shell by electrons and the result of charge-transfer is observed. Heterodimeric diatomic complexes are common in the construction of excimer lasers, which are used to produce higher energy (UV) radiation which is necessary in many biomedical applications (fluorescence lifetime imaging, DNA sequencing, protein analysis using matrix assisted laser desorption ionisation, etc). These lasers take advantage of the fact that excimer components have attractive interactions in the excited state and repulsive interactions in the ground state. In addition to the study of heterodimeric diatomic complexes, laser probing of charge transfer may also be used in sensor applications to study delocalised electrons in Rydberg matter clusters, where bonding is caused by delocalisation of the high-energy electrons to form an overall lower energy state. This is an emergent research topic of significant interest to the semiconductor and quantum optics communities because the observed lifetime of these states can exceed many seconds.

12.8 Energy Level Diagrams and Vibrational Sublevels

The representation in Figure 12.1 is an oversimplification of what happens in complex molecules because each electronic energy level is normally split into various discreet sublevels that vary in energy. Due to the existence of these vibrational sublevels, specific electronic states arise from atomic oscillations. The process of absorption of a photon enables the transition of that molecule from its lowest vibrational sublevel to a vibrational sublevel of an excited state.

The various bands observed in absorption spectra and fluorescence emission spectra are based on the Franck–Condon principle, which states that the nuclei change neither their separation nor their velocity during those transitions for which the absorption of a photon is more probable. A way to interpret the above principle is to assume that the light absorption event is so rapid that the nuclei do not have a chance to move during such short timescales. As stated earlier, the absorption of a photon required 10^{-15} s, whereas the period for one nuclear vibration along an oscillation requires more than 10^{-13} s.

Since nuclei can transfer energy to one another, the time for one cycle of a nuclear vibration is an estimate of the time in which excess vibrational energy can be dissipated as heat by the interaction of other nuclei. For many molecules the vibrational sublevels of both the ground state and the excited state are 10–20 kJ/mol apart, while the amount of light absorbed is maximal at a particular wavelength that corresponds to the most probable transition predicted by the Franck–Condon principle.

In an absorption process, transitions are most likely to originate from the point of maximum probability in the particular vibrational level of the lower state of the transition, and the relative intensities of transitions to vibrational levels of the upper state will be dependent on the probability of finding the upper state with that internuclear separation. Figure 12.3 illustrates the two different situations that arise: (a) when the upper and lower curves are similar in shape and size and (b) when the upper state is larger than the lower state. In case (a) it can be seen that (0, 0) transitions are the strongest in both absorption and emission, while in case (b) the strongest absorption band is (6, 0) and the strongest emission is (0, 4).

Vibrational sublevels can also be subdivided into rotational states. These occur in energy increments of about 1 kJ/mol, which correspond to approximately 3 nm in wavelength in the visible spectrum. Further broadening of absorption lines because of a continuum of translational energies of the whole molecule is about 0.1 kJ/mol, while the magnitude of shifts caused by intermolecular interactions when a molecule interacts with a solvent can be 5 kJ/mol. When all the above processes occur simultaneously, the result is an absorption band. The smoothness of absorption bands in most pigment molecules demonstrates that several photon

(a)

(b)

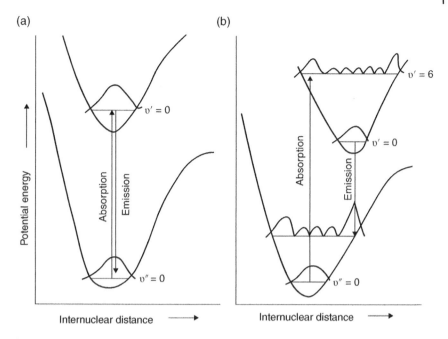

Figure 12.3 Electronic transitions of greatest probability for absorption and emission from lowest vibrational levels: (a) when both electronic states are of similar sizes and (b) when the upper state is larger than the lower one. *Source:* Adapted from [2]. Reproduced with permission of IEEE.

energies can produce an electronic excited state from the ground state. The above discussion clarifies that in optical fibre sensing experiments where an absorption spectrum is to be determined, details of the solvent used should be provided as this can have large effects on the inferred energy levels.

12.9 Distinction Between Absorption and Action Spectra

As stated earlier, the probability of a photon being absorbed by a molecule depends on its energy, and this is the absorption spectrum of that molecule. When, however, we use optical fibre sensors to observe the effect of light absorption as a function of wavelength, this is called the action spectrum. In photobiology, photochemistry, and photophysics, it is important to distinguish between the absorption and action spectrum of a molecule and evaluate the relative effectiveness of various wavelengths in producing a specific response. The Grotthuss–Draper law states that an action spectrum should resemble the absorption spectrum of the substance that absorbs the light, as it is responsible for the specific effect observed. By plotting the reciprocal of the number of photons required in different spectral bins to give a particular response and requiring the observed effect to be linear with photon flux density per spectral beam, it is possible to identify individual pigments responsible for an observed action spectrum.

12.10 Light Scattering Processes

There are several scattering processes that can be explored with fibre-optic sensing. One needs to firstly distinguish between elastic and inelastic scattering. An elastic scattering process refers to an interaction without a permanent exchange of energy between light and matter. This implies that molecules are in the same energy state as before and that the energy contained in the photon leaving the point of inter-action is also equal to the energy of the incoming photon. As a result, there is no change in frequency of the scattered photon. In contrast, inelastic scattering is associated with a permanent energy exchange between the photon and the molecule, and due to the principle of conservation of energy, both the scattered photon and the molecule have different energies. As a result, the scattered photon has a different wavelength when compared with that from the incoming photon. When light is incident in a volume of particles, scattering and absorption simultaneously take place as shown in Figure 12.4.

12.10.1 Elastic Scattering

Mie and Rayleigh scattering occur in volumes of randomly oriented particles and are used extensively in heat and mass transfer measurements. Bragg scattering can only be observed in structured arrangements of particles such as crystals (where

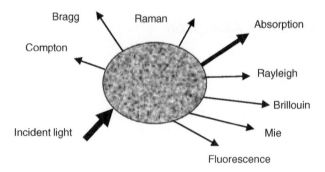

Figure 12.4 Scattering processes as well as absorption and fluorescence in a volume of particles.

the atoms are arranged in equidistant sets of planes) and if the distance between adjacent particles is of the same order as the incident wavelength. The light emitted from each atom due to the interaction with the incident light interferes and shows a characteristic intensity distribution similar to diffraction at regularly spaced intervals. The strongest intensity is observed if the Bragg condition is fulfilled, when the light impinges on the scattering plane at the Bragg angle. Since the distance in atoms in lattice structures is of the order of 1 Å, X-ray diffraction can be used to provide information regarding the structure of solid-state materials. Bragg scattering is associated with frequency shifts $\Delta\omega/\omega_0 = \pm(10^{-4} - 10^{-2})$ and is utilized in combination with inelastic Brillouin scattering in Bragg cells to perform laser Doppler velocimetry. Moreover, as discussed in other chapters, Bragg gratings are extensively used in several fibre-optic sensing modalities.

Mie scattering occurs when particles are large compared with the wavelength, whereas Rayleigh scattering occurs when particles are small compared with the wavelength (e.g. molecules, microscopic suspended particles, etc.) as shown in Figure 12.5.

Typical Mie scattering in the visible part of the spectrum are soot or dust particles with diameters in the region of 1–10 μm. Due to the large size of the particles, reflection and diffraction have to be considered besides the actual scattering by the particle. Changes in the phase amplitude and polarization of the incident light are associated with the Mie process (Figure 12.5). The transition region where the particle diameter is slightly larger than the incident wavelength ($q > 1$) is used in most applications of Mie scattering. In addition, one should note that the intensity of the light scattered in the forward direction is much larger than in the backward direction, the overall intensity increases with increasing diameter of the sphere, there is a strong angular dependence of the intensity distribution, and that intensity distributions can be very different for different incident light polarizations.

The assumption that the scatterers are spherical also does not hold often in most practical applications. Mie scattering has application in flow visualization and in the determination of particle size. Furthermore, there are other applications in

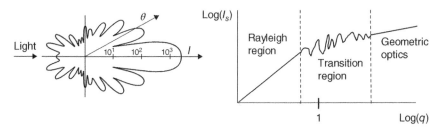

Figure 12.5 Typical polar plot associated with Mie scattering intensity illuminated by polarized light and scattering intensity I_s as a function of normalized particle diameter q (ratio of particle diameter to illumination wavelength).

particle image velocimetry. Scattering of particles often needs to be observed simultaneously from several directions in order to account for their non-spherical shape. The particle diameter is proportional to the following parameters: the experimentally observed phase difference, the wavelength of the incident light, a geometric factor associated with the experimental set-up, and the inverse value of the refractive index of the particle.

On the other hand, Rayleigh scattering occurs when the molecule absorbs one photon and moves in the excited state but then re-emits a photon of the same frequency returning to its ground state. Rayleigh scattering is normally used to obtain density or temperature measurements in gases or gaseous mixtures. The intensity of scattered light arriving at the detector is

$$I_R = E_D \left(\frac{d\sigma}{d\Omega} \right)_{\text{eff}} \Omega L N I_0 \tag{12.16}$$

where E_D is the detector efficiency, $(d\sigma/d\Omega)_{\text{eff}}$ is the effective differential scattering cross section, Ω is the solid angle of collection, N is the number of the molecules observed over a path length L associated with the solid angle, and I_0 is the incident light intensity. Observations of changes in scattered intensities as a function of time can provide additional information such as diffusion coefficients.

12.10.2 Inelastic Scattering

Compton scattering is observed when radiation of high frequency interacts with the electron that is closely bound in the atom. Since X-rays consist of photons with high energy, this large amount of energy cannot be absorbed by the electron as the electron movement is not controlled by the forces in the nucleus any more. As a result, the scattering process can be viewed as a collision between a photon and an electron. Some of the energy is used to dislocate the electron in a certain direction, while the photons deflected in the opposite direction contain less energy than before the collision. Compton scattering is associated with frequency shifts $\Delta\omega/\omega_0 = \pm(10^{-3} - 10^{-2})$, and the scattering process can be associated with molecular vibration and optical lattice vibration.

Brillouin scattering results from statistical density fluctuations due to acoustic vibrations in the scattering medium (liquids or crystals). These fluctuations travel at the local speed of sound, and the frequency of the scattered light is Doppler shifted. The relative velocity of the density fluctuation waves depends on the observation angle and ranges from the speed of sound (observed perpendicularly to the plane of the waves) to zero (in-plane observation). As a result, the amount of frequency shift varies with the observation angle. The highest intensity is observed if the Bragg condition is fulfilled (the angle between the direction of the incident light and the direction of the acoustic wave is equal to the Bragg angle). Since there is a frequency shift associated with Brillouin scattering, it can be classified as an inelastic process although there is no permanent exchange of energy associated with the process and as such could also be classified as an elastic process. Brillouin scattering is associated with frequency shifts $\Delta\omega/\omega_0 = \pm(10^{-6} - 10^{-5})$.

Raman scattering is a more complex process associated with rotational as well as vibrational states of the molecule as well as the electronic states. This is depicted more clearly in the energy diagram in Figure 12.6.

The photon absorption process lifts the molecule to a highly unstable virtual state and then it drops back to a stable energy level by emitting a photon of light. If this new state is energetically identical to the original state, Rayleigh scattering is observed. If the new level is higher or lower energetically than the original state, the scattered light has a different frequency than the incident light. The shift in frequency is referred to as a Raman shift and is proportional to the energy difference between the original state and the final de-excitation state. The energy difference between rotational levels is much smaller than vibrational levels, so the Raman lines from rotational transitions are much closer to the exciting line than those from vibrational transitions (Figure 12.7).

It is worth noting that laser-induced fluorescence described earlier is also useful for concentration and temperature measurements because the signals from this process are about 5–15 orders of magnitude stronger than those encountered in spontaneous Raman or Rayleigh scattering. Furthermore, since the transition of molecules from an excited state to the ground state is not always connected with the emission of light, there is collisional quenching. The quenching factors depend also on the pressure and temperature. In this sense, non-linearities in the observed signal due to quenching can be avoided by saturating the initial transition.

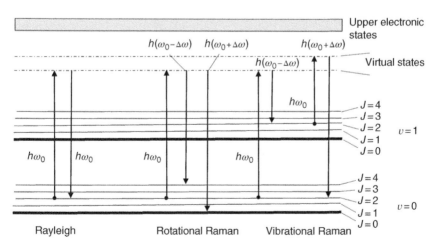

Figure 12.6 Energy level diagram for Rayleigh, rotational, and vibrational Raman scattering.

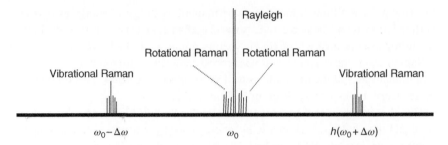

Figure 12.7 Typical Raman spectrum showing the rotational and vibrational Raman lines shifted to both sides of the excitation, which is identical with the Rayleigh line.

12.11 Induction of Non-linear Optical Processes

Two-photon absorption (TPA) is the absorption of two photons of identical or different frequencies in order to excite a molecule from one state (usually the ground state) to a higher energy, most commonly an excited electronic state. The energy difference between the involved lower and upper states of the molecule is equal to the sum of the photon energies of the two photons absorbed. TPA is a third-order process, typically several orders of magnitude weaker than linear absorption at low light intensities. It differs from linear absorption in that the optical transition rate due to TPA depends on the square of the light intensity; thus it is a non-linear optical process and can dominate over linear absorption at high intensities. Furthermore, the light intensity $I(x)$ versus distance x relation, which is normally given for single-photon absorption by the Bouguer–Lambert–Beer law $I(x) = I_0 e^{-ax}$, where I_0 is the initial light intensity and a is the one-photon absorption coefficient of the sample, changes to

$$I(x) = \frac{I_0}{1 + \beta x I_0} \tag{12.17}$$

where β is the TPA coefficient. Another parameter of interest is the molecular two-photon cross section. This is usually quoted in the units of Goeppert Mayer (GM) where 1 GM is 10^{-50} cm^4 s/photon. One can see that this unit results from the product of two areas (one for each photon, each in cm^2) and the time interval in which the two photons must arrive in order to be able to act together (the large scaling factor is introduced so that TPA cross sections of common dyes will have convenient values).

Induction of non-linear processes is often performed using seed pulses from Nd : YAG crystals, which are particularly useful for providing high pulse powers at nanosecond duration timescales, and Nd : YVO$_4$ crystals, which are used in low and moderate power applications providing pulses at picosecond duration timescales. Most commonly, however, Kerr-lens mode-locked Ti : sapphire lasers are used. They are tunable in the range of 700–1000 nm, with pulse energies in the range of 20–30 nJ, pulse durations under 100 fs, repetition rates around 100 MHz, and average powers of 2 W. Other possibilities include the use of Ti : Al_2O_3 crystals, which show remarkable versatility as they provide greater laser frequency tunability for ultrafast laser applications. Furthermore, there are other solid-state laser media that may also be used such as Nd : glass [3], Yb : glass [4], Yb : tungstate [5], and Cr : forsterite [6]. Since efficient multiphoton excitation requires peak intensities in the range 0.1–1 TW, in addition to tight spatial focusing, multiphoton processes typically require pulsed excitation sources that can provide efficient excitation at a practical average power. For example, a femtosecond laser with a pulse duration (τ) of 100 fs at a pulse repetition rate (f) of 100 MHz will enhance the two-photon excitation probability relative to that for continuous-wave light by the inverse of the duty cycle ($f \times \tau$), which is a factor of approximately 10^5.

Frequency conversion is possible using second-harmonic generation (SHG), sum-frequency generation (SFG), difference-frequency generation (DFG), optical parametric oscillation (OPO), and optical parametric amplification (OPA) in the visible and mid-IR spectral ranges. In addition, third-order harmonic stimulated Raman scattering (SRS) and four-wave mixing are also other promising techniques for frequency conversion. Nowadays, there are also several non-linear crystals such as KH_2PO_4 (KDP), $LiNbO_3$ (LN), $KTiOPO_4$ (KTP), β-BaB_2O_4 (β-BBO), LiB_3O_5 (LBO), and KBe_2BO_3F (KBBF), as well as a range of self-frequency doubling crystals [7] that can be used to induce non-linear phenomena.

Although the above techniques require the light pulses generated to be coupled to optical fibres (a process that often requires further compensation of the dispersion induced using pulse compression techniques), an alternative is to use fibre lasers to generate ultrafast pulses directly [8, 9]. In these lasers, pulses can be formed by balancing the non-linear gain and loss as well as non-linear refraction and dispersion (this process is called dissipative soliton generation) [10].

12.12 Concentrating Fields to Maximize Energy Exchange in the Measurement Process Using Slow Light

12.12.1 Slow Light Using Atomic Resonances and Electromagnetically Induced Transparency

A simple atomic resonance as the one depicted in Figure 12.8a is characterized by a resonant frequency ω_{12}, represented in Figure 12.8b by a spectrum of absorption coefficient and with a characteristic spectrum of its refractive index represented in Figure 12.8c. One can get a more intuitive picture of the physical phenomena

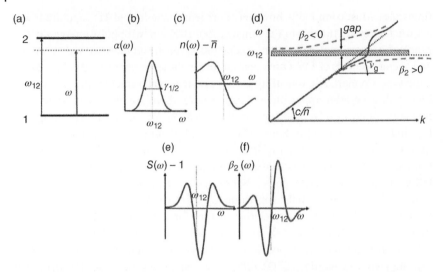

Figure 12.8 (a) Depiction of an atomic resonance (b) spectrum of absorption coefficient, (c) spectrum of refractive index, (d) dispersion diagram and group velocity in the vicinity of a resonance, (e) spectrum of slowdown factor, and (f) spectrum of group velocity dispersion (GVD). *Source:* Reproduced after [11]. Reproduced with permission of OSA publishing.

governing slow light propagation by plotting the dispersion curve in the absence of loss (i.e. $\gamma_{12} = 0$) as the dashed curve in Figure 12.8d. This curve corresponds to a well-known coupled mode model, also known as the polariton dispersion curve in solid-state physics. The first mode is a photon, which in the absence of an atomic transition is described by the linear dispersion curve $\omega_p = ck/\tilde{n}$. The second mode is the atomic polarization characterized by a resonance frequency ω_{12}. The dispersion curve of atomic polarization is a horizontal line $\omega_a = \omega_{12}$, indicating that it has zero group velocity, as atoms are assumed to be stationary in the timescales of the associated processes. In the vicinity of the resonance, two modes couple into each other, and the modified dispersion curve is split into two branches separated by the gap in which the light cannot propagate [11]. It can be observed that for each wave vector there are two coupled solutions characterized by two different group velocities. The one further away from the resonance has the higher group velocity and is usually referred to as 'photon-like', while the one closer to the resonance has lower group velocity and is usually referred to as 'atom-like'. One can then interpret the slow light propagation phenomenon as a process where the energy gets constantly coupled from the electromagnetic field to the atomic polarization and back. The longer the energy spends in the form of atomic excitation, the slower the coupled mode propagates. Thus slow light propagation in an atomic system can be understood as constant excitation and de-excitation of atoms in which coherence is preserved. Since the group velocity is also the velocity with which the energy propagates, the local energy density of the light becomes enhanced by a slowdown factor in the slow light medium.

In addition, the associated electric field does not become enhanced in the atomic slow light medium. This simply means that in atomic slow light schemes, all the additional energy compressed into the medium is stored in the atomic polarization. This is in contrast to photonic slow light schemes, where the electric field density does become enhanced substantially.

The aim in sensing applications of slow light is to take advantage of this enhanced electric field density to exchange more energy with the target molecules within a very localized area while at the same time maintaining a significant spectral bandwidth.

System losses, the dispersion of group velocity, and the dispersion of absorption are all parameters that need to be tailored to the application. One way of reducing dispersion is through the use of a double atomic resonance as encountered in electromagnetically induced transparency (EIT) experiments. This is qualitatively depicted in Figure 12.9. The dispersion curve is split into three branches, of which the central one, which is squeezed between two atomic resonances, is the one with a slow group velocity.

In the above double atomic resonant scheme, unfortunately it is not possible to maintain a variable bandwidth because the width of the passband cannot be changed. To change the passband width, one can adopt a spectral hole burning scheme using a strong pump in the inhomogeneously broadened transition as shown in Figure 12.10.

By changing the spectrum of the pump, using intensity or frequency modulation, one can change $\Delta\omega$ to achieve the maximum delay possible. The atomic oscillator in the absence of external modulation has just one resonant frequency ω_{12} in its response spectrum. If the wave is amplitude modulated with some frequency Ω, two sidebands at frequencies $\omega_{12} \pm \Omega$ in its spectrum will appear. By strongly modulating the strength of the atomic oscillator with some external frequency Ω, the absorption spectrum should show two absorption lines separated by 2Ω. As the resonant frequency increases, transparency is induced in the material.

A commonly used three-level 'Λ' EIT scheme can be visualized more clearly using the diagram in Figure 12.11. The ground-to-excited state transition ω_{12} is resonant with the frequency of the optical signal carrier ω_0 and has a dephasing rate of γ_{12}. In the absence of the pump, the absorption spectrum (dashed curve) is a normal Lorentzian line. There also exists a strong transition coupling the excited level 2 with the third level 3, but the transition between levels 1 and 3 is forbidden. When a strong resonant pump with intensity at frequency ω_{23} is turned on, the mixing of states 2 and 3 causes modulation of the absorption of the signal. The

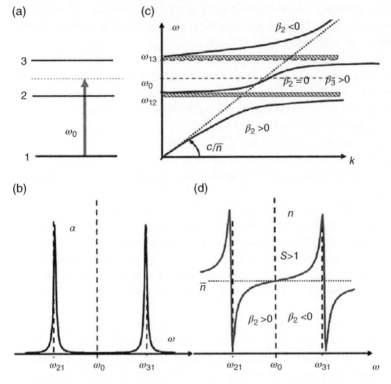

Figure 12.9 (a) Depiction of a double atomic resonance, (b) associated absorption spectrum, (c) dispersion diagram and group velocity in the vicinity of a resonance, and (d) refractive index spectrum. *Source:* Reproduced after [11]. Reproduced with permission of IEEE.

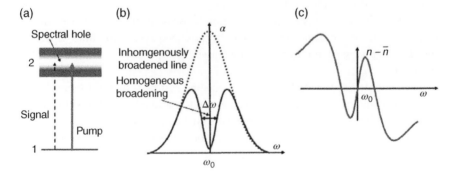

Figure 12.10 (a) Atomic transition with spectral hole burning, (b) associated absorption spectrum, and (c) refractive index spectrum. *Source:* Reproduced after [11].

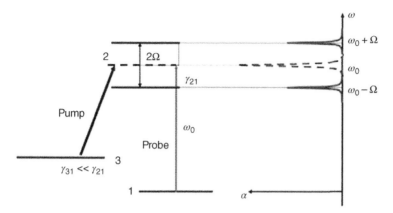

Figure 12.11 Energy levels in an EIT 'Λ' scheme. *Source:* Reproduced after [11]. Reproduced with permission of IEEE.

Lorentzian peak in the absorption spectrum splits into two smaller peaks at frequencies $\omega_0 \pm \Omega$, where Ω is the Rabi frequency. Changing the pump intensity allows one to achieve full tunability of the group velocity and to achieve very small group velocities with a tunable slowdown factor.

In EIT, the residual absorption rate at the resonant frequency is proportional to the dephasing rate of the intra-atomic excitation. Thus typically $\gamma_{31} \ll \gamma_{21}$, and the residual absorption is much weaker in the EIT than in the case of two independent resonances. This indicates that EIT is a coherent effect and that the reduction of absorption occurs due to the destructive interference of the absorption by two sidebands [11].

While in the simple single- or double-resonant slow light scheme, the energy is transferred from the electromagnetic wave to the atomic excitation and back, in the EIT scheme the signal photon propagates in the EIT medium, and then it transfers its energy to the excitation of the atomic transition between levels 1 and 2. Because of the presence of strong pump wave coupling between levels 2 and 3, the excitation is almost instantly transferred to the long-lived excitation between levels 1 and 3. Then the process occurs in reverse, and the energy is transferred back to the transition between levels 1 and 2 and finally back into the photon. Then the process repeats itself. Most of the time, the energy is stored in the form of 1–3 excitations, and thus it propagates with a very slow group velocity. Furthermore, the actual absorption event occurs only when the excitation 1–3 loses coherence and the energy cannot get back to the photon. Naturally, it is the dephasing rate of this excitation, i.e. $\gamma_{31} \ll \gamma_{21}$, that determines the residual absorption loss.

12.12.2 Slow Light Using Photonic Resonances

In photonic resonance, the energy is resonantly transferred between two or more modes of electromagnetic radiation. When the transfer takes place between a forward and backward propagating wave, the group velocity is reduced. The most common photonic resonance is the Bragg grating, which is a structure with its refractive index periodically modulated along its length z with period 'Λ' so that

$$n = \tilde{n} + \delta n \cos (2\pi/\Lambda z) \tag{12.18}$$

The bandgap appears in the vicinity of the Bragg frequency

$$\omega_B = \frac{\pi c}{\Lambda \tilde{n}} \tag{12.19}$$

and the dispersion law becomes

$$\frac{k - k_B}{k_B} = \sqrt{\left(\frac{\omega - \omega_B}{\omega_B}\right)^2 - \left(\frac{\delta n}{2\tilde{n}}\right)^2} \tag{12.20}$$

Close to the gap the group velocity reduces by a slowdown factor:

$$S = \frac{\left|\frac{\omega - \omega_B}{\omega_B}\right|}{\sqrt{\left(\frac{\omega - \omega_B}{\omega_B}\right)^2 - \left(\frac{\delta n}{2\tilde{n}}\right)^2}} \tag{12.21}$$

where the width of the forbidden gap is $\Delta\omega_{gap} = \omega_B \delta n / \tilde{n}$ and the index contrast $\delta n/\tilde{n}$ is called the 'strength' of the grating. This grating strength plays a role equivalent to that played by the oscillator strength of the atomic resonance. In this case, the slowdown effect in a photonic structure is the result of the transfer of energy between the forward and backward propagating waves (no energy is transferred to the medium). Hence, the strength of the electric field in photonic slow light structures becomes greatly enhanced.

It is possible to use simple photonic crystals whose dispersion curves are similar to Bragg gratings, but the problem of structures with a single photonic resonance is identical to the case of the single atomic resonance: the presence of a strong second-order dispersion. A solution is to consider structures with more than one photonic resonance with slightly different periods Λ_1 and Λ_2, ensuring a narrow passband $\Delta\omega$ between them. The slowdown factor in this scheme is

$$S(\omega_0) = \left[1 + \frac{\Delta\omega_{gap}}{2\Delta\omega}\right]^{1/2} \tag{12.22}$$

As shown in Figure 12.12, the dispersion curve becomes squeezed between two gaps, and the group velocity decreases with the passband, but the dependence is

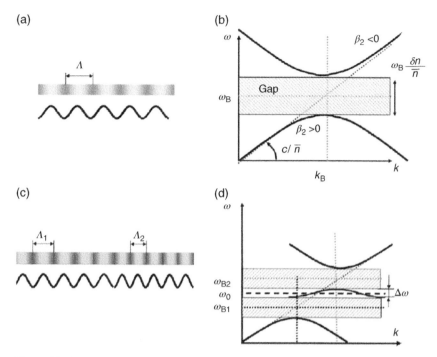

Figure 12.12 (a) Bragg grating and its index profile, (b) dispersion curves, (c) cascaded gratings of two different periodicities, and (d) associated dispersion curve. *Source:* Reproduced after [11]. Reproduced with permission of IEEE.

not as strong as in the case of atomic resonance; thus the photonic structures in general should have far superior performance at large bandwidths when compared with that achieved using an atomic medium.

A similar approach can be cascaded photonic crystal waveguides. Unfortunately, Bragg gratings have relatively small available index contrast and can be difficult to fabricate when the interferometric mask must provide different periods.

A periodic sequence of short Bragg grating segments can be considered as a new grating with periodically modulated properties (see, for instance, the Moiré grating in Figure 12.13a). In a Moiré grating the segments are not independent but interact coherently – hence its properties are different from the cascaded grating. A more efficient way of generating similar effects is through the use of coupled resonator structures (CRSs), such as those of Figure 12.13b and c, or by engineering defect modes in the photonic crystal, as described in Figure 12.13d. The existence of multiple resonances in Figure 12.13e indicates that the light propagating through the CRS can be considered a superposition of more than one forward and more than one backward propagating wave. Alternatively, one can also think about the resonators as 'photonic atoms' analogous to real atoms in EIT. The strength of the

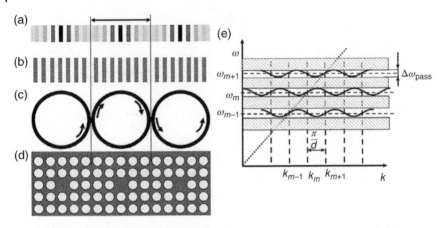

Figure 12.13 (a) Moiré grating, (b) coupled Fabry–Pérot resonators, (c) coupled ring resonators, (d) coupled defect modes, and (e) dispersion curves. *Source:* Reproduced after [11]. Reproduced with permission of IEEE.

electromagnetic field inside photonic structures increases by a factor of $S^{1/2}$, which is of particular interest to sensing applications.

Finally it is worth noting that since strong dispersion also takes place in the spectral vicinity of a resonant gain, it is possible to use gain media instead of passive structures. There are several advantages of using a gain medium: firstly, one can avoid attenuation, and secondly, the gain can be changed at will by changing the pump strength and the spectrum so that the delay and the bandwidth can be tuned. Tunable gain is commonly achieved using stimulated Brillouin scattering (SBS) or Raman scattering (SRS). An advantage of this approach of relevance to the fibre-optic community is that silica optical fibres can be used in this domain. In the SBS process, a high-frequency acoustic wave is induced in the material through an electrostriction process, and the density of the material increases in regions of high optical intensity. The process of SBS can be described classically as a non-linear interaction between the pump (at angular frequency ω_p) and a probe (Stokes) field ω_S through the induced acoustic wave of frequency Ω_A. The acoustic wave in turn modulates the refractive index of the medium and scatters pump light into the probe wave when its frequency is downshifted by the acoustic frequency. This process leads to a strong coupling among the three waves when this resonance condition is satisfied, which results in exponential amplification (absorption) of the probe wave. Efficient SBS occurs when both energy and momentum are conserved, which is satisfied when the pump and probe waves counter-propagate. This leads to a very narrow gain bandwidth.

An alternative way to achieve tunable delays is via SRS, which can also be achieved in optical fibres. SRS arises from exciting vibrations in individual molecules, also known as optical phonons – as opposed to exciting sound waves (acoustic phonons) as in the SBS process. The optical phonons, unlike acoustic phonons, are localized and have very short lifetimes, measured in picoseconds or fractions of picoseconds. In amorphous materials, such as glass, the frequencies of optical phonons are spread over a large interval (terahertz bandwidth). Therefore, the Raman gain is much broader than the Brillouin gain, albeit much smaller in absolute value.

12.13 Field Enhancement and Improved Sensitivity Through Whispering Gallery Mode Structures

One convenient way of improving light–matter interactions is through whispering gallery mode structures (WGMs), as these provide small mode volume and high-quality factor. They have been widely used in many applications such as low-threshold micro-lasers [12], non-linear optics [13, 14], narrow line width optical filters [15], and sensors. They can be implemented using various geometries such as microrings [16–18], microdiscs [19, 20], microtoroids [21, 22], and microspheres [23, 24]. These resonators normally have an axially symmetric structure that enables them to trap light in the form of WGMs that lead to optical resonances in the transmission spectrum. Microsphere resonators display a high coupling efficiency, high Q-factors, and low transmission loss when the light is coupled to the microsphere through a tapered fibre [25, 26]. These characteristics make them particularly useful in many applications such as trace-gas sensing, thermosensing, and biosensing [27]. When two microspheres are coupled to each other via their evanescent fields, extreme sensitivity can be achieved in the sensing process [28]. Coupled resonator optical waveguides (CROWs) in the form of microspheres or microdiscs can display coupled-resonator-induced transparency (CRIT) [29, 30], which is manifested as high transmittance over a narrow spectral range. CRIT occurs because of the destructive interference, and the phenomenon is in many respects similar to EIT [31]. The CRIT effect induces severe dispersion inside the transparency window and causes a low group velocity. The group velocity has an inverse relationship with dispersion and decreases further as the dispersion inside the transparency window increases. As a result, the speed of light decreases as the group velocity decreases. The slow light inside the microsphere cavity leads to a better light interaction with the surrounding environment, which is beneficial in sensing applications [32].

Light from a laser source is coupled inside the fibre taper. The transfer function of the coupled microspheres is defined as

$$T = \frac{E_2}{E_1} \tag{12.23}$$

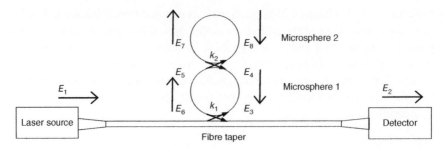

Figure 12.14 Generic diagram of a microsphere-based CROW structure. The input and output fields are E_1 and E_2; E_3, E_4, E_5, E_6, E_7, and E_8 are optical fields at several points inside the system.

where E_1 and E_2 are described in Figure 12.14.

Mode amplitude relations are as follows:

$$\begin{pmatrix} E_3 \\ E_2 \end{pmatrix} = \begin{pmatrix} t_1 & -ik_1 \\ -ik_1 & t_1 \end{pmatrix} \begin{pmatrix} E_6 \\ E_1 \end{pmatrix}$$
$$\begin{pmatrix} E_1 \\ E_5 \end{pmatrix} = \begin{pmatrix} t_2 & -ik_2 \\ -ik_2 & t_2 \end{pmatrix} \begin{pmatrix} E_8 \\ E_4 \end{pmatrix} \tag{12.24}$$

$$E_8 = a_2 \exp\left(i\phi_2\right)E_7 \tag{12.25}$$

$$E_6 = a_1 \exp\left(i\phi_1/2\right)E_5 \tag{12.26}$$

$$E_4 = a_1 \exp\left(i\phi_1/2\right)E_3 \tag{12.27}$$

where E_3, E_4, E_5, E_6, E_7, E_8, are also described in Figure 12.14; t_1, k_1 are transmission and coupling coefficients of the fibre taper and first sphere, respectively; t_2, k_2 are the transmission and coupling coefficients of the first microsphere and second microsphere, respectively; and a_1, a_2 are round-trip amplitude attenuation factors, which dramatically change when the surrounding environment changes. The two transfer functions for the structure can be obtained from these expressions:

$$T_1 = \frac{E_2}{E_1} = \frac{t_1 - a_1 \exp\left(i\phi_1\right)}{1 - t_1 a_1 \exp\left(i\phi_1\right)} \tag{12.28}$$

$$T_2 = \frac{E_5}{E_4} = \frac{t_2 - a_2 \exp\left(i\phi_2\right)}{1 - t_2 a_2 \exp\left(i\phi_2\right)} \tag{12.29}$$

while the effective phase shift of the structure is

$$\phi_{\text{eff}} = (\phi_1, \phi_2) = \arg(T_1) \tag{12.30}$$

and the associated group delay is

$$\tau(\omega) = \frac{d\phi_{\text{eff}}(\phi_1, \phi_2)}{d\omega} \tag{12.31}$$

whereas the group velocity is given from

$$V_g(\omega) = \frac{L}{\tau(\omega)} \tag{12.32}$$

where $L = L_1 + L_2$ corresponds to the overall path length across the structure (lengths are associated to the equatorial plane surroundings of the two microspheres).

In addition, the group index is given from

$$n_g(\omega) = \frac{c}{V_g(\omega)} \tag{12.33}$$

The round-trip phase delay for each sphere is given from $\phi_i = (2\pi n_{eff}L_i)/\lambda$, and the associated phase difference between two optical pathways is $\Delta\phi = \phi = \phi_1 - \arg(T_2)$:

$$\Delta\phi = \phi = \phi_1 - \left[\arctan\left(\frac{a_2 \sin\phi_2}{t_2 - a_2 \cos\phi_2}\right) - \arctan\left(\frac{a_2 t_2 \sin\phi_2}{1 - a_2 t_2 \cos\phi_2}\right) \right] \tag{12.34}$$

These two optical pathways destructively interfere when ϕ/π is an odd number, while a further requirement for CRIT to occur is that $a_1 < t_1 < t_2 < a_2$ [32].

12.14 Emergent Technological Trends Facilitating Multi-parametric Interactions of Light with Matter

12.14.1 Integration of Optical Fibres with Microfluidic Devices and MEMS

Through recent advances in femtosecond laser micromachining, it is nowadays possible to make microfluidic devices in transparent materials [33]. The development of optofluidic technology through the fusion of microfluidics and optics is an emergent interdisciplinary topic that promises to revolutionize sensing in the near future [34]. The technology promises to facilitate light–matter interactions in fluid–solid interfaces, purely fluidic interfaces, and colloidal suspensions, enabling the proliferation of portable devices for environmental monitoring and medical diagnostics. Low-cost manufacturing for these devices should also be possible using poly(dimethylsiloxane) (PDMS) [35]. So far, there have been several examples of integrated sensors with microfluidic chips, e.g. integrated microscopes using soft solid immersion lenses [36], compact tunable microfluidic interferometers [37], refractometers [38], surface plasmon resonance sensors [39], surface-enhanced Raman spectrometers using metallic nanostructures [40], tunable dye lasers with integrated mixer and ring resonators [41], photonic bandgap devices [42, 43], liquid-core/liquid-cladding optical waveguide structures for enhanced sensing [44], optical trapping devices [45], etc.

In addition to the above, integration of optical fibres with microelectromechanical systems (MEMS) provide new opportunities for manufacturing microspectrometers [46], endoscopic optical coherence tomography devices [47], fibre-optic non-linear optical microscopes and endoscopes [48], pressure sensors [49, 50], the sensing of displacement and vibration [51], etc. MEMS are also proving particularly useful when incorporated into more complex optical systems because they can provide improved alignment and thus coupling between optical fibres [52] as well as very controlled attenuation [53] and light chopping capability, e.g. using comb drive actuators [54].

Microfluidics and MEMS technologies can be combined to build systems with improved multi-parametric sensing capability. Furthermore, as it will be discussed in the following section, they can be used in conjunction with pulse shaping technologies to not only observe but also control molecular processes.

12.14.2 Pump–Probe Spectroscopy

Pump–probe spectroscopy using femtosecond duration laser pulses is routinely used nowadays to map the potential energy surface of chemical reactions. In pump–probe spectroscopy, a short-duration pump pulse is used to excite the molecular system under study from the ground state to an excited state. A synchronized, jitter-free probe pulse generated by splitting the pump pulse is then used to record the excitation/de-excitation spectra at a range of delay times imposed to the probe pulse using a scanning delay line. Phenomena evolving at picosecond or femtosecond timescales can be mapped using this technique. If the femtosecond pulses are amplified from 10 nJ to energies of 2 mJ each, the potential energy surfaces of the electron clouds surrounding the molecules are distorted, leading to a range of new phenomena.

12.15 Prospects of Molecular Control Using Femtosecond Fibre Lasers

12.15.1 Femtosecond Pulse Shaping

The general principle of pulse shaping is well established, and prototype devices have been constructed since the 1980s [55]. Nowadays, one can use an acousto-optic modulator to perform pulse shaping; alternatively, spatial light modulators in a 4f optical system [56] may be used to modulate the amplitude and phase of individual spectral bins. Systems for optical arbitrary waveform generation have been proposed [57–59], and several commercial vendors can provide the necessary hardware. Pulse shaping techniques are extensively discussed in the book by Weiner [60]. Most of these advances that stem from the ultrafast phenomena community have yet to migrate to the fibre-optic sensing community. An optical pulse synthesizer [61] and spectral line-by-line pulse shaping [62] can form an integral part of a fibre-based system. Transform-limited pulses (i.e. pulses of a wave that has the minimum possible duration for a given spectral bandwidth) may in principle be realized for any spectral phase signature arising from a coherent comb generation process, and advances in microresonator frequency combs [63] enable the integration of pulse shapers with optical fibre sensing systems. A typical system that is integrated with erbium-doped fibre amplifiers (EDFAs) is shown in Figure 12.15.

There are many ways that experiments in time-domain spectrometry can benefit from tailored pulse excitation. From a control theory and system identification point of view, for a system to be fully identified, one needs to be able to excite all the excitable modes of the system. Certain pulses (e.g. chirped or pseudo-binary sequences) ensure a more persistent excitation of the sample than what can be achieved using rectangular pulses. A higher pulse bandwidth achieved by shortening the pulse is also beneficial to the persistent excitation requirement, but, unfortunately, the energy available in each spectral bin diminishes as the duration

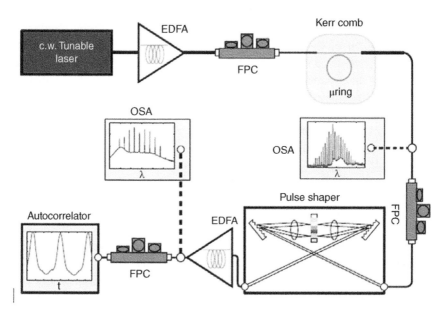

Figure 12.15 Line-by-line pulse shaping of a frequency comb from a silicon nitride microring. FPC, fibre polarization controller; µring, silicon nitride microring; OSA, optical spectrum analyser. The 4f optical system with the two gratings is also shown. *Source:* Adapted from [63]. Reproduced with permission of OSA Publishing.

of the pulse shortens (assuming the same pulse power). For a fixed measurement integration time, signal-to-noise ratio limitations in the observed spectra can be circumvented by either increasing the repetition rate of the laser or amplifying the pulses or alternatively concentrating the power at certain frequencies only as there is no point in probing away from molecular resonances where samples may be transparent. Such approach also fulfils another objective, which is the maximum possible exchange of power between the pump–probe pulses and the medium under study. Furthermore, in certain studies, there is scope to optimize the spectral content of the pulses so as to selectively excite only certain modes of the system. Such experiments require the use of tailored pulses.

12.15.2 New Opportunities for Coherent Control of Molecular Processes

From a femtochemistry point of view, the arrival time (phase) of certain components of the pulse is important. Appropriately designed pulses can excite a particular molecule to a spatial configuration that facilitates its interaction with another molecule by momentarily inducing additional steric effects, thus increasing or decreasing the reactivity of chemical reagents. In this sense, 'dressed states' of light in femtosecond laser pulses can have a catalytic or inhibitory function in chemical reactivity [64]. Such excitation also ensures that states of certain short-lived intermediates in a reaction can be either enhanced or suppressed, thus facilitating or inhibiting the progress of a reaction. If a robust methodology for producing high fidelity pulse shapes can be developed, appropriately chosen pulses should also be able to alter the dynamics of biochemical reactions by altering conformations in enzyme-substrate-based interactions, thus providing a versatile new tool for transcriptomics and proteomics researchers.

Instead of pursuing a purely combinatorial approach to design pulses for a particular experiment (e.g. the conversion of reactants A into products D as shown in Figure 12.16), we can use some of the chemometric information conveyed in the probe pulses of a time-domain spectrometer. Absorption lines or bands in the spectra are an indicator for the formation of the desired product, so it is possible to perform a closed-loop experiment where the absorption line/band strength is proportional to the concentration of the formed product D [65].

It is worth noting that as every feedback system requires a linear feedback signal to operate, the signal from the spectrometer must remain linear throughout the process. This is usually the case in most spectroscopic studies as long as care is taken to avoid saturating the absorbers by appropriately setting the optical flux density of the probe beam. Furthermore, appropriate shaping of the pulses using optimal control theory criteria may be performed so that a reaction may be driven along a certain trajectory ensuring preferential intermediate product formation

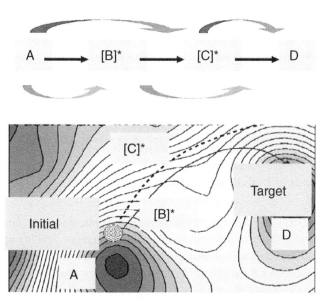

Figure 12.16 Potential energy surface of a reaction with short-lived intermediate state products B* and C* and final product D. Femtosecond pulses may be used to drive the reaction along different trajectories (dotted and solid lines) bypassing either intermediate B* or C*.

avoiding products that may be toxic or harmful to the environment. A selectivity in stereoisomer product formation is also possible using this approach.

In photoionization, molecular photodissociation, and atomic photoexcitation experiments of simple molecules, where accurate estimation of the Hamiltonian is possible, coherent control of quantum dynamics has been successfully performed by inverting the identified Hamiltonian. Problems are encountered, however, in strongly coupled systems such as large molecules/macromolecules in condensed phase, where it is impossible to obtain a complete knowledge of the entire Hamiltonian system [66]. Feedback learning algorithms overcome this limitation by using the physical system itself to explore its own quantum dynamics through an experimental search. In a typical search experiment, the goal is to compare the ability of several thousand different-shaped laser pulses to transform the system $|\psi\rangle$ from its initial state at $t = 0$ to some desired target state $|\chi\rangle$ at a later target time $t = T$. The pulse shapes are selected through a fitness-directed search protocol such as a genetic algorithm. The fitness is a measured quantity proportional to the objective functional $J[H;x_i]$, which is the square of the projection of χ onto ψ at the end of the experiment:

$$J[H;x_i] = |\langle \chi(T)|\psi(T)\rangle|^2 \tag{12.35}$$

The functional J that reaches its extreme value for the optimal pulse depends on the Hamiltonian H for the system evolution, which in turn depends on the laser electric field $E(t)$ determined by the settings of the n pulse shape control parameters x_i, $i = 1... n$. This pulse can be derived analytically using optimal control theory if H is known. Otherwise it must be discovered through the learning search. This is typically performed on a trial-and-error basis by searching the landscape of all the possible pump pulses that lead to the desired product formation. Instead of searching the entire space, however, an evolutionary approach may be used.

12.15.3 Developments in Evolutionary Algorithms for Molecular Control

Evolutionary algorithms are commonly used to find the proper phase for each frequency within the pulse in order to achieve constructive interference, leading to the desired outcome and destructive interference elsewhere [67, 68]. Evolutionary algorithms normally consist of three different parts: an objective function to be minimized, a population of amplitude and phase genes and operators. Algorithms are composed of three basic steps: (i) creation of a (random) population over the possible space of solutions, (ii) evaluation of the members of the population via the objective function (if any member minimizes the function, a stop criterion is used), and (iii) if no solution is found, then the software creates new members of the population, or an entirely new population via the operators and goes to the previous step.

The genetic algorithm population consists of a collection of waveforms, each made from two concatenated vectors, one specifying phases over a range of frequencies and another specifying corresponding magnitudes. An advantage of this representation is that the range of frequencies can be specified and the phases and amplitudes can be interpolated over them. This way the GA can map the entire frequency range defined by the pulse bandwidth with varying degrees of resolution. In addition, this scheme allows for amplitude to be encoded with less genes than the phase. Typically, waveforms have 20–60 phase genes but only 10–20 amplitude genes. Due to a very large pulse shape solution space, convergence is challenging, and the way the problem is parametrized matters.

One possible solution is the use of a differential evolution scheme in conjunction with new wavelet operators to solve problems of waveform matching and coherent control. There are three aspects to this – gene selection, fitness scaling and evolution:

Gene selection process: Roulette wheel selection (RWS), which is a probabilistic selection of breeding candidates where the degree of expression is proportional to fitness, may be adopted because members are never removed from the roulette wheel. Members with high fitness are virile, and, consequently, are chosen as parents more frequently than those with low fitness. The process can be coded into the algorithm by firstly creating a vector F of corresponding fitnesses for the population and then creating a cumulative summed vector (csv$F$) for all indices in F. Assuming N is the size of this vector, a random number between zero and max(csv$F$) may be chosen. It proves advantageous for the developed algorithm to also make use of elitists. Instead of creating a new population every single time, the best fit members of the population are reserved and directly added to the next generation. This ensures the best fitness of the population never decreases and lessens the computational load on the system; elitists do not need to have their fitness re-evaluated, as this must have been sampled in the previous generation.

Linear fitness scaling: As the population starts to converge to a solution, the average fitness of the population will also start to converge to the best fitness. To keep the selection pressure high, it is advantageous to employ fitness scaling and use the scaled fitnesses in RWS instead. By assigning the best fit member a scaled fitness of 2 and the average fitness a scaled fitness of 1, the best fit member is twice as likely to be chosen as the average fit member. Because the scaling is linear, some low fitness members will have negative scaled fitnesses. These members are assigned a zero fitness (i.e. given a null probability of being chosen as parents). The linear fitness scaling algorithm finds the best fitness F_{best} and the average fitness F_{avg} and then creates an array of scaled fitness from

$$F_{scaled}(n) = \frac{F_{best} - F(n)}{F_{avg} - F_{best}} + 2 \tag{12.36}$$

where $F_{scaled}(n) \geq 0$. After measuring and then linearly scaling the fitness, an operator is selected by RWS from a vector of operator weights. Depending on the operator chosen, either one or two parents are selected and then passed on to the operator, thereby creating the child. Operators originally proposed by Bucksbaum's group [68] can be used for this task.

Differential evolution (DE): This operator ensures that when new members of the population are created, they are only added to the population if they fit better than the least fit members of the previous generation. The *DE* only creates one new population member per generation (i.e. population size – elitists = 1). The operator also changes the way children are produced. Instead of directly selecting and using breeding candidates from the population, there is an intermediate step. Let each member of the population, sized n, be a vector of real numbers, $\mathbf{V}$, with the best fit member notated as $\mathbf{V}_{best}$. In *DE*, one initial vector, $\mathbf{V}_{init}$, would be selected using a traditional method, such as RWS. However, two other random vectors, $\mathbf{V}_{rand1}$ and $\mathbf{V}_{rand2}$, would also be selected, though in a manner with no regard to fitness. Then, the initial vector, $\mathbf{V}_{init}$, is added to a scaled difference of the two random vectors plus a scaled difference between the initial vector and the best fit vector, $\mathbf{V}_{\text{best}}$:

$$\mathbf{V}_{\text{parent}} = \mathbf{V}_{\text{init}} + \lambda(\mathbf{V}_{\text{best}} - \mathbf{V}_{\text{init}}) + F(\mathbf{V}_{\text{rand1}} - \mathbf{V}_{\text{rand2}}) \tag{12.37}$$

where $0.2 \leq F \leq 0.95$ and $0.6 \leq \lambda \leq 0.95$. Thus the parent vector is a linear difference of best fit and randomly chosen vectors. This randomness functions as a source of directed mutation creating a new vector with a possible domain larger than the domain of the original population. Self-organized (Kohonen) maps (SOM) may be used to assess the degree of expression of each operator during the convergence process of the developed meta-algorithm. Operators may outperform each other across different regions of the search space throughout the evolutionary process, confirming their respective utility. The above processes can be applied to perform pulse shaping in order to achieve the desired 'dressed' light states for tailored interactions of light with matter. An implementation of the above algorithms is discussed in [69], and further possibilities for chemical and biomolecular control are mentioned in [70, 71]. An issue with optical fibres when carrying out such experiments is the group velocity dispersion in the fibres. Photonic crystal fibres, however, may be tailored to compensate for this.

References

1 Nobel, P.S. (1991). *Physicochemical and Environmental Plant Physiology*. Academic Press.

2 Wayne, R.P. (1991). *Principles and Applications of Photochemistry*. Oxford University Press.

3 Aus der Au, J., Kopf, D., Morier-Genoud, F. et al. (1997). 60-fs pulses from a diode-pumped Nd:glass laser. *Opt. Lett.* 22: 307–309.

4 Hönninger, C., Morier-Genoud, F., Moser, M. et al. (1998). Efficient and tunable diode-pumped femtosecond Yb: glass lasers. *Opt. Lett.* 23: 126–128.

5 Druon, F., Balembois, F., and Georges, P. (2003). Laser crystals for the production of ultrashort laser pulses. *Ann. Chim. Sci. Mater.* 28: 47–72.

6 Seas, A., Petričević, V., and Alfano, R.R. (1992). Generation of sub-100-fs pulses from a cw mode-locked chromium-doped forsterite laser. *Opt. Lett.* 17: 937–939.

7 Yu, H., Pan, Z., Zhang, H., and Wang, J. (2016). Recent advances in self-frequency-doubling crystals. *J. Materiomics* 2: 55–65.

8 Fermann, M.E., Galvanauskas, A., Sucha, G., and Harter, D. (1997). Fibre-lasers for ultrafast optics. *Appl. Phys. B Lasers Opt.* 65: 259–275.

9 Fermann, M.E. and Hartl, I. (2013). Ultrafast fibre lasers. *Nat. Photonics* 7: 868–874.

10 Xu, C. and Wise, F.W. (2013). Recent advances in fibre lasers for nonlinear microscopy. *Nat. Photonics* 7: 875–882.

11 Khurgin, J.B. (2010). Slow light in various media; a tutorial. *Adv. Opt. Photon.* 2: 287–318.

12 Dai, L. et al. (2017). Effects of the slot width and angular position on the mode splitting in slotted optical microdisk resonator. *Photonics Res.* 5 (3): 194–200.

13 Beckmann, T. et al. (2011). Highly tunable low-threshold optical parametric oscillation in radially poled whispering gallery resonators. *Phys. Rev. Lett.* 106 (14): 143903.

14 Marquardt, C. et al. (2013). Nonlinear optics in crystalline whispering gallery resonators. *Opt. Photonics News* 24 (7): 38–45.

15 Vollmer, F. et al. (2008). Single virus detection from the reactive shift of a whispering-gallery mode. *Proc. Natl. Acad. Sci.* 105 (52): 20701–20704.

16 Iqbal, M. et al. (2010). Label-free biosensor arrays based on silicon ring resonators and high-speed optical scanning instrumentation. *IEEE J. Sel. Top. Quantum Electron.* 16 (3): 654–661.

17 Delezoide, C. et al. (2011). Vertically coupled polymer microracetrack resonators for label-free biochemical sensors. *IEEE Photon. Technol. Lett.* 24 (4): 270–272.

18 Khunnam, W. et al. (2018). Mode-locked self-pumping and squeezing photons model in a nonlinear micro-ring resonator. *Opt. Quant. Electron.* 50 (9): 343.

19 Wiersig, J. (2011). Structure of whispering-gallery modes in optical microdisks perturbed by nanoparticles. *Phys. Rev. A* 84 (6): 063828.

20 Kryzhanovskaya, N. et al. (2018). Enhanced light outcoupling in microdisk lasers via Si spherical nanoantennas. *J. Appl. Phys.* 124 (16): 163102.

21 Hunt, H.K. et al. (2010). Bioconjugation strategies for microtoroidal optical resonators. *Sensors* 10 (10): 9317–9336.

22 Lu, T. et al. (2011). High sensitivity nanoparticle detection using optical microcavities. *Proc. Natl. Acad. Sci.* 108: 5976–5979.

23 Farnesi, D. et al. (2014). Optical frequency conversion in silica-whispering-gallery-mode microspherical resonators. *Phys. Rev. Lett.* 112 (9): 093901.

24 Toncelli, A. et al. (2017). Mechanical oscillations in lasing microspheres. *J. Appl. Phys.* 122 (5): 053101.

25 Matsko, A.B. and Ilchenko, V.S. (2006). Optical resonators with whispering gallery modes I: basics. *IEEE J. Sel. Top. Quantum Electron.* 12 (3): 3–14.

26 Armani, A.M. et al. (2007). Label-free, single-molecule detection with optical microcavities. *Science* 317 (5839): 783–787.

27 Righini, G. et al. (2011). Whispering gallery mode microresonators: fundamentals and applications. *Rivista del Nuovo Cimento* 34 (7): 435–488.

28 Bo, F. (2015). Vertically coupled microresonators and oscillatory mode splitting in photonic molecules. *Opt. Express* 23 (24): 30793–30800.

29 Smith, D.D. (2004). Coupled-resonator-induced transparency. *Phys. Rev. A* 69 (6): 063804.

30 Smith, D.D. (2006). Coupled-resonator-induced transparency in a fibre system. *Opt. Commun.* 264 (1): 163–168.

31 Fleischhauer, M. et al. (2005). Electromagnetically induced transparency: optics in coherent media. *Rev. Mod. Phys.* 77 (2): 633–673.

32 Qian, K. (2016). Coupled-resonator-induced transparency in two microspheres as the element of angular velocity sensing. *Chin. Phys. B* 25 (11): 114209.

33 Osellame, R., Cerullo, G., and Ramponi, R. (2012). *Femtosecond Laser Micromachining: Photonic and Microfluidic Devices in Transparent Materials*. Springer.

34 Psaltis, D., Quake, S.R., and Yang, C. (2006). Developing optofluidic technology through the fusion of microfluidics and optics. *Nature* 442: 381–386.

35 Duffy, D.C., McDonald, J.C., Schueller, O.J.A., and Whitesides, G.M. (1998). Rapid prototyping of microfluidic systems in poly(dimethylsiloxane). *Anal. Chem.* 70: 4974–4984.

36 Gambin, Y., Legrand, O., and Quake, S.R. (2006). Microfabricated rubber microscope using soft solid immersion lenses. *Appl. Phys. Lett.* 88: 174102.

37 Grillet, C. et al. (2004). Compact tunable microfluidic interferometer. *Opt. Express* 12: 5440–5447.

38 Domachuk, P., Littler, I.C.M., Cronin-Golomb, M., and Eggleton, B.J. (2006). Compact resonant integrated microfluidic refractometer. *Appl. Phys. Lett.* 88: 093513.

39 Homola, J., Yee, S.S., and Gauglitz, G. (1999). Surface plasmon resonance sensors: review. *Sensors Actuators B Chem.* 54: 3–15.

40 Vo-Dinh, T. (1998). Surface-enhanced Raman spectroscopy using metallic nanostructures. *Trends Anal. Chem.* 17: 557–582.

41 Galas, J.C., Torres, J., Belotti, M. et al. (2005). Microfluidic tunable dye laser with integrated mixer and ring resonator. *Appl. Phys. Lett.* 86: 264101.

42 Domachuk, P., Nguyen, H.C., Eggleton, B.J. et al. (2004). Microfluidic tunable photonic band-gap device. *Appl. Phys. Lett.* 84: 1838–1840.

43 Shinn, A. and Robertson, W.M. (2005). Surface plasmon-like sensor based on surface electromagnetic waves in a photonic band-gap material. *Sens. Actuators B Chem.* 105: 360–364.

44 Wolfe, D.B. et al. (2004). Dynamic control of liquid-core/liquid-cladding optical waveguides. *Proc. Natl. Acad. Sci. U. S. A.* 101: 12434–12438.

45 Domachuk, P. et al. (2005). Application of optical trapping to beam manipulation in optofluidics. *Opt. Express* 13: 7265–7275.

46 Wolffenbuttel, R.F. (2005). MEMS-based optical mini- and microspectrometers for the visible and infrared spectral range. *J. Micromech. Microeng.* 15: S145–S152.

47 Xie, T., Xie, H., Fedder, G.K., and Pan, Y. (2003). Endoscopic optical coherence tomography with new MEMS mirror. *Electron. Lett.* 39: 1535–1536.

48 Fu, L. and Gu, M. (2007). Fibre-optic nonlinear optical microscopy and endoscopy. *J. Microsc.* 226 (Pt 3): 195–206.

49 Ge, Y., Wang, M., and Yan, H. (2008). Optical MEMS pressure sensor based on a mesa-diaphragm structure. *Opt. Express* 16: 21746.

50 Abeysinghe, D.C., Dasgupta, S., Boyd, J.T., and Jackson, H.E. (2001). A novel MEMS pressure sensor fabricated on an optical fibre. *IEEE Photon. Technol. Lett.* 13: 993–995.

51 Orłowska, K., Słupski, P., Świątkowski, M. et al. (2015). Light intensity fibre optic sensor for MEMS displacement and vibration metrology. *Opt. Laser Technol.* 65: 159–163.

52 Syms, R.R.A., Zou, H., Yao, J. et al. (2004). Scalable electrothermal MEMS actuator for optical fibre alignment. *J. Micromech. Microeng.* 14: 1633–1639.

53 Unamuno, A., Blue, R., and Uttamchandani, D. (2013). Modeling and characterization of a Vernier latching MEMS variable optical attenuator. *J. Microelectromech. Syst.* 22: 1229–1241.

54 Li, L. and Uttamchandani, D. (2004). Design and evaluation of a MEMS optical chopper for fibre optic applications. *IEE Proc.-Sci. Meas. Technol.* 151: 77–84.

55 Heritage, J.P., Weiner, A.M., and Thurston, R.N. (1985). Picosecond pulse shaping by spectral phase and amplitude manipulation. *Opt. Lett.* 10: 609–611.

56 Weiner, A.M. (2000). Femtosecond pulse shaping using spatial light modulators. *Rev. Sci. Instrum.* 71: 1929–1960.

57 Jiang, Z., Huang, C.-B., Leaird, D.E., and Weiner, A.M. (2007). Optical arbitrary waveform processing of more than 100 spectral comb lines. *Nat. Photonics* 1: 463–467.

58 Fontaine, N.K. et al. (2007). 32 phase × 32 amplitude optical arbitrary waveform generation. *Opt. Lett.* 32: 865–867.

59 Cundiff, S.T. and Weiner, A.M. (2010). Optical arbitrary waveform generation. *Nat. Photonics* 4: 760–766.

60 Weiner, A.M. (2009). *Ultrafast Optics.* Wiley.

61 Miyamoto, D. et al. (2006). Waveform-controllable optical pulse generation using an optical pulse synthesizer. *IEEE Photon. Technol. Lett.* 18: 721–723.

62 Jiang, Z., Seo, D.S., Leaird, D.E., and Weiner, A.M. (2005). Spectral line by line pulse shaping. *Opt. Lett.* 30: 1557–1559.

63 Ferdous, F., Miao, H., Leaird, D.E. et al. (2011). Spectral line-by-line pulse shaping of on-chip microresonator frequency combs. *Nat. Photonics* 5: 770–776.

64 Judson, R.S. and Rabitz, H. (1992). Teaching lasers to control molecules. *Phys. Rev. Lett.* 68: 1500–1503.

65 de Vivie-Riedle, R., Kurtz, L., and Hofmann, A. (2001). Coherent control for ultrafast photochemical reactions. *Pure Appl. Chem.* 73: 525–528.

66 Meshulach, D. and Silberberg, Y. (1999). Coherent quantum control of multiphoton transitions by shaped ultra-short optical pulses. *Phys. Rev. A* 60: 1287–1292.

67 Zeidler, D., Frey, S., Kompa, K.L., and Motzkus, M. (2001). Evolutionary algorithms and their application to optimal control studies. *Phys. Rev. A* 64: 023420.

68 Pearson, B.J., White, L.L., Weinacht, T.C., and Bucksbaum, P.H. (2001). Coherent control using adaptive learning algorithms. *Phys. Rev. A* 63: 63–74.

69 Shaver, A., Hadjiloucas, S., Walker, G.C., and Bowen, J.W. (2012). Femtosecond pulse shaping using a differential evolution algorithm and wavelet operators. *Electron. Lett.* 48 (21): 1357–1358.

70 Kim, Y.S. and Rabitz, H. (2002). Closed loop learning control with reduced space quantum dynamics. *J. Chem. Phys.* 117: 1024–1030.

71 Shaver, A., Hadjiloucas, S., Walker, G.C. et al. (2012). Femtosecond pulse shaping using a differential evolution algorithm and wavelet operators. *Electron. Lett.* 48 (21): 1357–1358.

13

Detection in Harsh Environments

Kamil Kosiel[1] and Mateusz Śmietana[2]

[1] *Łukasiewicz Research Network – Institute of Electron Technology, Al. Lotników 32/46, 02-668 Warsaw, Poland*
[2] *Institute of Microelectronics and Optoelectronics, Warsaw University of Technology, Koszykowa, Warsaw, Poland*

13.1 Introduction

Human activities develop more and more rapidly in industry, technology, and science domains while at the same time depend more and more on the control over environmental conditions, technological process parameters, stability of construction materials, etc. Therefore, there is increasing worldwide demand for various kinds of sensors allowing for gaining knowledge on a great variety of measurands [1–4].

Optical fibre sensors (OFSs) show many advantages, including possibility of remote, distributed or quasi-distributed [5] sensing, fast response, light weight, and small size, i.e. small diameters and volumes of the fibres (weight and volume of 1 km long, 200 μm diameter silica fibre are merely around 70 g and 31 cm^3, respectively). With their application, continuous monitoring or telemetry of multi-parameters is possible [5]. Additionally, they can be easily integrated into fibre-based large-scale networking and communication systems. In particular, in comparison with systems based on metal connections, which the most often include copper wires, application of the optical fibre systems offers many advantages, such as limited electromagnetic effect on the environment, minimal electromagnetic interference, broad bandwidth (OFSs offer the possibility of multiplexing a large number of individually addressed point sensors in a fibre network or distributed sensing), reduced risk of failure when exposed to water or other

Optical Fibre Sensors: Fundamentals for Development of Optimized Devices, First Edition.
Edited by Ignacio Del Villar and Ignacio R. Matias.
© 2021 The Institute of Electrical and Electronics Engineers, Inc.
Published 2021 by John Wiley & Sons, Inc.

conductive liquids, and lack of grounding concerns. Moreover, OFSs do not require electrical power at the sensing point. OFSs are intrinsically safe in combustible or explosive environments and reliable.

Another huge advantage offered by OFSs is the abundance of possible measurands that these sensors are able to monitor. By using a sensorial technique based on optical fibres, properties of physical [6], chemical [7–9], and biological [7, 9] nature can be analysed. In the previous chapters we saw the most often physical measurands: strain, stress or pressure, displacement, acceleration, temperature, magnetic field, electric current, and surrounding refractive index. In chemistry, the measurands are concentration of oxygen dissolved in water, hydrogen, pH, or humidity. Also a variety of biological molecules and even whole living cells were examined as analytes, and their detection with OFSs was elaborated. In both chemistry and biology domains, selective sensing of given analytes is also possible when functionalized OFSs are used.

Additional valuable feature of these sensors is their intrinsic resistance to many kinds of destructive factors and their robustness, especially during continuous measurements performed in various extreme conditions.

13.2 Optical Fibre Sensors for Harsh Environments

Optical fibres in harsh environment or harsh external conditions can be in general defined as devices operating in environments (surrounding conditions) that are different than normally applied to standard telecommunication fibres, i.e. conditions beyond the limits encountered for regular telecommunication and equipment for fibre data links [10–12]. In other words, any environmental condition requiring extra features or protection for a fibre or fibre cable to ensure reliability of the fibre sensor can be considered as the optical fibre harsh environment.

While for indoor communication the standard solutions comprise only relatively low level of protection (mainly aiming at basic protection of the fibres against mechanical destruction or safe collection of a number of fibres together within a common cable, e.g. simple-polymer-based fibre coatings and cable jackets, and aramid fibre strength members), the outdoor communication, in particular long-distance or submarine connections, requires consideration of more environmental hazards, e.g. water, additional mechanical stress, temperature changes, etc., and hence needs more robust cable assembly and additional fibre protection such as additional steel strength members, water-absorbing agents, etc.

However, advanced fibre sensors and sensorial systems have proved to be able to work in truly harsh conditions. In particular, they can withstand extreme high or low temperatures; extreme pressure from very high pressure to vacuum;

mechanical shocks, vibration, or stress imposed by tight bend configurations or linear movements; ionizing radiation such as UV, X-ray, or gamma; magnetic field of high strength; high electromagnetic radio-frequency interference; highly reactive chemicals; atmospheres containing dense vapour or liquid water, hydrogen, or other kinds of gas exhibiting detrimental impact on fibre component. In many cases some of these harsh factors occur together, exacerbating the sensor operation conditions and forcing the engineers to examine closely and optimize the solutions.

13.3 Need for Harsh Environment Sensing Based on Optical Fibres

A worldwide rapid increase in demand for harsh-environment-operating OFSs can be noticed nowadays. Today optical-fibre-based sensing is one of the fastest developing technologies, with numerous applications already existing in the market. Many market sectors, civil and military engineering, and several scientific fields have already increased use of harsh-environment-operating sensors for a few decades. As one can observe, the growth rate of the OFS world market accelerates during the last several years. This is mainly due to the rapid increase in need for harsh environment sensors, which fuels its progress.

Various measurands are of interest for different sectors. They are examined, i.e. measured or continuously monitored, in diversified environments, either 'natural' or anthropogenically impacted, as well as just related to technological or industrial processes. Various objects or properties of terrestrial (also underground), aqueous (e.g. marine or riverine), and air environments are accessible to fibre sensing technique – as well as environments beyond Earth (space and potentially other celestial bodies). The harmful factors being a threat to the sensors depend on the nature of these surroundings. The largest demand for harsh-environment-resistant sensors can be found in the industries of oil and gas (downholes, pipelines, etc.), energy (power plants – nuclear, coal-fired, etc.), chemical industry (reactors, gasifiers, etc.), transportation, telecommunication, structural engineering (e.g. for structural health monitoring), aerospace (civil and military), medicine and healthcare (health monitoring), military or defence, and geology (hydrogeology, control of geologically unstable areas, etc.). These applications correspond to both public and economic goals.

Among these numerous fields and stakeholders, one of the key beneficiaries of using harsh-environment-resistant optical fibre sensing systems is the oil and gas industry, which intensifies development of such sensors as it was stated in Chapter 6. This example perfectly shows the importance of progress in harsh environment fibre sensing. The main application here is downhole oil and gas exploration. The driving force there is that most of the discovered oil-based resources still remain unexcavated and unexploited in the ground after completion of the primary and secondary recovery stages (e.g. in the United States even up to approximately 65% of them, according to the US Department of Energy report [13]). For these two initial exploitation steps, the expected operating temperature range is around 100–150 °C; hence use of high-temperature special sensors is unavoidable for this condition monitoring, especially when expected pressure reaches up to 700 bar. However, after completion of these two initial recovery stages, quite different approach is necessary for further exploitation of the crude oil from the oil field. This also means much harsher conditions for the sensors. The possible further recovery stages [14], commonly termed as enhanced oil recovery (EOR) or tertiary recovery, can comprise various methods of exploitation. Essentially three main methods are used there, which are based on thermal, gas injection, and chemical injection treatment. In general, they are all focused on increasing mobility of the remaining oil in order to increase its extraction. However, it is crucial that economic justification for specific method of exploitation must always be verified on a case-by-case basis, and this should be achieved by means of appropriately accurate and reliable downhole monitoring performed *in situ* in harsh environments. In this case a quite common lack of robust instrumentation for such measurements is a crucial problem that prevents increased profit [10]. Therefore, there is an increasing need for such industrial monitoring, and it is important that it can be provided by the already existing fibre-optic based distributed measurement systems [15, 16]. Indeed, this type of sensing system is the most often used in this field. The most important data that is downhole monitored by the fibre sensors lowered into deep holes are real-time temperature and pressure. The measurements must take place in very harsh environment, in which temperatures or pressures are highly increased in comparison with standard well conditions and can reach over 300 °C and 1700 bar [17] in so-called high-pressure high-temperature (HPHT) wells. The harsh factors that the sensing in oil and gas industry has to face include not only the extremely high temperature, which ratings surpass even military specifications, or high pressure, but also harsh chemicals, which include oil, chemicals delivered for the exploitation process (e.g. mixture of surfactants, polymers, and alkali), and water, e.g. hot steam, which influences static fatigue of fibres. Additionally, the sensors must withstand intermittent shocks and vibration at different frequencies – the exemplary test protocols deal with 100 000 times repeated 2 ms shocks at 500 g. Such conditions much exceed those existing during the first two standard steps of downhole exploitation, i.e. the primary and secondary ones. The trend for dealing with expansion of HPHT wells is definitely going to be continued, because finding and exploitation of new hydrocarbon reserves involve facing increasingly harsh downhole conditions. The reserves continuously decline, however, advances in technology motivate the industry at the same time to drill deeper and in hotter geothermal regions.

High temperatures are also faced by the industry of catalytic processing, where around 200–400 °C can be expected. Even much higher temperatures in the range of around 800–1000 °C are generated in solid oxide fuel cells, which should be monitored. Gas turbines face operating temperature range of 1200–1400 °C, which can hardly be withstood by any existing sensing system dedicated for continuous monitoring. Extremely high temperature, even around 1450 °C, can be found at coal gasifier facilities (Figure 13.1).

Similarly to downhole operations, monitoring of process parameters in oil and gas burners, which can be achieved in them by spectral analysis of the flame at the temperature of hundreds of degrees Celsius, allows for better control and helps to increase burning process efficiency and to reduce the cost of the fuel.

High-temperature monitoring is highly desirable also in chemical industry – within various installations and processes.

Gas turbine engine operation and reliability can be improved by measurements of static and dynamic pressure at various locations [18]. The operating conditions for a pressure sensor are very demanding as not only the temperature can exceed 600 °C but also the pressure can reach around 6.9 bar.

Medical sensing applications place versatile requirements, according to the numerous needs, like autoclavability of sensors (and therefore need high temperature as well as high-pressure resistance), or their regenerability (and hence need of chemical resistance in alkaline environment [19, 20]). Regenerable biosensors

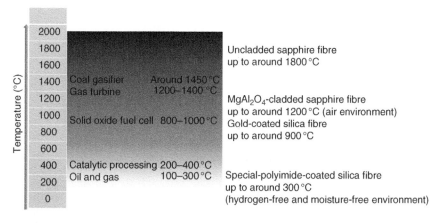

Figure 13.1 Graphical comparison of operating temperature ranges for some exemplary applications and maximum operational temperatures for sensors based on selected kinds of fibres.

with silane-based functionalization that are dedicated, e.g. for microbial sensing, antigen sensing or DNA detection, even if not used for sensing in harsh environment still require resistance to alkaline surrounding, at least up to pH 9, because of the standard procedure of their regeneration that employs alkali solution [19, 20] (alternatively an acidic surrounding, typically hydrochloric acid, can be used for regeneration). In addition, some medical or healthcare applications deal with monitoring of postural health of patients, even as wearable sensor systems [21, 22]. Thus they need sensors resistant to mechanical stress and strain. Medical applications also include imaging of ear, nose, throat, heart, etc. and optical catheters for intravenous sensing of oxygen, glucose, pressure, temperature, etc. [23].

Sensors used for structural health monitoring of various buildings, bridges, tunnels, hydraulic structures and for control of geological stability, as well as for structural monitoring of vehicles, aircraft, or spacecraft, are dedicated for remote real-time measurements of displacements, strain, convergences, cracks, etc. with sensitivity of around a micrometre. Structural health monitoring refers to the early damage detection of various man-made structures. Therefore, the sensing problem refers to changes to the material or geometric properties of a structural system that can adversely affect its performance. This can include changes of the boundary conditions and system connectivity. Fibre sensors can be mounted, for example, in or onto concrete or steel structures (Figure 13.2) in order to predict and detect small defects early. Distributed-strain-sensing OFSs mounted at critical junctures on, e.g. bridges, allow for movement and strain monitoring. Leakage monitoring of dams, pipes, soil levees, or embankments is performed by distributed temperature and strain monitoring. All such sensors need to be resistant to high mechanical loads, abrasion, and moisture (water) and additionally resistant to high

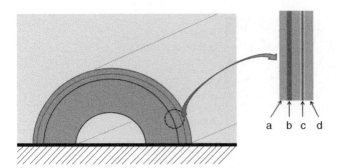

Figure 13.2 A scheme of an exemplary implementation of OFS for structural health monitoring of a tunnel, with tunnel structure shown on the left side and details of tunnel wall shown on the right side, where a, b, c, and d symbols denote a ferroconcrete secondary lining of the tunnel wall, waterproofing insulation, sensing optical cable within a braced girder, and primary lining, respectively. (The figure is based on a scheme shown in Fajkus et al. [24].)

accelerations or vibrations when mounted on various vehicles. Their proper functioning can save money and lives.

Because the average lifespan of typical civil engineering structures such as buildings, bridges, and dams is typically more than 70 years, the OFSs embedded for structural health monitoring of these objects must show similar robustness.

Another benefit of using optical fibre sensing in ships, trains, trucks, and heavy construction equipment deals with bearings pre-equipped with OFSs that are already commercially available. The efficiency of rotating equipment is improved thanks to continuous load sensing; at the same time damage to the bearing and, what is even more important, to expensive surrounding parts like gearboxes and engines is prevented by lifetime prediction and early failure detection.

Among other targets, OFSs are applied for continuous monitoring of wind turbines to predict, detect, and prevent failures before they lead to costly repairs – the most often they are installed on blades, bearings, gearbox, tower, and jacket of the turbine. OFSs can also be embedded in the blades for monitoring and by that increasing the efficiency of the turbine. In particular, fibre Bragg gratings (FBGs) embedded in wind turbine blades allow for monitoring of strain and counting the number of flex cycles in order to evaluate component fatigue.

Sensors dedicated to military (aircraft, missiles, tanks, etc.) or aerospace applications need matching special certifications [25, 26]. Nevertheless, it is expected the progressive application of OFS systems for structural health monitoring in both civil and military aircraft. One of the reasons is an increasing use of composite materials for aircraft constructions, which performance and durability need permanent monitoring (in Boeing 787 and Airbus 350 XWB already over 50% of structural components are made from composites [27]). This approach gives the advantage of decreasing weights, as well as fuel consumption and air pollution, while maintaining high strength of the construction and flightworthiness. This trend is definitely going to be continued, which means also more and more exploited distributed sensing systems, with preferred use of OFSs, based mainly on FBGs, as well as on Brillouin or Rayleigh scattering. The predominant concerns regarding the fibre performance are their resistance when exposed to radiation, large fluctuations in temperature, violent vibration, and vacuum. Since the fibres are embedded in the composite material, the critical locations for the fibres are the points of ingress and egress, as there the fibre is more vulnerable to intensive bending caused by the sharp gradients of pressure.

In space-flight field, similarly to aviation, the four most important factors that OFSs should withstand in order to avoid the sensing system failure are vacuum, vibration, excess temperatures, and radiation [25]. All non-metallic materials must present the property of not outgassing in vacuum in order to avoid surrounding contamination by any volatile and redepositing substances. The exemplary screening test is conducted at 125 °C for 24 hours below 10^{-6} Torr, and the criteria for this test are the total mass loss (TML) to be less than 1% and the total collected volatile condensable materials (CVCM) to be less than 0.1%. Regarding vibration durability tests, the overall acceleration (vibration) test level amounts 14.1 grms (the grms abbreviation refers to root mean square of the random acceleration), which is as a result of integrating the acceleration parameters over the entire spectral frequency range for the space flight, i.e. usually between 20 and 2000 Hz. Temperature range requirements are dependent on specific projects and generally are lower for LEO and GEO orbits, but some space missions may require that the OFSs withstand temperatures between −200 and 100 °C. The sensing system should also withstand radiation dose rate between around 0.01 and 0.04 rad/min during several-year-long mission.

OFSs in aerospace applications are implemented on a largest scale towards the above-mentioned structural health (strain) and temperature monitoring and as optical fibre gyroscopes.

Nuclear industry, i.e. nuclear storage and operation facilities and plants, and some research facilities such as particle accelerators or space applications, needs distributed and point sensors of temperature and strain, which are resistant to ionizing radiation.

Electric cable monitoring is also available remotely and over large lengths by optical fibres embedded into the power cables to look for hot spots along them as evidences of insulation breakdown or cable damage.

OFSs embedded within high-temperature superconductor (HTS) magnets have also been developed for real-time monitoring of strain, temperature, and irradiation and to detect quenches [28]. These crucial measurements aid the development of high-field magnets, where $Bi_2Sr_2CaCu_2O_x$ (Bi2212) and $YBa_2Cu_3O_y$ (YBCO) are the materials showing significant applicability for such magnets. Without application of suitable OFSs, which are high magnetic field resistant and radiation proof, development and operation of these magnets would have been limited, because any appropriate sensors and monitoring systems for manufacturing and operational processes had not been available. HTS-based magnets, developed for decades mainly for application in high energy physics, show great potential for high magnetic field or radiation environment applications, e.g. particle accelerators, nuclear magnetic resonance (NMR), and the plasma confinement systems for fusion reactors.

Monitoring of quality of potable water, i.e. the control over its biological and chemical purity, which in particular includes bacteria, nitrites, and organic or metal contamination, needs robust sensors resistant to long-term exposure to water.

Environmental monitoring sensors, employed in both the temporal and spatial domains, most often operate under harsh conditions (e.g. underground or subsea). They are required for detection of environmental changes in industrial, civil and manufacturing engineering, monitoring of by-products released into the environment, management of manufacturing processes, detection of chemical pollution, temperature changes, and also in agricultural engineering for understanding the environmental impact on agricultural productivity [4].

13.4 General Requirements for Harsh Environment OFSs

Generally, the above-mentioned harsh environments can be destructive when commercial standard sensors or sensorial systems available on the market are introduced to them or at least can dramatically shorten their lifetimes. Therefore, many specific applications dealing with such harsh conditions require sensors that are specially designed to operate safely there. Additionally, these extraneous factors should not limit or hinder the process of gathering information by the sensors. Harsh conditions in sensor application must be determined prior to development of such sensors and taken into account within engineering design process.

Different technological strategies and designs are in use, supporting reliability of measurements (sensing) and stability of the fibre sensors in harsh conditions. However, each real situation encountered in practice is different and therefore should be managed in a distinct manner. This often requires custom engineering or fine-tuning of the sensor system according to the local needs and towards operation within the specific constraints. Therefore, the fibre sensor systems intended to be used in harsh environments are only produced in low volumes (or short series); hence also their unit prices are distinctly higher than those for the standard products. However, it should be noticed that there are already companies offering complete systems for specialized harsh environment monitoring, e.g. the temperature-distributed sensing LIOS systems of NKT Photonics A/S [16].

The main component of such sensing systems, which plays a crucial role in the sensor, is the optical fibre itself. The fibres (and fibre cables) along with the necessary fibre connectors are often standard elements available from the market vendors, e.g. from Corning Inc., Fibrecore Ltd., Prysmian S.p.A., Alcoa Fujikura Ltd., NKT Photonics A/S, Fibreguide Industries Inc., Carlisle Interconnect Technologies Inc., Hitachi Cable America Inc., IRflex Corporation, CeramOptec GmbH, and others. Also various specialized types of fibres (or fibre cables) showing different demanded properties are already available on the market and affordable for systems engineering. Taking into account that the fibre (or fibre cable) must match specific operating conditions, depending on given application and kind of harsh environment, an appropriate initial selection of the most suitable fibre type is already crucial at the stage of sensing system design. For the purpose of the best selection of a fibre that best meets the system performance specifications while withstanding harsh environmental exposure, a system designer should be fully familiar with the relevant fibre properties, keeping also in mind an issue of cost targets meeting. While many types of specialty fibres are already commercially available, at the same time the rest of optic components of the sensing system should rather be specially designed from scratch and manufactured in order to match the specific requirements of the given project [11]. However, it should be noted that while the sensing fragment of the fibre is always vulnerable to harmful factors when introduced to harsh environment, the remaining components of the sensing system (e.g. an interrogator) can remain in safe conditions if the measurements are performed remotely.

There are a couple of different electromagnetic-radiation-transmitting materials, which have been used as a base for fabrication of optical fibres for harsh environment sensing. All they show specific advantages for the selected purposes, although they present also limitations for operation in harsh conditions, which to some extent can be mitigated by additional specialty protection measures (e.g. by special technology of fabrication, by appropriate specialty coatings, or by placing inside an optimized optical fibre cable or a probe). Therefore, these materials can be deployed for sensing beyond normal relatively safe conditions. The materials of optical fibres most commonly used for harsh environment sensing, or at least expected to be very promising, are silica glass, polymers, monocrystalline sapphire, chalcogenide glass, and polycrystalline silver halides. In contrast, fluorozirconate glass- or fluoroaluminate glass-based fibres show too many technological issues during the drawing process and are too mechanically sensitive and too sensitive to environmental influences (e.g. moisture) to be considered as good candidates for harsh environment sensing fibres. Therefore they probably could not compete with the just listed useful fibre materials, despite some very interesting and precious properties, like low attenuation (from around 0.01 dB/m for ZrF_4–BaF_2–LaF_3–AlF_3–NaF (ZBLAN) as for the exemplary fluorozirconate to around 0.1 dB/m for AlF_3-ZrF_4–BaF_2–CaF_2-YF_3 as for the exemplary fluoroaluminate, at a wavelength of approximately 3 μm).

It is important that in most applications the fibre core and a cladding (the two crucial components of a fibre ensuring the waveguiding effect) should be sufficiently protected against possible environmental threats by a fibre coating. Thus a coating must prevent from various types of damages that otherwise could weaken the fibre and compromise its performance. It is mostly the robustness of such coating what ensures safeness for the whole fibre in various harsh conditions. Therefore, its properties should appropriately match the given environmental conditions.

The concrete solutions regarding different coating types are described in further sections, but there are also some selected cases described further below, where only the specialty core (e.g. showing advanced core chemistries) can provide stable properties to the fibre. The coating can be composed of more than one layer. For example, in silica fibre coatings there are often two layers: an external layer of the coating is often referred to as buffer; though it should be noted, sometimes just the whole coating itself is referred to as coating buffer or just buffer. For extra protection in some fibre designs, an even more resistant jacket, e.g. made from special polymer, can be extruded over the initial coating. Also when a number of fibres are collected inside a cable, the fibre arrangement and the cable design, e.g. strength members, jackets, and additional fulfilment (e.g. moisture or hydrogen scavengers), which are used as specific protection, should correspond to the concrete harsh conditions.

Additionally, the structure of the light modulating region (interaction region), e.g. a Bragg grating [29] or long-period grating (LPG), a metallic coating for surface plasmon resonance (SPR) sensor or a dielectric coating for lossy mode resonance (LMR) sensor, a mirror structure for Fabry–Pérot resonator-based sensor, etc., depending on the applied sensing strategy, must also remain resistant to harsh conditions in operating environment (the special optical coatings for SPR and LMR sensors are fabricated exclusively within the interaction regions of their sensing fibres and should not be confused with protective fibre coatings that do not influence the optical properties of the fibres). Therefore, a method of fabrication of such interaction region should also match the type of harsh conditions for which the sensing fibre is dedicated. In particular, within a few methods of FBG fabrication for use in extreme environments, such as high temperature, pressure, or ionizing radiation, promising are femtosecond-pulse-duration infrared (fs-IR) laser-induced gratings (e.g. for silica and sapphire fibres) and the so-called grating regeneration process (the latter only for temperatures below 1000 °C).

13.5 Silica Glass Optical Fibres for Harsh Environment Sensing

In most cases of harsh-environment-resistant OFSs, the fibres based on silica glass (fused silica, fused quartz) are applied. In most applications, their cores and claddings are made exclusively from silica glass (all-silica fibres). Additionally, polymer-clad (or plastic-clad) silica (PCS) fibres are applied, in which low-refractive-index polymers are used for optical cladding fabrication. Silica fibres have been developed and used for decades in more and more improved forms for telecommunication. Hence their technology attained the most mature level of development in comparison with other fibre materials. They are widely available at reasonable cost. They show ability to be designed and fabricated while ensuring unique features, as well as tunable optical properties, matching wide variety of sensing applications. Nowadays, all-glass fibres have an undisputable advantage, which is very low attenuation, already only around 0.25 dB/km for single-mode fibres [30, 31]. Though the attenuation of multimode fibres is by an order of magnitude higher, it is low enough to allow for application of very long fibres for remote sensing (obviously their main application is still the connection of long-distance telecommunications).

PCS fibres show higher attenuation; hence they can be used for medium- or low-distance sensing applications in comparison with all-glass fibres whilst they show transmission properties that are intermediate between POFs and all-silica fibres.

The silica glass of the fibres shows quite large natural resistance to a wide spectrum of chemical environments, at least in the temperature range between room temperature to approximately 300 °C. However, it still shows susceptibility to water when it is subjected to mechanical stress and to some chemical agents, particularly when under the harshest possible conditions. Therefore, when high pressure or temperature comes into play, the standard fibre coatings have to be replaced by specialty versions.

Silica fibres are also relatively susceptible to mechanical abrasion. Therefore, fibre coatings are necessary even for the operating conditions comparable to standard telecommunication. A coating of a silica fibre, in particular, ensures its mechanical stability.

Indeed, the most obvious primary role of a coating is a mechanical protection of the fibre silica glass from scratching that without this protection would reduce the necessary strength of the fibre. Therefore, what is primarily demanded of a fibre coating, in terms of its own necessary mechanical properties, is its abrasion resistance, which should be adequate for the operation environment. It can also be noted here that the glass fibre coated by a polymer represents a kind of a simple fibreglass (i.e. a glass-fibre-reinforced plastic). The fibreglass by its nature shows mechanical strength even higher than aluminium, copper, or steel. Polymers have been used for over 35 years for fibre coatings in telecommunications as well as for specialty fibres that are more and more often used in other branches, like sensors. Obviously, a coating should also exhibit resistance to any other potentially dangerous factor present in an operating environment of a fibre. In harsh environment it must stay robust not only to any mechanical degradation (e.g. peeling, flaking) but also to thermal (e.g. melting, burning off) and chemical degradation, and in particular its dimensions and flexibility must remain the same in order to avoid attenuation of transmitted light within microbends. Therefore, in the specific environmental conditions and for the reasons described further below, the polymer of a coating must be replaced by other more relevant materials. In this sense, it must be remarked that the mere degradation of a coating, whatever the coating material, can easily lead to catastrophic decrease of a mechanical strength of the whole fibre and hence to its mechanical degradation.

At the same time, a major factor that limits the maximum operating temperature of a silica glass fibre is its coating, i.e. thermal stability of silica itself is significantly higher than that of the most types of applied coatings. Namely, one should take into account that silica fibreglass can be well applied at temperatures up to approximately 900 °C [32], although already above 500 °C embrittlement of a silica fibre is an issue [32], which induces using special protection and packaging inside a cable.

In any case, temperature as high as 900 °C is a way above degradation temperature of nearly any coating of a silica fibre. The only exception here is a gold coating. On the other hand, it should also be explained here that although the silica glass softening and deformation temperatures are as high as about 1600 [33] and 1700 °C [34], respectively, already at lower temperatures, some processes take place, limiting the long-term stability of silica fibre or influencing its properties. These are 'intrinsic' issues of silica glass at elevated temperatures and cannot be mitigated by any additional or 'external' measures (like better coatings or other means of protection). Namely, over 900 °C, instabilities in a glass appear, such as dopant diffusion and crystallization of the amorphous silica network [32]. Next, in the range of 1000–1100 °C, a process of internal strain relaxation in internal network of silica is initiated [32].

However, by using appropriate coatings, it is possible to enhance the thermal performance of the standard fibres, e.g. towards high-temperature sensing applications. For example, the standard telecommunication fibres, coated with typical acrylates, are designed only for a temperature range from about −55 up to 85 °C. The standard fibres degrade optically when used outside this range. On the other hand, a fibre with properly chosen coating allows for sensing applications at extended temperature ranges, which cannot be supported by the standard fibres. For instance, when a special high-temperature acrylate or silicone or polyimide coating is used, it allows for applications at up to 150, 200, or 300 °C, respectively [10, 35]. For the special polyimide coatings, even short temporary exposures to the temperature up to 400 °C are possible [35]. Such polyimide coatings remain tough with thicknesses of merely 10–15 µm [12]. Additionally, the special silicone and polyimide coatings can withstand high pressures – even over 1 kbar in the latter case. One example is the special polyimide-coated fibres fabricated by Fibretronix AB [10], in which the high-pressure resistance of the coating is achieved by its increased adhesion to the fibre. High pressures can lead to peeling off for an ordinary coating, followed by exposure of the vulnerable fibreglass surface. However, such high-temperature and high-pressure-resistant coatings give opportunities for OFS applications in oil and gas downhole industry, as well as in medicine, where autoclavable sensors are required. On the other hand, silicone can outgas volatile components at elevated temperatures [36–38]. Therefore, it may not match the requirements of some applications. In spite of this problem, special silicone coatings show additional resistance to ionizing radiation.

As it has already been mentioned, low-refractive-index polymers, with the notable example of fluoropolymers, are also used for fabrication of claddings in PCS fibres. Fluoropolymers are particularly useful as polymer claddings, as fluorination in general decreases their refractive index value but at the same time increases hardness and thermal and chemical resistance. Such fluoropolymer-clad fibres are known also as hard-clad silica (HCS) or hard-polymer-clad silica (HPCS) fibres.

Anyway, it should be emphasized here that by far the most heat-resistant coatings are made from metals. These coatings, mostly fabricated in the thickness range of 15–60 μm [12], withstand temperatures exceeding 300 °C, which are unavailable for polymer-coated fibres. For example, the operating temperatures for aluminium-coated (Figure 13.3) or copper-coated fibres reach 400 °C [10, 39], while for copper-alloy-coated fibres they reach 600 °C [40], and for gold-coated ones (Figure 13.3), they significantly surpass all the above-mentioned temperatures reaching even 700 °C [39]. Even around 1000 °C as a safe temperature for some applications was reported [41].

Another valuable feature of metallic coatings is that they do not ignite, which is a valuable feature particularly in oil and gas industry. These properties of metal-coated fibres are particularly precious for downhole oil and gas industry, turbine-machinery or jet-engine monitoring, high-temperature manufacturing (like in monitoring of deposition processes used for semiconductor or solar panel fabrication or in steel or glass manufacturing), oil or gas burner (furnace) operation monitoring, and sensing for military and aerospace. In these fields extremely high temperatures come into play, and any type of polymer-coated fibre cannot withstand them.

However, it should be mentioned here that using metals for fibre coatings causes also numerous technological problems, being, for example, the result of large difference of thermal expansion coefficients of glass and metal and hence a reason of undesired enhanced microbend losses in metal-coated fibres [10]. Additionally, the process of coating by aluminium or copper requires strict oxygen-free conditions, because of high chemical affinity of these metals for oxygen. It is also necessary to ensure strict temperature control over the coating processes, particularly for the gold coatings, in which fabrication takes place at ultra-high temperature (approximately 1100 °C) and requires using special non-metallic crucibles and handling equipment. Moreover, the coating process must be performed just after fibre drawing by application of the metal from its melt, which makes the overall fabrication process very complicated.

However, an advantage of metallic coating technology is that it can be widely applied for all kinds of fibres, regardless of their diameter (for single-mode as well

(a)

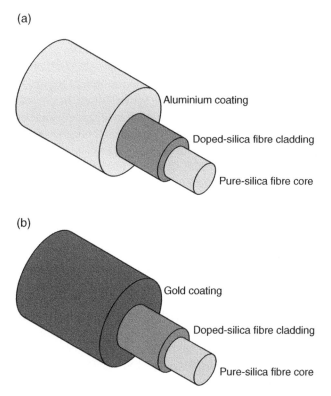

Aluminium coating

Doped-silica fibre cladding

Pure-silica fibre core

(b)

Gold coating

Doped-silica fibre cladding

Pure-silica fibre core

Figure 13.3 Schematic cross sections of an exemplary gold-coated silica fibre (a) and aluminium-coated silica fibre (b). The figure is based on schemes shown in [39]. *Source:* Reproduced with permission of IEEE.

as for different multimode fibres, e.g. step-index and graded-index ones), while a serious drawback of gold-coated fibres is the high price of their coating material.

At the other extreme of the temperature range available with specially coated fibres, special polyimide-coated fibres and cables allow for cryogenic applications, as they withstand even the temperatures dropping to approximately −180 °C. This is a valuable feature widely used by liquid natural gas facilities and pipelines for their temperature-monitoring-based leakage control. Even much lower temperatures of optical fibre operation, reaching around −270 °C, are available with the use of metal-coated fibres, with coatings based on aluminium or gold [39]. Additionally, metal-coated fibres show the advantage of not outgassing in vacuum, which is beneficial, e.g. for aerospace sensor applications.

A proper coating should also protect the fibre against any aggressive chemicals present in its operating environment. However, it is important to notice that even water (high moisture) would act as a destructive factor for a fibreglass coated by polymer, as it accelerates its damage by crack propagation under mechanical stress conditions (i.e. tensile load or bending stress) – a phenomenon known as stress corrosion, static fatigue, or delayed failure [42]. Such increasing cracks, enhanced by moisture that penetrates the polymer coating and attacks the fibreglass surface, lead to eventual fibre breakage [10, 43–45] and sensing system failure. In general, polymers do not allow for fabrication of hermetic enough coatings to water or hydrogen. An exemplary way of effective fibre protection used in this case is a hermetic coating based on amorphous carbon [43, 46–48] (like in the case of specialty fibres fabricated by Fibretronix AB [10]), which allows for longer lifetime of the sensing system when fibre stays in moist surrounding conditions. Additionally, following the necessary further optimization, such a tight carbon coating was demonstrated to be effective as a protection against hydrogen diffusion into the fibreglass (also this kind of fibres was elaborated by Fibretronix AB). The carbon coatings are fabricated by chemical vapour deposition (CVD) technique with thicknesses typically below 50 μm [12]. In addition, their pinhole-free nature is crucial for a hermetical protection.

The ability to protect against hydrogen and water is also a well-known property of metallic coatings, which are used for some specialty optical fibres [39]. Like in the case of amorphous carbon coatings, they operate in contact with water and hydrogen as hermetic seals. In wet surrounding, even under high mechanical stress, the metallic coating does not allow for water diffusion [39]. Regarding hydrogen diffusion into the fibre core, it is a well-known reason of strongly enhanced optical background losses in the case of standard optical fibres, being a serious problem in fibre sensing of downhole oil and gas exploration and exploitation. Therefore, use of exclusively hermetically coated fibres is a valuable option for the sensing in downhole gas and oil extractive industry, similarly as in any other case of sensing in hydrogen-rich environment. The appropriately coated fibres are able to safely operate in hydrogen-rich environment, and the specialty carbon-based or metallic coatings block diffusion of concentrated hydrogen even at elevated temperatures. For example, the carbon-coated fibres of Fibretronix AB do not show any enhancement of radiation attenuation that would be associated with molecular hydrogen (H_2) or Si-OH-related absorption (the broad absorption bands centred at the wavelengths of 1240 and 1385 nm, respectively) even after exposure to hydrogen pressure of 1 bar at the temperature of 170 °C [10]. Such optical fibres hermetically coated by carbon film are commercially offered nowadays, but it should be mentioned their promising properties have already been known for about 30 years [47].

Though such carbon or metallic coatings are normally called 'hermetic', it looks that a more adequate name for them would rather be 'quasi-hermetic'. The point is that their water and hydrogen resistance is only a matter of kinetics of diffusion and incorporation of these chemical species into the coating (in other words they are kinetically resistant). The process of diffusion of water or hydrogen is only slowed down by an appropriate coating, as a result of a large activation energy for incorporation process, which for hydrogen is nearly twice larger in carbon than in polymer. Actually, the elevated temperature exacerbates the problem of diffusion, leading to faster eventual core crack (by water) or deterioration of the core transmittance (by hydrogen). Therefore, the water and hydrogen resistance of such specially coated fibre sensors cannot be considered separately from the surrounding temperature, as well as from the surrounding pressure. Hence such an approach relying on kinetic stability rather than on thermodynamics still remains an issue for the long-term use at the highest possible operating temperatures. As a result, very long-term application of specially coated fibre sensors in hydrogen- or moisture-rich conditions is possible but only at temperatures lower by a few hundred degrees Celsius referring to those working in dry air or inert environment. Actually, in the case of hydrogen-rich environment, this limit of safe temperature is only around 150 °C [12], but for water surrounding it is slightly higher.

Regarding carbon coatings, although they help in suppressing water-related delayed failure (at least at medium temperatures), their drawback is that they promote brittle fracture to silica fibres and, therefore, they are a cause of approximately 50% decrease of tensile strength compared with non-coated fibres [12].

Consequently, another alternative strategy for fibre core protection against selected harsh factors (e.g. hydrogen) is to adequately modify the properties of the core itself, instead of using a special coating. In this sense, one possibility is to fabricate a fibre with a pure silica core (PSC). By a strict control of its chemical composition, it is devoid of any germanium doping, unlike the contemporary standard core fibres. It must be pointed out that hydrogen reacting with germanium is a reason of an additional and broad absorption band of around 200 nm with sharp increase of intensity already below approximately 1200 nm observed in conventional germanium-doped fibres. Therefore, making the fibre core germanium-free nearly eliminates this type of absorption issue [10], even if the exposure to hydrogen is under 1 bar pressure and the temperature is elevated to 250 °C for up to a few weeks. However, as a result of a lack of any effective barrier, the hydrogen diffusion to the PSC still takes place when the fibre is located in a hydrogen-rich surrounding, still resulting in some absorption at the 1240 and 1385 nm bands, which is particularly important for distributed optical fibre sensing of temperature, e.g. for downhole control in oil and gas industry, where the used sensorial signal wavelength is often equal to 1064 nm.

The already mentioned carbon-based and metallic coatings, which maintain tightness in moist and hydrogen-rich environments, are also able to protect the fibreglass against other chemicals more aggressive than pure water. Apart from not being tight enough in moist surrounding and allowing water to diffuse through, polymer coatings can be dissolved or degraded by a large number of chemical substances, e.g. oil, acids, alkali, sodium chloride solution (e.g. seawater), and many others. Such a complicated issue is found in downhole oil industry where, apart from oil itself and hot water, also many other chemicals are present at elevated temperatures. A mixture of these chemicals is used towards the described so-called 'enhanced recovery' industrial processes. Moreover, the susceptibility of polymer-coated fibres to seawater precludes their applications in marine environment. Carbon-coated fibres, with amorphous carbon coatings deposited directly on the fibreglass surface, with a thickness merely below 500 nm, proved to be resistant to 49% hydrofluoric acid aqueous solution at 23 ° C [47] or 5% sodium hydroxide aqueous solution at 95 °C [47]. The protection is effective even for fibres tested under high tensile stress exceeding 17 kbar, where reference standard polymer-coated fibres degrade at such conditions within a few hours.

Coming back to PCS fibres, they are also used for manufacturing some types of fibre sensors, e.g. for the sensors which operation is based on LMRs [49, 50]. This is in particular because of relatively low cost of these fibres and relative simplicity of LMR technology. In LMR sensor [51] manufacturing, the original cladding is removed from the selected section of the fibre, and a dielectric functional coating with specially matched optical properties, i.e. mainly the real part of refractive index, which should be higher than that of the fibre core, is deposited directly on the core [50]. In this way the LMR sensing section is made in the fibre, and the application of PCS fibres gives the advantage of a simple process of cladding removal. Specially designed LMRs can show resistance to selected harsh factors, e.g. harsh chemicals. This can be achieved by using a functional coating matching not only the necessary optical requirements but also additionally showing some special properties, e.g. chemical resistance. For example, tantalum oxide, which is known as a relatively chemically resistant material, can be used for such a coating fabrication [50]. The appropriately deposited chemically resistant coating can cover not only the sensing section but also the whole surface of the fibre, together with the native polymer coating of the fibre, thus ensuring chemical resistance in the whole fibre.

As it has already been mentioned, ensuring chemical resistance to alkaline solutions is necessary for biosensors with silane-based functionalization when the sensors are to be regenerated and reused in multiple measurements. The biosensor regeneration based on the removal of the 'old' silane functionalization layer and biolayers after measurement is one of the sensor regeneration methods and allows for multiple repeated cycles of covering the fibre by the new functionalization and performing measurements with the same biosensor. The regeneration procedure used for functionalization removal employs pH 9 solution. Such reusable biosensors based on fibres with induced LPGs were fabricated with tantalum oxide optical coatings. The coatings appropriately matched the sensorial properties of the fibres and, at the same time, ensured their resistance in alkali solutions [19, 20].

Ionizing radiation is another factor that increases optical losses in the fibre core, as well as degradation of polymer-based fibre coatings [52]. Therefore, it is destructive for the fibre sensors and reduces the sensor system lifetime, which is obviously a serious issue in the case of some special sensorial applications, e.g. in nuclear industry, particle accelerators, or space applications. Again, the problem can be mitigated by the employment of special fibre coatings, as a tight external protection, or by using special cores, which intrinsic properties, mainly the composition, determine their ionizing-radiation hardness. Usefulness of special metallic coatings, instead of polymer-based ones that are vulnerable in ionizing radiation, was demonstrated by Fibretronix AB [10]. Aluminium coatings proved to slow down the damage caused to sensorial fibres used in nuclear plants by two orders of magnitude, even at the operation temperature of 400 °C [39]. As it has already been mentioned, metallic coatings are also a method of mitigating the issues of a very high temperature, vacuum, hydrogen-rich, or moist environments. Also the already mentioned PSC fibres, which require extreme chemical purity of the silica core, are a type of special core fibres showing increased robustness in ionizing radiation, as they are highly resistant to induction of the radiation-based optical losses. On the other hand, suitably chosen dopants introduced to the fibre core glass can also do the job in specific conditions. For example, a silica-based core can be protected against gamma radiation, which is a reason of its darkening, by doping with cerium because this element acts as an annihilator of colour centres in silica.

An overview and classification of basic types of OFSs and basic sensing principles that are used for different measurement strategies has already been presented in the first chapters of the book. Regarding distributed sensing, it is worth remembering and highlighting that this technology permits to measure different parameters in harsh environment with a single system. This fact simplifies the collection and monitoring of diverse measurements. For temperature or strain monitoring of spatially extensive objects around the world – buildings, dams, subsea umbilical cords, pipelines, downhole applications, wind turbine blades, etc., the most often is performed using quasi-distributed or distributed sensing, either through FBG, LPG, or Brillouin scattering-based sensors. In such applications the quasi-distributed sensing (by grating-based sensors) or distributed measurements (based on distributed sensors) are used to replace interferometric sensors, which are much more suitable for single-point detection [27]. For distributed sensing of temperature or strain used for unstable geological terrains and phenomena, e.g. soil erosion, expansive soils, slope failures, ground subsidence, river channel changes, glaciers, and coastal erosion, whether human induced or of natural origins, the most often distributed sensing based on Brillouin, Raman, or Rayleigh scattering is applied [27, 53]. Fibre grating generated in an optical fibre is a periodic perturbation along given section of the optical fibre core. This perturbation in the fibre may be made either in the form of refractive index, thickness, or density of material. Most of the fibre gratings have periodic refractive index modulation in the core, which poses perturbation to the optical path. FBGs have already become a key enabling technology for distributed strain and temperature sensors. Such distributed sensing using optical fibre allows for monitoring over many-kilometre distances (e.g. even tens of kilometre-long tunnels) with a few-millimetre-dense spatial resolution, thus replacing thousands of traditional point sensors – a feature quite unique and unavailable for any other sensing systems. For geohydrological monitoring of levees or embankments also geotextiles or geogrids, embedded with silica or POF-based sensing fibres have been investigated [54]. However, the tightly bonded polymer protective sleeves can be a useful protection of FBG-induced optical fibres (that are more generally fragile than the fibres without gratings) when embedded in geogrids, additionally ensuring strain transfer from the geogrid to FBG. For multi-kilometre-long monitoring scales, the sensing fibres can only be based on silica in order to ensure high enough transmission. As a commercial example dealing with such applications, Fibrecore Ltd. offers two ranges of highly photosensitive (PS) fibre for FBG fabrication. The first one is the PS range of high-germania boron co-doped fibre for writing high-reflectivity gratings without the need to hydrogen-load the fibre. The second one is the high-germania, bend-insensitive SM1500 fibres, containing fivefold enhanced germania concentration in comparison with standard telecommunication fibres – this enables gratings to be written with or without hydrogen loading while maintaining low attenuation around 1550 nm. Additionally, the erbium-doped fibres offered by Fibrecore Ltd. (IsoGain™ and MetroGain™) are capable of supporting FBG fabrication for distributed feedback lasers and light sources.

On the other hand, NKT Photonics A/S offers the complete system of distributed temperature and strain sensing based on Raman scattering, called LIOS [55]. The monitoring with LIOS is possible with 1 m spatial resolution. The LIOS sensing system is cost effective in large installations of many-kilometre sizes and durable in harsh environments. Based on the statistical field analysis, mean time between failures (MTBF) of over 45 years is declared for LIOS. Hence it matches the requirements of critical applications such as oil and gas exploration, fire detection in road and rail tunnels, special hazardous buildings, power cable and aerial transmission line monitoring, and for industrial induction furnace surveillance, in which it has already been installed worldwide. The NKT company provides full solutions from the early design phase, over installation and commissioning, to service and maintenance once the system is deployed. Over 5000 LIOS sensing systems based on OFSs have already been installed worldwide.

Even the traditional downhole cable, targeted to withstand temperatures elevated up to 150 °C (or as low as −40 °C), pressure up to 1400 bar, and corrosive or hydrogen-rich environments, is equipped with a loose-tube design, hydrogen scavenging gel that surrounds the collected fibres, inner stainless steel tube, polypropylene inner jacket, outer stainless steel tube, and outer encapsulation.

For LIOS sensing systems, the fibre-in-metal-tube (FIMT) concept is employed, which is a hermetically sealed and rugged construction for very long sensor lengths that increases their mechanical stability. It also particularly effectively protects the fibres against hydrostatic pressures, high-temperature effects, and corrosive environments, just like in EOR downhole applications at temperatures reaching 450 °C in hydrogen-rich surroundings. Additionally, as the fibre sensor is to be exposed to drastic and rapid changes in temperature and pressure, the excess fibre length (EFL) loaded into the metal tube during manufacturing is adjusted to the sensing system requirements. It is done by a kind of a loose-tube design, which protects the fibres from tension due to temperature changes even better than the tight-buffered cable. The purpose of this is to allow for relieving stresses that are related to different coefficients of thermal expansion of the materials in the structure of FIMT. A wide variety of high-quality FIMT designs is offered within LIOS systems.

NBG Holding GmbH offers special harsh environment sensing cables that are protected by multiplied layers of steel, up to four layers. Various grades of steel can be applied according to the customer need, as well as special steel grades (316L, 825, 625), with cost-effective steel-grade mixing [56]. Special high-temperature scavenger gels are used as well. The cables can withstand temperatures in the range of between −190 and 600 °C, which depends mostly on the fibre cladding material (carbon and metallic claddings are also offered).

For high-temperature measurements up to 1000 °C with fused silica-based fibres, apart from distributed sensing systems, also various point sensors have been demonstrated with different kinds of optical structures such as Mach–Zehnder interferometers [57] or Michelson interferometers [58].

13.6 Polymer Optical Fibres for Harsh Environment Sensing

The restricted flexibility of silica fibres generates some issues in specific applications, in particular special sensing, where particularly intensive bending or stretching of the fibre has to be taken into account. In such cases plastic (polymer) optical fibres (POFs) have been investigated for several years and are more and more often preferred vs. glass-based fibre sensors in specific conditions, where high temperatures or harsh chemicals do not interact at the same time. In contrary to silica fibres, POFs show high flexibility. However, POFs typically have considerably higher attenuation than glass fibres (typically 1 dB/m or higher). Even for the special low-attenuation perfluorinated graded-index POFs, their attenuation is still by an order of magnitude higher than that of multimode silica fibres [59]. Such higher attenuation of POFs limits the range of POF-based applications. Namely, POFs can be chosen instead of glass fibres when rather short distances are expected, and hence the attenuation is only a minor issue. In the case of sensing, they are very convenient, e.g. for medical tests or monitoring (breathing cycles or respiration in wider terms [60, 61], joint movements [21], or spinal posture [22]). In particular, POFs are perfectly suitable for wearable sensing, wearable health monitoring (Figure 13.4) [21, 22, 61], etc. This was discussed in detail in Chapter 7, devoted to biomechanics.

However, POF sensors are also applied or at least seriously considered beyond medicine and healthcare sensing in diverse fields including structural health monitoring, automotive industry, aeronautics and aerospace or robotics, etc. Many of the applications are within flexible and stretchable foils or skins [62]. They are also

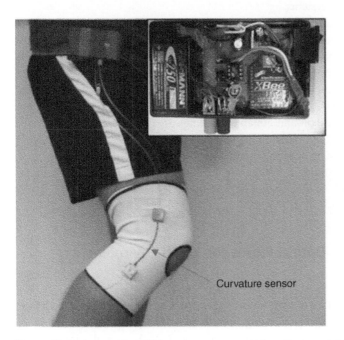

Figure 13.4 A wearable POF-based sensor mounted on the knee joint brace, with electronics placed in the box (see the inset in the picture) and attached to the patient's belt – in the bottom picture. *Source:* from [21]. Reproduced with permission of IEEE.

convenient for sensing (though still most often used just for data transmission) when the gained data are transmitted only locally and particularly when the fibre must be tightly packed in a limited space. So, they have been considered for application, e.g. in aircraft, helicopters, and ships, particularly for military and aerospace needs like tanks, missiles, and spacecraft. The elasticity of POFs and also their resiliency to mechanical shock or vibration, easy handling, and lightweight, are the advantages that justify their use in the above specific sensing applications. Their cores are most often made from poly(methyl methacrylate) (PMMA), and their claddings from some fluorinated polymers. Such polymers show satisfactory transparency, which allows for attenuation losses not exaggerated on short distances, though as it has already been mentioned their transmission performance is not able to surpass that of all-glass-based fibres. Their fluoropolymer-based claddings show very conveniently low refractive index value, and, therefore, they can ensure satisfactory optical confinement [63]. For short-distance and low-speed sensing applications, PMMA-core POFs are generally a lower-cost solution than silica-based fibre systems. However, technology of POFs is still under development. For example, certain fibre manufacturers, e.g. Omron, Asahi Kasei, and others, have already developed the multicore high-flex fibre structures [64, 65] that are equipped with multiple independent cores. Thanks to this solution, the attenuation related to fibre bending was practically reduced to zero, even for bending radius as small as 1 mm, which demonstrates the outstanding mechanical flexibility of POFs. Such a small bending radius is at least several times smaller than the radius of some silica fibres [66]. In this sense, it should be taken into account that, even if a silica fibre can be bent to some extent without mechanical damage, its bending-related attenuation already appears for much larger bending radius. However, at the same time, POFs still cannot surpass, for example, the thermal resistance of silica-based fibres (particularly when the latter are coated with an appropriately resistant material). For example, the Asahi Kasei high-temperature POF-based fibre cables can only operate up to 105 °C [65], while the lowest operating temperature is −55 °C. The relatively heat-resistant POF cables are offered by Hitachi Cable America Inc. Their fibre cores are not made from PMMA, but they are based on silicone or acrylic, giving these alternative materials the maximum operating temperatures of 150 and 130 °C (for 3000 hours), respectively [67]. At the same time, however, they show increased values of the minimum bending angle, i.e. 35° and 20°, respectively [67], which is a drawback in comparison with the above-described PMMA-core POFs.

Because fluoropolymers show increased chemical and thermal stability in comparison with most of not fluorinated polymers, they are also often used for coatings in various types of optical fibres or for buffers of optical fibre cables. In the case of fluoropolymer-cladded fibres (e.g. for POFs and for PCS fibres), their claddings can be intentionally thickened, and also the appropriate coatings can be obtained in this way. Alternatively, even more chemically and thermally robust fluoropoly-mers can be applied as coatings [35], e.g. polytetrafluoroethylene (PTFE), a per-fluoropolymer showing strong chemical inertness and thermal resistance at least to approximately 200 °C, which is typically referred to under the brand name Teflon [68], or ethylene tetrafluoroethylene (ETFE), a fluoropolymer showing strong chemical inertness and thermal resistance at least to approximately 150 °C, and even higher toughness than PTFE. Like PTFE, also ETFE is typically referred to under a brand name, in this case Tefzel [69].

Anyway, the glass fibres, particularly equipped with robust enough metallic or carbon coatings, are generally still preferred to POFs when the sensing fibre is to be directly exposed to high temperatures or to elevated temperatures and reactive chemicals (including harsh solvents) at the same time. This is true particularly because the thermal resistances of PTFE and ETFE are far from those of metallic coatings.

Hence, the above-described drawbacks of POFs hinder wide applications of POF-based sensors in military and aerospace, where specific attestations and cer-tificates are necessary, before the product can start to operate commercially. The elevated-temperature-operating, flame-retardant flexible fibres are the minimum goal for such specialty POFs [26]. The LITEflight® POF Fibre Optic Cable, pro-duced by Carlisle Interconnect Technologies Inc., is probably the only aero-space-qualified POF-based structure offered on the market [70].

There is also an important economical aspect of POF sensor technology. Because POFs have rather larger diameters (around 1 mm) than silica glass fibres (typically a fraction of millimetres), their components, e.g. connectors, are less complex and cheaper and show lower probability of damage than their counter-parts used for glass optical fibres. The lower complexity of POF-related connec-tors stems from different level of tolerances in comparison with silica fibre solutions. Hence, such components can be fabricated from inexpensive plastics rather than the precision-machined stainless steel or ceramics that are required by silica glass fibres [26].

13.7 Chalcogenide Glass and Polycrystalline Silver Halide Optical Fibres for Harsh Environment Sensing

Chalcogenide glass-based [71] and silver halide polycrystalline-based [72] optical fibres have been developed for years as they show a valuable feature of broadband infrared (IR) transmission. Chalcogenide glasses show very wide range of trans-mission wavelengths, transmitting quite well (with low energy losses) typically in the range between around 2 and 20 μm, depending on their chemical compo-sition [73]. For example, the $Te_2As_3Se_5$ (TAS) glass shows the minimum of attenuation below 1 dB/m in the wavelength range of 6.5–9 μm. Silver halide polycrystalline fibres show also broadband IR transmission, in the range from a few micrometres up to several micrometres. In this sense, it is important to remark that the transmission windows of these materials, extending far in the mid-infrared (MIR) region, cover usually the two important atmospheric windows: 3–5 and 8–12 μm. Therefore, they match the spectral requirements of specific applications that need such long operating wavelengths, such as near-infrared (NIR) or MIR range sensing or thermal imaging. The OFSs based on these materials operating in the IR region, where optical signatures of chemical molecules and in particular biomolecules are located, are already applied and still show a high potential of development as analytical techniques giving *in situ* and real-time information about chemical or technological processes, medical or biological samples, metabolic patterns, etc. For the same reason, they are valuable for public and environmental safety in detection and monitoring of a variety of hazardous chemicals such as toxic gases, atmospheric pollution, gas emissions from various industrial processes, explosives, industrial chimneys and greenhouse gases. There are also many other potential fields of application for sensors based on such IR-transmitting fibres, such as aerospace, military, etc. However, in comparison with the mature technology of silica, these fibres are just emerging from their infancy.

In the group of chalcogenides, the requirements for drawing are particularly satisfactorily matched by binary and ternary sulphides, selenides, and tellurides [71], and their numerous chemical compositions can be considered for optical fibre fabrication. They quite often can contain arsenic, germanium, or antimony, and they are relatively stable and durable (though still much more delicate than the oxide glass fibres), with the additional advantage that they lack a susceptibility to moisture. However, their drawback is that they transmit rather poorly in the visible (VIS) spectral range. Additionally, their refractive indices are high, which generate issues regarding appropriately matched optical claddings for the fibres.

The polycrystalline silver halide fibres became a more convenient alternative for halides of thallium, which were invented earlier but occurred to be toxic. Silver halides show generally higher temperature resistance than chalcogenides and are generally high pressure and water resistant [74]. Nonetheless, they still have some serious drawbacks, compared with silica glass fibres. Their optical quality is relatively good but remains strongly susceptible to air contamination – a reason of transmission loss. They are also capable of undergoing photodegradation under exposure to ultraviolet (UV) and VIS radiation. Additionally, they are corrosive to many metals, so special connectorization of the silver halide fibre cores using titanium, gold, or ceramic materials is necessary. Regarding the ageing effects, the fibre transmission losses typically increase with time. Moreover, the extrusion method, which is used for their fabrication, is expensive and hard for implementation. Chalcogenide and silver halide fibres are commercially offered by companies such as IRflex Corporation [75] and CeramOptec GmbH [76].

Some of the most promising implementations of both chalcogenide-based and silver halide-based fibres are in optical fibre evanescent wave spectroscopy (FEWS) and for attenuated total reflection (ATR)-based probes. Both methods, as optical-fibre-based analyses, allow for real-time *in situ* remote control. The principle of this FEWS is based on an evanescent wave generated directly at the interface between the unclad (core-only) fibre and the surrounding area. The transmission of light propagated in the optical fibre is reduced selectively, being partially absorbed at each reflection. This wavelength selective reduction of the fibre transmission depends on absorption bands in the IR spectral region, shown by chemical or biological species being in direct physical contact with the fibre. Therefore, it carries information about the surrounding composition measured and analysed within the FEWS method. What is important is that in FEWS the uncladded fibre serves as both the optical waveguide transmitting radiation and the internal reflection element (IRE) at the same time. In ATR probes this basic concept is developed towards using special cone-shaped high-refractive-index IREs. Such probes have already been commercialized and are available from art photonics GmbH as harsh-environment-operating ATR immersion fibre-optic probes [77, 78]. They are able to work at temperatures elevated up to 250 °C and at high pressure, even up to 300 bar. Their wide range of operation temperatures starts from −100 °C. However, their thermal resistance cannot compete with silica-based fibre or sapphire-based sensors. The robustness of these probes in harsh environments is ensured by their special design, namely, they are constructed using special shaft made of hastalloy, with airflow cooling inside it, which protects the ATR tip placed inside the shaft. Some different types of the probes with patented design are available, with operating parameters depending on the particular type of the probe. They are equipped with ATR tips made from different materials (diamond, ZnSe, and silicon for the silver halide fibres or ZrO_2 for the chalcogenide fibre), cone-shaped for immersion in liquid flow without dead zone. The fibres and the shafts are available with optional total lengths of 1–5 m and 300–700 mm, respectively. The fibre minimal bending radius is 130 mm. The probes are flexible for industrial applications of NIR and MIR fibre-based ATR and are compatible with all spectrometers and automated process interfaces produced for any type of FT-NIR, FT-IR and other IR spectrometers, photometers, and IR LED-based or QCL-based spectral sensors. They are suitable for real-time reaction monitoring in laboratories and pilot plants and for full-automated process control, being offered as a particularly useful tool in remote polymerization control, crystallization process screening *in situ* IR spectroscopy, and process analytical technologies (PATs) in chemical, petrochemical, atomic, biopharmaceutical, and food industry.

Based on chalcogenide as well as silver halide, optical fibres can be used as representatives of IR waveguides for thermometry [79, 80]. The measurements can be performed at temperatures elevated up to a few hundred degrees Celsius. As an example, a silver halide two-channel thermopile-amplifier sensor was proposed [74], where the thermometer was based on two AgCl–AgBr solution core–clad fibres placed in two different probes (the signal and the reference probe) equipped with heads showing different emissivity values and operating at temperatures up to 250 °C when immersed in the medium under analysis and showing the potential also for measurements at temperatures down to −200 °C. The surrounding temperature was measured with an accuracy of approximately 1.5 °C by monitoring the difference between the IR signals of the two sensing probes. The measurement was based on the previously determined relationship between the temperature of the environment and the difference in IR signals. The halide fibres, though naturally water resistant, were embedded hermetically in stainless steel within the probes in order to allow for application in even more harsh chemical environments.

13.8 Monocrystalline Sapphire Optical Fibres for Harsh Environment Sensing

Sapphire fibres are another type of fibres that show great potential for harsh environment sensing. They show quite broadband transmission in the VIS and IR wavelength range (approximately 0.3–5 μm). Sapphire fibres are significantly more expensive than silica fibres, and their technology still suffers from poorly elaborated subject of appropriate fibre optical claddings [32]. These reasons still hinder the otherwise very attractive sensing applications [81, 82] of sapphire fibres. In the form of core-based fibres (i.e. as sapphire core-only fibres), they are resistant to extremely high temperatures, pressures, and harsh chemicals, showing even much higher robustness in these conditions than more popular and technologically mature silica fibre-based sensors. They easily can operate at temperatures significantly exceeding 1000 °C, at least up to 1800 °C [32]. The true high-temperature operation potential they offer is probably even higher as their melting temperature reaches around 2054 °C [32]. Such extremely high operating temperatures make the sapphire fibres practically the only candidates for the future monitoring of gas turbines (1200–1400 °C) or coal gasification processes (around 1450 °C). Cores of sapphire fibres are made of monocrystalline α-Al_2O_3 material (in contrary to amorphous silica fibres). Therefore, they are even more chemically resistant to reactive chemicals than the silica ones. The c-axis of the monocrystal is oriented along the length of the fibre. Additionally, sapphire fibres show high flexibility [83]. These properties are a prerequisite of successful special applications of the sapphire fibre-based sensors in harsh environments, e.g. for oil and gas industry, catalytic processing, chemical plants, or monitoring of solid oxide fuel cells, gas turbine, and coal gasification process. However, the most problematic question here is lack of durability of an appropriate cladding material that would match stability of sapphire cores at extremely high temperatures or in harsh chemicals. Also core–cladding interface stability is an issue. In contrary to the extreme thermal and chemical resistance of the unclad sapphire fibre, fibres equipped with claddings show much poorer robustness although they may still surpass harsh environment abilities of silica fibre-based sensors.

Apart from possible deposition of an additional outer material as a cladding layer, which covers most of the applied strategies, there are some other methods for sapphire fibre cladding fabrication aiming at generation of an appropriate core–cladding refractive index contrast. They are the modification of a sapphire surface by implantation (so far mostly by hydrogen ions) or doping or the surface microstructurization (the creation of a pattern by a controlled removal of surficial sapphire), all leading to the surficial refractive index decrease [32]. The general problem here concerns obtaining a cladding that would match the necessary optical requirements, e.g. low enough in comparison with sapphire real part of refractive index and low enough losses, and at the same time to ensure the enough thermochemical and mechanical long-term stability, additionally showing similar to sapphire thermal expansion coefficient. So far, probably the most promising high-temperature claddings for sapphire fibres can be fabricated using films made of three different materials: $MgAl_2O_4$ spinel [32, 84], zirconium oxide [32, 85, 86], and polycrystalline aluminium oxide [32, 87, 88]. The film-cladded sapphire fibres can withstand temperatures up to around 1200 °C under ambient air conditions, up to 1400 °C, and up to 1600 °C for the spinel, zirconium oxide, and aluminium oxide claddings, respectively. By using appropriate methods, they all can be fabricated with relatively low cost. Their drawbacks are thermodynamic instability at higher temperatures for the spinel and requirement of additional microstructurization as well as deterioration of the surface morphology for the other two oxides. In particular, because the refractive index of zirconium oxide is higher than that of sapphire, additional microstructurization of zirconium oxide claddings is necessary for controlled decrease of their mass density and for the refractive index value reduction obtained by this method. As an alternative method of cladding fabrication for sapphire fibres, the implantation process by hydrogen ions would be a prospective solution, particularly taking into account that it allows for maximum operating temperature of up to 1700 °C. However, its cost is relatively high. What is more, the long-term stability at high temperatures of such claddings is kinetically limited because of hydrogen diffusion further activated in such conditions [32, 89].

Some further solutions for claddings of sapphire fibres are supposed to be tested in the near future [32]. Anyway, only unclad sapphire fibres have already been available on the market. On the basis of unclad fibres, some promising sensors have been demonstrated, with a relatively large number of evanescent-field-based applications. Based on unclad fibre coated on its end by a ZrO_2 film, also a Fabry–Pérot interferometer-based temperature sensor was demonstrated up to the temperature of 1200 °C. Moreover, generation of FBGs has been demonstrated in sapphire fibres, and even several FBGs were written in the same sapphire fibre for operating in multiplexed mode at temperature up to 1745 °C. In such a device also the Bragg grating pattern has to show high enough thermal stability in order to withstand the harsh operation conditions. Hence, femtosecond infrared (fs-IR) laser-induced gratings were used here [32].

The sapphire FBGs are suitable for harsh combustion environments such as jet engines, coal gasification reactors, and natural gas turbines for electrical power generation. One of the issues of unclad sapphire sensors is that they show too large refractive index contrast between sapphire and its surrounding, which is particularly large in the case of unclad sapphire fibre sensor in air, being 1.77 to 1 for the VIS radiation range, leading eventually to shallow surrounding penetration by the evanescent field. This property hinders the evanescent-field-based sensing solutions. Another issue of such unclad fibres is the possible fibre surface contamination by adsorbed microparticles (the effect was stated for a few-micron-diameter carbon particles [90]) coming from the fibre surrounding and being the cause of severe absorption and scattering losses, hence hindering the operation of the sensor due to the consequent attenuation of the optical fibre transmission. This again shows that the elaboration of appropriate optical claddings for sapphire fibres is necessary, since it would also protect their surfaces against detrimental contamination, which is still a crucial point in this sensor technology. It is also one of the main conditions for true commercialization of the sapphire fibre sensors.

13.9 Future Trends in Optical Fibre Sensing

Some future trends can be recognized in the field of harsh-environment-resistant fibre sensors. The harsh-environment optical-fibre-based sensing has already become indispensable in more and more industry and engineering fields. Apart from monitoring of EOR processes in petroleum industry and already widely employed structural health monitoring, harsh-environment optical-fibre-based systems have experienced increased demand in aerospace applications. This trend is going to be continued.

It can be noticed that large companies, leading the global market in the design and manufacturing of such special fibre devices, try to consolidate their importance and influence on the market by integration of a wide product range, gaining ability to better respond to consumer demands. The acquisition of Fibretronix AB by Fibrecore Ltd., which took place in 2014, can serve as an example of such a trend [91]. Fibretronix AB is a Swedish company with production focused on harsh environment fibres and special fibre coatings, in particular based on hermetic carbon and high-temperature-resistant polyimide. Fibrecore Ltd. is a UK-based company, a specialized optical fibre and cables global market leader and high-volume manufacturer, that consequently implements its long-term strategy of support for oil and gas industry. Certainly, the crucial factor here is the interest in sensing tools of the main customer in the market, i.e. the oil and gas industry. What is important here is an accelerating trend of generation of HPHT wells, i.e. the industry whose success strongly depends on permanent monitoring of downhole process parameters, as already emphasized. For example, in 2012 the created HPHT wells represented only around 1.5% of the total number of wells drilled in this period, while a 10-fold increase of this rate can already be expected in 2020.

What is more, already custom-designed, multi-parameter monitoring systems are available on the market, by which one is able to remotely monitor large-scale objects or areas with unrivalled precision, like the above-described LIOS system sold already in thousands. Though the sensing by LIOS is performed in harsh environments with a variety of destructive factors, its declared MTBF reaches decades.

Additionally, with prospective future success of sapphire fibre technology, one can expect even increased abilities of monitoring in extremely harsh environments, like temperatures exceeding 2000 °C in harsh chemical surroundings.

Equally exciting are, still under development, biomedical applications of optical-fibre-based monitoring and imaging.

References

1 Kim, D.-S. and Tran-Dang, H. (2019). *Industrial Sensors and Controls in Communication Networks: From Wired Technologies to Cloud Computing and the Internet of Things*. Springer International Publishing.

2 Soloman, S. (2010). *Sensors Handbook*. New York: McGraw-Hill.

3 Distributed Sensing Cable in Industrial Environments|Corning [Online]. https://www.corning.com/worldwide/en/products/communication-networks/products/fibre-optic-cable/distributed-sensing-cables/distributed-sensing-cable-industrial-environments.html (accessed date 3 May 2019).

4 Joe, H.-E., Yun, H., Jo, S.-H. et al. (2018). A review on optical fibre sensors for environmental monitoring. *International Journal of Precision Engineering and Manufacturing-Green Technology* 5 (1): 173–191.

5 Kersey, A.D. and Dandridge, A. (1988). Distributed and multiplexed fibre-optic sensors. In: *Optical Fibre Sensors*. New Orleans, LA: OSA Publishing paper WDD1.

6 Pinet, É. (2009). Fabry–Pérot fibre-optic sensors for physical parameters measurement in challenging conditions. *Journal of Sensors* 2009: 1–9.

7 M. A. Pérez, O. González, and J. R. Arias, 'Optical fibre sensors for chemical and biological measurements', in *Current Developments in Optical Fibre Technology*, S. W. Harun, Ed. InTech, 2013.

8 Norris, J.O.W. (2000). Optical fibre chemical sensors: fundamentals and applications. In: *Optical Fibre Sensor Technology: Advanced Applications — Bragg Gratings and Distributed Sensors* (eds. K.T.V. Grattan and B.T. Meggitt), 337–378. Boston, MA: Springer US.

9 Wang, X. and Wolfbeis, O.S. (2016). Fibre-optic chemical sensors and biosensors (2013–2015). *Analytical Chemistry* 88 (1): 203–227.

10 Rehman, S. and Norin, L. (2019). Specialty Optical Fibres for Harsh Environments [Online]. https://www.photonics.com/Articles/ Specialty_Optical_Fibres_for_Harsh_Environments/a44541 (accessed 8 March 2019).

11 Pirich, R. and D'Ambrosio, K. (2011). Fibre optics for harsh environments. *2011 IEEE Long Island Systems Applications and Technology Conference*, Farmingdale, NY (6 May 2011), pp. 1–4.

12 Chris Emslie (2019). Fibre-Optic Components: Harsh-Environment Optical Fibre Coatings: Beauty is Only Skin Deep [Online]. https://www.laserfocusworld.com/ articles/print/volume-51/issue-04/features/fibre-optic-components-harsh-environment-optical-fibre-coatings-beauty-is-only-skin-deep.html (accessed 14 April 2019).

13 Enhanced Oil Recovery, Energy. Gov. [Online]. https://www.energy.gov/fe/ science-innovation/oil-gas-research/enhanced-oil-recovery (accessed 28 March 2019).

14 Enhanced Oil Recovery (EOR)|Schlumberger [Online]. https://www.slb.com/ services/technical_challenges/enhanced_oil_recovery.aspx (accessed 6 May 2019).

15 Enhanced Oil Recovery (EOR) Processes, LIOS SENSING [Online]. https://www. nktphotonics.com/lios/en/application/enhanced-oil-recovery-eor-processes/ (accessed 19 March 2019).

16 Distributed Temperature Sensing, LIOS SENSING [Online]. https://www. nktphotonics.com/lios/en/technology/distributed-temperature-sensing/ (accessed 19 March 2019).

17 Haywood, R. and Gore, W.L. (2017). Downhole Tools in the Oilfield Services Industry, High Temperature [Online]. https://www.psma.com/sites/default/files/ uploads/tech-forums-capacitor/presentations/is186-downhole-tools-oilfield-services-industry-transformation-improve-reliability.pdf (accessed 8 April 2019).

18 Gautam, A.K. (2019). A Design of High Temperature High Bandwidth Fibre Optic Pressure Sensors [Online]. https://www.academia.edu/20872948/ A_Design_of_High_Temperature_High_Bandwidth_Fibre_Optic_Pressure_ Sensors (accessed 8 April 2019).

19 Kosiel, K., Dominik, M., Ściślewska, I. et al. (2018). Alkali-resistant low-temperature atomic-layer-deposited oxides for optical fibre sensor overlays. *Nanotechnology* 29 (13): 135602.

20 Piestrzyńska, M., Dominik, M., Kosiel, K. et al. (2019). Ultrasensitive tantalum oxide nano-coated long-period gratings for detection of various biological targets. *Biosensors and Bioelectronics* 133: 8–15.

21 Stupar, D.Z., Bajic, J.S., Manojlovic, L.M. et al. (2012). Wearable low-cost system for human joint movements monitoring based on fibre-optic curvature sensor. *IEEE Sensors Journal* 12 (12): 3424–3431.

22 Dunne, L., Walsh, P., Smyth, B., and Caulfield, B. (2006). Design and evaluation of a wearable optical sensor for monitoring seated spinal posture. *2006 10th IEEE International Symposium on Wearable Computers*, Montreux, Switzerland (11 October 2006–14 October 2006), pp. 65–68.

23 Carotenuto, B., Ricciardi, A., Micco, A. et al. (2018). Smart optical catheters for epidurals. *Sensors* 18 (7): 2101.

24 Fajkus, M., Nedoma, J., and Mec, P. (2017). Analysis of the highway tunnels monitoring using an optical fibre implemented into primary lining. *Journal of Electrical Engineering* 68 (5): 364–370.

25 Ott, M.N. (2005). Validation of commercial fibre optic components for aerospace environments. *Smart Structures and Materials 2005: Smart Sensor Technology and Measurement Systems* 5758: 427–439.

26 Polishuk, P. (2006). Plastic optical fibres branch out. *IEEE Communications Magazine* 44 (9): 140–148.

27 Di Sante, R. (2015). Fibre optic sensors for structural health monitoring of aircraft composite structures: recent advances and applications. *Sensors* 15 (8): 18666–18713.

28 Schwartz, J., Johnson, R.P., Kahn, S.A., and Kuchnir, M. (2008). Multi-purpose fibre optic sensors for HTS magnets. *11th European Particle Accelerator Conference, EPAC 2008*, Genoa, Italy (23 June 2008–27 June 2008), pp. 2458–2460.

29 Mihailov, S.J. (2012). Fibre Bragg grating sensors for harsh environments. *Sensors* 12 (2): 1898–1918.

30 Wandel, M.E. (2006). *Attenuation in Silica-Based Optical Fibres*. Technical University of Denmark.

31 Werneck, M.M. and Allil, R.C.S.B. (2011). *Optical Fibre Sensors*. Modern Telemetry.

32 Chen, H., Buric, M., Ohodnicki, P.R. et al. (2018). Review and perspective: Sapphire optical fibre cladding development for harsh environment sensing. *Applied Physics Reviews* 5 (1): 011102.

33 Silica Glass (SiO_2) Optical Material [Online]. https://www.crystran.co.uk/optical-materials/silica-glass-sio2 (accessed 10 April 2019).

34 Silica Glass – Characteristics. https://www.tosoheurope.com [Online]. https://www.tosoheurope.com/our-products/silica-glass/silica-glass-characteristics (accessed 10 April 2019).

35 Polymer_Coatings_for_Silica_Optical_Fibre_Nov_2009.pdf [Online]. https://www.molex.com/mx_upload/superfamily/polymicro/pdfs/Polymer_Coatings_for_Silica_Optical_Fibre_Nov_2009.pdf (accessed 3 April 2019).

36 von Münchhausen, H. and Schittko, F.J. (1963). Investigation of the outgassing process of silicone rubber. *Vacuum* 13 (12): 549–553.

37 Rothka, J., Studd, R., Tate, K., and Timpe, D. (2002). Outgassing of Silicone Elastomers. ISC, pp.1–9.

38 Villahermosa, R.M. and Ostrowski, A.D. (2008). Chemical analysis of silicone outgassing. *Presented at the Optical System Contamination: Effects, Measurements, and Control 2008,* San Diego, CA, USA (13 August 2008–14 August 2008), vol. 7069, article number 706906.

39 K. L. Industries Fibreguide (2019). Metal-Coated Fibres [Online]. https://www.photonics.com/Articles/Metal-Coated_Fibres/a31635 (accessed 31 March 2019).

40 Metal Coated Silica Fibres, AZoM.com, 16 August 2017 [Online]. https://www.azom.com/article.aspx?ArticleID=14354 (accessed 4 April 2019).

41 Hsu, K., Csipkes, A., and Jin, T. Gold and Steel Protected FBGs Enable Robust Sensing in Harsh and High Temperature Environments [Online]. https://technicasa.com/gold-steel-protected-fbgs-enable-robust-sensing-harsh-high-temperature-environments (accessed 4 April 2019).

42 Wiederhorn, S.M. and Bolz, L.H. (1970). Stress corrosion and static fatigue of glass. *Journal of the American Ceramic Society* 53 (10): 543–548.

43 Glaesemann, G.S. (July 2017). Optical Fibre Mechanical Reliability: Review of Research at Corning's Optical Fibre Strength Laboratory. White Paper, WP8002, July 2017

44 Anurag Dwivedi, Scott Glaesemann, G. (2011). Optical Fibre Strength, Fatigue and Handleability After Aging in a Cable. TR3290, January 2011.

45 Klingsporn, P.E. (2011). Characterization of Optical Fibre Strength Under Applied Tensile Stress and Bending Stress. KCP-613-6655, 1054754, August 2011.

46 Huff, R.G. and DiMarcello, F.V. (1988). Hermetically coated optical fibres for adverse environments. *Presented at the 1987 Symposium on the Technologies for Optoelectronics,* Cannes (19–20 November 1987), vol. 867, pp. 40–47.

47 Lu, K.E., Glaesemann, G.S., VanDewoestine, R.V., and Kar, G. (1988). Recent developments in hermetically coated optical fibre. *Journal of Lightwave Technology* 6 (2): 240–244.

48 Kurkjian, C.R., Simpkins, P.G., and Inniss, D. (1993). Strength, degradation, and coating of silica lightguides. *Journal of the American Ceramic Society* 76 (5): 1106–1112.

49 Villar, I.D., Zamarreño, C.R., Sanchez, P. et al. (2010). Generation of lossy mode resonances by deposition of high-refractive-index coatings on uncladded multimode optical fibres. *Journal of Optics* 12 (9): 095503.

50 Kosiel, K., Koba, M., Masiewicz, M., and Śmietana, M. (2018). Tailoring properties of lossy-mode resonance optical fibre sensors with atomic layer deposition technique. *Optics & Laser Technology* 102: 213–221.

51 Del Villar, I., Hernaez, M., Zamarreño, C.R. et al. (2012). Design rules for lossy mode resonance based sensors. *Applied Optics* 51 (19): 4298.

52 West, R.H. (1988). A local view of radiation effects in fibre optics. *Journal of Lightwave Technology* 6 (2): 155–164.

53 Schenato, L. (2017). A review of distributed fibre optic sensors for geo-hydrological applications. *Applied Sciences* 7 (9): 896.

54 Wang, Z., Wang, J., Sui, Q. et al. (2015). Development and application of smart geogrid embedded with fibre Bragg grating sensors. *Journal of Sensors* 2015: 1–10.

55 Why LIOS Sensing Systems? LIOS SENSING [Online]. https://www.nktphotonics.com/lios/en/technology/why-lios-sensing-systems/ (accessed 2 May 2019).

56 Sensing Cables, NBG Systems GmbH [Online]. https://www.nbg.tech/sensing-overview/ (accessed 6 May 2019).

57 Hu, X., Shen, X., Wu, J. et al. (2016). All fibre M-Z interferometer for high temperature sensing based on a hetero-structured cladding solid-core photonic bandgap fibre. *Optics Express* 24 (19): 21693.

58 Yin, J., Liu, T., Jiang, J. et al. (2016). Assembly-free-based fibre-optic micro-Michelson interferometer for high temperature sensing. *IEEE Photonics Technology Letters* 28 (6): 625–628.

59 Graded-Index Polymer Optical Fibre (GI-POF) [Online]. https://www.thorlabs.com/catalogPages/1100.pdf (accessed 9 April 2019).

60 Sartiano, D. and Sales, S. (2017). Low cost plastic optical fibre pressure sensor embedded in mattress for vital signal monitoring. *Sensors (Basel)* 17 (12): 2900.

61 Grillet, A., Kinet, D., Witt, J. et al. (2008). Optical fibre sensors embedded into medical textiles for healthcare monitoring. *IEEE Sensors Journal* 8 (7): 1215–1222.

62 Photonic Skins For Optical Sensing|Projects|FP7|CORDIS|European Commission [Online]. https://cordis.europa.eu/project/rcn/86613/factsheet/en (accessed 5 April 2019).

63 Drobny, J.G. (2008). *Technology of Fluoropolymers*. CRC Press.

64 An Introduction to Fibre-Optic Sensors, Sensors Magazine [Online], 1 December 2001. https://www.sensorsmag.com/components/introduction-to-fibre-optic-sensors (accessed 3 April 2019).

65 Asahi Multi-Core Cable – Asahi-Kasei Offered by Industrial Fibre Optics, Inc. [Online]. http://i-fibreoptics.com/multi-core-fibre-cable.php (accessed 8 April 2019).

66 Glaesemann G. S., Castilone, R. J., (2002). The Mechanical Reliability of Corning Optical Fibre in Bending. WP3690, September 2002.

67 Heat Resistant Plastic Optical Fibre Cable (POF): Hitachi Cable America [Online]. http://www.hca.hitachi-cable.com/products/hca/products/materials/heat-resistant-plastic-optical-fibre.php (accessed 8 April 2019).

68 PTFE Coating, Extrusion, Molding|Teflon TM PTFE [Online]. https://www.chemours.com/Teflon_Industrial/en_US/products/product_by_name/teflon_ptfe/index.html (accessed 8 April 2019).

69 ETFE Resin|Tefzel TM ETFE Fluoroplastic [Online]. https://www.chemours.com/Teflon_Industrial/en_US/products/product_by_name/tefzel_etfe/index.html (accessed 8 April 2019).

70 LITEflight® Fibre Optic Cable [Online]. https://www.carlisleit.com/products/wire-cable/liteflight-fibre-optic-cable/ (accessed 3 April 2019).

71 Harrington, J.A. (2004). *Chalcogenide glass fibre optics.* In: *Infrared Fibres and their Applications*, 83–105. SPIE Press.

72 Damin, C.A. and Sommer, A.J. (2013). Characterization of silver halide fibre optics and hollow silica waveguides for use in the construction of a mid-infrared attenuated total reflection Fourier transform infrared (ATR FT-IR) spectroscopy probe. *Applied Spectroscopy* 67 (11): 1252–1263.

73 Bureau, B., Boussard, C., Cui, S. et al. (2014). Chalcogenide optical fibres for mid-infrared sensing. *Optical Engineering* 53 (2): 027101.

74 Yoo, W.J., Jang, K.W., Seo, J.K. et al. (2011). Development of a 2-channel embedded infrared fibre-optic temperature sensor using silver halide optical fibres. *Sensors* 11 (10): 9549–9559.

75 Chalcogenide Glass Fibres|FAQs|IRFlex Corporation [Online]. https://irflex.com/about/faqs/ (accessed 9 April 2019).

76 Fibres/CeramOptec [Online]. https://www.ceramoptec.com/products/fibres.html (accessed 23 April 2019).

77 Fibre optic ATR-Probes, Art Photonics [Online]. https://artphotonics.com/product/fibre-optic-atr-probes/ (accessed 4 April 2019).

78 Fibre Optic ATR-Probes for Harsh Environment, Art Photonics [Online]. https://artphotonics.com/product/fibre-optic-atr-probes-for-harsh-environment/ (accessed 14 April 2019).

79 Saito, M. and Kikuchi, K. (1997). Infrared optical fibre sensors. *Optical Review* 4 (5): 527–538.

80 Harrington, J.A. (2004). *Infrared Fibres and Their Applications*. SPIE.

81 Wang, A., Gollapudi, S., May, R.G. et al. (1995). Sapphire optical fibre-based interferometer for high temperature environmental applications. *Smart Materials and Structures* 4 (2): 147–151.

82 Wang, J., Dong, B., Lally, E. et al. (2010). Multiplexed high temperature sensing with sapphire fibre air gap-based extrinsic Fabry–Perot interferometers. *Optics Letters* 35 (5): 619.

83 Sapphire Fibres – Other Fibres [Online]. http://www.micromaterialsinc.com/specsFibre.html (accessed 22 May 2020).

84 Jiang, H., Cao, Z., Yang, R. et al. (2013). Synthesis and characterization of spinel $MgAl_2O_4$ thin film as sapphire optical fibre cladding for high temperature applications. *Thin Solid Films* 539: 81–87.

85 Wang, J., Lally, E.M., Wang, X. et al. (2012). ZrO_2 thin-film-based sapphire fibre temperature sensor. *Applied Optics* 51 (12): 2129.

86 Davis, J.B., Lofvander, J.P.A., Evans, A.G. et al. (1993). Fibre coating concepts for brittle-matrix composites. *Journal of the American Ceramic Society* 76 (5): 1249–1257.

87 Nubling, R.K., Kozodoy, R.L., and Harrington, J.A. (1994). Optical properties of clad and unclad sapphire fibre. *Presented at Biomedical Fibre Optic Instrumentation 1994*, Los Angeles, USA (23 January 1994–29 January 1994), vol. 2131, 28 July 1994, pp. 56–61.

88 Shen, Y., Tong, L., and Chen, S. (1999). Study on performance stability of the sapphire fibre and cladding under high temperature. *Presented at the Harsh Environment Sensors II*, Boston, MA, USA (19 September 1999–19 September 1999), vol. 3852, pp. 134–142.

89 Spratt, W., Huang, M., Murray, T., and Xia, H. (2013). Optical mode confinement and selection in single-crystal sapphire fibres by formation of nanometer scale cavities with hydrogen ion implantation. *Journal of Applied Physics* 114 (20): 203501.

90 Zhang, J., Xiong, F., and Djeu, N. (2009). Sapphire fibre evanescent wave absorption in turbid media. *Applied Spectroscopy* 63 (8): 932–935.

91 Fibrecore – News Article – Fibrecore Acquires Fibretronix, Fibrecore [Online], 1 September 2014. https://www.fibrecore.com/news-article/fibrecore-acquires-fibretronix (accessed 17 March 2019).

14

Fibre-Optic Sensing

Past Reflections and Future Prospects

Brian Culshaw[1] and Marco N. Petrovich[2]

[1] *Department of Electronic and Electrical Engineering, University of Strathclyd, Glasgow, Scotland, UK*
[2] *Optoelectronics Research Centre, University of Southampton, Southampton, UK*

14.1 Introductory Comments

The previous chapters in this book have covered the basic concepts of fibre sensors and have presented some insights into current interests in research into prospective fibre sensor architectures. There has also been some discourse on applications, notably for distributed sensing and structural monitoring. Intriguingly, while applications have concerned advances predominately in sensing physical parameters with much activity in strain and temperature measurements, the emerging interest lies predominantly in (bio)chemical and medical applications.

This chapter is essentially about future prospects, but in order to gain some insight into the future, we shall first of all reflect upon the past and on the achievements that fibre sensors have made over the past half-century. There is a huge and immediately obvious contrast between fibre optics in communications and fibre optics in sensing. The former has precipitated the words 'fibre optics' into everyday language with fibre to the home becoming readily and widely available. The latter has made but modest impact on the sensor industry, currently laying the claim to around 1% of the total marketplace [1]. The profound difference lies in the observation that sensing can be addressed by multiple technologies for the vast majority of applications. To exemplify this, a book [2] published towards the end of the last century lists several pages of techniques to make chemical measurements – none of which mentions fibre optics.

Optical Fibre Sensors: Fundamentals for Development of Optimized Devices, First Edition.
Edited by Ignacio Del Villar and Ignacio R. Matias.
© 2021 The Institute of Electrical and Electronics Engineers, Inc.
Published 2021 by John Wiley & Sons, Inc.

In the remainder of this chapter, we shall first attempt to explain the simple observation that market presence of this technology is modest at best. The technology has had a half-century in which to make its impact. Paradoxically, the technology still enjoys energetic attention from within the research community and indeed, by implication, from the miscellaneous diverse funding agencies. So how can this happen? Here, it is critical to understand what it is that makes the sensor that works in the practical world, outside the research community. We also need to fully appreciate the technological tools and scientific principles that are currently emerging and how these may contribute to solving fibre sensing problems in the future. What follows endeavours to provide at least some insight into these critical factors. This points towards an optimistic future for the technologies that are currently evolving and for the all-important young researchers involved.

14.2 Reflections on Achievements to Date

Fibre-optic sensors have been with us for at least half a century, possibly more. As mentioned in the first chapter, Harold Hopkins' fibre endoscope [3] from 1954 could be viewed as the first example of a fibre sensor. Certainly it conveys remote information to the user that is the basic function of a sensor system. The other contender, the Fotonic sensor [4], emerged a decade later and is still sold [5] in its original format to perform specialized precise measurements of surface finish. The fibre endoscope, however, while certainly still used, also exemplifies the impacts of changes over half a century in technological development. Some endoscopic examinations now involve a remote CCD camera on a lead, and the 'endoscopic pill', embracing energy supply, radio transmitter for Wi-Fi linkages and camera has also proved its efficacy. This very brief discourse exemplifies much of the background towards the modest penetration of fibre-optic sensing overall especially from the perspective of the absolutely critical final operator, whose interests are in obtaining reliable information fitted to the technical and financial needs of the application, regardless of technological novelty. We need to recognize not only the fibre-optic sensor itself but also the consumers' needs and impacts of other rapidly improving technologies, both complementary to the fibre sensor evolution and competing in cost–performance ratio.

First of all, it is useful to reflect on what a fibre sensor actually is and relate this to the user that the sensor might want. Figures 14.1 and 14.2 attempt to summarize this. There is, of course, the sensor head itself within which light emanating from the input source is modulated in one of its many properties – phase, intensity, spectral distribution, polarization state, etc. – and thereafter hopefully retains this modulation until it is detected. The light should ideally be modulated only by the parameter of interest, but in reality there will almost always be interfering phenomena. Finally, the end user seeks an answer that is often along the lines of 'yes, everything is fine' or 'maybe there are problems'. For the latter further detail may

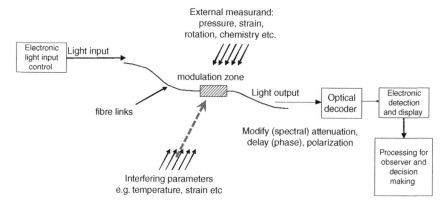

Figure 14.1 The elements of the fibre-optic sensor indicating the multiple interfaces – all of which need to be calibrated and stabilized. The modulation zone can impose linear (same spectrum, different distribution) or non-linear (different spectrum – luminescence, Brillouin, Raman scatter) changes on the input. The modulation zone can also either be intrinsic – i.e. the light is modulated while still in the fibre – or extrinsic implying optical launch and collection.

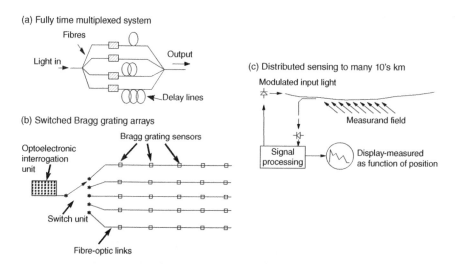

Figure 14.2 Some examples of multiplexed fibre sensor architectures. The time multiplexed system shown in (a), (b) a time-switched Bragg grating array exemplifies wavelength multiplexing (often used single channel), while (c) illustrates distributed sensing, a technique unique to fibre sensing.

be required. Additionally, the user needs to be assured that the sensor system meets all the necessary regulatory and operational requirements. All this is glaringly obvious but often forgotten. Indeed, most of the preceding chapters focus entirely upon the modulation zone and often seek to highlight one particular measurement rather than considering interfering phenomena, optical input, transmission, and detection requirements. The key to this is that everything involving all the many interfaces needs to be stable throughout the operating regime of the sensor system. Figure 14.2 illustrates one of the major potential (and often realized) benefits of fibre sensors – namely, the ability to sense many points simultaneously, either through arrays of extrinsic sensors or through the single intrinsic sensor shown for distributed measurements. Either format can function over lengths of many tens of kilometres, but again the factors mentioned above must all be taken into account – sometimes a major challenge! For the occasional cooperative measurand, self-correction can be accomplished [6]. A real sensor is far more than simply light in at one end and out at the other. In practice, along the path from concept to reality, by far the bulk of the investment is dedicated to all these peripheral activities.

There are also many competing technologies for the vast majority of measurement functions. This then leads into the question – why should the user chose a particular approach? Additionally, for most technologies, there is a well-established alternative as opposed to the 'new boy' (even after half a century) of fibre optics. Figure 14.3 presents a flow chart that seeks to highlight the necessary

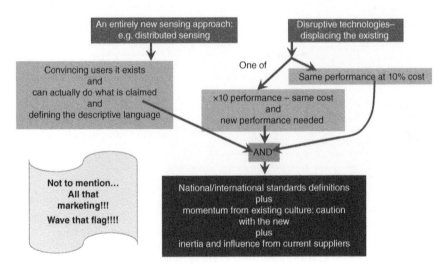

Figure 14.3 A representative of the processes needed to establish a new sensor technology – the original paper is just the starting block!

features for any new technology to be adopted. In essence the basic necessities are significantly improved performance (which is also needed for the application) and very significant cost savings. A vital, often ignored, necessity in the process is researcher communication with developers and buy-in from end users throughout the process, as encapsulated in the diagram. A little thought on this soon confirms that two decades can easily pass by between the initial idea and the final emergence into the marketplace. This process is certainly not constrained to fibre-optic sensors but is common to the vast majority of technological innovation, as embodied in the generalized format of Porter's five forces [7].

With this background it is interesting to examine the current fibre sensor successes. The intrinsic and the extrinsic sensor has been described in the first chapters of this book. In the context of Figure 14.1, intrinsic enjoys the benefit of a much relaxed stability requirements within the sensing region itself – the region that is almost always exposed to the greatest environmental changes. Therefore, the benefit of no optical interface ensures stability that otherwise can be very difficult to attain. However, even (as exemplified in Figure 14.4) for the intrinsic sensor, there are many transfer points between the measurand and the light being transmitted in the fibre. Some measurands, however, can penetrate to and interact with the propagating light without relying significantly on any of the interfaces between the core of the fibre – where the modulated light is located – and the world outside. Three spring to mind, namely, temperature (for which there can be a slight delay due to thermal conductivity effects), magnetic fields, and gravitational linkages exemplified in the Sagnac interferometer. Mechanical strain and acoustic signals, with sufficient care in the design of the fibre cladding to 'world outside' interface, can also be made extremely reliable.

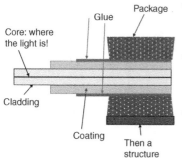

• Direct interactions with light in core:
  • Thermal (phase, Raman & Brillouin scatter)
  • Magnetic field (Faraday rotation)
  • Sagnac loop interferometer – phase
• Interactions with light in core via packaging and coatings:
  • Strain, quasi static and dynamic (including acoustic/ultrasonic)
  And slowly gaining increasing user acceptance:
• Interactions totally outside fibre core coupled via end or side:
  • Typically refractive index via colour, phase, spectral distribution, for (bio) chemical, thermal, target motion etc.

Figure 14.4 Illustrating the numerous interfaces in fibre sensing between the parameter to be measured and the fibre core that carries the light to be modulated. Modulators external to the fibre itself – either at the end or side coupled – have to date gained modest adoption.

It is then, perhaps, not surprising with the wisdom of hindsight that these direct interfaces feature in what are currently the majority of commercially viable applications of fibre sensors. Additionally, even though all of these intrinsic sensor interfaces are sensitive to whatever is being measured throughout their entire length, they can be configured to make very reliable, repeatable measurements that can be ascribed to a particular section of the fibre either through carefully designing the interaction or through carefully interrogating the fibre itself. All, however, do have competing alternative technologies, but in all cases the 'order of magnitude benefit in performance or price' convincingly applies.

In the strain and temperature case, distributing sensing or multiplexed fibre Bragg gratings (FBGs) offer very significant benefits in terms of installation (assuming the necessary transfer between the core and the world outside is in place), communication over a measurement range that can extend to 10 km or more, and versatility in application. These have been discussed in detail in Chapters 5 and 6. For magnetic field detection, the principal application to date has been current measurement in high voltage, high current transmission systems [8] (Figure 14.5). Faraday rotation is the basic interaction mechanism, and so care must be taken, and any other spurious non-reciprocal polarization drift (for example, birefringence introduced through fibre bending, polarization mode

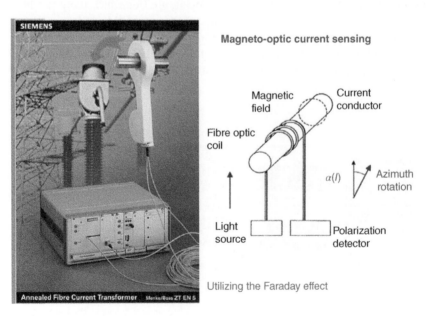

Figure 14.5 Fibre-optic current monitoring, concept published 1973 [9], uses Faraday rotation inducing current-dependent circular birefringence for current monitoring.

dispersion, etc.) should be avoided. This, in turn, implies careful attention to fibre design (frequently involving 'spun' fibre) and to the mounting process for the fibre loop. The enormous benefits are, however, that current to magnetic field to polarization rotation relationships are very well characterized and can be stabilized and, probably even more important, the whole system needs no electrical contact with the current-carrying conductor.

In the gravitational field context, the fibre-optic gyroscope has made immense inroads in rotation measurement (Figure 14.6) and has demonstrated unprecedented ranges of application capability [10]. In the gyroscope context, integrated optics [11] has also played an important part, and systems ranging from space-qualified guidance units for deep space exploration to inertial navigation units for autonomous vehicles have all emerged based upon fibre gyro technology. Of course, in this context, there are numerous competitive approaches: the laser gyro based upon essentially the same Sagnac interferometer principle and many variations upon the theme of mechanical rotation detection gyroscopes. All of these, however, have complex and critical mechanical interfaces, whether in the rotation of air bearings as in mechanical systems or in the alignment of laser cavities and the need for mechanical dither. All present reliability and stability problems. The fibre gyroscope then, thanks to specifically produced polarization-maintaining single-mode fibre formats, ingenious design approaches, and careful signal processing, delivers immensely superior performance at a very competitive cost.

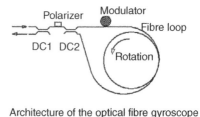

Architecture of the optical fibre gyroscope

Image from KVH web site.

Figure 14.6 The fibre-optic gyroscope showing the basic architecture of the fibre Sagnac interferometer and two very different production versions.

There are, of course, a few other success stories that do not rely upon intrinsic systems. Of these a very early example is a temperature measurement system that relies upon phosphor decay time measurement [12] with the phosphor skilfully attached to the end of the fibre. There are also other examples – a disposable intra-cranial pressure sensor for use [13] in neurosurgery is one – but in terms of fibre sensing, overall these extrinsic devices represent a relatively small market fraction.

So to summarize the current situation with fibre sensor commercial activity, by far the majority of successful sensors are based on intrinsic devices and address specialized niche areas where either long, electromagnetic interference free access as, for example, in distributed sensing and fibre Bragg grating arrays is needed in the application, or they displace previous difficult technologies as in current sensing and the fibre gyroscope. However, when it comes to displacing established generic systems, direct electronic sensing rules in the vast majority of cases (at least for the time being). It is straightforward to see why. The only sensor interface in these all electronic sensing systems links directly into the electronic circuit (invariably an integrated circuit, readily designed – to custom needs if necessary – and fabricated using standard tools), and the output to the final user is also via the same electronic circuit. The interfaces are, therefore, minimized, so by implication reliability is significantly enhanced. So how might this change in the future? And what is it that the excites those of us working with fibre-optic sensors as a technology?

14.3 Photonics: How Is It Changing?

The essentials of a photonic system are illustrated diagrammatically in Figure 14.7, which in parallel indicates the key parameters relevant to fibre sensors.

This perspective also points towards areas where new sensor research can be productive. These can be briefly summarized as:

- Finding more ways to modulate the light while it is guided in the fibre or indeed an optical waveguide.
- Finding more ways to modulate the light in an extrinsic sensor when the light is outside the fibre.
- Evolving approaches to ensure that the interfaces between the input light and the measurand, the measurand and the modulation system for the light, between the fibre that carries the light that has been modulated and the receiver, and between the detected light and the observer all remain stable and predictable over long periods of time.
- Making signal processing systems more versatile to ensure effective interface to the user and also, when possible, to provide a self-correction for any drifts within

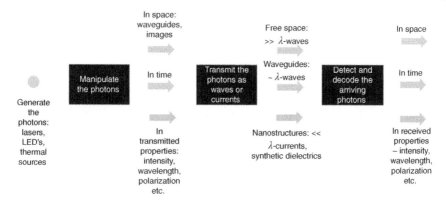

Figure 14.7 What is photonics? And where are the tools for fibre sensing? There are in all three stages, and there are also essential aspects in generation and detection. Ideally, the fibre sensing engineer is well versed in this vast range of possibilities. Quite a challenge!!!

the measurand interface and/or photonic system that precedes the detection processing. Such advances have figured prominently in realizing distributed sensing, especially for acoustic signals, in fibre gyroscopes and many fibre-optic-based spectroscopically dependent chemical measurement systems.

The past few decades have seen striking advances in technological capabilities that may, in principle, contribute to improving all or some of the above factors in fibre-optic sensing systems. By far the majority, indeed perhaps all of these tools, are associated with the increasing acceptance of the prefix of 'nano' in both conceptual design and materials application. The striking advances in semiconductor technology that have facilitated ever more versatile circuit functionalities are undeniably by far the most prevalent. Spin-offs from this technology into photonics have impacted source and detector design: the quantum dot and single-photon detector, transmission media (including the photonic crystal fibre with its huge versatility), and the photonic integrated circuit, most notably in silicon, are but a few examples. There are also important tools emerging in possible measurand interfaces with the modulation processes: nanoscale plasmonics, 'metamaterials' (synthetic materials functionality enabled through structure), control of inherent molecular properties, and transmitter and receiver interfaces as in, for example, the photonic scale Yagi antenna. Not to mention the ever-increasing versatility and precision of 3D printing.

Other factors are also at work. There is no doubt that societal needs will evolve rapidly over the coming decades. The self-driving car will almost certainly benefit from the fibre gyroscope. Infrastructure requirements will become more demanding thanks to more people and ever-increasing lifespan expectations. This natural aspiration towards ever-increasing longevity will inevitably also produce biochemical, biomedical, and environmental monitoring and measuring requirements. There are also numerous examples within our society where the availability of a new tool creates the demand. Perhaps the Internet itself is the most pervasive and obvious.

So there is a fascination both technologically and in potential application opportunities for fibre-optic sensing in the future and for the young researcher to benefit immensely from fibre sensing as a research activity. Some indication of the interests and activity may be gleaned from a quick examination of the proceedings of the 26th International Conference on Optical Fibre Sensors that took place in Lausanne in September 2018. This meeting was, by far, the best attended to date with around 600 people coming to Switzerland for the meeting and around 400 papers presented. The meeting certainly featured the well-established: ongoing improvements in distributed sensing, fibre gyroscopes, and current sensors, together with a significant presence on the theme of the fibre Bragg grating. However, the conference also demonstrated an expanding evolution of sensors technologies targeted towards biomedical, biochemical, and environmental applications. There was even one submission discussing production techniques for medical and industrial applications based on Fabry–Pérot sensing systems [14], indicating a very definite trend towards these extrinsic modulation sensors and their emerging applications sectors. Interestingly, the prefix 'nano' also had a very strong presence. Metamaterials though have yet to make an impact, but silicon photonics is gently emerging. The next section of this chapter is dedicated to exploring some of these prospects – and others – in a little more detail.

14.4 Some Future Speculation

Photonics as a discipline is most certainly arousing strategic interest with, for example, targeted research and development programmes emerging in both the United States and the EU. UK government figures indicate that the photonics industry in the United Kingdom (about 7% of manufacturing currently) makes comparable contributions to pharmaceuticals and is likely to expand significantly. New functionalities enabled through advances in materials will play a major part in this future evolution. For certain, some of these functionalities will enable, often currently unimagined, radically new approaches and capabilities in fibre-optic sensing.

Many of these advances rely upon the continuing improvements in the precision with which structures can be defined and the associated patterning can be reliably reproduced. Similar progressions in materials handling capabilities have enabled the major changes in the past. Indeed, the telephone in your pocket exemplifies such a development over the past half-century or so, in parallel with the arrival of the ubiquitous Internet and countless other changes in our society. At the basic level, many, possibly most, of these developments have been enabled by increasingly precise materials handling capabilities. In this section of the chapter, we will very briefly describe a few of these emerging techniques and present some thoughts on how these may contribute to future sensor initiatives.

14.4.1 Photonic Integrated and Plasmonic Circuits

The basic ideas of integrated optics have been around for roughly half a century. Initially fabrication technologies were well beyond photolithographic capabilities, and furthermore the only substrate material that had been, at that stage, identified – namely, lithium niobate – was incompatible with straightforward processing technologies. An early pioneer of the subject notably commented that 'integrated optics is now and always will be a technology for the future' [15].

However, two major factors contributed to a significant change. The first involves all those improvements in photolithographic precision driven initially through the silicon semiconductor industry. The second was the result of the pressures from silica – as in optical fibres. The optical losses in silica fall dramatically in the near infrared (IR), most notably around 1.5 μm wavelength. Early optical communication systems worked on semiconductor lasers then limited to the 850 nm band. However, the demand to venture into the longer wavelength region triggered a host of materials handling capabilities and the emergence of heterojunctions in compound III–V semiconductors to produce lasers at 1.5 μm and thereabouts – a wavelength at which silicon is also conveniently transparent and around which all the other fibre communication system technologies have been developed. Additionally, with a little subtle juggling, III–V and II–VI semiconductor sources and detectors in this wavelength band can be integrated on to the silicon platform. Furthermore, expanding into germanium is beginning to enable the technology to be applicable further towards the mid-IR region – which can often be important in a sensing context.

Research ventures into silicon photonics were emerging towards the end of the last century and are now becoming mainstream. There are now books on the subject [16], and a number of major corporations are investing heavily in the technology. Both predominately target optical communications, as indeed with optical fibre technology. Sensing, however, has benefited from, indeed relied upon, advances in communications in the past and is likely to continue doing so.

In the fibre-optic sensing context, there has to date been relatively little penetration of the integrated photonic ideas (with the exception of the high-performance fibre-optic gyroscope). There are however many obvious possibilities in integrating the optical systems at the source end of the fibre sensor where the modulation needs may be very demanding but compatible with integrated optical circuits and the complementary optical domain receiver end decoding systems and detection into optical integrated circuits rather than the often custom discrete optics. Silicon integrated optics custom circuit foundries are now beginning to establish themselves so that concepts can be evaluated without local investment in processing and some early sensor system prototypes are emerging. Also, in the longer term, there is the tantalizing prospect for competitively priced larger-scale production.

There have also been the examples of silicon photonics being utilized as ingenious chip sensor systems with optical inputs and outputs attached to or even incorporated within the chip itself. Examples include coupled resonators operating at 'exceptional points' where the slightest disturbance produces a rapid change in optical output [17]. There is also another recent demonstration of a 'nanophotonic optical gyroscope' based upon two coupled loops on a chip of an area less than $2 \, mm^2$ [18]. Ingenious design within the system overall produced a phase sensitivity of a few nanoradians. At the decoding end of the network, there have also been ventures into the optical integrated circuit signal processing concept exemplified in an interrogation system for de-multiplexing fibre Bragg gratings [19].

Photonics integration has another aspect to it, namely, what has become known as 'plasmonics,' which is in effect utilizing the electrical current carrying properties of metals at optical frequencies. The skin effect penetrates but nanometres at optical frequencies implying huge losses, so any current carrying can only go a very short distance – a fraction of the optical wavelength. However, somewhat along the lines of an exceptionally tiny lightening conductor, these 'current carrying capacities' can be used as an approach to significantly enhancing optical fields in a dielectric region, and hence by implication if this region contains, for example, a chemical sample, then the propagation characteristics of light around the structure become modified in response to the refractive index of the sample.

The basic idea has been around for a very long time in the form of the Kretschmann configuration [20] (Figure 14.8) where one face of a triangular prism is coated with a very thin metal layer (tens of nanometres at most). This then produces an enhanced field in the far side of the layer, but this field can only be accessed when the phase velocities in the prism and the material outside are matched at which point total internal reflection from the prism ceases, resulting in this refractive index sensor system. The basic concept has also been around as the principle for in-fibre polarizers [21] for many years. More recently, the basic idea of the 'lightening conductor' has been adopted to the 'lab on the end of a fibre' concept also shown in Figure 14.8. Yet again, controlling material properties and the material dimensions to nanometre tolerances is an essential contribution to this process. In the Kretschmann version only thickness was involved so that

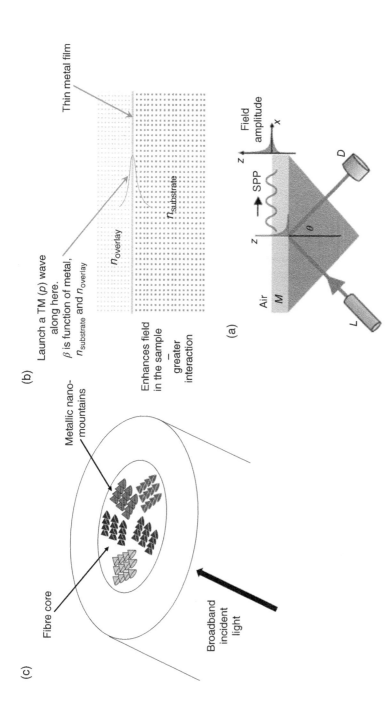

(b)

Thin metal film

Launch a TM (p) wave along here. β is function of metal, $n_{substrate}$ and $n_{overlay}$

$n_{overlay}$

$n_{substrate}$

Enhances field in the sample – greater interaction

(a)

Air

M

z

SPP

Field amplitude

x

z

θ

D

L

(c)

Fibre core

Metallic nano-mountains

Broadband incident light

Figure 14.8 The well-established Kretschmann configuration is shown bottom right (a) with the guided wave version at (b) as used in both fibre-optic and integrated optic polarizers and some chemical (really refractive index) sensors. (c) A 'lab-on-fibre tip' concept wherein each group of 'nanometre metallic mountains' is coated with a different reactive material to select different chemicals and with a specific optical wavelength associated with each mountain group.

simple evaporation processes sufficed. Taking this to three dimensions does, however, present challenges and also leads into the more general question of metamaterials briefly addressed in the next section.

14.4.2 Metamaterials in Sensing

A 'metamaterial' can be viewed as one within which the structure of the material itself can introduce novel physical or chemical properties that dominate significantly over the bulk properties of the original material components. In many applications, the physical properties concerned focus on the electromagnetic impact of the structure at optical frequencies. However, if we generalize the thought into, for example, acoustic wave properties, then the idea of dimensions and/or loading impacting upon wave motion properties is evident in guitar strings, organ pipes, and a host of other domains. At another dimensional level the quantum dot dictates the quantum wave properties of a material via dimensional control to, for example, produce lasing phenomena impossible in the bulk.

Metamaterials have created immense interest in optics and photonics triggered by the concept of the perfect diffraction-free imaging using material with refractive index of −1 (Figure 14.9) that is featured in the very early applications [22]. The intervening years have seen much research interest, notably in optics and terahertz portions of the spectrum, and the publication of several texts (see, for example, [23]).

The idea of perfect imaging by incorporating such a 'perfect' lens on the end of a fibre emerged very soon after the concept was initially published. In principle, this could, for example, undertake a molecule by molecule chemical analysis of a sample positioned on the distal side of the meta-lens. This basic concept of the meta-lens on the end of a fibre has been recently demonstrated [24].

The other much publicized aspect of metamaterials centres on 'cloaking', which enables a coating to be placed on an object that directs incident radiation around the object without distortion so the observer can see beyond this particular coated object as if it were totally absent. It is less clear how this might be utilized in a fibre sensor context. However, since cloaking is very much wavelength dependent (a general trend among metamaterials since many of the facilitating concepts require a structurally determined resonance to be realized in a suitable well-controlled periodic lattice), perhaps we have here a tool for directing light of specific colours to specific points within a target from a single incoming fibre – the ultimate flexible highly dispersive lens structure.

Metamaterials – like silicon photonics – rely on highly precise fabrication processes to realize different structures. Although basic lithographic concepts are sometimes used in this context, there are a whole host of other tools available. For example, precise fabrication tools such as electron and ion beam etching/

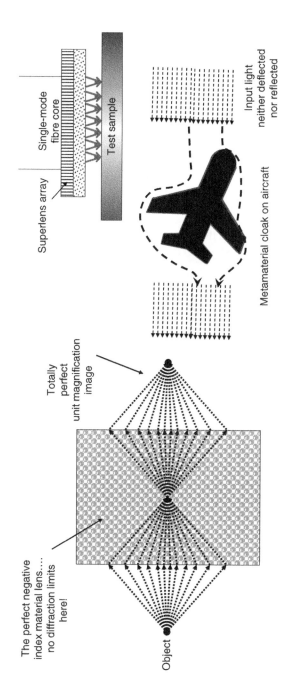

Figure 14.9 Metamaterials and examples of the concepts, which in principle can enable using these structured synthetic materials. In the fibre sensor concept, the idea of the meta-lens with super-resolution capability on a fibre has already been demonstrated.

milling have important roles in optical metamaterial realization. Self-assembly is another often explored route – indeed perhaps this is the most appealing. Again conceptually, we are accustomed to these ideas. Growing a crystal from a concentrated solution onto a carefully chosen substrate – namely, the seed crystal – results in a carefully organized larger crystal precipitated from the solution. On a broader front, self-assembly produces what could also be regarded as metamaterials. There is a very subtle and rapidly evolving range of techniques and material sources to bring this into fruition. In this context zinc oxide [25] has received a very great deal of attention. When doped with other materials, zinc oxide has proved to have an almost confusing array of optical and electrical properties after this self-assembled system has been put together – sensing functionality among them.

14.4.3 More Variations on the Nano Story

The possibilities afforded through the ability to control nanoscale structures of one form or another appear to be endless. Of the many, a few are beginning to make a presence.

Variations on the theme of the gold nanowire enable a vast range of functionalities. In the context of chemical sensing [26], the wire can be coated with one or a few layers of appropriate chemical indicator. The presence of the selected chemical species changes the optical properties of the nanowire, into which light has been launched through a distal end. Coating and assembling these nanowire arrays onto a suitable optical substrate does most certainly present challenges, but the promise of very compact, highly selective, and very sensitive chemical sensing systems based on optical interrogation continues to stimulate considerable interest. However, the nanowire is by definition very tiny, so much care is needed to maintain the necessary levels of cleanliness.

These nanowire systems are based upon similar principles to the much more well-established operation of the fibre taper used extensively in the scanning near-field optical microscope. However, in the sensor context, the design trade-offs are considerably more delicate. For the scanning near-field microscope, the taper needs to be metal coated to prevent the optical field spreading into the surrounding air. Losses in the coating are a major issue, but provided the tiny sub-wavelength end spot is maintained, and the system is clean, all is well. However, in the fibre sensor context, the metal coating is definitely no longer needed. Here the evanescent field needs to spread, in controlled, predictable, and stable way, into the sample. This has recently shown promising results for sensitive hydrogen detection using Raman spectroscopy [27] (Figure 14.10). As a basic spectroscopic concept, the Raman approach has considerable appeal over the use of intermediary chemical indicator-based detection systems and over direct absorption spectroscopy for many species, since the Raman shift is independent of excitation wavelength.

Single-mode fibre

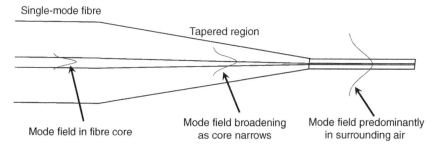

Mode field in fibre core

Mode field broadening
as core narrows

Mode field predominantly
in surrounding air

Figure 14.10   An idealized sketch of a tapered single-mode optical fibre illustrating the expanding mode field as the taper narrows down until most of the optical field is in the surrounding air. Since the taper typically has micron dimensions, very high optical fields can penetrate into the surrounding air facilitating strong, potentially very useful non-linear interactions.

Consequently, this shift can be excited ('pumped' is the most commonly used terminology) using sources of any convenient optical wavelength. Additionally, since for many species the Raman shifts are well separated, the same pump source can address many different species. It does though need very high localized optical power densities to excite the relatively weak Raman response and a reliable means to collect the shifted spectrum. The fibre taper can provide both of these functions, but consistency in shaping the taper and absolute cleanliness are also basic necessities.

Much of the above discussion on nanotechnologies has essentially considered ways to increase the localized electric field within the sample. Plasmonics – already mentioned – has already had a role to play here, but the optical losses in metals remain a substantial issue. However, graphene [28] has emerged over the past half-century or so as very low-loss optical conductor offering enormous promise in a diversity of optical functionalities. Once more, the precision fabrication tolerances developed in relative recent times have ensured that controlled graphene-based structures are realistically available. The ability to control deposition on a molecular layer-by-layer basis coupled to the quite extraordinary optical and electrical (and mechanical too) properties of the material is beginning to show a presence in fibre sensing in biochemical applications in particular. There will undoubtedly be more to come.

14.4.4   Improving the Signal-to-Noise Ratio

Signal-to-noise ratio in an optical system is inherently limited by the fact that our detectors to date all detect photons so the statistical variations in arrival rate present a basic limit.

However, in its simplest form, this limit, when viewed in terms of the wave transmission equivalent, is statistically equally split between phase noise and amplitude noise. Sensors operate by detecting one or the other of these. Consequently, if the light can be squeezed – an idea that has been around for more than 20 years [29] – for example, to push the noise in signal phase into the amplitude dimension, then a phase-sensitive sensor will exhibit better performance. Thus far, this has entailed exotic optics along the lines of those found in LIGO [30], and gravitational waves would not have been detected without the squeezing (and meeting lots of other engineering challenges). More recently, we have seen attempts to squeeze light on a photonic integrated circuit chip, typically based on coupling into very high Q-factor resonators. To date, the improvements have been modest – a decibel or so [31] – but who knows what may happen in the future?

The optical frequency comb [32], originally developed as ultra-precise optical frequency reference tool for metrology application, is stimulating growing interest in optical sensing. Its output comprises hundreds and even thousands of finely, evenly spaced lines, obtained by stabilizing a mode-locked laser to an optical or radio-frequency clock. This combination of broad wavelength range and fine resolution is very attractive for high-sensitivity and high-selectivity spectroscopic sensing of gases and molecules, particularly as optical combs can be designed to operate in the near- to the mid-IR wavelength range. Dual comb spectroscopy [33, 34] can resolve both the tooth-by-tooth attenuation and phase shift caused by molecular absorption and has been proposed for applications in chemical sensing and breath analysis [34]. Recently, an all-fibre system incorporating an optical fibre spectrometer has been used to achieve spectral fingerprinting of acetylene in the near IR though acquisition of 500 comb lines [35]. The optical frequency comb has also been applied to high-sensitivity, high-resolution interrogation of FBGs for strain and temperature measurements [36], as well as intra-cavity laser strain [37] and refractive index [38] sensing. It should be noted that this area is very much in its infancy and at the lab demonstration level but certainly fast developing.

To enhance the availability of the frequency comb, the integrated optic version recently made its appearance [39], so perhaps this is a prospective contributor to further evolving remote spectroscopy in fibre. There have even been all-fibre versions of the frequency comb demonstrated [40] – one less interface to worry about! At the other end of the system, the single-photon avalanche detector (SPAD) has become readily available. To date little if anything has happened in the fibre sensing domain exploiting the prospects of single-photon detection.

14.4.5 Quantum Sensing, Entanglement, and the Like

Quantum communications and quantum computing are making the headlines [41]. There is also much interest and considerable promise in quantum memory

– an essential aspect of both computing and communications. In the fibre context, the quantum clock based on a gas atomic vapour loaded in the core of hollow core (HC) fibres (see Section 14.4.6) has also appeared. Quantum sensing is occasionally mentioned though to the authors' knowledge little, if anything, has, at time of writing, occurred with fibres. But it probably will....

The claims with quantum sensing all centre upon high sensitivity. Intuitively the observation that the entangled pair is either there or not there seems to indicate a high sensitivity is feasible. Such sensitivity has been demonstrated in an open path Sagnac interferometer [42]. There are clearly questions on the extent to which this 'high sensitivity' is reproducible, based on the sensitivity to mechanical and other surrounding phenomena. There are also questions concerning potential linearity and dynamic range. The possible benefits of a fibre-based system may be that at least some of the mechanical alignment issues could be considerably simplified and perhaps (there are precedents for this philosophy for sure!) something can be borrowed from quantum communications. It will be intriguing to see how this might evolve [43].

14.4.6 The Many Prospects in Fibre Design and Fabrication

We have already reflected on the observation that most of the current effective fibre sensor technologies rely on modulating the light within the fibre rather than taking the light out, interacting with the world outside and then putting it back into the fibre. Initially, fibres were solid structures based on ultra-pure silica with a doped higher index core, the dimensions of which, together with the core–cladding index difference, controlled the mode propagation characteristics of the fibre. The potential in communication systems offered by these fibres stimulated huge investments in their fabrication in order to minimize losses, optimize the dispersion characteristics, and control the modal characteristics [44]. As a result, the single-mode optical fibre is now produced in hundreds of millions of kilometres per year, and the total deployed length around the globe is estimated to be several billion kilometres. However, in the sensing context, the fibre was restricted to detecting phenomena that can alter the propagation characteristics of the light carried in the core but without having direct access to the core, albeit over impressively long distances. For example, in excess of 100 km has been demonstrated for many distributed systems.

A little over two decades ago, the microstructured fibres (MOFs), also known as photonic crystal fibres [45], dramatically increased the variety of geometries and optical responses available to fibre-based sensing. These fibres have arrays of wavelength-scale air holes in their transverse cross section, which extend along the full fibre length and define the waveguide properties. The holes surround a light guiding core, which can be either solid, or, more intriguingly, can itself be an oversized air hole. HC MOFs are arguably the most innovative development in optical fibre technology since the advent of single-mode optical fibre [46]. Consequently, immense functionality in combined optical and mechanical properties can be realized.

While mode propagation in conventional fibres is controlled by the core dimension and the core–cladding index difference, in MOFs, it can be tailored via the overall fibre geometry, resulting in a substantially broader range of optical properties than can be achieved in the conventional fibres. The number, size, shape, and arrangement of the air holes can be specified with a wide degree of freedom.

Solid core MOFs (Figure 14.11a–c) work on a principle akin to total internal reflection, but by tailoring the design of the holey cladding, they can achieve properties that are simply not possible in conventional optical fibres. These properties include, among others, broadband single-mode guidance, by which the fibre supports a single degenerate pair of transverse optical modes at all wavelengths, and unique dispersion properties, by which the chromatic dispersion can be designed to be extremely flat over extended portions of the spectrum, or indeed to have zero points at wavelengths below the silica zero-dispersion wavelength at 1.3 μm, again something that is not possible in conventional fibres. Solid core MOFs also facilitate very high power densities within a single spatial mode, and hence high effective non-linearity, which, in association with the tailorable dispersion, makes all manners of non-linear optical effects readily possible. These have made the biggest impact as the essential enabler in the supercontinuum 'white light' lasers [53] that can extend from the ultraviolet to the mid-IR, providing unprecedented sources for spectroscopy-based sensing, advanced microscopy, detection of fluorophores in biosensing, etc. Index-guiding MOFs can be made to have very small core and a substantial fraction of evanescent field, which can be used to detect gases by direct absorption [54, 55].

Arguably however, it is the HC MOFs that have to date made the biggest impact in the context of optical fibre sensing [56, 57]. These fibres guide light by virtue of either photonic bandgap effects or anti-resonant effects, or a combination of the two. Both mechanisms are radically different from total internal reflection in conventional fibres. Photonic bandgap fibres (PBGFs) (Figure 14.11d–f) comprise an array of high-aspect-ratio air holes in the cladding and a core formed by replacing some elements at the centre by an oversized air hole. The photonic crystal cladding produces optical bandgaps through which specific wavelengths are reflected into the HC and therefore guided. Anti-resonant hollow fibres (Figure 14.11g–i), in some cases referred to as Kagome fibres or negative curvature fibres, comprise simpler structures, in which light confinement is achieved by designing the thin glass boundaries around the core to be in anti-resonance with the light guided in the core: light is thus expelled from the silica regions and confined in the central HC.

Figure 14.11 Examples of microstructured optical fibres (MOFs), also known as photonic crystal fibres [47]. (a)–(c) Index-guiding MOFs: large mode area, highly non-linear, suspended core fibre; fibres (b) and (c) can be used for evanescent sensing [48, 49] as a substantial fraction of the mode is guided in the holes surrounding the core (d)–(f): photonic bandgap hollow core fibres: 3, 7, and 19 cell fibre [45]; (g)–(i) anti-resonant hollow core fibres: hexagram [50], Kagome [51], and tubular [52] fibre. With >99% light guided in the hollow region, hollow core fibres provide and ideal platform for light–matter interaction over long path lengths and in a very compact format. *Source:* Reproduced with permission of IEEE.

Despite the structural complexity, requiring precise multistage fabrication procedures that are currently still far less mature when compared with conventional fibres, both types of HC MOFs have been demonstrated to attain losses as low at ~1 dB/km [58, 59] and have been fabricated in lengths exceeding 10 km [60]. The anti-resonant variety is believed to still be quite far from its intrinsic limits, which are predicted to be lower than solid fibre counterparts [61]. The promise of ultra-low loss in association with order-of-magnitude lower non-linearity as compared with solid fibres has raised substantial interest in such fibres for telecom applications as well as for long-distance laser power delivery. In the sensing context, HC fibres can extend their operation window well beyond the transparency window of silica glass and in particular towards the mid-IR region of the spectrum [62, 63], which is obviously of great interest for direct absorption sensing.

Interest in sensing applications of HC MOFs has steadily grown since the turn of the century. A few potential application areas have emerged, although these remain at the level of lab-based prototypes. The most obvious prospective application area is absorption or Raman-based spectroscopic gas sensing. With >99% of the optical power guided in the HC, losses as low as a few decibels per kilometre, and negligible macrobending losses, HC-PBGFs open up the possibility for extremely long effective interaction lengths. Additionally, sensor heads based on PBGFs can be made extremely compact by coiling the fibre and additionally require minimal amounts of test gas. The long inevitably slow diffusion of the gas measurand into the HC is however a significant drawback. Filling time is prohibitively long (minutes to hours) even for relatively short lengths of ~1 m, unless the source of gas under test is somehow pressurized. The anti-resonant Kagome fibres [64] have significantly larger core diameters and comparatively faster filling times. Additionally, the response time can be reduced to ~10 s of seconds by having periodic apertures in order to speed up the in- and out-diffusion. This can be realized by segmenting the fibre [65] or, more elegantly, via femtosecond laser machining [66] (see Figure 14.12a and b). These options have been demonstrated very convincingly achieving hitherto impossible sensitivities in a lab environment. However, the long-term reliability of such access points needs to be thoroughly assessed before any actual application. Also, while the response time is far from critical in some applications, users are invariably far more comfortable with seconds rather than minutes! For liquid sensing, surface tension and viscosity also come into play whether for in-core or surface access. A closely related issue is that of contamination and dust. A very thin layer of almost any contaminant will totally change the interaction mechanisms, and a few grains of dust will have enormous impact.

This ultra-low non-linearity, thanks to the low overlap between the guided optical field and the silica, also provides other benefits for sensor applications, such as optical fibre gyroscopes (FOGs) [70]. Conventional fibres introduce Kerr effect non-linearities. Additionally, thermal fluctuations and magnetic fields (via Faraday rotation) likewise introduce errors through non-reciprocal propagation [10]. HC-PBGFs are orders of magnitude less sensitive to all these effects mechanisms, and this may prove crucial to developing future FOG sensors. There are however some properties of HC fibres that can cause their own spurious effects

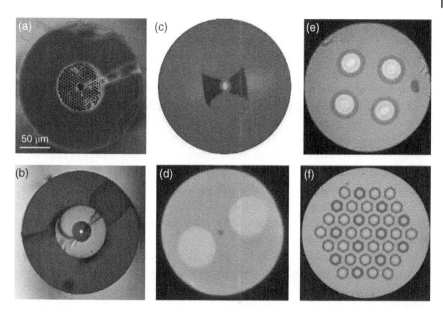

Figure 14.12 (a, b) Hollow core PBGFs with fs laser-machined microchannels to provide faster diffusion channels for gas [66]. (c, d) Bow-tie and panda polarization-maintaining fibres; (e, f) [67] examples of multicore optical fibres: 4 [68] and 37 cores [69]. *Source:* Reproduced with permission of AIP.

in sensing – of which perhaps the most evident is the transmission of several modes rather than just one. Also high birefringence is hard to achieve in these fibres. Another important issue is that for HC-PBGFs to achieve their full potential, the low-loss splices and interconnections with conventional fibres, long-term stability against gas and moisture ingression, and long-term mechanical stability all need to be thoroughly investigated. Initial results are encouraging – it seems probable that appropriate technological solutions will emerge in due course.

The scope of new fibres also extends beyond the photonic crystal into what is in effect a complementary domain whereby the traditional components – core and cladding – can be assembled in different formats. The highly birefringent elliptical core fibre has been with us for many years and find extensive application in fibre gyroscopes, for example [10]. The stress-induced 'panda' or 'bow-tie' fibre has very similar properties (Figure 14.12c and d).

More recently multicore fibres have emerged as a means to spatially multiplex fibre communication channels [71], with record number of cores now routinely integrated in a single fibre cross section (see Figure 14.12f). Unsurprisingly, multicore fibres have also been applied in the sensing context, particularly as bend and twist sensitive systems utilizing differential phase modulation among the various cores [72]. This observation points towards another prospect for future exploration, which is, in effect, the combination of the location and compositions of the core or cores and cladding within a fibre structure. Additionally, there are options for carving slots along a fibre [73] into which either chemicals for detection can be allowed to enter or injected or, alternatively, cladding materials can be incorporated as intermediary chemically sensitive transformers.

There are also rapidly expanding prospects for the application of multimode fibres. These often involve finding approaches to maintaining a modal infrastructure in transmission through the fibre and ensuring the input pattern at one end of the fibre can be recovered at the other. This concept has been applied for what is effectively endoscopic imaging both directly and for Raman spectral analysis [74]. It is an obvious extrapolation that this same basic process could be applied, for example, to fluorescence detection from chemical samples to perform remote spatially resolved analysis.

There remains a huge potential for the application of versatile fibre formats in fibre-optic sensing. The biggest benefit is probably the ability to keep the light in the fibre while being able to engineer the interactions with the outside world. There are particular prospects in bend and torsion sensing with multicore fibres and in a variety of options for biochemical sensing with HC fibres and indeed the similar 'side access' variants.

Meanwhile the fascination remains, and despite the relatively modest exploitation of the photonic crystal fibre concept to date (notably in optical sources especially the supercontinuum), the research continues, stimulated by tantalizing prospects encountering new, thus far undreamt of, functionalities and investigating fascinating physics in the process.

14.4.7 Technologies Other than Photonics

There is much, much more to making a fibre sensor that actually operates than simply the optics for which we have mentioned a few of the emerging concepts outlines above. What really matters in the end is that the operator gets a consistent and reliable answer to his questions concerning parameters to be measured. The operator needs confidence. Indeed, for the average fibre sensor system (or many other new developments), the investments in time, effort, and indeed technologies required to establish this operator confidence comfortably exceed those involved in demonstrating the sensing principle.

There are ongoing innovations that will help to contribute in the future. Some examples might include packaging technologies. It is essential that all except the sensing part of the system are appropriately protected from environmental changes. Many of the concepts we have discussed above require complex optics at the sensing end that is frequently in an unpleasant environment. Here technologies such as precision injection moulding and high-definition 3D printing will find their place. Yet again very high precision fabrication will make significant contributions.

Machine learning, the Internet of Things, and big data access will also have their part to play in, for example, interpreting potentially erroneous data in sensor arrays and distributed sensing. Similarly, techniques of this nature will continue to improve the quality and ease of assimilation of data as presented to the final user. Appropriately designed and implemented electronic systems can also contribute substantially to maintaining sensor system stability through, for example, establishing protocols whereby sensor system can be reliably reset to zero and in some cases recalibrated using the same process.

The principal purpose of this very short discussion is to highlight this very important point. There is much, much more involved in fibre-optic sensor than simply the optics. The contributing technologies that facilitate all this additional functionality are for sure continuing to become more versatile and more capable.

14.4.8 Societal Aspirations in Sensor Technology

There are countless surveys and studies that have assessed the areas within which new engineering techniques can contribute to solving imminent problems facing society. These documents can be viewed as having two objectives. The first is to give hints into the engineering community as to where they may focus their efforts. The second is of course to raise awareness in the community outside that engineering is doing a good job on their behalf. In the photonics context, Photonics 21 in Europe and the National Photonics Initiative in the United States have both produced reports along these lines and with the same general conclusions, namely, environment, safety, agricultural efficiency, optimization of transportation technologies, and health all need considerable attention and that photonics can help in this regard [75] (see Figure 14.13). Sensors feature very strongly in this since all of these societal parameters need to be measured and photonics can measure all of them with fibre-optic technology having potentially critically important role to play.

14.4.9 The Future and a Quick Look at the Sensing Alternatives

Applying sensor technology to the needs mentioned above (and indeed to a huge variety of other requirements) has over the past decade or so established its own vocabulary of which the Internet of Things [76] features probably the most strongly. The concept of using an established communication infrastructure with standard interfaces to facilitate sensory inputs has obvious appeal. Perhaps the camera on the mobile phone is the most familiar example – and this is also evident in a much enhanced version enabling medical practitioners to view X-ray images

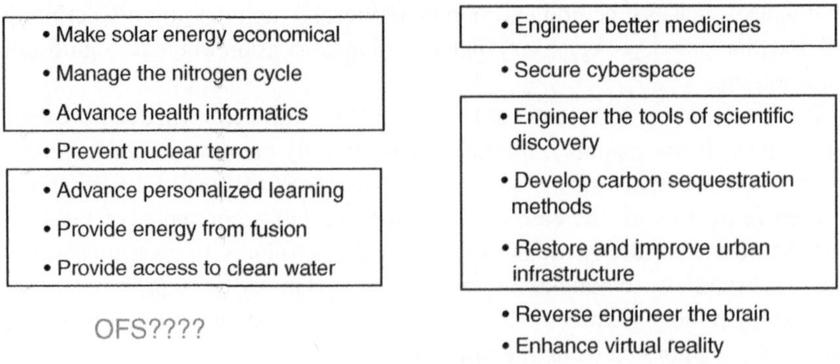

- Make solar energy economical
- Manage the nitrogen cycle
- Advance health informatics
- Prevent nuclear terror
- Advance personalized learning
- Provide energy from fusion
- Provide access to clean water

OFS????

- Engineer better medicines
- Secure cyberspace

- Engineer the tools of scientific discovery
- Develop carbon sequestration methods
- Restore and improve urban infrastructure

- Reverse engineer the brain
- Enhance virtual reality

Figure 14.13 A representation of the principal aspirations in society (representing an overview of the opinions from many organizations – see text) and highlighting where OFS may make a future contribution.

and CT scans in real time when miles from the instrument. There is also the access to home heating controls from anywhere in the internet – something now taken for granted! Much is also said of 'Industry 4.0,' and in the context of all these initiatives, a considerable electronics-based infrastructure is evolving [77].

This immediately leads into major questions. Electronics-based sensory systems abound. Here a sensor element, for which the conductivity, capacitance, or combinations thereof varies with the parameter to be measured, is interfaced directly into an electronic circuit. And this can all be built into the chip. Users are very comfortable with this, and extensive systems have been built around this basic idea. For example, many households have carbon monoxide detectors based on the principle – here a direct chemical interface to an electronic circuit. Sensory systems that require an interface, optical, ultrasonic, or whatever are less familiar and are therefore invariably greeted with scepticism unless they have been around for a very long time and/or have been utilized by a technically erudite community, for example, oil exploration and recovery and also, more familiarly, in medical imaging – the ultrasonic scan of the unborn child. Even this technically erudite community is inherently suspicious of new technology and, with the exception of the domestically familiar (e.g. fibre to the home, TV and phone displays, digital cameras, and light-emitting diodes), needs to be carefully persuaded the viability of new technology. This persuasion process is often lengthy, involving numerous stages (Figure 14.14), and can typically take up to two decades, maybe more, in most of the sectors within which fibre-optic sensors can usefully contribute. There is a paradoxical exception in that the healthcare community sometimes respond to innovation in half the time of sectors like oil exploration and recovery. This phenomenon is quite generic, but the time constants can vary significantly. It was first suggested to a wider community through the previously mentioned Porter's five forces analysis [7] over 40 years ago.

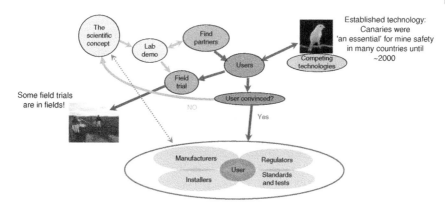

Figure 14.14 A representation of the communities involved in adopting new technologies indicating also the essential commutations channels and user cultures into which the researcher needs to effectively interface. Hence – this can go to two decades – or more!

The fibre-optic sensor community requires skills and aptitudes well beyond simply presenting the technical paper with the great idea to achieve the ever-expanding presence that the technology warrants. A principal requirement is effective empathetic communications with the eventual user audience in particular through an appreciation of what is required for an application to become real. This requirement goes well beyond the often vaunted 'low cost', which appears in research papers and invariably refers optimistically to the cost of the components in the sensor head rather than the system overall. Its implications go well beyond technology into the definition and acceptance of standards and procedures and significant investment in market development, customer, and regulator persuasion. Always there is the need to recognize the very definite influence of the cautious user and established technologies. So communications from the researcher throughout this chain are an essential aspect of eventual external acceptance.

14.4.10 So What Has Fibre Sensing Achieved to Date

When simply viewed in terms of market penetration figures, fibre sensing occupies but a very tiny blip in a market dominated by electronic devices for industrial and domestic applications. However, the tiny blip that is occupied does solve critical problems, and this is especially true for distributed and quasi-distributed systems operating within safety system applications. Nothing else can observe, e.g. what is happening many tens of kilometres away along a pipeline, continuously and totally reliably. Perhaps somehow the fibre sensor community should in the future do much more to emphasize these benefits to the community at large and expose these unique and critical achievements to the widest possible audience. Remember that fibre sensors have also made a presence in seismic surveying, in hydrophone systems for safety and security, in perimeter surveillance, in clinical applications e.g. intracranial surgery, in navigational systems through the fibre gyro, and in a host of other specialized, high value, technologically critical niches.

Fibre sensing from a research student perspective also has significant, indeed very helpful, challenges associated with it. The topic is very much interdisciplinary, unlike the vast majority of research themes that are 'single track'. The student needs to appreciate at least some elements of the application, needs to be knowledgeable of the diversity of photonic technologies that can solve the application, needs to be aware of the material processing, needs required to put the sensor together, and needs to be aware of how to make realistic test procedures to measure what has emerged against what was expected. Consequently, graduates from these programmes do have a greater diversity of opportunity than many others. Yes, several do go into further academic research with varying degrees of success, many go into industry and several establish spin-out companies, a few end up in government laboratories and similar environments, and, from personal experience, at least one has become a church minister while another publishes a specialized comic book on a regular basis. For sure the research graduates can apply themselves to a huge diversity of possibilities.

14.5 Concluding Observations

Half a century of fibre sensing has provided some exciting research opportunities that invariably require interdisciplinary skills and applications awareness. The interests at the research level will, for sure, continue, exemplified in the record attendance of OFS26 in 2018. Eventual careers of graduates reflect this broad exposure – a wide diversity of opportunities are available, and the necessary technical curiosity has been certainly aroused through the research experience. Fibre-optic sensing will also continue to provide not only exciting research opportunities but also intriguing (and profitable) application niches in the future. While these niches are difficult to identify at the moment, future societal demands will ensure fibre sensing will have its role to play.

References

1 See for example: https://www.alliedmarketresearch.com/sensor-market for world sensors market at ~US$150Bn per annum and https://www.marketstudyreport.com/request-a-sample/1175913 on current fibre sensors markets at ~US$1 Bn per annum.

2 Janata, J. (1989). *Principles of Chemical Sensors*. Springer.

3 Hopkins, H.H. and Kapany, N.S. (1954). A flexible fibrescope, using static scanning. *Nature* 173: 39–41.

4 Menadier, C., Kissinger, C., and Adkins, H. (1967). The fotonic sensor. *Instrum. Control Syst.* 40 (6): 114;
And for interferometry: Snitzer, E. Apparatus for controlling the propagation characteristics of coherent light within an optical fibre. US Patent 3,265,589, filed 19 August 1969.

5 https://www.mtiinstruments.com/products/non-contact-measurement/fibre-optic-sensors/fibre-optic-mti-2100-fotonic-sensor. Accesses the latest version of the Fotonic sensor.

6 Such self-correction approaches are often jealously guarded! However, the 'OptoSniff' and 'OptoMole' have successfully incorporated such self-correction procedures into every single gas sensing measurement. https://www.optosci.com.

7 Porter, M.E. (1979). How competitive forces shape strategy. *Harv. Bus. Rev.* 59 (2): 137–145. https://en.wikipedia.org/wiki/Porter%27s_five_forces_analysis.

8 Bohnert, K., Frank, A., Müller, G.M. et al. Fibre optic current and voltage sensors for electric power transmission systems. *Proeedings of SPIE 10654, Fibre Optic Sensors and Applications XV*, 1065402, Orlando, USA (17 April 2018–18 April 2018).

9 Rogers, A. (1973). Optical technique for measurement of current at high voltage. *Proc. IEEE* 120 (2): 262–267.

10 LeFevre, H.C. (2014). *The Fibre Optic Gyroscope*, 2e. Boston and London: Artec House.

11 See, for example: Tran, M.A., Hulme, J.C., and Komljenovic, T. The first integrated optical driver chip for fibre optic gyroscopes. *4th IEEE International Symposium on Inertial Sensors and Systems, INERTIAL 2017,* Kauai, USA (27 March 2017–30 March 2017). Article number 7935659, pp. 47–49.

12 Wickersheim, K.A. Optical temperature measurement technique utilizing phosphors. US Patent 4,215,275, filed 15 February 1978. See https://lumasenseinc.com/EN/home/home-lumasense-technologies.html.

13 Crutchfield, J.S., Narayan, R.K., Robertson, C.S., and Michael, L.H. (1990). Evaluation of a fibreoptic intracranial pressure monitor. *J Neurosurg.* 72 (3): 482–487.

14 Inaudi, D. and Pinet, E. (2018). Large-volume Fabry–Pérot fibre-optic sensors production for medical devices and industrial applications. *26th International Conference on Optical Fibre Sensors, OSA Technical Digest* (Optical Society of America), paper TuE32, Lausanne, Switzerland (24–28 September *2018*).

15 This comment was made during an invited overview presentation from a Bell Laboratories presenter at ECOC 1978 in Genova, Italy. One of your authors happened to be there – and it stuck!!!

16 Zimmermann, H.K. (2016). *Integrated Silicon Optoelectronics*, Springer Series in Optical Sciences 2010. Berlin Heidelberg: Springer-Verlag; See also: Komljenovic, T., Davenport, M., Hulme, J. et al. (2016). Heterogeneous silicon photonic integrated circuits. *J. Lightwave Technol.* 34: 20–35; And: Reed, G.T. et al. (2018) Advancing silicon photonics by germanium ion implantation into silicon. *Proceedings of the SPIE 10536, Smart Photonic and Optoelectronic Integrated Circuits XX*, 105361T, San Francisco, USA (29 January 2018–1 February 2018). doi: https://doi.org/10.1117/12.2292199.

17 Hodaei, H., Hassan, A.U., Wittek, S. et al. (2017). Enhanced sensitivity at higher-order exceptional points. *Nature* 548: 187–191; See also: Osborne, I.S. (2019). Exceptional points in optics. *Science* 363 (6422): 39–41.

18 Khial, P.P., White, A.D., and Hajimiri, A. (2018). Nanophotonic optical gyroscope with reciprocal sensitivity enhancement. *Nat. Photonics* 12: 671–675.

19 Marin, Y., Celik, A., Faralli, S. et al. (2018). Silicon photonic chip for dynamic wavelength division multiplexed FBG sensors interrogation. *26th International Conference on Optical Fibre Sensors, OSA Technical Digest* (Optical Society of America), paper ThE45, Lausanne, Switzerland (24–28 September *2018*).

20 See for example: https://en.wikipedia.org/wiki/Surface_plasmon_resonance for introductory material or Li, L., Liang, Y., Guang, J. et al. (2017). Dual Kretschmann and Otto configuration fibre surface plasmon resonance biosensor. *Opt. Express* 25 (22): 26950–26957, as an example of a sensor configuration.

21 See for example: Willsch, R. (1990). High performance metal-clad fibre-optic polarisers. *Electron. Lett.* 26 (15): 1113–1115.

22 Shekhar, P., Atkinson, J., and Jacob, Z. (2014). Hyperbolic metamaterials: fundamentals and applications. *Nano Convergence* 1: 14.

23 Shalaev, V.M. and Cai, W. (2009). *Optical Metamaterials, Fundamentals and Applications*. Springer.

24 Yang, J., Ghimire, I., Wu, P.C. et al. (2019). Photonic crystal fibre metalens. *Nanophotonics* 8 (3): 443–449.

25 Wang, Z.L. (2004). Nanostructures of zinc oxide. *Mater. Today* 7 (6): 26–33.

26 Talataisong, W., Ismaeel, R., and Brambilla, G. (2018). A review of microfibre-based temperature sensors. *Sensors* 18 (2): 1–26.

27 Qi, Y., Zhao, Y., Bao, H. et al. (2019). Nanofibre enhanced stimulated Raman spectroscopy for ultra-fast, ultra-sensitive hydrogen detection with ultra-wide dynamic range. *Optica* 6: 570–576.

28 Falkovsky, L.A. (2008). Optical properties of graphene. *J. Phys. Conf. Ser.* 129: 012004.

29 See for example: *J. Modern Opt.* 24: 6/7(June/July 1987) – this was a special issue on squeezed light.

And also Tombesi, P. and Pike, E.R. (1989). *Squeezed and Non-Classical Light*, NATO ASI Series B Physics, vol. 190. Plenum Press.

30 Aasi, J., Abadie, J., Abbott, B.P. et al. (2013). Enhnaced sensitivity of the LIGO gravitational wave detector by using the squeezed states of light. *Nat. Photon.* 7: 613–619.

31 Dutt, A., Luke, K., Manipatruni, S. et al. (2015). On-chip optical squeezing. *Phys. Rev. Appl.* 3: 044005.

32 Picqué, N. and Hänsch, T.W. (2019). Frequency comb spectroscopy. *Nat. Photon.* 13: 146–157.

33 Zolot, A.M., Giorgetta, F.R., Baumann, E. et al. (2012). Direct-comb molecular spectroscopy with accurate, resolved comb teeth over 43 THz. *Opt. Lett.* 37 (4): 638–640.

34 Coddington, I., Newbury, N., and Swann, W. (2016). Dual-comb spectroscopy. *Optica* 3: 414–426.

35 Coluccelli, N., Cassinerio, M., Redding, B. et al. (2016). The optical frequency comb fibre spectrometer. *Nat. Commun.* 7: 12995.

36 Kuse, N., Ozawa, A., and Kobayashi, Y. (2013). Static FBG strain sensor with high resolution and large dynamic range by dual-comb spectroscopy. *Opt. Express* 21: 11141–11149.

37 Minamikawa, T., Ogura, T., Nakajima, Y. et al. (2018). Strain sensing based on strain to radio-frequency conversion of optical frequency comb. *Opt. Express* 26: 9484–9491.

38 Oe, R., Taue, S., Minamikawa, T. et al. (2018). Refractive-index-sensing optical comb based on photonic radio-frequency conversion with intracavity multi-mode interference fibre sensor. *Opt. Express* 26: 19694–19706.

39 Okawachi, Y., Saha, K., Levy, J.S. et al. (2011). Octave-spanning frequency comb generation in a silicon nitride chip. *Opt. Lett.* 36: 3398–3400.

40 Poiana, D.A., Garcia-Souto, J.A., Posada-Roman, J.E., and Acedo, P. (2018). All-fibre electro-optic dual optical frequency comb for fibre sensors. *Proceedings OFS 26*, Paper WF82, Lausanne, Switzerland (24–28 September 2018).

41 This is a field that is constantly making news! See for example: https://en.wikipedia.org/wiki/Quantum_information_science.

42 Gustavson, T.L., Landragin, A., and Kasevich, M.A. (2000). Rotation sensing with a dual atom interferometer Sagnac gyroscope. *Class. Quantum Grav.* 17 (12): 2385.

43 Dowling, J.P. (2013). *Schrodinger's Killer App*. CRC Press.

44 Hecht, J. (1999). *City of Light – The Story of Fibre Optics*. Oxford University Press Revised 2004. https://www.jeffhecht.com/city.html.

45 Russell, P.S.J. (2003). Photonic crystal fibres. *Science* 299 (5605): 358–362.

46 Poletti, F., Petrovich, M.N., and Richardson, D.J. (2013). Hollow-core photonic bandgap fibres: technology and applications. *Nanophotonics* 2 (5–6): 315–340.

47 All fibres shown in Fig. 13.11 have been fabricated at the Optoelectronics Research Centre, University of Southampton. Credits: M.N. Petrovich, J.R. Hayes, K. Furusawa, N.V. Wheeler, N.K. Baddela, Y. Chen, T. Bradley, S.R. Sandoghchi.

48 Hoo, Y.L., Jin, W., Ho, H.L. et al. (2002). Evanescent-wave gas sensing using microstructure fibre. *Opt. Eng.* 41: 8–9.

49 Petrovich, M.N., van Brakel, A., Poletti, F. (et al.) (2005). Microstructured fibres for sensing applications. *SPIE Proceedings, Photonic Crystals and Photonic Crystal Fibres for Sensing Applications*, Bruges, Belgium (23 May 2005–27 May 2005), vol. 6005, 60050E.

50 Hayes, J.R., Poletti, F., Abokhamis, M.S. et al. (2015). Anti-resonant hexagram hollow core fibres. *Opt. Express* 23: 1289–1299.

51 Bradley, T.D., Wheeler, N.V., Jasion, G.T. et al. (2016). Modal content in hypocycloid Kagomé hollow core photonic crystal fibres. *Opt. Express* 24: 15798–15812.

52 Hayes, J.R., Sandoghchi, S.R., Bradley, T.D. et al. (2017). Antiresonant hollow core fibre with an octave spanning bandwidth for short haul data communications. *J. Lightwave Technol.* 35 (3): 437–442.

53 Ranka, J.K., Windeler, R.S., and Stentz, A.J. (2000). Visible continuum generation in air–silica microstructure optical fibres with anomalous dispersion at 800 nm. *Opt. Lett.* 25: 25–27;
Also see: Dudley, J.M. and Taylor, J.R. (2009). Ten years of nonlinear optics in photonic crystal fibre. *Nature Photon.* 3: 85.

54 Monro, T.M., Belardi, W., Furusawa, K. et al. (2001). Sensing with microstructured optical fibres. *Meas. Sci. Technol.* 12 (7): 854–858.

55 Jin, W., Ho, H.L., Cao, Y.C. et al. (2013). Gas detection with micro- and nano-engineered optical fibres. *Opt. Fibre Technol.* 19 (6): 741–759.

56 Ritari, T., Tuominen, J., Ludvigsen, H. et al. (2004). Gas sensing using air-guiding photonic bandgap fibres. *Opt. Express* 12: 4080–4087.

57 Austin, E., van Brakel, A., Petrovich, M.N., and Richardson, D.J. (2009). Fibre optical sensor for C2H2 gas using gas-filled photonic bandgap fibre reference cell. *Sens. Actuators B* 139 (1): 30–34.

58 Mangan, B.J., L. Farr, A. Langford et al. (2004). Low loss (1.7 dB/km) hollow core photonic bandgap fibre. *Optical Fibre Communication Conference, OFC 2004*, Los Angeles, CA (February 2004), vol. 2, 3.

59 Bradley, T.D., J. R. Hayes, Y. Chen et al. (2018). Record low-loss 1.3 dB/km data transmitting antiresonant hollow core fibre. *2018 European Conference on Optical Communication (ECOC)*, Rome (23–27 September 2018), pp. 1–3.

60 Chen, Y., Liu, Z., Sandoghchi, S.R. et al. (2016). Multi-kilometer long, longitudinally uniform hollow core photonic bandgap fibres for broadband low latency data transmission. *J. Lightwave Technol.* 34 (1): 104–113.

61 Poletti, F. (2014). Nested antiresonant nodeless hollow core fibre. *Opt. Express* 22: 23807–23828.

62 Wheeler, N.V., Heidt, A.M., Baddela, N.K. et al. (2014). Low-loss and low-bend-sensitivity mid-infrared guidance in a hollow-core–photonic-bandgap fibre. *Opt. Lett.* 39: 295–298.

63 Yu, F. and Knight, J.C. (2013). Spectral attenuation limits of silica hollow core negative curvature fibre. *Opt. Express* 21: 21466–21471.

64 Couny, F., Benabid, F., and Light, P.S. (2006). Large-pitch kagome-structured hollow-core photonic crystal fibre. *Opt. Lett.* 31: 3574–3576.

65 Lazaro, J.M., Cubillas, A.M., Silva-Lopez, M. (et al.) (2008). Methane sensing using multiple-coupling gaps in hollow-core photonic bandgap fibres. *Proceedings of SPIE 7004, 19th International Conference on Optical Fibre Sensors*, 70044U, Perth, WA, Australia (15 April 2008–18 April 2008).

66 van Brakel, A., Grivas, C., Petrovich, M.N., and Richardson, D.J. (2007). Micro-channels machined in microstructured optical fibres by femtosecond laser. *Opt. Express* 15: 8731–8736.

67 Images of bowtie and Panda fibres courtesy of Fibrecore. http://www.fibrecore.com

68 Matsui, T., T. Sakamoto, Y. Goto et al. (2015). Design of 125 μm cladding multi-core fibre with full-band compatibility to conventional single-mode fibre. *European Conference on Optical Communication (ECOC)*, Valencia, Spain (27 September 2015–1 October 2015), pp. 1–3.

69 Sasaki, Y., Takenaga, K., Aikawa, K., Miyamoto, Y. and Morioka, T. (2017). Single-mode 37-core fibre with a cladding diameter of 248 μm. *Optical Fibre Communications Conference and Exhibition (OFC)*, Los Angeles, CA (March 2017–23 March 2017), pp. 1–3.

70 Kim, H.K., Digonnet, M.J.F., and Kino, G.S. (2006). Air-core photonic-bandgap fibre-optic gyroscope. *J. Lightwave Technol.* 24: 3169–3174.

71 Richardson, D.J., Fini, J.M., and Nelson, L.E. (2013). Space-division multiplexing in optical fibres. *Nat. Photon.* 7: 354.

72 Flockhart, G.M.H., MacPherson, W.N., Barton, J.S. et al. (2003). Two-axis bend measurement with Bragg gratings in multicore optical fibre. *Opt. Lett.* 28: 387–389.

73 Zhang, Y., Zhao, Y., and Wang, Q. (2013). Multi-component gas sensing based on slotted photonic crystal waveguide with liquid infiltration. *Sens. Actuators B Chem* 184: 179–188.

74 See http://www.osa.opn.org/news.endoscopy-fibre Feb 2019. And for example: Chen, M., Mas, J., Forbes, L.H. et al. (2018). Depth-resolved multimodal imaging: wavelength modulated spatially offset Raman spectroscopy with optical coherence tomography. *J. Biophoton.* 11 (1): e201700129.

75 See for example: reports from Photonics 21 in the EU and the US National Photonics Initiative. These are available for free download and a readily located.

76 A good overview of the Internet of Things is available on https://en.wikipedia.org/wiki/Internet_of_things.

77 Similarly, https://en.wikipedia.org/wiki/Industry_4.0 presents a good overview on industry 4.0.

Index

Optical Fiber Sensors: Fundamentals for Development of Optimized Devices, First Edition.
Edited by Ignacio Del Villar and Ignacio R. Matias.
© 2021 The Institute of Electrical and Electronics Engineers, Inc.
Published 2021 by John Wiley & Sons, Inc.

IEEE Press Series on Sensors

Series Editor: Vladimir Lumelsky, Professor Emeritus, Mechanical Engineering, University of Wisconsin-Madison

Sensing phenomena and sensing technology is perhaps the most common thread that connects just about all areas of technology, as well as technology with medical and biological sciences. Until the year 2000, IEEE had no journal or transactions or a society or council devoted to the topic of sensors. It is thus no surprise that the IEEE Sensors Journal launched by the newly-minted IEEE Sensors Council in 2000 (with this Series Editor as founding Editor-in-Chief) turned out to be so successful, both in quantity (from 460 to 10,000 pages a year in the span 2001–2016) and quality (today one of the very top in the field). The very existence of the Journal, its owner, IEEE Sensors Council, and its flagship IEEE SENSORS Conference, have stimulated research efforts in the sensing field around the world. The same philosophy that made this happen is brought to bear with the book series.

Magnetic Sensors for Biomedical Applications
Hadi Heidari, Vahid Nabaei

Smart Sensors for Environmental and Medical Applications
Hamida Hallil, Hadi Heidari

Whole-Angle MEMS Gyroscopes: Challenges, and Opportunities
Doruk Senkal and Andrei M. Shkel

Optical Fibre Sensors: Fundamentals for Development of Optimized Devices
Ignacio Del Villar and Ignacio R. Matias.

Printed and bound by CPI Group (UK) Ltd, Croydon, CR0 4YY
19/05/2022
03125027-0004